# Amorphous Solids and the Liquid State

# PHYSICS OF SOLIDS AND LIQUIDS

**SUPERIONIC CONDUCTORS**
Edited by Gerald D. Mahan and Walter L. Roth

**HIGHLY CONDUCTING ONE-DIMENSIONAL SOLIDS**
Edited by Jozef T. Devreese, Roger P. Evrard, and Victor E. van Doren

**ELECTRON SPECTROSCOPY OF CRYSTALS**
V. V. Nemoshkalenko and V. G. Aleshin

**MANY-PARTICLE PHYSICS**
Gerald D. Mahan

**THE PHYSICS OF ACTINIDE COMPOUNDS**
Paul Erdös and John M. Robinson

**THEORY OF THE INHOMOGENEOUS ELECTRON GAS**
Edited by Stig Lundqvist and Norman H. March

**POLYMERS, LIQUID CRYSTALS, AND LOW-DIMENSIONAL SOLIDS**
Edited by Norman H. March and Mario Tosi

**AMORPHOUS SOLIDS AND THE LIQUID STATE**
Edited by Norman H. March, Robert A. Street, and Mario Tosi

*Forthcoming*
**CHEMICAL BONDS OUTSIDE METAL SURFACES**
Norman H. March

A Continuation Order Plan is available for this series. A continuation order will bring delivery of each new volume immediately upon publication. Volumes are billed only upon actual shipment. For further information please contact the publisher.

# Amorphous Solids and the Liquid State

Edited by

## Norman H. March
*University of Oxford*
*Oxford, England*

## Robert A. Street
*Xerox Palo Alto Research Center*
*Palo Alto, California*

and

## Mario Tosi
*International Center for Theoretical Physics*
*Trieste, Italy*

**Plenum Press • New York and London**

5282519X

Library of Congress Cataloging in Publication Data

Main entry under title:

Amorphous solids and the liquid state.

(Physics of solids and liquids)
Bibliography: p.
Includes index.
1. Liquids. 2. Amorphous substances. I. March, Norman H. (Norman Henry), 1927–
. II. Street, Robert A. III. Tosi, Mario, 1932–    . IV. Series.
QC145.2.A46   1985                     530.4                     85-12031
ISBN 0-306-41947-5

©1985 Plenum Press, New York
A Division of Plenum Publishing Corporation
233 Spring Street, New York, N.Y. 10013

Printed in the United States of America

# Contributors

*H. Beck*, Institut de Physique, Université de Neuchâtel, Rue A.-L. Breguet 1, CH-2000 Neuchâtel, Switzerland.

*P. N. Butcher*, University of Warwick, Coventry CV4 7AL, England.

*J. M. D. Coey*, Department of Pure and Applied Physics, Trinity College, Dublin 2, Ireland.

*E. A. Davis*, Department of Physics, University of Leicester, Leicester LE1 7RH, England.

*J. E. Enderby*, H. H. Wills Physics Laboratory, University of Bristol, Royal Fort, Tyndall Avenue, Bristol BS8 1TL, England.

*J. L. Finney*, Birkbeck College, Malet Street, London WC1E 7HX, England.

*M. Gerl*, Laboratoire de Physique du Solide (L.A. au C.N.R.S. n° 155), Faculté des Sciences, B.P. 239, 54506 Vandoeuvre Les Nancy Cedex, France.

*J. Hafner*, Institut für Theoretische Physik, Technische Universität Wien, Karlsplatz 13, A 1040 Wien, Austria; and Laboratoire de Thermodynamique et Physico-Chimie Métallurgiques, Domaine Universitaire, B. P. 75, F 38402 Saint Martin d'Hères, France.

*J. P. Hansen*, Laboratoire de Physique Théorique des Liquides, Université P. et M. Curie, 75230 Paris Cedex 05, France.

*P. Lagarde*, LURE, Université de Paris-Sud, Bâtiment 209 C, 91405 Orsay Cedex, France.

*N. H. March*, Theoretical Chemistry Department, University of Oxford, 1 South Parks Road, Oxford OX1 3TG, England.

*A. E. Owen*, Department of Electrical Engineering, School of Engineering, University of Edinburgh, Edinburgh EH9 3JL, Scotland.

*W. A. Phillips*, Cavendish Laboratory, Department of Physics, University of Cambridge, Madingley Road, Cambridge CB3 0HE, England.

*G. Rickayzen*, The Physics Laboratory, University of Canterbury, Kent CT2 7NR, England.

*M. P. Tosi*, Istituto di Fisica Teorica dell'Università and International Centre for Theoretical Physics, Trieste, Italy.

# Preface

This book has its origins in the 1982 Spring College held at the International Centre for Theoretical Physics, Miramare, Trieste.

The primary aim is to give a broad coverage of liquids and amorphous solids, at a level suitable for graduate students and research workers in condensed-matter physics, physical chemistry, and materials science. The book is intended for experimental workers with interests in the basic theory.

While the topics covered are many, it was planned to place special emphasis on both static structure and dynamics, including electronic transport. This emphasis is evident from the rather complete coverage of the determination of static structure from both diffraction experiments and, for amorphous solids especially, from model building. The theory of the structure of liquids and liquid mixtures is then dealt with from the standpoint of, first, basic statistical mechanics and, subsequently, pair potentials constructed from the electron theory of simple metals and their alloys. The discussion of static structure is completed in two chapters with rather different emphases on liquid surfaces and interfaces. The first deals with the basic statistical mechanics of neutral and charged interfaces, while the second is concerned with solvation and double-layer effects.

Dynamic structure is introduced by a comprehensive discussion of single-particle motion in liquids. This is followed by the structure and dynamics of charged fluids, where again much basic statistical mechanics is developed. Electronic-transport properties are then examined, first with reference to the nonlocalized electron states in liquid metals and then with emphasis on amorphous solids, the rate-equation approach describing hopping conduction being treated in detail.

A number of specialized topics are dealt with in the final part of the book: local coordination as determined by the Exafs technique, thermodynamics and kinetics of the glass transition, ferromagnetism in amorphous materials, thermal properties of amorphous solids interpreted in terms of two-level systems, and finally a brief chapter on the nature of defects in amorphous materials.

Valuable background reading can be found in the cited books by Mott and Davis on electrons in noncrystalline solids, by Hansen and MacDonald and by March and Tosi on liquids. Some of the material in the book *Ill-Condensed Matter* is also very relevant in this context.

*N. H. March*
*R. A. Street*
*M. P. Tosi*

# Contents

## 3. Structure and Forces in Liquids
## and Liquid Mixtures ............................................... 53

*N. H. March*

M. P. Tosi

## 8. Structure and Dynamics of Charged Fluids ........................ 229

*J. P. Hansen*

# I

# Mainly Static Structure, Including Interfaces

# Diffraction Studies of Liquids

## J. E. Enderby

### 1.1. Introduction

Simple liquids are those for which the intermolecular potential-energy function $\Phi(\mathbf{r}_1, \mathbf{r}_2, \ldots, \mathbf{r}_N)$ has the special form given by

$$\Phi(\mathbf{r}_1, \mathbf{r}_2, \ldots, \mathbf{r}_N) = \sum_{i<j} \phi(\mathbf{r}_k, \mathbf{r}_j) \qquad (1.1.1)$$

where $\mathbf{r}_k$ is the position vector of the $k$th atom or ion in the liquid, $N$ is the total number of atoms, and $\phi$ is the so-called pair potential. We also assume that $\phi$ is spherically symmetric, i.e., it is a function of $|r| = |\mathbf{r}_k - \mathbf{r}_j|$. For a simple mixture composed of, say, $a$ and $b$ atoms, the interactions are characterized by three pair potentials, each spherically symmetric, given by $\phi_{aa}(r)$, $\phi_{bb}(r)$, and $\phi_{ab}(r)$. There is a considerable body of evidence which suggests that $\phi$ for the liquid form of the rare gases (Ar, Kr, Ne, and Xe) approximates very closely a sum of spherically symmetric pair potentials. An example of the potential $\phi$ used for liquid rare gases is given in Figure 1.1. More surprisingly, a model based on pair potentials can also be justified for liquid nontransition metals like Na and Al. In these cases, the form of $\phi$ is not so well defined, but the general features presented in Figure 1.1 do not probably involve a serious error. A review of the usefulness of pair potentials for metallic systems lies outside the scope of this treatment; the reader is referred to the book by Faber[1] and to the references cited therein. Pair potentials are also very useful in

**J. E. Enderby** · H. H. Wills Physics Laboratory, University of Bristol, Royal Fort, Tyndall Avenue, Bristol BS8 1TL, England.

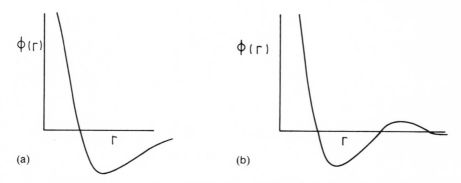

Figure 1.1. Schematic pair potentials: (a) insulators (e.g., argon), (b) metals (e.g., sodium).

molten salts. This topic has been reviewed by Sangster and Dixon.[2] Huggins and Mayer[3] have expressed the general form of $\phi_{ij}(r)$ as

$$\phi_{ij}(r) = \frac{Z_i Z_j e^2}{r} + B_{ij} \exp(-\alpha_{ij} r) - \frac{C_{ij}}{r^6} - \frac{D_{ij}}{r^8} + \cdots \qquad (1.1.2)$$

with $i, j = +, -$.

The choice of parameters appearing in equation (1.1.2) differentiates one potential from another. For example, Fumi and Tosi[4] allow $Z_1$ and $Z_2$ to assume the full aionic values (i.e., $\pm 1$ for alkali halides) and use the Mayer[5] values of the coefficients $C_{ij}$ and $D_{ij}$. The determination of $B_{ij}$ and $\alpha_{ij}$ is too complicated to discuss here and reference should be made to the article by Sangster and Dixon.[2] Potentials of this general form can be classed as "rigid-ion" potentials and it is clearly important to know whether any structural feature in molten salts reflects the fact that in reality the ions are deformable. To achieve this, a new set of polarizable ion potentials that incorporate ideas derived from the lattice dynamic shell model have been developed by Sangster and Dixon.[2]

In aqueous solutions of electrolytes the unusual nature of water presents a serious complication. In *primitive* models, the molecular nature of the water is neglected and the properties of the solvent are simulated by an effective dielectric constant $\varepsilon$. The interaction between ions is then represented by a pair potential of the form

$$\phi_{ij}(r) = A_{ij}(r) + \frac{Z_i Z_j e^2}{\varepsilon r} \qquad (1.1.3)$$

where $A_{ij}(r)$ represents the repulsive (overlap) part of the ion–ion poten-

**Table 1.1. Parameters for the Various Primitive Models of Electrolytes
[Equation (1.1.3)]**

| $A_{++}$ | $A_{--}$ | $A_{+-}$ | Name |
|---|---|---|---|
| 0 | 0 | 0 | Debye–Hückel |
| $\infty\ (r \leq \sigma)$ | $\infty\ (r \leq \sigma)$ | $\infty\ (r \leq \sigma)$ | Restricted |
| $0\ (r \geq \sigma)$ | $0\ (r \geq \sigma)$ | $0\ (r \geq \sigma)$ | |
| $\infty\ (r \leq \sigma_1)$ | $\infty (r \leq \sigma_2)$ | $\infty \left( r \leq \dfrac{\sigma_1 + \sigma_2}{2} \right)$ | Extended |
| $0\ (r \geq \sigma_1)$ | $0\ (r \geq \sigma_2)$ | $0 \left( r \geq \dfrac{\sigma_1 + \sigma_2}{2} \right)$ | |
| $A(r)$ contains terms which simulate quantum mechanical overlap and polarization effects | | | Refined |

tial. We specify in Table 1.1 various forms of $A_{ij}(r)$ frequently encountered in the literature.

One system of considerable heuristic value is a hypothetical one composed of a dense assembly of hard spheres and represents a useful reference system. A hard-sphere liquid is clearly a simple liquid wi the above definition and the form of $\phi$ is given in Figure 1.2a; the corresponding functions $\phi$ for a binary mixture are illustrated in Figure 1.2b.

Two rather extreme pictures have emerged for amorphous solids. In the dense random packing of hard spheres (DRPHS), the interaction can be represented by pair potentials like those in equation (1.1.2) above. If the small sphere is a metalloid like phosphorus, this approach needs to be modified[6] to ensure that each metalloid atom is surrounded by metal atoms.

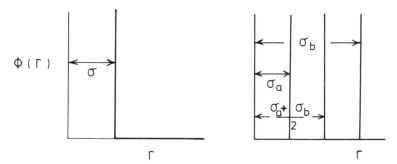

Figure 1.2. Hard-sphere potentials.

For chalogenide glasses, the 8-$N$ rule is often a guide to coordination. In these cases, a simple pairwise potential would not be sufficient to ensure that the coordination/valence requirements are properly satisfied and chemical arguments become very important.

## 1.2. Structure Factor $S(k)$ and Its Fourier Transform $g(r)$

### 1.2.1. Determination by Experiment

The scattering of radiation from a system containing identical atoms can, in the Born approximation, be most usefully discussed in terms of the so-called scattering law $S(k, \omega)$. The quantity measures the probability that energy $\hbar\omega$ will be transferred to the neutron if momentum to $\hbar k$ is imparted to the system.

In a conventional diffraction experiment (Figure 1.3) an integrated intensity $I(\theta)$ is measured at a given angle of scatter $2\theta$. The integration over the energy transfer is performed by the detector so that an effective differential scattering cross-section $(d\sigma/d\Omega)_{\text{eff}}$ is measured. It was Placzek[7] who first showed that

$$\left(\frac{d\alpha}{d\Omega}\right)_{\text{eff}} \propto \int_{-E_0/\hbar}^{\infty} \frac{k_1}{k_0} S(k, \omega) f(\omega) d\omega \tag{1.2.1}$$

where $E_0$ is the incident energy, $k_0$ and $k_1$ are the incident and final wave numbers, and $f(\omega)$ measures the energy dependence of the detector efficiency. The momentum and energy transfers are related by

$$\frac{\hbar k^2}{2m} = 2E_0 + \hbar\omega - 2(E_0^2 + \hbar\omega E_0)^{1/2} \cos 2\theta$$

Thus an exceedingly complex cross-section is determined in a conventional experiment: when $E_0$ greatly exceeds the energy transfer, the so-called

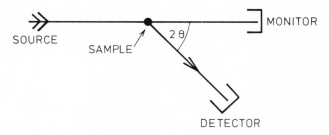

Figure 1.3. Conventional layout for diffraction studies.

static approximation applies and the right-hand side of equation (1.2.1) reduces to the static structure factor $S(k)$. This is defined in the usual way by an equation of the form

$$Nf^2 S(k) = \frac{d\sigma}{d\Omega} \qquad (1.2.2)$$

where

$$(d\sigma/d\Omega) = f^2 \left\langle \sum_k \sum_j \exp i\mathbf{k} \cdot (\mathbf{r}_k - \mathbf{r}_j) \right\rangle$$

and $f$ is an appropriate scattering factor. The angle brackets $\langle \ \rangle$ denote a time (or ensemble) average.

In practice, the observed scattered intensity $I(\theta)$ contains substantial contributions from incoherent effects and multiple scattering. These must be corrected and means found to put the experimental data on an absolute scale and allow for sample absorption.

Quite generally

$$I(\theta) = \alpha(\theta) \left[ \frac{d\sigma}{d\Omega} + \delta(\theta) \right] \qquad (1.2.3)$$

and the challenge which faces experimentalists is to determine the calibration parameters $\alpha(\theta)$ and $\delta(\theta)$ by theory, experiment, or both.

In Table 1.2 we show the physical origins of $\alpha(\theta)$ and $\delta(\theta)$ for two principal types of radiation used, X-rays and neutrons. We shall not dwell on the determination of the calibration parameters, except to note that it is not a trivial matter; for example, the contribution to $\delta(\theta)$ from inelastic effects for the case of neutron scattering from hydrogenous liquids needs particular care.[8]

**Table 1.2. Contributions to the Calibration Parameters**

|  | $\alpha(k)$ | $\delta(k)$ |
|---|---|---|
| X-rays | Sample geometry<br>Apparatus geometry<br>Polarization corrections<br>Absorption in the sample | Compton scattering<br>Multiple scattering |
| Neutrons | Sample geometry<br>Apparatus geometry<br>Absorption in the sample<br>Absorption by the sample container | Multiple scattering<br>Scattering by the sample container<br>Incoherent scattering<br>Placzek (inelastic) effects |

Summaries of the principal methods used for the reduction of $I(\theta)$ to $S(k)$ have been given by Enderby[9] and Pings[10] for neutrons and X-rays, respectively.

We conclude this section with a brief discussion of the $f$-factors for the two principal types of radiation, X-rays and neutrons. The essential difference can be seen by reference to Figures 1.4 and 1.5.[11] For X-rays, $f$

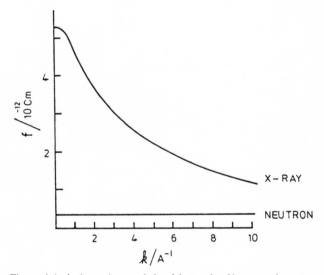

Figure 1.4. $k$-dependence of the $f$-factor for X-rays and neutrons.

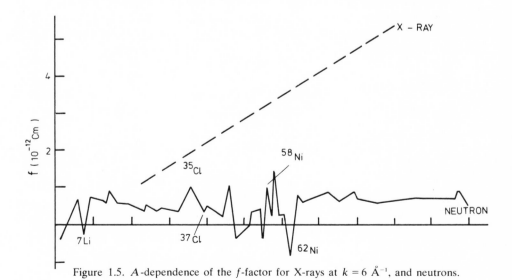

Figure 1.5. $A$-dependence of the $f$-factor for X-rays at $k = 6$ Å$^{-1}$, and neutrons.

Table 1.3. Examples of Coherent Scattering Lengths ($10^{-12}$ cm)

| Element or isotope | $f$ | Element or isotope | $f$ |
|---|---|---|---|
| H | −0.372 | Fe | 0.951 |
| D | 0.670 | $^{54}$Fe | 0.42 |
| $^{6}$Li | 0.18 | $^{56}$Fe | 1.01 |
| $^{7}$Li | −0.21 | $^{57}$Fe | 0.23 |
| $^{35}$Cl | 1.17 | | |
| $^{37}$Cl | 0.308 | Ni | 1.03 |
| | | $^{58}$Ni | 1.41 |
| Ca | 0.49 | $^{60}$Ni | 0.282 |
| $^{40}$Ca | 0.49 | $^{62}$Ni | −0.87 |
| $^{44}$Ca | 0.18 | $^{64}$Ni | −0.037 |

(usually referred to as the form factor) is $k$-dependent and, at a given value of $k$, increases with atomic weight $A$. The $f$-factor for neutrons, the so-called coherent scattering length, is isotropic, isotope-dependent, and bears no simple relationship to $A$. These properties make neutrons a particularly useful tool with which to study ionic liquids. For example, the ability of light elements to scatter neutrons allows unambiguous information to be obtained about ion–water conformations in aqueous solution. However, the most significant property of the coherent scattering length so far as structure is concerned is its isotopic dependence (Table 1.3) and we shall return to this theme in due course.

Note that the method of structure determination and the interpretation in terms of $S(k)$ and $g(r)$ (see Section 1.2.2 below) applies equally well to both amorphous solids and liquids. The application of scattering methods will be illustrated in this chapter by reference to liquids, and Chapter 2 will focus particular attention on amorphous solids.

### 1.2.2. Relationships between $g(r)$ and $S(k)$

The structure factor introduced in equation (1.2.2) was defined by

$$S(k) = \frac{1}{N} \left\langle \sum_k \sum_j \exp i\mathbf{k}(\mathbf{r}_k - \mathbf{r}_j) \right\rangle$$

$$= \frac{1}{N} \left\langle \sum_{k=j} e^{-i\mathbf{k}\cdot(\mathbf{r}_k - \mathbf{r}_j)} + \sum_{k \neq j} e^{i\mathbf{k}(\mathbf{r}_k - \mathbf{r}_j)} \right\rangle$$

$$= 1 + \frac{1}{N} \left\langle \sum_{k \neq j} e^{-i\mathbf{k}(\mathbf{r}_k - \mathbf{r}_j)} \right\rangle$$

$$= 1 + \frac{1}{N} \left\langle \sum \sum \cos \mathbf{k} \cdot (\mathbf{r}_k - \mathbf{r}_j) \right\rangle \qquad (1.2.4)$$

We now define the quantity $n_2(\mathbf{r}_k, \mathbf{r}_j)d\mathbf{r}_k d\mathbf{r}_j$ as the probability that both volume elements $d\mathbf{r}_k$ and $d\mathbf{r}_j$ are occupied. We may therefore write

$$S(k) = 1 + \frac{1}{N} \int_V \int_V \cos k(\mathbf{r}_k - \mathbf{r}_j) n_2(\mathbf{r}_k, \mathbf{r}_j) d\mathbf{r}_k d\mathbf{r}_j \qquad (1.2.5)$$

where $V$ is the volume of the sample. In a liquid $n_2(\mathbf{r}_k, \mathbf{r}_j)$ depends only on $|(\mathbf{r}_k - \mathbf{r}_j)| = r$. Let $g(r)$, the radial distribution function, be given by $g(r)\rho^2 = n_2(r)$ where $\rho$ is the mean density of atoms. With this definition, $g(r)$ tends to unity at large $r$ because the occupancy of two distant volume elements is uncorrelated. It follows that (1.2.5) may be written, in the limit $V \to \infty$, as

$$S(k) = 1 + \rho \int_{\text{all space}} \cos(\mathbf{k} \cdot \mathbf{r})[g(r) - 1]d\mathbf{r} + \int_{\text{all space}} \cos(\mathbf{k} \cdot \mathbf{r})d\mathbf{r} \qquad (1.2.6)$$

The last term in equation (1.2.6) is a delta function situated at $k = 0$ and may be disregarded. Thus

$$S(k) = 1 + \rho \int_0^\infty \int_{\phi=0}^\pi \cos(kr \cos \phi)[g(r) - 1]2\pi r^2 dr \sin \phi d\phi$$

$$= 1 + \rho \int \sin kr[g(r) - 1] \frac{4\pi r^2}{kr} dr$$

$$= 1 + \frac{4\pi\rho}{k} \int dr[g(r) - 1]r \sin kr \qquad (1.2.7)$$

It is a straightforward matter to invert (1.2.7) and hence obtain an explicit expression for $g(r)$ in the form

$$g(r) = 1 + \frac{1}{2\pi^2\rho r} \int dk[S(k) - 1]k \sin kr \qquad (1.2.8)$$

Thus measurements of $S(k)$ can, in principle, be used to determine $g(r) - 1$. A difficulty in practice, however, is that the upper integration limit cannot exceed approximately 12 Å$^{-1}$ so that some truncation errors are unavoidable. A further problem is that the random errors in $S(k)$, though usually small for any given value of $k$, give rise to a cumulative error in the transform that appears as a sharp increase in the magnitude of $[g(r) - 1]$ at values of $r$ less than about 1 Å. Thus, within the hard-core radius where strictly $g(r) \equiv 0$, considerable "experimental" noise is often in evidence.

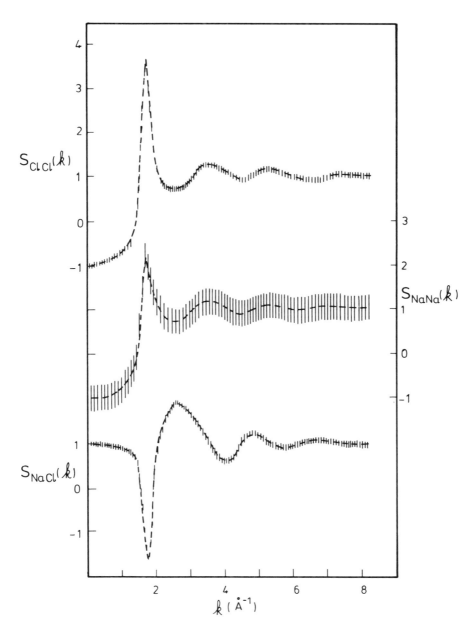

Figure 1.7. Partial structure factors for molten NaCl at 875 °C (after Edwards *et al.*[15]).

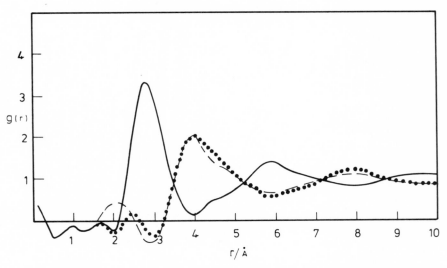

Figure 1.8. Pair correlation function for molten NaCl at 875 °C derived by Biggin and Enderby[17] from the original data of Edwards *et al.*[15] Full curve, $g_{NaCl}(r)$; broken curve, $g_{NaNa}(r)$; dotted curve, $g_{ClCl}(r)$.

## 1.4. Structure of Molten Salts

Ten molten salts have been studied by the isotopic substitution method so far. A summary of the results is presented in Table 1.4. This experimental work has enabled definite conclusions to be drawn about the structure of molten salts and has allowed, for the first time, detailed comparisons to be made with theory.

So far as alkali-metal/chloride salts are concerned the following conclusions are worth special mention:

1. The $k$-space data are dominated by a "coulomb" peak in $S_{--}(k)$ and $S_{++}(k)$ at a value of $k$ equal to 2 Å$^{-1}$ and a corresponding "coulomb" dip in $S_{+-}(k)$ at the same value of $k$. The existence of a small amount of penetration by the like ions into the first coordination shell is clearly visible (see Figures 1.7 and 1.8).
2. The position and magnitude of the principal peaks in $g_{++}$, $g_{+-}$, and $g_{--}$ generally coincide with those predicted by the ion–ion interactions represented by equation (1.2.1).
3. The tendency for charge cancellation first noted theoretically by Woodcock and Singer[18] is evident in the experimental data. Beyond 5 Å, the cancellation is almost complete.

## Table 1.4. Data on Ten Molten Salts Studied by the Isotopic Substitution Method

| Melting temperature °C | Salt | Ref. | Position of 1st max/Å | | | Position of 1st min/Å | | | Coordination number | | | Method for $n$ | Measuring temperature (°C) | Solid-state values (25 °C) | | |
|---|---|---|---|---|---|---|---|---|---|---|---|---|---|---|---|---|
| | | | $+-$ | $++$ | $--$ | $+-$ | $++$ | $--$ | $n_{+-}$ | $n_{++}$ | $n_{--}$ | | | $r_{+-}/$Å | $n_{+-}$ | Structure |
| 801 | NaCl | a | 2.8 | 3.9 | 3.9 | 3.9 | 6.0 | 5.8 | 5.3 | 13.0 | 13.0 | (iv) | 875 | 2.81 | 6 | B1 |
| 770 | KCl | b | 3.1 | 4.8 | 4.8 | 4.9 | 7.0 | 7.0 | 6.1 | 15.9 | 16.2 | (iv) | 800 | 3.41 | 6 | B1 |
| 718 | RbCl | c | 3.2 | 4.9 | 4.9 | 5.4 | 7.4 | 7.2 | 6.9 | 13.0 | 14.0 | (iv) | 750 | 3.27 | 6 | B1 |
| 645 | CsCl | d | 3.4 | 3.9 | 3.9 | 5.4 | 4.3 | 4.3 | 6.0 | — | — | (iv) | 700 | 3.51 | 6 | B1 |
| 430 | CuCl | d | 2.3 | 3.7 | 3.9 | 3.7 (broad) | 5.2 | 5.3 | — | — | — | — | 500 | 2.35 (445 °C) | | Zinc blend |
| 455 | AgCl | e | 2.6 | 3.4 | — | 3.4 | 3.9 | — | 4.3 | 3.1 | — | ? | 510 | 2.77 | 6 | B1 |
| | | | 2.6 | 3.2 | 3.2 | 3.1 | 3.9 | 3.7 | 2.7 | 4.1 | 2.7 | ? | 850 | | | |
| 963 | BaCl$_2$ | f | 3.1 | 4.9 | 3.9 | 4.1 | 7.7 | 5.1 | 6.4 | 14.0 | 7.0 | (ii) | 1025 | 3.18 (920 °C) | 8 | Fluorite |
| 283 | ZnCl$_2$ | g | 2.3 | 3.8 | 3.7 | 3.2 | 4.8 | 4.7 | 4.3 | 4.7 | 8.6 | (ii) | 323 | 2.27 | 4 | Irregular |
| 782 | CaCl$_2$ | h | 2.8 | 3.6 | 3.7 | 3.5 | 4.8 | 4.7 | 5.3 | 4.2 | 7.8 | (i) | 820 | 2.78 | 6 | Rutile |
| 875 | SrCl$_2$ | i | 2.9 | 5.0 | 3.8 | 4.0 | 7.0 | 4.8 | 6.9 | 13.6 | 9.3 | (iv) | 925 | 3.03 | 8 | Fluorite |

[a] S. Biggin and J. E. Enderby, *J. Phys. C* **15**, L305 (1982) (see also Edwards *et al.*[(15)]).
[b] Y. Derrien and J. Dupuy, *J. Phys. (Paris)* **36**, 191 (1975).
[c] E. W. J. Mitchell, P. F. J. Poncet, and R. J. Stewart, *Phil Mag.* **34**, 721 (1976).
[d] S. Eisenberg, J.-F. Jal, J. Dupuy, P. Chieux, and W. Knoll, *Phil. Mag. A* **46**, 195 (1982).
[e] Y. Derrien and J. Dupuy, *Phys. Chem. Liq.* **5**, 71 (1976).
[f] F. G. Edwards, R. A. Howe, J. E. Enderby, and D. I. Page, *J. Phys. C* **11**, 1053 (1978).
[g] S. Biggin and J. E. Enderby, *J. Phys. C* **14**, 3129 (1981).
[h] S. Biggin and J. E. Enderby, *J. Phys. C* **14**, 3577 (1981).
[i] R. McGreevy and E. W. J. Mitchell, *J. Phys. C* **15**, 5537 (1982).

The situation for 2–1 molten salts is more complicated and reference should be made to the original papers for an adequate discussion. Small cations ($Mg^{2+}$, $Zn^{2+}$, $Ca^{2+}$) tend to occupy tetrahedral sites produced by the relatively close-packed $Cl^-$ subsystem and there is very little penetration (Figure 1.9). On the other hand, as the cation size increases the amount of like-ion penetration becomes very significant (Figure 1.10).

In the past there has been considerable controversy about the correct value of the coordination number $n_{+-}$ for molten salts. Biggin and Enderby[17] have shown that this controversy arises from: (1) the sensitivity to definition, particularly marked for ionic liquids; and (2) inherent ambiguity in the weighted distribution function $G(r)$. They emphasize that, in statistical mechanics, there is no special significance attached to the coordination number, thermodynamic properties being related to integrals over $g_{\alpha\beta}(r)$ for all $r$. However, in approximate treatments of ionic melts, such as the quasi-chemical theory, a physically reasonable number must be available for near-neighbor interactions. Table 1.5 presents values of $n_{+-}$ for two molten salts, NaCl and $BaCl_2$, derived by four different methods[21] and using input data $g_{+-}(r)$ and $G(r)$.

We conclude that for systems in which the cation is small (such as $Li^+$, $Na^+$, $Mg^{2+}$, $Zn^{2+}$, and $Mn^{2+}$), method (i) applied to $G(r)$ is satisfactory. The fundamental reason for this is the absence of significant penetration of like

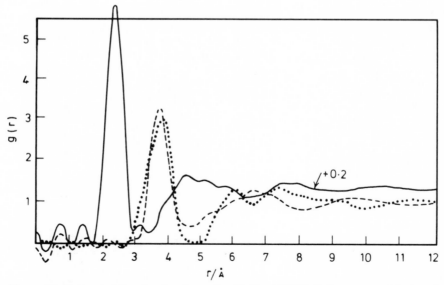

Figure 1.9. Pair correlation function for molten $ZnCl_2$ at 327 °C. Full curve, $g_{ZnCl}(r)$; dotted curve, $g_{ZnZn}(r)$; broken curve, $g_{ClCl}(r)$.[19] For the sake of clarity $g_{ZnCl}(r)$ has been displaced by +0.2.

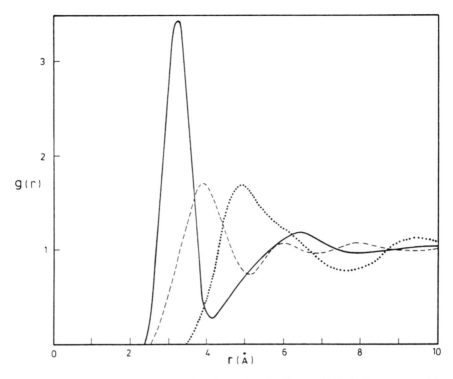

Figure 1.10. Radial distribution functions for molten $BaCl_2$ at 1298 K. Full curve, $g_{BaCl}(r)$; broken curve, $g_{ClCl}(r)$; dotted curve, $g_{BaBa}(r)$.[20]

## Table 1.5. Coordination Numbers for NaCl and BaCl₂

| Melt | Method[a] | (i) $[rg(r)]_{symm}$ | (ii) $[r^2g(r)]_{symm}$ | (iii) 1st minimum in $r^2(r)$ | (iv) 1st minimum in $g(r)$ |
|---|---|---|---|---|---|
| NaCl | $g_{+-}(r)$ | $3.9 \pm 0.2$ | $4.0 \pm 0.2$ | $5.2 \pm 0.4$ | $5.3 \pm 0.4$ |
| (Na⁺ radius | $G_{total}(r)$ | 4.0 | 4.1 | 4.1 | 4.5 |
| 0.95 Å) | (X-ray) | | | | |
| BaCl₂ | $g_{+-}(r)$ | $6.4 \pm 0.2$ | $6.3 \pm 0.2$ | $7.0 \pm 0.4$ | $7.0 \pm 0.4$ |
| (Ba²⁺ radius | $G_{total}(r)$ | 15 | 16 | 19 | 26 |
| 1.35 Å) | (neutron) | | | | |

[a] The four methods are described and discussed by Waseda.[21]

### Table 1.6. Coordination Numbers for Various Molten Salts

| Melt | Cationic radius (Å) | Equation (4.1) | $n_{+-}$ Method (i) applied to $g_{+-}(r)$ |
|------|------|------|------|
| $ZnCl_2$ | 0.74 | 3.5 | 4.3 |
| NaCl | 0.95 | 4.0 | 3.9 |
| $CaCl_2$ | 0.99 | 5.8 | 5.3 |
| $SrCl_2$ | 1.12 | 6.5 | 5.1 |
| KCl | 1.33 | 4.1 | 4.1 |
| $BaCl_2$ | 1.35 | 7.5 | 6.4 |
| RbCl | 1.47 | 3.8 | 3.5 |

ions into the first shell. For large cations there is substantial penetration of the first shell by like ions. Thus $G(r)$ will, by any of the methods, produce a very misleading value for $n_{+-}$, as is evident from the example of $BaCl_2$ given in Table 1.5.

A final noteworthy point is that method (i) can be theoretically justified in terms of solid-like local order[22] and should therefore yield coordination numbers in good overall agreement with the semiempirical formula proposed by Ohno and Furukawa[23]

$$n_{+-}(\text{liquid}) = n_{+-}(\text{solid}) \left(\frac{V_s}{V_l}\right) \left(\frac{r_{+-}(\text{liquid})}{r_{+-}(\text{solid})}\right)^3 \qquad (1.4.1)$$

where $V_l$ and $V_s$ are the molar volumes of the liquid and solid states, respectively. The results in Table 1.6 show that, provided $g_{+-}(r)$ is used to obtain $n_{+-}$, equation (1.4.1) is a very useful approximation for a wide variety of molten salts.

## 1.5. Structure of Concentrated Ionic Solutions

### 1.5.1. Introduction

Consider a salt $MX_p$ dissolved in $H_2O$. From the arguments outlined in Section 1.2.3, it is clear that there are *ten* radial distribution functions and ten corresponding partial structure factors.

(i) The distribution functions $g_{MO}$, $g_{MH}$, $g_{XO}$, and $g_{XH}$ tell us about the coordination of water molecules around the cations (M) and the anions (X), and are of interest to scientists working on ion hydration.

(ii) The distribution functions $g_{MM}$, $g_{XX}$, and $g_{MX}$ describe ion–ion correlations and can be compared with those derived from theoretically with those derived from effective (Mayer–MacMillan) pair potentials similar to those described in Table 1.1.

(iii) Finally, the distribution functions $g_{OO}$, $g_{HH}$, and $g_{OH}$ describe the water structure in the solution and can therefore be used to discuss changes in the water brought about by the presence of ions.

The function $F(k)$ for $MX_p$ in $H_2O$ can be written explicitly in the form

$$F(k) = \tfrac{1}{9}(1 - c - pc)^2 f_O^2(S_{OO} - 1) + \tfrac{4}{9}(1 - c - pc)^2 f_H^2(S_{HH} - 1)$$
$$+ \tfrac{4}{9}(1 - c - cp)^2 f_O f_H(S_{OH} - 1) + c^2 f_M^2(S_{MM} - 1) + p^2 c^2 f_X^2(S_{XX} - 1)$$
$$+ 2pc^2 f_X f_M(S_{XM} - 1) + \tfrac{2}{3}c(1 - c - pc)f_M f_O(S_{MO} - 1)$$
$$+ \tfrac{4}{3}c(1 - c - pc)f_M f_H(S_{MH} - 1) + \tfrac{4}{3}pc(1 - c - pc)f_X f_H(S_{XH} - 1)$$
$$+ \tfrac{2}{3}pc(1 - c - pc)(S_{XO} - 1)$$

where $c$ is the atomic fraction of the cation (see below).

The difficulties involved in using diffraction data to determine the local order in solutions are now evident and can be summarized as follows:

1. The above equation illustrates that $F(k)$ comprises ten partial structure factors each weighted by appropriate scattering and concentration factors. Even for very concentrated solutions the ion–water contributions to the total scattering pattern are typically only 30% (X-rays) and 10% (neutrons), and it is therefore exceedingly difficult to disentangle unambiguously the ion–water and ion–ion terms. As the electrolyte concentration is reduced, the situation becomes progressively less favorable (Figure 1.11).

2. Both neutron and X-ray methods suffer from special disadvantages for hydrogenous liquid. We have already referred to the inelastic corrections so far as neutrons are concerned. X-ray methods suffer from the disadvantage that the contribution from ion–hydrogen correlations to the total pattern is essentially zero.

The method developed by our group to overcome these difficulties has been fully discussed in the literature (see, for example, Enderby and Neilson[24]). Essentially, by changing the isotope systematically, one obtains information about ion hydration and ion–ion correlations.

## 1.5.2. Ion Hydration

Suppose diffraction experiments have been conducted on two solutions, which are identical chemically but differ isotopically. For example, the two solutions might be 1 molal $K^{35}Cl$ and 1 molal $K^{37}Cl$ and we tabulate

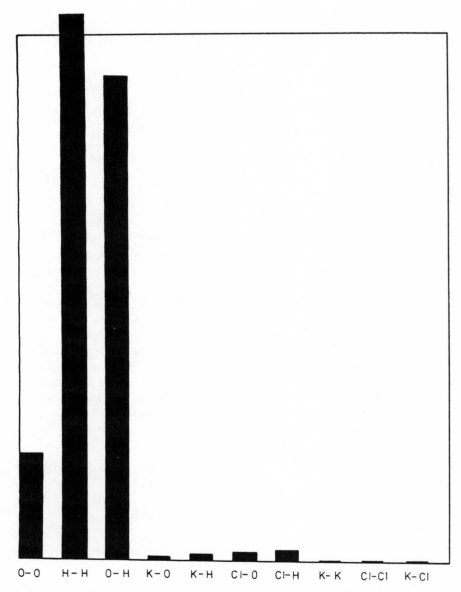

Figure 1.11. Weighting of the various contributions to a total neutron pattern for a 1 molal solution of KCl in water.

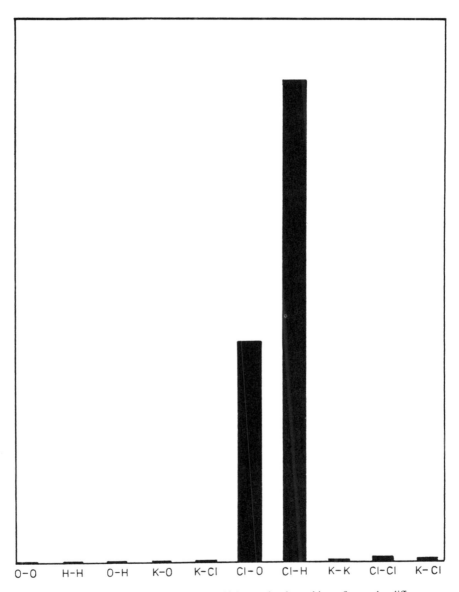

Figure 1.12. Weighting of the four terms which remain after taking a first-order difference.

the two intensity patterns $I_1(k)$ and $I_2(k)$. The first-order difference $\Delta(k) = I_2(k) - I_1(k)$ will contain only the four contributions

$$2N_0 N_{Cl} f_0 \Delta f_{Cl}[S_{ClO}(k) - 1 - 1] + 2N_D N_{Cl} f_D \Delta f_{Cl}[S_{ClD}(k) - 1]$$
$$+ 2N_K N_{Cl} f_K \Delta f_{Cl}[S_{KCl}(k) - 1] + N_{Cl}^2 \Delta f_{Cl}^2[S_{ClCl}(k) - 1] \qquad (1.5.1)$$

(We usually carry out neutron experiments with heavy water, $D_2O$, rather than light water, $H_2O$; the subscript D in place of H indicates this fact.) When the numerical factors are substituted into equation (1.5.1), we find that only the first two really matter (Figure 1.12). Thus for practical purposes, $\Delta(k)$, and hence by equation (1.2.8), $\Delta G(r)$, yields the information needed to construct our pattern of solutions — the distribution of water molecules around the anion $Cl^-$. Similarly, if the isotope of the cation is changed, the distribution of water molecules around $K^+$ can be deduced directly.

Many systems have been studied by this method and a few examples are shown in Table 1.7. Let us consider in more detail one of them, the monovalent ion $Li^+$. The form of $\Delta G(r)$ for one of the solutions studied is shown in Figure 1.13. It is clear from the value of $\Delta G(r)$ near $r = 3.2$ Å that

**Table 1.7. Results of Ion Hydration**

| Ion and solute | Molality | Ion–oxygen distance/Å | Ion–deuterium distance/Å | $\theta/\deg^a$ | Coordination number |
|---|---|---|---|---|---|
| $Ni^{2+}$ | | | | | |
| $NiCl_2$ | 0.086 | $2.07 \pm 0.03$ | $2.80 \pm 0.03$ | $0 \pm 20$ | $6.8 \pm 0.8$ |
| | 0.46 | $2.10 \pm 0.02$ | $2.80 \pm 0.02$ | $17 \pm 10$ | $6.8 \pm 0.8$ |
| | 0.85 | $2.09 \pm 0.02$ | $2.76 \pm 0.02$ | $27 \pm 10$ | $6.6 \pm 0.5$ |
| | 1.46 | $2.07 \pm 0.02$ | $2.67 \pm 0.02$ | $42 \pm 8$ | $5.8 \pm 0.3$ |
| | 3.05 | $2.07 \pm 0.02$ | $2.67 \pm 0.02$ | $42 \pm 8$ | $5.8 \pm 0.2$ |
| | 4.41 | $2.07 \pm 0.02$ | $2.67 \pm 0.02$ | $42 \pm 8$ | $5.8 \pm 0.2$ |
| $Ni(ClO_4)_2$ | 3.80 | $2.07 \pm 0.02$ | $2.67 \pm 0.02$ | $42 \pm 8$ | $5.8 \pm 0.2$ |
| $Ca^{2+}$ | | | | | |
| $CaCl_2$ | 1.0 | $2.46 \pm 0.03$ | $3.07 \pm 0.03$ | $38 \pm 9$ | $10.0 \pm 0.6$ |
| | 2.8 | $2.39 \pm 0.02$ | $3.02 \pm 0.03$ | $34 \pm 9$ | $7.2 \pm 0.2$ |
| | 4.5 | $2.41 \pm 0.03$ | $3.04 \pm 0.03$ | $34 \pm 9$ | $6.4 \pm 0.3$ |
| $Cu^{2+}$ | 4.32 | $2.05 \pm 0.02$ | $2.56 \pm 0.10$ | $56 \pm 5$ | $2.3 \pm 0.3$ |
| $CuCl_2$ | | | | | |
| $Cu(ClO_4)$ | 3.0 | $1.97 \pm 0.02$ | $2.60 \pm 0.02$ | $39 \pm 8$ | $6 \pm 1$ |

$^a$ $\theta$ is the tilt angle (see Figure 1.13).

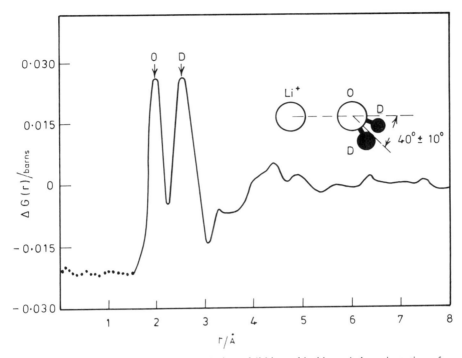

Figure 1.13. $\Delta G(r)$ for a 3.57 molal solution of lithium chloride and the orientation of a water molecule close to $Li^+$.

$Li^+$ forms, in solution, a very well hydrated ion. (An independent measurement suggests a residence time of $10^{-11}$ s for a water molecule close to a lithium ion, a long time compared with the rearrangement time for ions in electrolyte solutions.) The number of water molecules around each $Li^+$ is six for solutions of concentration 3 molal or less, rather than four as had been previously supposed. The average orientation of each water molecule is also shown in Figure 1.13. Note that the tilt angle between the Li–O axis and the plane of the water molecule is about 40°. The detail that the new method uncovers in the ion–water conformation provides a sharp experimental test for the interaction parameters given by quantum theory,

## 1.6. Structure of Liquid Semiconductors

Liquid Te frequently forms one element in a wide range of liquid semiconductors.[25] When Tl, Ag, Cu, or Ni are added to liquid Te, a variety

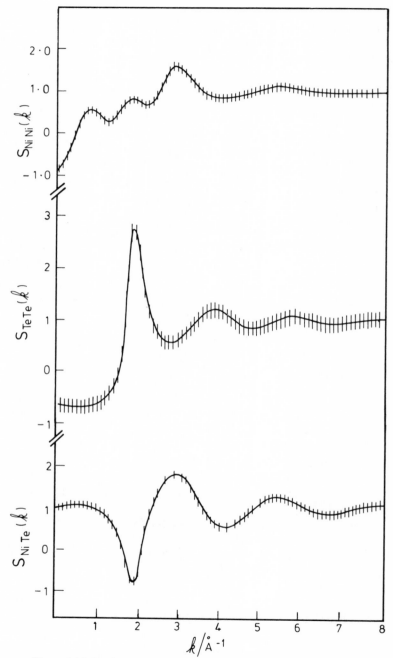

Figure 1.14. Partial structure factor $S(k)$ for liquid NiTe at 930 °C.

of electrical behavior occurs at low concentration. The addition of copper to tellurium hardly changes the conductivity until the concentration CuTe is reached, while silver decreases it by approximately 35 $\Omega cm^{-1}$/at% over most of the composition range up to $Ag_2Te$. The behavior of Tl is intermediate between Cu and Ag. All three systems show a sharp minimum in conductivity at the stoichiometric compositions $Tl_2Te$, $Cu_2Te$, and $Ag_2Te$.[25] The conductivity of liquid Te, on the other hand, *increases* with the addition of nickel and is doubled by as little as 10 at%.[26]

A series of structural measurements has been made on these alloy systems and we refer especially to the recent work on liquid NiTe and $NiTe_2$.[27] In $k$-space (Figure 1.14) the dip in the cross term $S_{NiTe}(k)$ when $k$ equals about 2 $\overset{\circ}{A}^{-1}$ corresponds to maxima in $S_{NiNi}(k)$ and $S_{TeTe}(k)$. This behavior is a signal that the charge transfer from the Ni to the Te has taken place. From the real-space data shown in Figure 1.15 it is possible to extract the coordination numbers and peak positions, and these are shown in Table 1.8 where we include, for comparison, some earlier data on liquid CuTe.[28]

It is clear that both the Te–Te structure factor and pair correlation function are very different from those characteristic of pure liquid Te. At

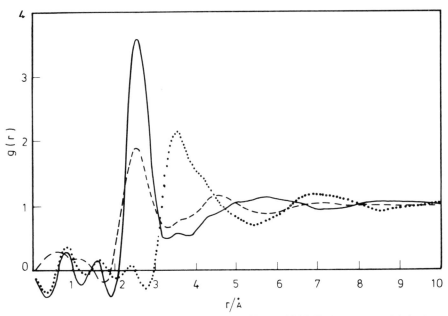

Figure 1.15. Pair correlation functions for liquid NiTe at 930 °C. Full curve, $g_{NiTe}(r)$; broken curve, $g_{NiNi}(r)$; dotted curve, $g_{TeTe}(r)$.

**Table 1.8. Average Coordination Numbers and Distances (M = Ni or Cu)**

|  | Average coordination numbers | | | Distances (Å) | | |
|---|---|---|---|---|---|---|
|  | M–M | Te about M | Te–Te | M–M | Te–M | Te–Te |
| NiTe | $2.9 \pm 0.4$ | $4.5 \pm 0.5$ | $10 \pm 3$ | $2.54 \pm 0.03$ | $2.56 \pm 0.03$ | $3.56 \pm 0.04$ |
| NiTe$_2$ | $1.6 \pm 0.4$ | $4.6 \pm 0.3$ | $11 \pm 3$ | $2.81 \pm 0.03$ | $2.56 \pm 0.03$ | $3.55 \pm 0.04$ |
| CuTe$^2$ | $3.6 \pm 0.4$ | $2.9 \pm 0.3$ | $10 \pm 3$ | $2.82 \pm 0.03$ | $2.55 \pm 0.03$ | $3.66 \pm 0.05$ |

NiTe, for example, all traces of the unusual structural properties of the liquid Te have been lost. It is particularly significant that $r_{TeTe}$ for the alloy is some 20% *greater* than in the pure liquid and $n_{TeTe}$ has changed from 3 to 11. These facts strongly support the notion that tellurium in Ni–Te alloys behaves as an anion; in molten NaCl, for example, $n_{ClCl}$ is 13 and $r_{ClCl}$ is 3.8 Å. The sum of the conventional ionic radii $r_{Ni}^{2+} + r_{Te}^{2-}$ is 2.80 Å, which is in reasonable agreement with $r_{NiTe}$ (2.56 Å).

If we focus attention on $g_{NiTe}(r)$ and $g_{NiNi}(r)$ we conclude that, on the average, a nickel ion is surrounded by about four $Te^{2-}$ ions at 2.56 Å and about two $Ni^{2+}$ ions at 2.54 Å for NiTe. This local order provides the essential basis for understanding the electrical behavior of Ni–Te alloys. The $Te^{2-}$ ions form a close-packed structure in much the same way as $Cl^-$ does. However, in the molten-salt case once the ratio $r_-/r_+$ exceeds approximately 2.4, essentially no cations can penetrate the first coordination shell.[29] By contrast, the cations in these alloys clearly do penetrate the first coordination shell and are able to form a threefold coordinated structure with a Ni–Ni spacing sufficiently short to allow significant $d$-band overlap. Similar considerations apply to liquid CuTe. We therefore conclude that the local order characterizing both $Ni^{2+}$ and $Cu^{2+}$ in the melt is sufficient to give an unfilled $d$-band and, in accordance with the principles set out by Wilson,[30] electrons in this band are on the delocalized side of the Mott transition.* In this sense both liquid NiTe and CuTe are "metallic" and the sign of the thermopower, for example, simply reflects the location of the Fermi energy with respect to the hybridized $d$-band. These liquids represent examples where a purely ionic description of the interaction needs to be modified to accommodate the specific nature of $d$–$d$ bonding.

---

* In the solid state CuTe and NiTe are metallic.[31] For NiTe the cation–cation distance in the liquid is smaller than in the solid (2.54 Å compared with 2.68 Å). The reverse is true for CuTe (2.82 Å compared with 2.63 Å) and this probably accounts for the higher resistivity.

Let us consider $Cu_xTe_{1-x}$ and $Ni_xTe_{1-x}$ for $x > 0.5$. For Cu–Te, the resistivity increases and becomes a maximum at $Cu_2Te$. By contrast, NiTe continues its gradual progression to metallic behavior. The presence of monovalent $Cu^+$ in the $d^{10}$-configuration implied by the stoichiometry of $Cu_2Te$ will clearly lead to low conductivities; no such behavior will occur for Ni–Te and the model therefore gives a natural explanation of the contrasting behavior of the two alloy systems for $x > 0.5$. We also considered liquid $Tl_2Te$ from the same point of view and concluded that the most appropriate structural model will again be based on $Tl^+$ and $Te^{2-}$ ions. Here, the complication of $d$-band effects can be neglected so that the conductivity of the stoichiometric liquid ($Tl_2Te$) will be low.

## 1.7. Conclusions

The structure of complex liquids and amorphous solids, though at an early stage in its development, is clearly benefiting from the method of multipattern diffraction. We anticipate that the increased availability of isotopes and high-intensity neutron sources (such as the SNS at the Rutherford Appleton Laboratory) will further enlarge the scope of the isotope method in the future. Of particular interest will be the isolation of the partial structure factors for systems in which a substantial degree of covalency is anticipated, such as liquid or amorphous $GeSe_2$. At the same time, methods based on the anomalous scattering of X-rays will also become feasible when powerful light sources like those at Daresbury and Brookhaven become fully operational.

ACKNOWLEDGMENTS. I wish to acknowledge my colleagues at Leicester and Bristol for helpful discussions and for performing many of the experiments described here. The Science and Engineering Research Council is also thanked for its continued support of the method of isotopic substitution.

## References

1. T. E. Faber, *Introduction to the Theory of Liquid Metals*, Cambridge University Press (1972).
2. M. J. L. Sangster and M. Dixon, *Adv. Phys.* **25**, 247 (1976).
3. M. L. Huggins and J. E. Mayer, *J. Chem. Phys.* **1**, 643 (1933).
4. F. Fumi and M. P. Tosi, *J. Phys. Chem. Solids* **25**, 31 (1964).
5. J. E. Mayer, *J. Chem. Phys.* **1**, 270 (1933).
6. D. E. Polk, *J. Non-Cryst. Solids* **11**, 381 (1973).

7. G. Placzek, *Phys. Rev.* **86**, 377 (1952).
8. J. G. Powles, *Mol. Phys.* **36**, 1161 (1978).
9. J. E. Enderby, in: *Physics of Simple Liquids* (H. N. V. Temperley, J. S. Rawlinson, and G. S. Rushbrooke, eds.), North-Holland, Amsterdam (1968).
10. C. N. J. Pings, in: *Physics of Simple Liquids* (H. N. V. Temperely, J. S. Rawlinson, and G. S. Rushbrooke, eds.), North-Holland, Amsterdam (1968).
11. G. Bacon, *Neutron Diffraction*, Clarendon Press, Oxford (1975).
12. G. Vineyard, *Liquid Metals and Solidification*, American Society for Metals, Cleveland, Ohio (1968).
13. D. T. Keating, *J. Appl. Phys.* **34**, 923 (1963).
14. J. E. Enderby, D. M. North, and P. A. Egelstaff, *Phil. Mag.* **14**, 961 (1966).
15. F. G. Edwards, J. E. Enderby, R. A. Howe, and D. I. Page, *J. Phys. C* **8**, 3483 (1975).
16. J. R. Westlake, *A Handbook of Numerical Matrix Inversion and Solution of Linear Equation*, Wiley, New York (1968).
17. S. Biggin and J. E. Enderby, *J. Phys. C* **15**, L305 (1982).
18. L. V. Woodcock and K. Singer, *Trans. Faraday Soc.* **67**, 12 (1971).
19. S. Biggin and J. E. Enderby, *J. Phys. C* **14**, 3129 (1981).
20. F. G. Edwards, R. A. Howe, J. E. Enderby, and D. I. Page, *J. Phys. C* **11**, 1053 (1978).
21. Y. Waseda, *Structure of Non-Crystalline Materials: Liquids and Amorphous Solids*, McGraw-Hill, New York (1980).
22. C. A. Coulson and G. S. Rushbrooke, *Phys. Rev.* **56**, 1216 (1939).
23. H. Ohno and K. Furukawa, *J. Chem. Soc., Faraday Trans. 1* **77**, 1981 (1981).
24. J. E. Enderby and G. W. Neilson, *Rep. Prog. Phys.* **44**, 593 (1981).
25. M. Cutler, *Liquid Semiconductors*, Academic Press, New York (1977).
26. R. J. Newport, R. A. Howe, and J. E. Enderby, *J. Phys. C* **15**, 4635 (1982).
27. V. T. Nguyen, M. Gay, J. E. Enderby, R. J. Newport, and R. A. Howe, *J. Phys. C* **15**, 4627 (1982).
28. I. Hawker, R. A. Howe, and J. E. Enderby, *Proc. Int. Conf. Electrical and Magnetic Properties of Liquid Metals*, p. 262, UNAM, Mexico (1975).
29. S. Biggin and J. E. Enderby, *J. Phys. C* **14**, 3577 (1981).
30. J. A. Wilson, *Adv. Phys.* **21**, 143 (1972).
31. F. Hulliger, *Struct. Bonding (Berlin)* **4**, 83 (1968).

# Modeling of Liquids and Amorphous Solids

## J. L. Finney

This chapter is concerned with the development and application of techniques used to model the structures of noncrystalline dense assemblies in the laboratory and on the computer. Stress is laid on amorphous solids and glasses, where the technical problems are greatest.

## 2.1. Introduction: Laboratory Models of Ideal Single-Component Assemblies

### 2.1.1. Crystals, Gases, and Liquids: The Nature of Models

We wish ideally to be able to derive from a theory or a model of a given assembly (such as a real crystal or a simple liquid) the partition function

$$Z = \sum_{\substack{\text{all} \\ \text{configurations}}} \exp(-\mathcal{H}/kT) \tag{2.1.1}$$

where $\mathcal{H}$ is the Hamiltonian for the assembly. The kinetic-energy terms in the Hamiltonian may be integrated out to yield

$$Z = Z_k \cdot Q \tag{2.1.2}$$

**J. L. Finney** · Crystallography Department, Birkbeck College, Malet Street, London WC1E 7HX, England.

where $Q$, the configurational integral, contains structural or positional information on the $N$ particles in a volume $V$ through the potential energy $U(q_1, \ldots, q_N)$, where $q_j$ are the position coordinates of particle $j$. Integral $Q$ can then be written as

$$Q = \frac{1}{N!} \int \cdots \int \exp[-U(q_1, \ldots, q_N)/kT] dq_1 \cdots dq_N \qquad (2.1.3)$$

Given $Q$, and hence $Z$, we can evaluate the thermodynamics of the assembly by formal manipulation. The central problem is to evaluate $U(q_1, \ldots, q_N)$, for which a model is required. Such a model could be a set of sample coordinates, which could be termed a structural model. It is with such models that we are concerned here. Note that we consider only a structural model — an instantaneous snapshot from which all movements have been frozen out. The separation of the Hamiltonian [equation (2.1.2)] allows this.

### 2.1.2. Models of Gases and Crystals

In an ideal gas, the instantaneous positions are determined solely by statistics. There is no intermolecular potential, no molecular size, and no structure: the radial distribution function is a constant. Potential energy $U(q_1, \ldots, q_N)$ is zero for all configurations, and $Q$ becomes simply $V^N/N!$ The thermodynamics of the ideal gas follows. For a real or imperfect gas, the intermolecular forces and molecular size result in a departure of the structure from complete randomness. The structural consequences of the small number of weak interactions in an imperfect gas can be considered in terms of pairs, triplets, and larger molecular clusters as the density increases. Unfortunately, such a cluster expansion fails to converge for even moderately dilute gases and hence is not easily applied to liquids.

In contrast to the gas, where we must consider a small number of weak interactions, the crystal presents us with the problem of handling a large number of strong interactions. In an ideal crystal, however, we know the location of every atom (all atomic positions can be referred to a lattice), hence we can write down $U(q_1, \ldots, q_N)$ and evaluate $Q$. Real (imperfect) crystals can be regarded as perturbations from the ideal structure. The mean position of each atom is known, and the problem is now essentially one of characterizing the lattice vibrations.

The liquid presents us with an intermediate case. The high (approximately crystal) density means we must cope with a large number of strong interactions. This precludes the simple approach from the gas side (too many interactions; the cluster expansion does not converge). Moreover, the absence of an underlying lattice prevents us making use of the crystal-lattice concept. We must look for other ways to approach the problem.

### 2.1.3. Models of Ideal Single-Component Liquids

By analogy with the ideal crystal (Section 2.1.2), we try to construct a model — initially in the laboratory — of an ideal liquid. Real liquids may then be referred to this ideal in a way similar to the real–ideal crystal relationship. To be acceptable, such a model must be consistent with known structural information from experiment, in particular the radial distribution function (RDF) and the density. If the RDF is correctly normalized, the density constraint is automatically included.

One of the earliest attempts to model a liquid structure was made by Bernal and Fowler[1] in relation to water. Although their discussion of variable topology in silicate and ice structures laid the basis for later work on water, Bernal argued that we needed a better understanding of simple liquids before full understanding of more complex ones could be achieved. Rice[2] in 1944 made a physical model of a simple van der Waals liquid by mixing colored gelatin spheres in a gelatin bath. Light-scattering measurements confirmed that the model could reproduce the main characteristics of liquid scattering. In the late 1950s, Bernal returned to the problem and constructed a laboratory ball-and-spoke model, using a distribution of spoke lengths that was consistent with the known RDF.[3]

This kind of approach was formalized in the 1960s by the construction of hard-sphere packing models,[3-5] in an attempt to examine the detailed geometric consequences of intermolecular interactions. The crystal was known to be a structure that was consistent with experimental density, the potential function, and the existence of a lattice. The liquid structure could be considered similarly as being also of the correct density and consistent with the potential function: the lattice constraint is, however, completely absent.

Within this conceptual framework, the ideal liquid can be realized by focusing on the main structure-determining characteristics of the potential function. Ideal-crystal structures were based on the use of hard spheres, resultant structures (fcc, hcp) being consequences of volume exclusion effects. The ideal liquid can be approached similarly by trying to realize a structure that is controlled by volume exclusion in the absence of an imposed lattice. Such models have been built in the laboratory by several methods; details of the construction, which stress the prevention of crystallinity and the achievement of adequate density and the measurement of the individual sphere coordinates, have been described

Figure 2.1. Part of a hard-sphere random packing. Note the lack of a crystalline lattice.

elsewhere.[3-8] The resultant structure, a "random close packing of hard spheres" (Figure 2.1), appears to have approximately the correct packing density (0.637) compared to that of the close-packed crystal (0.7405), giving a fractional density change on melting of about 15%, as observed for inert-gas crystals. The RDF was also generally consistent with experiment results on real liquids.[3]

This random close packing of hard spheres, a "homogeneous, coherent, and essentially irregular assemblage of molecules containing no crystalline regions,"[3] is taken as our ideal-liquid model. This beautifully simple idea is seen by Ziman as "the key to any qualitative or quantitative understanding of the physics of liquids," and "supersedes various other theoretical approaches based on phenomenological constructs such as 'holes in lattice', 'paracrystals', 'significant structures', 'dislocations', etc."[9] It is a model that is "homologous to that of the crystalline solid, as well as radically different in kind, and [has] a general quality of homogeneity."[3] Qualitatively, it accounts for the essential fluidity and high entropy of the liquid, and explains qualitatively both the discontinuity of state at the melting point, and the contrasted continuity with the vapor phase. It is consistent with the existence of a liquid-gas critical point and the absence of a crystal–liquid critical point. Moreover, it can be demonstrated[10] that the simple model and a computer-simulated liquid can be referred to each other in a simple way, effectively validating the ideality of the model with respect to more realistic liquids (cf. real and ideal crystals). Other, more recent molecular-dynamics calculations allow us to identify the random close-packed hard-sphere model with a hard-sphere glass at absolute zero.[11]

### 2.1.4. Summary

Even though a crystal of an assembly of atoms has a structure consistent with experimental density, the intermolecular interactions, and the existence of a lattice organization, a liquid or amorphous solid structure must be consistent with the same constraints except that of a lattice. All modeling work is thus concerned with realizing structures consistent with these constraints, either in the laboratory or on the computer. The constraints may differ quantitatively (e.g., silicates are network structures where dense packing of spheres are not relevant), but the underlying philosophy is unchanged.

## 2.2. Short-Range Order in Spherical and Network Glasses

### 2.2.1. Short-Range Order in the Random Close-Packed Model

The random close-packed model is our ideal model of a single-component glass of spherical atoms. Its detailed local structure, as described by the RDF (Figure 2.2), is a consequence of the packing constraints exerted in the hard-sphere assembly. It is noteworthy that the high-resolution RDF shown in Figure 2.2 exhibits a clear splitting of the second peak. Such a splitting is not observed for a simple liquid, but is a characteristic of metallic glasses; indeed, it is this RDF feature that first encouraged the application of the random packing as a model of metallic glass structure.[12,13]

The RDF peaks show those interatomic distances that are more probable (i.e., that occur more frequently) than others: this short-range order (SRO) is a consequence of the packing constraints in the structure. We can identify approximately the local geometries corresponding to these

RDF features as follows:

1. The large first peak at 1.0 diameters corresponds to first-neighbor contacts. The area under the peak relates to the first-neighbor coordination number (see Chapter 1 concerning this point).
2. The first component of the split second peak occurs at about 1.73 diameters, which is close to $\sqrt{3}$ diameters. This distance corresponds to that across two edge-sharing equilateral triangles.
3. The second component of the split second peak is at 2.0 diameters, and corresponds to second neighbors in a straight line, i.e., separated by a common first neighbor along the line of centers. Such "collineations" of three (and more) spheres were noted by Bernal[3] in his early work on laboratory-built models.

Note that these distances occur also in close-packed crystals: we must *not*, however, be tempted to draw the incorrect conclusion that the crystalline and noncrystalline structures are similar. For not only do these distances occur with different frequencies in crystal and noncrystal, but also (1) the spreads are qualitatively different and (2) characteristic *crystal* distances occur relatively rarely in the noncrystalline packing. Of particular interest is the absence of the $\sqrt{2}$ diameter peak found in fcc crystals; in the random-packing model, in fact, there is a minimum at this distance (Figure 2.2). Thus, although both crystal and glass models show some of the same

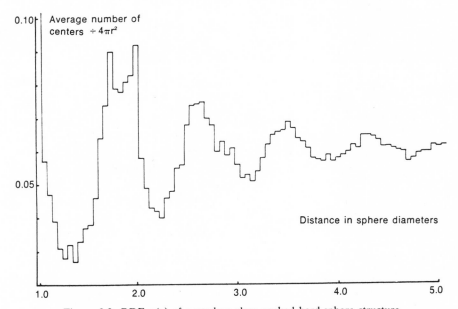

Figure 2.2. RDF $g(r)$ of a random close-packed hard-sphere structure.

characteristic distances, this does *not* imply the structures are in any sense "similar." Rather *both* crystal and glass show local configurations consistent with the dense packing of equal spheres. The relative frequencies with which these local configurations occur are characteristic of the short-range order of the structure.

In three dimensions, the most frequently-occurring such clusters — the dominant polyhedra — are the tetrahedron and the octahedron. Being natural ways of arranging spheres, both polyhedra occur in close-packed crystals and the "random" close-packed ideal-glass model. There are two main distinctions, however, between their arrangement in the ideal crystal and ideal glass:

1. Both the relative frequency and relative arrangements of the tetrahedra and octahedra are different. In the crystal the relative arrangement is regular, consistent with a lattice, while in the glass this restriction does not apply. In the ideal glass, pentagonal and helical arrangements of face-sharing tetrahedra are observed.[3] The same building blocks are used in both crystal and glass, but the topology or connectivity of the units is different.

2. The glass model contains *small* fractions of other, more complex polyhedra where the strict hard-sphere packing constraints prevent even distorted tetrahedra and octahedra from fitting together.[3] There is a sense in which these additional polyhedra could be considered as "defects" in the otherwise "perfect" structure of tetrahedra and octahedra.*

### 2.2.2. Short-Range Order in Network Glasses: Vitreous Silica

When developing the hard-sphere glass model above, we built a model to (1) have the correct density, (2) be consistent with the potential function, and (3) be *not* referable to an underlying lattice. In this case, the potential function, being spherically symmetrical, was particularly simple. This, together with the required high density, meant the structure was determined largely by packing constraints. Put slightly differently, the chemistry of the system was in the prevention of orbital overlap; the structural consequence of the chemistry was a model controlled by volume exclusion and therefore by packing constraints.

We can follow the same path in modeling a very different kind of

---

* It is noteworthy here that in *soft*-sphere models (see Section 2.3.3 below) these more complex polyhedra tend to become squeezed out. In such cases, both crystal and glass models can be considered, to a good approximation, as made up largely of tetrahedra and octahedra only. Again, the topology and relative frequencies of these two building blocks differ in the crystal and glass cases.[14]

amorphous solid, namely vitreous silica. As in the spherical case above, we try to assess the geometric consequences of the chemistry of the system. The problem is to build into a model the structural consequences of the potential functions operating in the assembly. A possible starting point is to examine the wide range of silicate crystal structures, and note that they can all be rationalized in terms of the linking of $SiO_4$ tetrahedra; in fact simple crystal chemical arguments based on radius-ratio considerations suggest the strong stability of this tetrahedral unit. The model-building problem for the glass is to decide how to connect together the tetrahedra through corner-linked oxygens in a way that is consistent with the RDF.

We thus proceed to consider the peaks in the RDF in conjunction with what we know of the local "chemical ordering" — the local short-range order which, in the silicate case, is determined by the chemistry rather than the packing constraints of the hard-sphere model. We draw the following conclusions (Figure 2.3)[15]:

1. Each Si is surrounded by four oxygens, as expected if the $SiO_4$ tetrahedra are preserved.
2. The first-neighbor O–O peak shows the tetrahedra are not significantly distorted.
3. The first-neighbor Si–Si peak tells us the distribution of the Si–O–Si angle in the glass; this distribution can then be used in building up a laboratory model. Without this information, we might well go badly wrong when connnecting neighboring tetrahedral units.
4. Other higher-order correlations (higher peaks in the RDF) can be examined to obtain other information, which can be usefully used as constraints when building a model.[15]

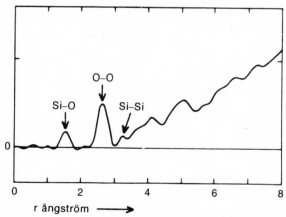

Figure 2.3. RDF of vitreous silica showing the decomposition of the various components in terms of atom–atom distances. $r^2 g(r)$ is the quantity plotted in this figure.

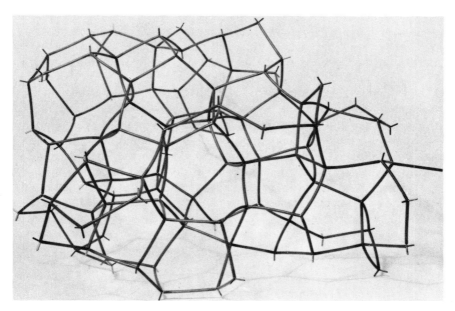

Figure 2.4. An ideal fully coordinated tetrahedral random-network model. For clarity, only the linkages between tetrahedral centers (Si) are shown.

The result of this network model-building process is a fully four-coordinated random-network model (Figure 2.4).[16–18] Just as the hard-sphere model serves as an ideal reference structure for *packed* glasses, so this random network is an ideal reference structure for *network* glasses, where the intermolecular interactions are strongly directional. For example, amorphous Si and Ge can be modeled in a similar way, although the much higher constraints on bond angles in these systems implies some deviation from full four-coordination may be necessary. As a model of vitreous silica itself, the random network fits experiment remarkably well, and can be used in further studies, such as its vibrational spectroscopic properties.[18,19] Note, however, that the model is not perfect. For example, the resultant bond-angle distribution of constructed models is not identical to that fed in: the competing constraint of complete four-coordination always seems to force some departure from the target angle distribution.

### 2.2.3. Summary: Ideal Glass Models

The two structural models so far discussed: (1) the random close-packed hard-sphere model; and (2) the random four-coordinated network model, can be considered as ideal structures for two extreme noncrystalline assemblies, namely (respectively) (1) the ideal hard-sphere glass (at absolute zero); potential function spherically symmetrical; (2) the ideal four-coordinated network glass, with tetrahedral directional bonding, but with some variation in bond angles possible.

In type 1 (packed) glasses, the short-range order is determined by packing constraints, the only significant chemistry being the prevention of overlap by the Pauli exclusion principle. In type 2 (network) glasses, there is strong (essentially perfect) chemical ordering leading to a tetrahedral building block. The linking of these units through the Si–O–Si angle in silica determines the intermediate-range order in the structure.

Real glasses are more complex than either reference structure. In proceeding to consider real glasses, these two extreme ideal structures should be borne in mind. In many cases, other glasses may be considered as variations of either the packed structure, or the network structure, or a mixture of both.

## 2.3. Computer Modeling of Spherical One-Component Glasses

### 2.3.1. Introduction

The ideal hard-sphere dense random-packing model reproduces qualitatively the current experimental density (and hence is consistent with packing constraints) and the RDF. There are, however, major problems in using the ideal hard-sphere model as a true model of real assemblies. When we compare the RDF in detail with that of amorphous (slightly impure) Ni or Co, we find the following small but significant discrepancies (Figure 2.5):

1. The relative positions of the split second-peak components are incorrect by 3% to 5%. This is outside experimental error.
2. The relative intensities of the two components are reversed.

In addition, there is the apparently trivial point that most amorphous metallic systems are alloys, and thus contain a second component. Although in the early X-ray experiments the contribution of the second component to the scattering is small (about 8% in $Ni_{76}P_{24}$), it is not negligible.

This underlines the points made above that our hard-sphere model should be treated rather as an ideal reference system for packed glasses, and *not* as an accurate model of the details of a real alloy glass. For real systems, the model requires refining with respect to (1) the potential functions (e.g., using a Lennard-Jones 6:12 potential), and (2) the presence of the second component.

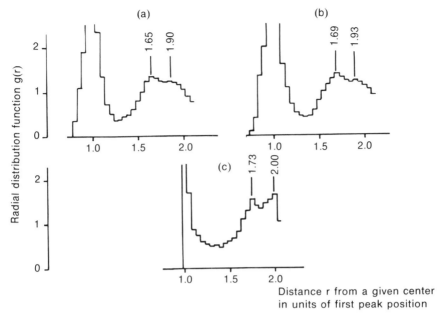

Figure 2.5. RDFs of (a) amorphous $Ni_{76}P_{24}$;[13] (b) amorphous (slightly impure) cobalt.[19] The figures indicate relative positions of the split second-peak components in terms of the first-neighbor peak position. (c) is the $g(r)$ for the random close-packed hard-sphere model, shown for comparison.

To make progress with such refinements, we must move away from relatively inflexible laboratory model-building procedures and try to develop less *ad hoc* methods of "model building" to handle these subtelties. Attempts have therefore been made to build (initially hard-sphere) models on the computer. These attempts have forced us to try to rationalize the decision-making processes used partly subconsciously in laboratory model building, and in addition has focused attention on the importance of adequate control of density (and hence packing constraints) and boundary conditions, in addition to the exact nature and consequences of the potential function used.

We consider this process of "model refinement" in three parts: (1) hard-sphere model building by computer; (2) refinement using soft potentials; (3) the use of Monte-Carlo and molecular-dynamics simulation techniques for modeling liquids and (possibly) glasses.

## 2.3.2. Hard-Sphere Construction Methods

### 2.3.2.1. Bennett Method and Derivatives[20]

Using a preexisting seed of $n$ atoms (usually three spheres in contact, or possibly a planar surface), the $(n + 1)$th sphere is added to a tetrahedral pocket (i.e., in contact with three other spheres). This "sequential-addition" procedure is continued until the model is of the desired size.

At any stage in the building process, many possible tetrahedral pockets are available and a choice must be made between them. The following criteria have been used:

1. *Bennett global.* Choose that pocket closest to the center of mass of the existing model. This procedure is approximately equivalent to adding at the lowest energy site for a gravitational or other long-range potential.
2. *Bennett local.* Choose the deepest available pocket. This is perhaps related to the lowest energy site in a short-range potential.
3. *Ichikawa.* Choose a pocket on the basis of a prespecified degree of "tetrahedral perfection" $k$, where $k_{123} \equiv \max(r_{ij})/(R_i + R_j)$, $i, j = 1, 2, 3$; $i > j$; $i \neq j$, and $R_i$ is the radius of sphere $i$. Note that $k = 1$ corresponds to allowing only perfect tetrahedra, while $k \leq 2$ is identical to the Bennett criterion, which makes no distinction on the basis of tetrahedral perfection. A "perfection window" (e.g., $k \leq 2$) is usually defined, and only pockets within these limits are accepted.

Consider the following properties of Bennett-type models:

1. The density decreases as the model size increases. This reflects the incomplete exertion of packing constraints during the construction process, and suggests that isolated clusters of spheres, which by definition have a free boundary, can attain higher densities than "infinite" packings, where the packing constraints are fully effective. This difference in behavior between isolated and embedded clusters has been termed the "embedding problem";[22] it places severe restrictions on the validity of "cluster" models.

The projected density of an infinite model built by this technique is about 0.61, significantly lower than that achievable by laboratory-built models. This is a basic problem with sequential-addition methods. To achieve higher densities, we need to program into the building procedure the "collective rearrangements," which are achieved by shaking, vibrating, or kneading when building laboratory models.

2. The models are strongly anisotropic, a consequence of the fact that, during building, only the packing constraints to one side (toward the

6. *Very* soft potentials are required to bring in the first component closer to experiment (Figure 2.6e).
7. The result of relaxation can depend on the starting point chosen, although the RDF may not be sensitive enough to show these differences. Figures 2.6a, f show RDFs of two Lennard-Jones relaxations using different density starting points, and there is little difference between them. There is some evidence from a detailed Voronoi analysis (see Appendix[26]) that there are differences in the internal local structures of these two assemblies.

## 2.3.4. Construction by Computer Simulation

These techniques are particularly suitable for modeling liquids. In principle, assuming a potential function, we can set up in the computer assemblies of model interacting particles at a given state point (e.g., volume and temperature) that are sample Gibbs ensembles of those model particles.

There are two main methods.

1. Monte Carlo (MC) techniques are essentially a method of estimating ensemble averages (multidimensional integrals) using random numbers. The standard "Metropolis sampling"[32] is usually used. The method is in principle very flexible if different sampling distributions are chosen depending on the aim of the investigation. Full details are given elsewhere.[33]

Application of the Monte Carlo method to simple liquids has been well explored. For more complex liquids, such as water, there are not insignificant ergodic problems, and there is no sure way of telling whether or not the ensemble is at equilibrium. These problems are even more severe for glasses, which are not even equilibrium ensembles. The application to glasses may therefore be limited, and great care must be taken in any such attempt.

2. Molecular dynamics (MD) follows the time evolution of a box of interacting particles through repeated solution of Newton's laws of motion. In addition to the equilibrium thermodynamic and structural information given also by MC, MD allows time-dependent information, such as transport properties, to be extracted from the ensemble. Ergodic problems occur with MD also; it might be argued that the deterministic nature of the method makes it less flexible than MC for handling nonequilibrium and ergodic difficulties. For early liquid-state applications of the technique, see Rahman's paper.[34]

Both techniques have been applied to glasses, but in addition to the above reservations it is unclear how far the results can be taken as referring to a *glass* — which, except in the hard-sphere and possibly the Lennard-Jones system, would be expected to relax its structure at least locally on cooling — or to a "*configurationally arrested liquid*," which is of less interest.

Consider, for example, an MD simulation of a soft potential simple "glass." Starting from the liquid configuration, we progressively reduce the temperature following all the time the molecular motions of the system. Ideally, we would like to reduce the temperature at a rate consistent with that used in preparing the corresponding real glass in the laboratory. Typically, cooling rates of up to $10^6$ K $s^{-1}$ are used experimentally: such slow rates are computationally inaccessible, and much faster rates *must* be used in the simulation. There is therefore no guarantee that our simulated configuration will bear an adequately strong resemblance to the real glass we are trying to model.

Work to date on simple systems (hard spheres, Lennard-Jones)[11,35,36] gives results that are effectively the same as the modeling and relaxation procedures discussed in earlier sections, perhaps implying that computer and "laboratory" quench rates for these simplest systems are commensurate. There is evidence for quench-rate dependence of properties in the Lennard-Jones glass,[36] but this has not as yet been explored fully.

In summary, computer simulation techniques can be used very successfully for modeling liquids, though for more complex molecules (e.g., with strong directionality in the potential, or for mixed systems of widely different characteristic time scales of motion) there may be awkward ergodic problems. It must also be remembered that the result will depend strongly on the assumed potential function.

The problems are greater in application to glasses, except perhaps for the simplest systems in which the distinction between the glass and the configurationally arrested liquid may be minor or even nonexistent. Laboratory quench rates cannot be even approached in the computer. The use of judicious variation of quench rates, together with energy minimization procedures from different starting points taken during the quench, might be explored profitably.[26,28]

## 2.4. Modeling Real Binary Amorphous Metals

We have discussed above the construction of ideal and slightly nonideal models for single-component *packing* glasses, in which the short-range order is determined by packing constraints. At the other end of the "chemical order spectrum" we discussed the building of models, such as the random four-coordinated network for vitreous silica. In building this latter ideal structure of a *network* glass, we understood the local chemical ordering, and what possible ways of joining together the basic chemically ordered building blocks were consistent with experiment.

Real binary-alloy glasses fall somewhere in between; their structures must be consistent with some (often incompletely known) degree of local chemical ordering, together with the full exertion of packing constraints consistent with the known high experimental density. We saw in Section 2.3 that this reconciling of short-range order with high density was not trivial even when the chemical ordering amounted to zero, merely disallowing the overlap of neighbors. For binary alloys, the problem is at least an order of magnitude more difficult.

It is probably fair to say that, because of the problem of reconciling local short-range chemical ordering with the constraints of space-filling, we have not as yet been able to construct a single adequate model of a binary-alloy glass. The problem is complex and is discussed in detail elsewhere.[28] The two methods that have been tried can be regarded as starting from the opposite ends of the packing-network spectrum defined by the ideal hard sphere and random-network reference structures, respectively.

The first method due to Boudreaux[23,37] starts from the packing end of the spectrum, emphasizes packing constraints first, and tries to feed in chemical ordering by applying additional constraints to the Bennett addition procedure (modified for two components). Subsequent relaxation may or may not use empirical potential functions designed to enhance — or at least to try to retain — the desired chemical ordering. The second method, due to Gaskell,[38] starts from the *network*, or strongly chemically ordered end of the model spectrum, and builds a network model of the elementary local chemically ordered units thought to exist in the glass from other considerations (such as trigonal prisms in amorphous $Pd_3Ge$). Thus an attempt is made to design-in the chemical ordering at the expense of packing constraints, leaving a subsequent relaxation to bring in the necessary packing constraints.

It is not presently clear how successful either of these methods is in reproducing the structures of the real alloy glass. Partly this is because of inadequate experimental data with which to test the models. For single-component glasses and liquids, only one RDF is needed to describe the average structure. For binary systems, the corresponding total RDF is an inadequate test. As stressed in Chapter 1, we need information on the three partial RDFs to test a structural model. Such information is only now becoming available from neutron-scattering measurements using isotope substitution.

Other approaches to modeling binaries can be envisaged, either using model-building or liquid-state simulations as starting points for relaxation calculations. Some of the possibilities are discussed more fully elsewhere.[28]

## Appendix 2.1. A Note on the Voronoi Polyhedron

The RDF gives information on the average structure of a liquid or glass, projected from three dimensions onto one dimension. Although this may be all we require for the thermodynamics and may be the most we can expect from scattering experiments, it is useful to be able to compare in detail the *local* three-dimensional structures in noncrystalline packed models. This is particularly important with respect to alloy glasses in which chemical ordering is expected: here we need a sensitive probe for the degree of chemical ordering and its variation throughout the structure.

The concept of the Voronoi polyhedron is often used for this purpose. This construction allows us to partition the space occupied by a model into polyhedral regions, one of which is associated with each "atom." The construction, which is totally nonarbitrary, draws perpendicular bisectors through lines joining atom centers. The smallest closed polyhedron so formed about each atom contains all space closer to that atom than to any other. The set of polyhedra completely fills all the space occupied by the assembly.

Various metric and topological characteristics of the polyhedral assembly can be used in comparing different packed structures, for example:

1. Polyhedron volumes and their distributions.
2. Numbers of faces per polyhedron: average and distribution.
3. Number of edges per face: average and distribution.
4. Polyhedron type index $(n_3, n_4, n_5, n_6, \dots)$, where $n_i$ is the number of faces to the polyhedron with $i$ edges.

The procedure, although undoubtedly useful, must be used with care. For example, certain polyhedron types relate to fcc, hcp, bcc, or icosahedral local structures. The technique can therefore be used as a diagnostic of, e.g., crystallization or degree of icosahedrality in a given structure. Examples are given elsewhere.[22,26,28,31,39,40]

A generalization of the procedure (the radical plane polyhedron[41]) can be used for multicomponent systems, where the existence of atoms of more than one size leads to problems in using the simple Voronoi construction.

## References

1. J. D. Bernal and R. H. Fowler, *J. Chem. Phys.* **1**, 515 (1933).
2. O. K. Rice, *J. Chem. Phys.* **12**, 1 (1944).
3. J. D. Bernal, *Proc. R. Soc. London, Ser. A* **280**, 299 (1964).

4. J. D. Bernal, *Nature* **188**, 910 (1960).
5. G. D. Scott, Nature **188**, 908 (1960).
6. J. L. Finney, *Proc. R. Soc. London, Ser. A* **319**, 479 (1970).
7. J. D. Bernal, I. A. Cherry, J. L. Finney, and K. R. Knight, *J. Phys. E* **3**, 388 (1970).
8. J. L. Finney, Ph.D. thesis, Univ. of London (1968).
9. J. M. Ziman, *Models of Disorder*, Chapter 2, Cambridge University Press (1979).
10. J. L. Finney, *Proc. R. Soc. London, Ser. A* **319**, 495 (1970).
11. L. V. Woodcock, *J. Chem. Soc., Faraday Trans. 2* **72**, 1667 (1976).
12. J. D. Bernal and J. L. Finney, *Acta Crystallogr., Sect. A* **25**, part S3, S89 (1969). G. S. Cargill III, *Acta Crystallogr., Sect. A* **25**, part S3, S95 (1969).
13. G. S. Cargill III, *J. Appl. Phys.* **41**, 12 (1970).
14. J. L. Finney and J. Wallace, *J. Non-Cryst. Solids* **43**, 165 (1981).
15. R. L. Mozzi and B. E. Warren, *J. Appl. Cryst.* **2**, 164 (1969).
16. W. Zacharaisen, *J. Am. Chem. Soc.* **54**, 3841 (1932).
17. D. L. Evans and S. V. King, *Nature* **212**, 1353 (1966). R. J. Bell and P. Dean, *Nature* **212**, 1354 (1966).
18. P. Dean and R. J. Bell, *New Sci.* **45**, 104 (1970).
19. P. K. Leung and J. G. Wright, *Phil. Mag.* **30**, 185 (1974).
20. C. H. Bennett, *J. Appl. Phys.* **43**, 2727 (1972).
21. T. Ichikawa, *Phys. Status Solidi A* **29**, 293 (1975).
22. J. L. Finney, *Nature* **266**, 309 (1977).
23. D. S. Boudreaux and J. M. Gregor, *J. Appl. Phys.* **48**, 152 (1977).
24. J. F. Sadoc, J. Dixmier, and A. Guinier, *J. Non-Cryst. Solids* **12**, 46 (1973).
25. J. L. Finney, *J. Phys. (Paris)* **36**, Colloque C2, 1 (1975).
26. J. L. Finney, in: *Diffraction Studies of Non-Crystalline Substances* (I. Hargittai and W. Y. Orville-Thomas, eds.), p. 440, Elsevier, Amsterdam (1981).
27. J. L. Finney, *Mater. Sci. Eng.* **23**, 199 (1976).
28. J. L. Finney, in: *Amorphous Metallic Alloys* (F. E. Luborsky, ed.), Chapter 4, Butterworth, London (1983).
29. G. Mason, *Discuss. Faraday Soc.* **43**, 75 (1967).
30. W. S. Jodrey and E. M. Tory, *Powder Technol.* **30**, 111 (1981).
31. J. A. Barker, J. L. Finney, and M. R. Hoare, *Nature* **257**, 120 (1975). L. von Heimendahl, *J. Phys. F* **5**, L141 (1975).
32. M. Metropolis, A. W. Metropolis, M. N. Rosenbluth, A. H. Teller, and E. Teller, *J. Chem. Phys.* **21**, 1087 (1953).
33. J. P. Valleau and S. G. Whittington, in: *Modern Theoretical Chemistry*, Vol. 5 (Statistical Mechanics part A: Equilibrium Techniques) (B. J. Berne ed.), Chapter 4, Plenum, New York (1977). J. P. Valleau and G. M. Torrie, in: *Modern Theoretical Chemistry*, Vol. 5 (Statistical Mechanics part A: Equilibrium Techniques) (B. J. Berne, ed.), Chapter 5, Plenum, New York (1977).
34. A. Rahman, *J. Chem. Phys.* **45**, 2585 (1966).
35. A. Rahman, J. J. Mandell, and J. P. McTague, *J. Chem. Phys.* **64**, 1564 (1976).
36. W. Damgaard Kristensen, *J. Non-Cryst. Solids* **21**, 203 (1976).
37. D. S. Boudreaux and J. M. Gregor, *J. Appl. Phys.* **48**, 5057 (1977).
38. P. H. Gaskell, *J. Non-Cryst. Solids* **32**, 207 (1979).
39. M. Tanemura, Y. Hiwatari, H. Matsuda, T. Ogawa, N. Ogita, and A. Ueda, *Prog. Theor. Phys.* **58**, 1079 (1977). C. S. Hsu and A. Rahman, *J. Chem. Phys.* **71**, 4974 (1980).
40. J. N. Cape, J. L. Finney, and L. V. Woodcock, *J. Chem. Phys.* **75**, 2366 (1981).
41. B. J. Gellatly and J. L. Finney, *J. Non-Cryst. Solids* **40**, 313 (1982).

**3**

# Structure and Forces in Liquids and Liquid Mixtures

## N. H. March

### 3.1. Introduction

In this chapter, we shall discuss first the statistical mechanical theory of liquids,[1] and then of liquid mixtures. The discussion will include the theory of freezing of some simple liquids. This is closely related to the theory of the bulk liquid structure, the main topic of the first part of this chapter. Here, the so-called Born–Green–Yvon hierarchy is the starting point of all structural theories of simple liquids. This hierarchy was also used in the pioneering studies of Kirkwood and his school[2] on the theory of freezing. Modern trends in the theory of both liquid structure and freezing have emphasized the Ornstein–Zernike direct correlation function $c(r)$, which is more directly connected with the interparticle interaction than the liquid pair function $g(r)$, to be defined precisely below. This function $c(r)$ [see equation (3.4.2) below] will therefore be a focal point in this discussion. However, first we shall give an operational definition of the structure factor $S(k)$ of a simple liquid like argon, by summarizing the results for the intensity of X-ray scattering (see Chapter 1 for full details of the diffraction experiments).

---

**N. H. March** · Theoretical Chemistry Department, University of Oxford, 1 South Parks Road, Oxford OX1 3TG, England.

## 3.2. Structure Factor of a Simple Liquid

Suppose X-rays are scattered from a simple liquid through an angle $2\theta$, the X-ray wavelength being $\lambda$. By setting $k = 4\pi \sin \theta / \lambda$, the intensity $I$ of scattered X-rays can be expressed in the form

$$I(k) = N |f(k)|^2 S(k) \qquad (3.2.1)$$

where $N$ is the number of atoms in the liquid scatterer, $f(k)$ is the atomic scattering factor, which in argon, say, is simply the Fourier transform of the electron density $\rho(r)$ in the free argon atom, i.e.,

$$f(k) = \int \rho(r) \exp(i\mathbf{k} \cdot \mathbf{r}) d\mathbf{r} \qquad (3.2.2)$$

while $S(k)$ is defined as the liquid-structure factor (cf. Figure 4.10 of Chapter 4). Its Fourier transform $g(r)$ is the liquid pair function introduced above, which gives a quantitative measure of the short-range order in the liquid.

## 3.3. Thermodynamics in Terms of the Pair Function for Two-Body Interactions

If the total potential energy $\Phi(\mathbf{r}_1, \mathbf{r}_2, \ldots, \mathbf{r}_N)$ for $N$ atoms at positions $\{\mathbf{r}_i\}$, $i = 1, \ldots, N$, can be expressed as a sum of pair interactions:

$$\Phi(\mathbf{r}_1, \mathbf{r}_2, \ldots, \mathbf{r}_N) = \sum_{i<j} \phi(r_{ij}) \qquad (3.3.1)$$

then the internal energy $E$ and the pressure $p$ are readily written down.

### 3.3.1. Internal Energy

As a first application of the pair function $g(r)$, these quantities will be expressed in terms of it. This is possible provided the intermolecular potential energy $\Phi$ can be decomposed as in equation (3.3.1). For the moment, we can best treat $\phi(r)$ as essentially the potential energy of interaction of two (say argon) atoms in vacuo.

The internal energy $E$ can be written in the form

$$E = \tfrac{3}{2} N k_B T + \langle \Phi \rangle \qquad (3.3.2)$$

where the kinetic energy follows from the equipartition-of-energy principle for a classical system in thermal equilibrium at temperature $T$. The mean potential energy is simply

$$\langle \Phi \rangle = \frac{\int \cdots \int e^{-\Phi/k_B T} \Phi d\mathbf{r}_1 \cdots d\mathbf{r}_N}{\int \cdots \int e^{-\Phi/k_B T} d\mathbf{r}_1 \cdots d\mathbf{r}_N} \tag{3.3.3}$$

For pair forces, all the $N(N-1)/2$ terms contribute equally to $\langle \Phi \rangle$ and hence

$$\langle \Phi \rangle = \frac{N(N-1)}{2} \int \phi(r_{12}) \left( \frac{\int \cdots \int e^{-\Phi/k_B T} d\mathbf{r}_3 \cdots d\mathbf{r}_N}{Z} \right) d\mathbf{r}_1 d\mathbf{r}_2 \tag{3.3.4}$$

where $Z$ is the partition function. However, from the definition of the distribution functions, the quantity in parentheses on the right-hand side of equation (3.3.4) is simply the two-body function $g(\mathbf{r}_1 \mathbf{r}_2)$ and hence, utilizing the isotropy of the liquid,

$$\langle \Phi \rangle = \frac{N(N-1)}{2} \iint \phi(r_{12}) \frac{(N-2)!}{N!} \rho^2 g(r_{12}) d\mathbf{r}_1 d\mathbf{r}_2$$

$$= \frac{\rho^2 \Omega}{2} \int_0^\infty \phi(r) g(r) 4\pi r^2 dr : \rho = N/\Omega \tag{3.3.5}$$

where one of the integrations yields immediately the volume $\Omega$ of the fluid. Hence, from equations (3.3.2) and (3.3.5), we derive the expression

$$E = \tfrac{3}{2} N k_B T + \frac{N\rho}{2} \int_0^\infty \phi(r) g(r) 4\pi r^2 dr \tag{3.3.6}$$

In fact, the potential energy could have been written down directly on physical grounds, since the number of molecules on the average within a distance between $r$ and $r + dr$ from a given molecule is $\rho g(r) 4\pi r^2 dr$ and the factor $\tfrac{1}{2}$ is present to avoid counting interactions twice.

### 3.3.2. Equation of State

We start from the well-known virial relation

$$2K + \langle \mathbf{r} \cdot \mathbf{F} \rangle = 3p\Omega \tag{3.3.7}$$

where $K$ is the kinetic energy while $\mathbf{r} \cdot \mathbf{F}$ denotes the virial of the forces $\mathbf{F}$.

Of course, in a classical liquid the kinetic energy $K$ is immediately determined by the temperature $T$, and hence

$$3p\Omega = 3Nk_{\mathrm{B}}T + \langle \mathbf{r} \cdot \mathbf{F} \rangle \qquad (3.3.8)$$

In a system with pair forces, this equation can be readily expressed as

$$p = \rho k_{\mathrm{B}}T - \frac{\rho^2}{6} \int d\mathbf{r}\, g(r) r \frac{\partial \phi}{\partial r} \qquad (3.3.9)$$

which is the desired result for the equation of state in terms of $g(r)$ and the pair potential. We shall see later that there is an alternative route to the equation of state via the compressibility; one that involves only the Ornstein–Zernike direct correlation function and not the assumption that a pair-potential description is possible.

## 3.4. Ornstein–Zernike Direct Correlation Function

In the force equation (3.5.2) below, the liquid pair function $g(r)$ is expressed in terms of the potential of the mean force $U(r)$ [cf. equation (3.5.1)]. It is of considerable interest to note that there is a further correlation function, $c(r)$ say, introduced by Ornstein and Zernike, and commonly referred to as the direct correlation function, which is much more intimately related to the pair potential $\phi(r)$ than is $g(r)$. As in equation (3.5.2), the mean force is decomposed into a pair force, plus a part from the "other atoms," so the total correlation function $h(r)$ defined by

$$h(r) = g(r) - 1 \qquad (3.4.1)$$

can be written as the direct correlation function $c(r)$, plus another part expressed as a convolution of $h$ and $c$. We shall motivate this convolution form a little further below when we consider briefly approximate structural theories of liquids, but for the moment we simply introduce the definition of $c(r)$ through the relation

$$h(r) = c(r) + \rho \int h(|\mathbf{r} - \mathbf{r}'|) c(\mathbf{r}') d\mathbf{r}' \qquad (3.4.2)$$

By a Fourier transform, with

$$\tilde{c}(k) = \rho \int c(r) e^{i\mathbf{k} \cdot \mathbf{r}} d\mathbf{r} \qquad (3.4.3)$$

we obtain almost immediately

$$\tilde{c}(k) = \frac{S(k) - 1}{S(k)}$$

If this appears to be a minor change from $S(k)$, note that in liquid argon around the triple point, the long-wavelength limit $S(0)$ of the structure factor, to be discussed below, is about 0.06, and hence $\tilde{c}(0)$ equals about $-16$. For liquid metals just above the freezing point, $S(0)$ is even smaller and $\tilde{c}(k)$ is, for simple metals, dominated by its short-range behavior. We shall see shortly that, at large $r$, $c(r)$ is directly proportional to the pair potential, the proportionality constant being simply $-\beta = -(k_B T)^{-1}$.

We show in Appendix 3.1 how $S(0)$, and hence $\tilde{c}(0)$, can be related to the isothermal compressibility $K_T$, thereby providing an alternative route to the equation of state (3.3.9).

Thermodynamic consistency with "density-independent" pair potentials then yields the result

$$-k_B T \rho \int c(r) d\mathbf{r} = \frac{-\partial}{\partial \rho} \left( \int d\mathbf{r} r g \frac{\partial \phi}{\partial r} \frac{\rho^2}{6} \right) \tag{3.4.4}$$

which implies, after an integration by parts,[3]

$$-\rho c(r) r^2 = \phi(r) \frac{\partial^2}{\partial \rho \partial r} \left[ \frac{g r^3 \rho^2}{6 k_B T} \right] + F \tag{3.4.5}$$

where $\int_0^\infty F dr = 0$.

The assumption that $F$ falls off more rapidly than $\phi$, far from the critical point, leads back to the asymptotic result $c(r) = -\phi(r)/k_B T$ at sufficiently large $r$. This is consistent with experimental results[4] on low-angle scattering from liquid argon (cf. Figure 3.1 below).

## 3.5. Three-Atom Correlation Function

So far we have been concerned (1) with the definition of $S(k)$ and $g(r)$ and the way they can be obtained from experiment, (2) with the expression of the thermodynamics of a liquid with pairwise interactions in terms of $g(r)$ and the pair potential $\phi(r)$, and (3) with an alternative route to the equation of state via the compressibility relation to the direct correlation function $c(r)$.

Nonetheless, of course, a theory of liquid structure must provide a route whereby one can calculate $g(r)$ from a given pair potential $\phi(r)$.

This, as we explain below, involves us in a quite fundamental way with the three-atom correlation function $\rho^{(3)}(\mathbf{r}_1\mathbf{r}_2\mathbf{r}_3)$. To proceed, the pair function $g(r_{12})$ in a classical liquid is expressed in terms of the potential of the mean force $U(r_{12})$ through the Boltzmann factor:

$$g(r_{12}) = \exp[- U(r_{12})/k_{\mathrm{B}}T] \tag{3.5.1}$$

Now we set up the so-called force equation. The total force $- \partial U(r_{12})/\partial \mathbf{r}_1$ on atom 1 is made up of two parts: a direct part $- \partial \phi(r_{12})/\partial \mathbf{r}_1$ and an indirect part. To set up the indirect part, suppose there is a third atom at $\mathbf{r}_3$ when atoms 1 and 2 are at $\mathbf{r}_1$ and $\mathbf{r}_2$. Then one must weight the force $- \partial \phi(r_{13})/\partial \mathbf{r}_1$ with the appropriate probability, which is not just $\rho^{(3)}(\mathbf{r}_1\mathbf{r}_2\mathbf{r}_3)$ but $\rho^{(3)}(\mathbf{r}_1\mathbf{r}_2\mathbf{r}_3)/\rho^2 g(r_{12})$, to take account of the fact that there are certainly atoms at positions $\mathbf{r}_1$ and $\mathbf{r}_2$. Thus we are led to the force equation

$$\frac{- \partial U(r_{12})}{\partial \mathbf{r}_1} = \frac{- \partial \phi(r_{12})}{\partial \mathbf{r}_1} - \int \frac{\partial \phi(r_{13})}{\partial r_1} \frac{\rho^{(3)}(\mathbf{r}_1\mathbf{r}_2\mathbf{r}_3)d\mathbf{r}_3}{\rho^2 g(r_{12})} \tag{3.5.2}$$

by integrating the indirect contribution over all positions of atom 3. This is an exact equation of classical statistical mechanics for a liquid with pairwise interactions. However, to obtain an explicit theory of structure one must clearly make an approximation for $\rho^{(3)}(\mathbf{r}_1\mathbf{r}_2\mathbf{r}_3)$.

### 3.5.1. Kirkwood Approximation: Its Consequences and Limitations

The earliest and still useful approach is due to Kirkwood, who decoupled $\rho^{(3)}$ into a product of pair functions in the form

$$\rho^{(3)}(\mathbf{r}_1\mathbf{r}_2\mathbf{r}_3) = \rho^3 g(r_{12})g(r_{23})g(r_{31}) \tag{3.5.3}$$

Though it is important to assess its limitations, this assumption leads to a complete, though of course approximate, theory of structure: the Born–Green theory This theory is seldom used nowadays; nonetheless one of the currently favored structural theories, called the hypernetted chain theory because of its basis in diagrammatic analysis, is closely related to it in a manner to be outlined now.

If equation (3.5.3) is inserted into equation (3.5.2), the resulting equation can be integrated following Rushbrooke[5] to yield

$$\frac{U(r) - \phi(r)}{k_{\mathrm{B}}T} = \rho \int E(\mathbf{r} - \mathbf{r}')h(\mathbf{r}')d\mathbf{r}' \tag{3.5.4}$$

where the quantity $E(r)$ is defined by

$$E(r) = \int_r^\infty \frac{g(s)\phi'(s)ds}{k_B T} \tag{3.5.5}$$

Since $g \to 1$ at large distances, it follows from equation (3.5.5) that $E(r)$ has the asymptotic form $-\phi(r)/k_B T$ and therefore equals the direct correlation function $c(r)$ in that limit.

While equation (3.5.4) is the basic equation of the Born–Green theory, the hypernetted chain theory is obtained by replacing $E$ in equation (3.5.4) by the direct correlation function $c(r)$. By using the convolution equation defining $c(r)$ the hypernetted chain equation is derived in the form

$$h(r) - c(r) = \frac{\phi(r) - U(r)}{k_B T} \tag{3.5.6}$$

Since $h(r)$ equals approximately $-U(r)/k_B T$ at sufficiently large $r$ from equation (3.5.1) because $h = g - 1$, we regain $c(r) = -\phi(r)/k_B T$ at large $r$. However, if we retained the Born–Green theory, we would find $c(r) = -\alpha\phi(r)/k_B T$, where $\alpha$ depends on the thermodynamic state and can deviate substantially from unity.[6]

Thus the first limitation of the Kirkwood approximation (3.5.3) is that it does not yield the correct quantitative asymptotic relation between $c(r)$ and $\phi(r)$. We have already seen in Section 3.4 that this asymptotic equation $c(r) = -\phi(r)/k_B T$ bears a close relation to thermodynamic consistency and it is then not surprising that the Born–Green theory can, under some circumstances, display very substantial thermodynamic inconsistencies. We stress that the above discussion is valid far from the critical point: other longer-range critical fluctuations dominate there.

### 3.5.2. Inverse Problem of Extracting the Pair Potential from the Measured Structure

Johnson and March[7] proposed that one should extract the pair potential from measured structure factors. One evidently had to appeal at that time to the approximate structural theories discussed above. However, with computer experiments some check of such theories becomes possible, and we shall briefly outline the conclusions to which one is then led.

Essentially, one can construct $U - \phi$ from computer experiments and hence derive an essentially exact estimate of the term involving the three-body function $\rho^{(3)}$ in the force equation. If one wishes, one can then compare the result with that obtained using the Kirkwood approximation. This procedure has been implemented by Ebbsjö et al.,[8] whose principal conclusion is that the Kirkwood approximation is more successful for liquid Na than for liquid argon, the latter being poorly described. There is clearly much, long overdue work to be done in making further progress with the inverse problem of extracting $\phi(r)$ from $g(r)$.* In addition to the theoretical aspect discussed above, the determination of $c(r)$ from experiment still remains unsolved due to the need for very-small-angle data. This again is important for the future: the theory of small-angle scattering is well established, at least in the absence of a collective mode. Clearly, $\phi(r) = -k_B Tc(r)$ if $c(r)$ can be determined from experiment at sufficiently large $r$.

### 3.5.3. Difference of Specific Heats in Terms of the Three-Body Correlation Function

Schofield[9] appears to have been the first to show that $c_p$ and $c_v$, the specific heats at constant pressure and constant volume, respectively, involve separately three- and four-particle correlation functions.

The method can be developed for $c_p - c_v$[10] with the result

$$\frac{c_p - c_v}{S(0)} = 1 - \frac{2\pi\rho}{3k_B T} \int dr r^3 g(r) \frac{d\phi}{dr}$$
$$- \frac{\rho}{2k_B TS(0)} \left[ \int d\mathbf{r} g(r)\phi(r) + \rho \int d\mathbf{r} d\mathbf{s} \left\{ \frac{\rho^{(3)}(\mathbf{r},\mathbf{s})}{\rho^3} - g(r)g(s) \right\} \phi(r) \right] \tag{3.5.7}$$

To estimate $c_p - c_v$, the Kirkwood approximation is inadequate. However, the replacement

$$\frac{\rho^{(3)}(\mathbf{r},\mathbf{s})}{\rho^3} - g(r)g(s) \cong g(r)h(|\mathbf{s}-\mathbf{r}|) \tag{3.5.8}$$

inside the integral in equation (3.5.7) gives $c_p - c_v$ equal to approximately $3k_B$ at the triple point of argon, compared with the measured value of $2.8k_B$.

For simple liquid metals, however, a more physical picture of the specific heats proves possible using density-fluctuation arguments (see Section 3.6 below).

### 3.5.4. Pressure Dependence of the Pair Correlation Function

In place of $Z_G$ in equation (A3.1.1) of Appendix 3.1, it is convenient to

---

* *Note added in proof.* Since this was written, an important contribution to the solution of this inverse problem has appeared [D. Levesque, J. J. Weis and L. Reatto, *P. Rev. Lett.* **54**, 451 (1985)].

introduce a thermodynamic property $z$, called the fugacity, given by

$$z = \left(\frac{mk_B T}{2\pi\hbar^2}\right)^{3/2} \exp(\mu/k_B T) \tag{3.5.9}$$

where $\mu$ is the chemical potential. In the grand ensemble, the dependence of $g(r)$ on density $\rho$ at constant $T$ is then expressed only in terms of the fugacity and we can write

$$\left\{\frac{\partial}{\partial\rho}[\rho^2 g(r)]\right\}_T = V\left\{\frac{\partial[\rho^2 g(r)]}{\partial\langle N\rangle}\right\}_T = V\left\{\frac{\partial[\rho^2 g(r)]}{\partial z}\right\}_T \bigg/ \left(\frac{\partial\langle N\rangle}{\partial z}\right)_T \tag{3.5.10}$$

In a similar manner to that used in relating $S(0)$ to the isothermal compressibility $K_T$ one derives

$$z\left\{\frac{\partial[\rho^2 g(r)]}{\partial z}\right\}_T = \langle N\rho^{(2)}(\mathbf{r}_1\mathbf{r}_2)\rangle - \langle N\rangle\langle\rho^{(2)}(\mathbf{r}_1\mathbf{r}_2)\rangle \tag{3.5.11}$$

which evidently represents fluctuation in the product $N\rho^{(2)}$, where $\rho^{(2)}$ is two-body distribution function. Some tedious, but straightforward manipulation allows equation (3.5.11) to be expressed in terms of the three-atom correlation function as

$$z\left\{\frac{\partial[\rho^2 g(r)]}{dz}\right\}_T = 2\rho^2 g(r) + \int d\mathbf{r}_3[\rho^{(3)}(\mathbf{r}_1\mathbf{r}_2\mathbf{r}_3) - \rho^3 g(r)] \tag{3.5.12}$$

This equation involves only the "$s$" term of $\rho^{(3)}$, while the force equation requires information only on the "$p$" term. The above analysis shows that useful, though naturally incomplete information on the three-atom correlation function can be obtained by observing the pressure dependence of the pair function $g(r)$. This has been skillfully exploited experimentally by Egelstaff and his colleagues.[11] The theory of the pressure dependence of $g(r)$,* based on the hierarchy, has been considered by de Angelis and March.[12]

## 3.6. Density Fluctuations and Collective Modes in Simple Liquid Metals

The theory of independent phonons in crystals leads to the standard result for the frequencies of the longitudinal phonons (see, for example, Pines[13]):

$$\omega_k^2 = \frac{\rho}{M}\sum_G\left\{\left[\frac{\mathbf{k}}{k}\cdot(\mathbf{k}+\mathbf{G})\right]^2\tilde{\phi}(\mathbf{k}+\mathbf{G}) - \left(\frac{\mathbf{k}\cdot\mathbf{G}}{k}\right)^2\tilde{\phi}(k)\right\} \tag{3.6.1}$$

* Note added in proof. For a review of the pressure dependence of $g(r)$, and in particular the applicability of the one-component plasma model to simple liquid metals, see the book by N. H. March and M. P. Tosi on Coulomb Liquids (Academic: London, 1984).

where $\tilde{\phi}$ denotes the Fourier transform of the interparticle pair interaction and **G** denotes a reciprocal lattice vector.

### 3.6.1. Model of Independent Harmonic Oscillators

In a fluid, one works with the density fluctuations $\rho_k$ and canonically conjugate momenta $P_k$. In terms of $P_k$, we can construct a Hamiltonian representing uncoupled harmonic oscillators, this time with frequencies given by

$$\omega_k^2 = (\hbar k^2/2M)^2 + \frac{k^2 \rho}{M} \tilde{\phi}(k) \qquad (3.6.2)$$

where $\tilde{\phi}$ is the Fourier transform of the pair potential.

We shall not give details here, but rather introduce an elementary model for a classical liquid showing the origin of the second term on the right-hand side of the above equation. To anticipate our brief discussion of atomic dynamics in liquids at the end of this chapter, we introduce the dynamical generalization (see also Section 3.9 below) of the structure factor $S(k)$, namely $S(k\omega)$, such that first

$$S(k) = \int_{-\infty}^{\infty} S(k\omega)d\omega \qquad (3.6.3)$$

and second, we note the physical interpretation of $S(k\omega)$ as the probability that a neutron impinging on the liquid will transfer momentum $\hbar k$ and energy $\hbar\omega$ to the liquid. For a classical liquid, the zeroth moment in equation (3.6.3) can be supplemented by the known second moment[1]

$$\int_{-\infty}^{\infty} \omega^2 S(k\omega)d\omega = \frac{k_B T k^2}{2M} \qquad (3.6.4)$$

Bearing in mind the "phonon" model introduced above, note that a simple collective-mode model can be based on the assumption

$$S(k\omega) = \tfrac{1}{2}[\delta(\omega - \omega_k) + \delta(\omega + \omega_k)]S(k) \qquad (3.6.5)$$

which evidently already satisfies the zeroth moment. The second-moment equation (3.6.4) then suffices to relate the collective-mode frequency $\omega_k$ to the structure factor $S(k)$ through (cf $\tilde{c}(k) \sim -1/S(k) \sim -\rho k_B T \tilde{\phi}(k)$)

$$\omega_k^2 S(k) = k_B T k^2/2M \qquad (3.6.6)$$

That a collective mode of the kind discussed here exists in liquid Rb has been demonstrated in the neutron studies of Copley and Rowe,[14] who extracted a dispersion relation for the collective mode by plotting the variation of $S(k\omega)$ against $\omega$ for various values of $k$, and then plotting the positions of the $\omega$ peaks as functions of $k$. This has the general shape predicted by equation (3.6.6) when we insert the structure factor $S(k)$, though the theory is not quantitative.

### 3.6.2. Small-Angle Scattering from Liquid Metals

Matthai and March[15] have compared and contrasted the behavior of the small-angle scattering in liquid argon and liquid rubidium. Their work amounts to studying the small-$k$ expansion of the liquid-structure factor $S(k)$. In liquid argon, with no collective mode, it is well established that $S(k)$ has the form[16]

$$S(k) = S(0) + a_2 k^2 + a_3 k^3 + \cdots \qquad (3.6.7)$$

where $a_3$ can be calculated as $\pi^2 \rho [S(0)]^2 c_6 / 12 k_B T$ from the result $c(r) = -\phi(r)/k_B T = (c_6/r^6)/k_B T$ in terms of the van der Waals constant $c_6$ at sufficiently large $r$; $a_2$ has also been studied by Gaskell et al. using the random phase approximation. Matthai and March[15] confirm (see Figure 3.1) that the above small-angle scattering theory fits the neutron data of Yarnell et al. on liquid argon to the experimental accuracy ($a_{2\,\text{theory}} = -0.08\ \text{Å}^2$; $a_{2\,\text{exp}} = -0.12\ \text{Å}^2$; $a_{3\,\text{theory}} = 0.375\ \text{Å}^3$; $a_{3\,\text{exp}} = 0.35\ \text{Å}^3$). However, for liquid rubidium Matthai and March[15] modified the above expansion to

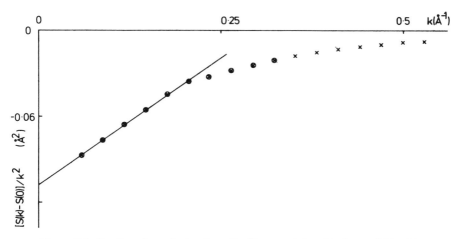

Figure 3.1. Small-angle scattering from liquid argon (after Matthai and March).

**Table 3.6.1. Specific Heat of Liquid Metals at Melting Point**[a]

| Element | Na | K | Rb | Zn | Cd | Ga | Tl | Sn | Pb | Bi |
|---|---|---|---|---|---|---|---|---|---|---|
| $\gamma = c_p/c_v$ | 1.12 | 1.11 | 1.15 | 1.25 | 1.23 | 1.08 | 1.21 | 1.11 | 1.20 | 1.15 |
| $c_v/Nk_B$ | 3.4 | 3.5 | 3.1 | 3.1 | 3.1 | 3.2 | 3.0 | 3.0 | 2.9 | 3.1 |

[a] Measured values from Kleppa.[18]

read

$$S(k) = S(0) + a_1 k + a_2 k^2 + a_3 k^3 + \cdots \qquad (3.6.8)$$

where they attribute the nonzero value of $a_1$ to the dispersion of the collective mode discussed above for Rb. Further studies, both experimental and theoretical,* are clearly required in this area for those simple liquid metals believed to display a collective mode. However, thermal properties give strong support to a model of independent density fluctuations in simple liquid metals, such a model yielding a ratio of specific heats $\gamma = 1$ and $c_v = 3R$ in the zeroth order. Refinements of these values are strongly suggested by the data in Table 3.6.1 and are discussed elsewhere.[17,10]

## 3.7. Solution Theories of Thermodynamics of Liquid Mixtures

After this introduction to the thermodynamics and structure of simple monatomic liquids, the remainder of this chapter, until the final section 3.10 on the theory of freezing, will be devoted to binary liquid mixtures. This area is well reviewed from the point of view of thermodynamics and phenomenology in the book by Rowlinson.[19] A natural starting point here is to discuss solution theories, especially regular solution theory, and a generalization due to Flory, which allows for the possibility of a large size difference between the components of the mixture. However, we shall start with some basic discussion of concentration and number fluctuations in liquid mixtures.

### 3.7.1. Concentration and Number Fluctuations

We shall begin by relating the fluctuations $\langle (\Delta c)^2 \rangle$, etc., to thermodynamic quantities, following the work of Bhatia and Thornton.[20]

Let $N_1$ and $N_2$ denote the mean number of atoms of two types in a given macroscopic volume $\Omega$. Furthermore, $\Delta$ shall denote the instantaneous deviation from the mean. The probability of these deviations (see, for

---

* *Note added in proof.* Y. Waseda (*Z. Naturforsch.* **A38**, 509, 1983) gives diffraction data in support of equation (3.6.8) for liquid metals. Relevant theoretical work is that of R. Kumaravadivel and M. P. Tosi (*Nuovo Cim.* **4D**, 39, 1984).

example, the book by Tolman[21] or Hill[22]) is given by

$$w = w_0 \exp\left(-\sum_{i,j} \frac{F_{ij}\Delta N_i \Delta N_j}{2k_BT}\right), \qquad i,j = 1,2 \tag{3.7.1}$$

where $w_0$ is the normalization constant and

$$F_{ij} = \frac{\partial^2 F}{\partial N_i \partial N_j}\bigg|_{\Omega,T} = \left(\frac{\partial \mu_i}{\partial N_j}\right)_{T,\Omega,N}$$

$$= \left(\frac{\partial \mu_j}{\partial N_i}\right)_{T,\Omega,N} = F_{ji} \tag{3.7.2}$$

As usual, the chemical potential $\mu_i$ of species $i$ is given by

$$\mu_i = \frac{\partial F}{\partial N_i}\bigg|_{T,\Omega,N'} = \frac{\partial G}{\partial N_i}\bigg|_{T,P,N'} \tag{3.7.3}$$

the subscript $N'$ indicating that an $N_i$ not being differentiated is held fixed. From the property of the Gaussian distributions, we have

$$\langle \Delta N_i \Delta N_j \rangle = k_B T (F^{-1})_{ij} \tag{3.7.4}$$

where $(F^{-1})_{ij}$ is evidently the $ij$ element of the inverse matrix of $F_{ij}$. Now

$$\left(\frac{\partial \mu_i}{\partial N_j}\right)_{T,\Omega,N'} = \left(\frac{\partial \mu_i}{\partial N_j}\right)_{T,P,N'} + \frac{v_i v_j}{\Omega K_T} \tag{3.7.5}$$

where $v_i = (\partial\Omega/\partial N_i)_{T,P,N'}$ is the partial molar volume for the species $i$ and

$$\Omega = N_1 v_1 + N_2 v_2 \tag{3.7.6}$$

Hence one obtains

$$\sum_{ij} F_{ij}\Delta N_i \Delta N_j = \frac{1}{\Omega K_T}(v_1\Delta N_1 + v_2\Delta N_2)^2 + \sum_{ij}\left(\frac{\partial \mu_i}{\partial N_j}\right)_{T,p,N'}\Delta N_i \Delta N_j \tag{3.7.7}$$

At this point the Gibbs–Duhem relations in the form appropriate for two components,

$$\sum_{i=1}^{2} N_i \left(\frac{\partial \mu_i}{\partial N_j}\right)_{T,p,N'} = 0, \qquad j = 1,2 \tag{3.7.8}$$

is used to simplify the above summation. Remembering that

$$\Delta N = \Delta N_1 + \Delta N_2 \tag{3.7.9}$$

and

$$N\Delta c = (1 - c)\Delta N_1 - c\Delta N_2 \qquad (3.7.10)$$

one obtains

$$\sum_{ij} F_{ij}\Delta N_i \Delta N_j = \left(\frac{\Omega}{K_T N^2}\right)(\Delta N + N\Delta c\delta)^2 + B(\Delta c)^2 \qquad (3.7.11)$$

where the size difference $\delta$ is given by

$$\delta = \frac{v_1 - v_2}{cv_1 + (1 - c)v_2} \qquad (3.7.12)$$

and

$$B = \left(\frac{N^4}{N_2^2}\right)\left(\frac{\partial\mu_1}{\partial N_1}\right)_{T,p,N'} = \left(\frac{\partial^2 G}{\partial c^2}\right)_{T,p,N} \qquad (3.7.13)$$

In this way, the fluctuations that are of considerable interest in binary liquid mixtures, namely $\langle(\Delta c)^2\rangle$, $\langle(\Delta N)^2\rangle$, and $\langle\Delta N\Delta c\rangle$, can be written in terms of the size factor $\delta$, the isothermal compressibility $K_T$, and the second derivative with respect to the concentration of the Gibbs function $G$ (see also Appendix 3.2.)

### 3.7.2. Regular and Conformal Solutions

The work of Bhatia et al.[23] uses the approach of Longuet-Higgins,[24] which he refers to as the model of conformal solutions. The idea is to express the thermodynamic properties of the mixture in terms of those of a reference species $L_0$, say.

The basic equation derived by Longuet-Higgins for the molar Gibbs free energy in terms of the mole fractions $x_r$ and $x_s$ of species $L_r$ and $L_s$ in the solution is

$$G = \sum_r x_r(G_r + RT \ln x_r) + \sum\sum_{r<s} x_r x_s E_0 d_{rs} \qquad (3.7.14)$$

Here $G_r$ is the molar free energy of species $L_r$ at given $T$ and $p$, while $E_0$ is the molar configurational energy of the reference species, and is explicitly $RT - Q_0$, where $Q_0$ is the latent heat of vaporization of $L_0$ at temperature $T$ and pressure $p$; $d_{rs}$ is an interaction parameter for each pair of components.

The dependence of volume on composition follows in the form

$$V = \sum_r x_r V_r + V_0(p\beta_0 - T\alpha_0) \sum\sum_{r<s} x_r x_s d_{rs} \qquad (3.7.15)$$

where $\alpha_0$ and $\beta_0$ are, respectively, the thermal expansion coefficient and the isothermal compressibility of the reference liquid $L_0$.

These two equations express the thermodynamic properties of a conformal solution in terms of those of its components, together with a single interaction parameter for each pair of components. Therefore, it is clear that, when this model is applicable, the thermodynamic properties of a solution of more than two components can be determined from a study of the appropriate possible binary systems.

The theory is closely related to the theory of regular solutions. In fact, the only assumption made in the model of Longuet-Higgins that is not involved in the theory of regular solutions is that the intermolecular forces are approximately equal in magnitude for different pairs of species.

A generalization of this model has been effected by Byers Brown (see Rowlinson[19] for a summary and further references).

### 3.7.2.1. Application to a Binary Alloy, Especially Na–K

When specializing to two components, the conformal solution has a Gibbs free energy given by

$$G = (1 - c)G_1 + cG_2 + RT[c \ln c + (1 - c)\ln(1 - c)] + c(1 - c)w \qquad (3.7.16)$$

where $G_1$ and $G_2$ refer to the pure species 1 and 2, while $w \equiv E_0 d_{12}$ depends on $p$ and $T$ but is independent of concentration.

The molar volume $V$ is immediately found to have the form

$$V \equiv \left(\frac{\partial G}{\partial p}\right)_T = (1 - c)V_1^0 + cV_2^0 + c(1 - c)w', \qquad w' = \left(\frac{\partial w}{\partial p}\right)_T \qquad (3.7.17)$$

$V_1^0$ and $V_2^0$ evidently being the molar volumes of the pure species 1 and 2.

The concentration fluctuations are readily obtained from the result

$$\left(\frac{\partial^2 G}{\partial c^2}\right)_{T,p} = RT\left[\frac{1}{c(1 - c)}\right] - 2w \qquad (3.7.18)$$

the quantity $S_{cc}(0)$ being given in Appendix 3.2. Also given in that Appendix are expressions for the long-wavelength limit of the structure factors $a_{ij}$ $(k = 0)$. The quality of the predictions of these partial structure factors in the long-wavelength limit is shown by comparing the conformal-solution theory with the experimentally derived results of McAlister and Turner[25] in Figure 3.2. The salient features of the experiments are all faithfully reflected by the conformal-solution theory. For a further discussion of solution theories applied to liquid metals, the reader is referred to the review by Bhatia.[26]

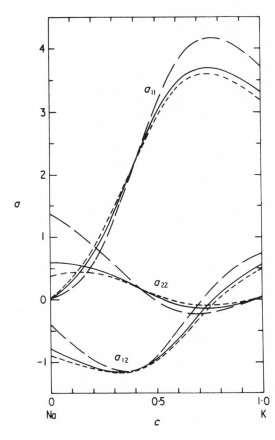

Figure 3.2. Long-wavelength limit of structure factors in liquid Na–K alloys for various sets of parameters in conformal solution theory.[23]

We stress that conformal-solution theory should not be used in the case of large size differences. It fails, for instance, for liquid Na–Cs, where the ratio of atomic volumes is 3. This case is discussed separately below by a modified approach based on Flory's work.

### 3.7.3. Phase Diagrams and Concentration Fluctuations

As a further application of conformal-solution theory for liquid metal alloys, we shall now outline the calculation of the phase diagram of Na–K.[27]

The liquidus curve of an ideal solution has long been known to have the approximate form

$$\ln(1 - c_2) = (L_{10}/R)(T_1^{-1} - T^{-1}) \qquad (3.7.19)$$

$c_2$ being the concentration of element 2 and $L_{10}$ the latent heat at the freezing temperature $T_1$ of pure liquid 1.

### 3.7.3.1. Thermodynamic Equations along the Liquidus

In terms of chemical potentials

$$\mu_{10}^s(T) = \mu_1(T, c_2) \qquad (3.7.20)$$

with the differential form

$$\frac{\Delta T}{\Delta c_2} = -\left(\frac{\partial \mu_1}{\partial c_2}\right)_{p,T} \bigg/ \left[\left(\frac{\partial \mu_1}{\partial T}\right)_{c_2,p} - \left(\frac{\partial \mu_{10}^s(T)}{\partial T}\right)_p\right]$$

$$= \left(\frac{\partial \mu_1}{\partial c_2}\right)_{p,T} \bigg/ (L/T) = -c_2\left(\frac{\partial^2 G}{\partial c_2^2}\right)_{p,T} \bigg/ (L/T)$$

$$= -\frac{RT^2 c_2}{S_{cc}(0)L} \qquad (3.7.21)$$

where subscript zero refers to a pure substance and superscript s denotes solid. Hence it can be seen that the liquidus curve depends crucially on the concentration fluctuations $\langle(\Delta c)^2\rangle$. In the above equation for the slope of the liquidus curve, $L$ is a generalized concentration-dependent latent heat defined by

$$\frac{L}{T} = \frac{L_{10}}{T_1} + \int_{T_1}^{T} \frac{\Delta c_{p_{10}}}{T} dT - \frac{\partial}{\partial T} \{RT \ln[\gamma_1(1-c_2)]\} \qquad (3.7.22)$$

$\gamma_1$ being the activity, while $\Delta c_{p_{10}} = c_{p_{10}}^L - c_{p_{10}}^S$.

If we now expand $\Delta c_{p_{10}}$ around $T_1$ and neglect higher terms than the first, and calculate the activity $\gamma_1$ from conformal-solution theory ($RT \ln \gamma_1 \div wc^2$), then the equation of the liquidus curves is obtained in the form

$$c(t) - 1 = (c_{ideal} - 1)\exp(-Wc^2/t) \qquad (3.7.23)$$

where $t = T/T_1$ and $W = w/RT_1$. If the value $w^{Na}/RT_1 = 1.1$ is used, the liquidus curves of the eutectic Na–K mixture are found to be in good agreement with experiment, as shown in Figure 3.3.

### 3.7.4. Flory's Theory and Large Size Differences

In the work of Flory,[28] the Gibbs free energy of mixing has the form

$$\Delta_m G = Nk_B T[c \ln \phi + (1-c)\ln(1-\phi)] + Ng(c)w \qquad (3.7.24)$$

where the last term is the energy of mixing, the concentration dependence not being written out explicitly.

The modification in the entropy term is that $\phi$ is the concentration by volume of species 1. If $v_1$ and $v_2$ are the partial molar volumes, then $\phi$ is given by

$$\phi = \frac{cv_1}{cv_1 + (1-c)v_2} \equiv c + c(1-c)\delta \qquad (3.7.25)$$

This model must be used in Na–Cs because the size difference is large, the ratio of the molar volumes being such that $v_2/v_1 = 3$. After a short calculation, the result for the concentration fluctuations is found to be

$$S_{cc}(0) \doteq c(1-c)\left[1 + c(1-c)\delta^2 + c(1-c)g''(c)\frac{w}{k_{\mathrm{B}}T}\right]^{-1} \qquad (3.7.26)$$

where $g(c)$ is a known function entering equation (3.7.24). The above treatment has been used with some success by Bhatia and March[29] for

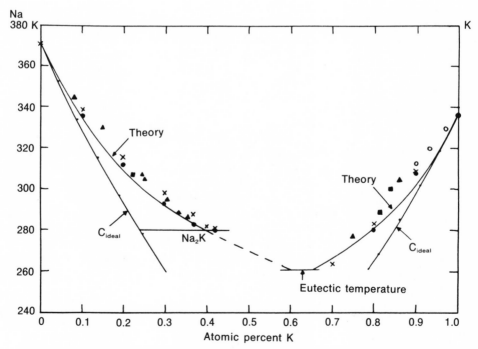

Figure 3.3. Eutectic curves of liquid Na–K alloys. Points refer to measurements and curves are from conformal-solution theory (after Bhatia and March[27]).

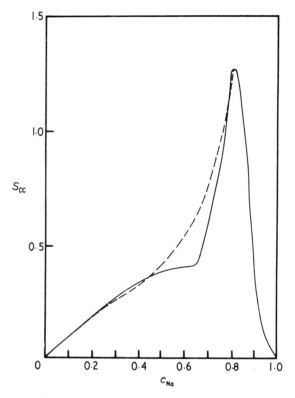

Figure 3.4. Concentration fluctuations in liquid Na–Cs alloy. The solid curve refers to measurements and the dashed curve to Flory's method (after Bhatia and March[28]).

comparison with the measurements of Ichikawa and Thompson[30] on liquid Na–Cs. There is a large asymmetry in the measured value of $S_{cc}(0)$, with a large peak around $c_{Na}$ equal to about 0.75. Bhatia and March conclude that the position of the peak depends on a balance between the size ratio $v_2/v_1$ and the interchange energy $w$, which they estimate as $w/k_B T = 1.14$ at the temperature of the Ichikawa–Thompson measurements. The overall agreement between theory and experiment is quite reasonable, as shown in Figure 3.4.

It is noteworthy that the liquidus curves of Na–Cs are also described quite reasonably by the model of Flory. The slope of the liquidus curve was earlier seen to be given by

$$\frac{\Delta T}{\Delta c} = -\frac{RT^2 c}{S_{cc}(0)L} \tag{3.7.27}$$

so it is evident that a rather flat portion of the liquidus curve will be found where there is a huge peak in $S_{cc}(0)$ and, of course, this is observed near $c = 0.75\text{--}0.8.$*

### 3.7.5. Relation of Miedema's Work to Solution Theories

We shall see later in this chapter that the interchange energy $w$ in conformal-solution theory is calculable from the properties of the reference liquid [in fact, from $\phi(r)$ and $g(r)$]. However, how to find the reference liquid for a given alloy system remains somewhat of a problem.

Therefore, it is of interest here to point out that we can use the work of Miedema and his colleagues (see, for example, the review by Miedema, de Châtel, and de Boer[31]) to express $w$ in an a–b alloy of terms of (1) the Wigner–Seitz cell boundary densities $n_a$ and $n_b$ in pure metals a and b, and (2) the electronegativity difference (see also Chapter 1).

Since we have seen that $w$ can be estimated from thermodynamic (activity) data and from phase diagrams, a link can therefore be forged between solution theories of liquid metal mixtures and the electron theory of metals.[32] This link is more direct, even though at present less fundamentally based, than that via alloy pair potentials, discussed fully in Chapter 4 (see also Section 3.8.2).

If $V_a$ and $V_b$ denote the atomic volumes of pure metals a and b, then from Miedema's calculations of the heat of solution of $i$ in $j$ [say $\Delta H_s$ ($i$ in $j$)], we obtain

$$w = \tfrac{1}{2}[\Delta H_s(\text{a in b}) + \Delta H_s(\text{b in a})] \qquad (3.7.28)$$

Using Miedema's results this yields

$$w = \frac{V_a^{2/3} + V_b^{2/3}}{n_a^{-1/3} + n_b^{-1/3}} [-P(n_a^{1/3} - n_b^{1/3})^2 + Q(x_a - x_b)^2] \qquad (3.7.29)$$

where $x_a - x_b$ is the electronegativity difference while $P$ and $Q$ are almost constant, through a wide class of alloys. This equation therefore provides the desired link between $n_a$ and $n_b$, which can be obtained from the electron theory of pure metals, the electronegativity difference $(x_a - x_b)$, and the interchange energy $w$ of solution theory.

Table 3.7.1 summarizes numerical consequences of the above equation for liquid mixtures of two alkalis. Actually, $w$ as obtained by this method is about a factor of 2 too large for Na–K; this is an unacceptable error in calculating the concentration fluctuations $S_{cc}(0)$. Refinements of Miedema's work are therefore clearly needed, but the gross trends of the measured interchange energy are illustrated by this treatment.

---

* *Note added in proof.* In relation to NaCs and to Flory's formula, we refer to (a) F. E. Neale and N. E. Cusack, *J. Phys. F* **12**, 2839 (1982) and (b) R. N. Singh and A. B. Bhatia, *J. Phys. F* **14**, 2309 (1984).

**Table 3.7.1. Interchange Energy $w$ for Liquid Alkali Mixtures as Extracted from Experiment Compared with That Obtained Using Miedema's Work (after Alonso and March[32])**

| Alloy | $w$ (units are kJ/g) at solute | Exp |
|---|---|---|
| NaK | 5.5 | 2.9 |
| CsK | 0.0 | 0.45 |
| (NaCs)[a] | 10.0 | 5.0 |
| RbCs | 0.0 | −0.5 |
| KRb | 0.0 | 0.5 |
| NaRb | 7.5 | 5.4 |

[a] NaCs is not a conformal solution, for reasons discussed in Section 3.7.4.

## 3.8. Structure of Binary Liquid Mixtures

It is clear that to develop a comprehensive microscopic theory of a liquid a–b mixture, we require a way of determining the partial radial distribution functions $g_{aa}(r)$, $g_{bb}(r)$, and $g_{ab}(r)$, which have been discussed from the standpoint of diffraction measurements in Chapter 1 and from the point of view of models in Chapter 2. Though progress in the microscopic theory is still somewhat limited, we shall generalize the approach given in Section 3.5 for monatomic liquids to binary liquid mixtures.

### 3.8.1. Force Equations

Consider again a binary liquid mixture with $N$ atoms of type a at $\mathbf{R}_1, \ldots, \mathbf{R}_N$ and $n$ atoms of type b at $\mathbf{r}_1, \ldots, \mathbf{r}_n$. For a configuration in which atom a is situated at $\mathbf{R}_1$ and atom b at $\mathbf{r}_1$, let us write down an equation for the mean force $-\partial U_{ab}(\mathbf{R}_1, \mathbf{r}_1)/\partial \mathbf{R}_1$ acting on atom a. We have the direct interaction, say $-\partial \phi_{ab}(\mathbf{R}_1, \mathbf{r}_1)/\partial \mathbf{R}_1$, and second [cf. equation (3.5.2)] a contribution from the "rest" of the system namely the indirect interaction.

#### 3.8.1.1. Calculation of Indirect Contribution

Consider a second atom of type a at $\mathbf{R}_2$ acting with a force $-\partial \phi_{aa}(\mathbf{R}_1 - \mathbf{R}_2)/\partial \mathbf{R}_1$ and a second atom of type b at $\mathbf{r}_2$, with resulting force $-\partial \phi_{ab}(\mathbf{R}_1, \mathbf{r}_2)/\partial \mathbf{R}_1$. We must then sum these contributions, multiply by the probability of this four-atom configuration:

$$\rho_{abab}^{(4)}(\mathbf{R}_1 \mathbf{r}_1 \mathbf{R}_2 \mathbf{r}_2)/\rho_{ab}^{(2)}(\mathbf{R}_1 \mathbf{r}_1)$$

and finally integrate over the positions of the second type of atom. The result is

$$-\frac{\partial U_{ab}(\mathbf{R}_1\mathbf{r}_1)}{\partial \mathbf{R}_1} = -\frac{\partial \phi_{ab}(\mathbf{R}_1\mathbf{r}_1)}{\partial \mathbf{R}_1}$$

$$-\int \frac{\rho_{abab}^{(4)}(\mathbf{R}_1\mathbf{r}_1\mathbf{R}_2\mathbf{r}_2)}{\rho_{ab}^{(2)}(\mathbf{R}_1\mathbf{r}_1)} \left[\frac{\partial \phi_{aa}(\mathbf{R}_1 - \mathbf{R}_2)}{\partial \mathbf{R}_1} + \frac{\partial \phi_{ab}(\mathbf{R}_1\mathbf{r}_2)}{\partial \mathbf{R}_1}\right] d\mathbf{R}_2 d\mathbf{r}_2 \tag{3.8.1}$$

Since three-atom correlations $\rho^{(3)}$ are related to $\rho^{(4)}$ by

$$\rho_{aba}^{(3)}(\mathbf{R}_1\mathbf{r}_1\mathbf{R}_2) = \int \rho_{abab}^{(4)}(\mathbf{R}_1\mathbf{r}_2\mathbf{R}_2\mathbf{r}_2)d\mathbf{r}_2 \tag{3.8.2}$$

and

$$\rho_{abb}^{(3)}(\mathbf{R}_1\mathbf{r}_1\mathbf{r}_2) = \int \rho_{abab}^{(4)}(\mathbf{R}_1\mathbf{r}_1\mathbf{R}_2\mathbf{r}_2)d\mathbf{R}_2 \tag{3.8.3}$$

we obtain finally

$$-\frac{\partial U_{ab}}{\partial \mathbf{R}_1}(\mathbf{R}_1\mathbf{r}_1) = -\frac{\partial \phi_{ab}(\mathbf{R}_1\mathbf{r}_1)}{\partial \mathbf{R}_1} - \int \frac{\rho_{aab}^{(3)}(\mathbf{R}_1\mathbf{r}_1\mathbf{R}_2)}{\rho_{ab}^{(2)}(\mathbf{R}_1\mathbf{r}_1)} \frac{\partial \phi_{aa}(\mathbf{R}_1\mathbf{R}_2)d\mathbf{R}_2}{\partial \mathbf{R}_1}$$

$$-\int \frac{\rho_{abb}^{(3)}(\mathbf{R}_1\mathbf{r}_1\mathbf{r}_2)}{\rho_{ab}^{(2)}(\mathbf{R}_1\mathbf{r}_1)} \frac{\partial \phi_{ab}(\mathbf{R}_1\mathbf{r}_2)d\mathbf{r}_2}{\partial \mathbf{R}_1} \tag{3.8.4}$$

A similar force equation can be written for two atoms of type a.

### 3.8.2. Approximate Pair Potentials from Structure Factors

While the above approach to a classical binary liquid is formally exact, the situation parallels that of pure liquids in that one cannot make further progress without invoking approximations. Again, one possibility is the Kirkwood-type approximation. As before, we can then invert the problem and ask how $\phi_{aa}$, etc., might be extracted from measured structure data. If, as in simple liquid metals, $\phi_{aa}$ and $U_{aa}$ are not very different, one can replace $\phi_{aa}$ in the "indirect" term by $U_{aa}$. One can then derive the following equations with, admittedly, a limited range of validity:

$$\phi_{aa}(R) = U_{aa}(R) + \frac{k_B T \rho_a}{8\pi^3 \rho^2} \int [S_{aa}(k) - 1]^2 \exp(i\mathbf{k} \cdot \mathbf{R})d\mathbf{k}$$

$$+ \frac{k_B T \rho_b}{8\pi^3 \rho^2} \int [S_{ab}(k) - 1]^2 \exp(i\mathbf{k} \cdot \mathbf{R})d\mathbf{k} \tag{3.8.5}$$

and

$$\phi_{ab}(R) = U_{ab}(R) + \frac{k_B T \rho_a}{8\pi^3 \rho^2} \int [S_{aa}(k) - 1][S_{ab}(k) - 1] \exp(i\mathbf{k} \cdot \mathbf{R}) d\mathbf{k}$$

$$+ \frac{k_B T \rho_b}{8\pi^3 \rho^2} \int [S_{ab}(k) - 1][S_{bb}(k) - 1] \exp(i\mathbf{k} \cdot \mathbf{R}) d\mathbf{k} \qquad (3.8.6)$$

It is of interest that one can, at least in principle, use measured structure data to estimate $\phi_{aa}$, $\phi_{ab}$, and $\phi_{bb}$. However, such an approach could at best only be quantitative if $\phi_{aa} \sim U_{aa}$, etc. One example in which this approach has been pushed through is the work of Page *et al.*[33] on a 50–50 liquid Na–K alloy. The pair potentials thus obtained show sensible features, even though they can hardly be fully quantitative. These aspects are discussed further in Chapter 4 with the aid of different methods.

### 3.8.3. Partial Structure Factors from Conformal-Solution Theory

Parrinello *et al.*[34] have obtained the concentration–concentration partial structure factor $S_{cc}(k)$ in a conformal solution in terms of the properties of the reference liquid. Even though this treatment clearly has all the limitations of conformal-solution theory, it has the merit that it is derived precisely from statistical mechanics and satisfies the important sum rule given in Appendix 3.1.

We shall not attempt to give the argument here, but it is worth remarking that conformal-solution theory is based on a (monatomic) reference liquid pair potential $\phi(r)$, which is scaled to yield the potentials $\phi_{ij}(r)$ in a binary liquid mixture by the relation

$$\phi_{ij}(r) = A_{ij}\phi(\lambda_{ij}r) \qquad (3.8.7)$$

where the deviations of $A_{ij}$ and $\lambda_{ij}$ from unity are treated as perturbations. As a result of this scaling, it is worth asking how the scaling of the potential in a monatomic liquid, namely

$$\phi(r) \rightarrow \phi_m(r) \equiv A\phi(\lambda r) \qquad (3.8.8)$$

affects the liquid properties.

### 3.8.3.1. Scaling of Thermodynamic Quantities

The chemical potential $\mu$ scales according to

$$\mu_m(\beta, \Omega) = A\mu(A\beta, \lambda^3\Omega) + \frac{3}{\beta}\ln\lambda : \beta = (k_B T)^{-1} \qquad (3.8.9)$$

so that the fugacity becomes

$$z_m(\beta, \Omega) = \lambda^3 A^{3/2} z(A\beta, \lambda^3 \Omega) \qquad (3.8.10)$$

### 3.8.3.2. Scaling of Pair Function $g(r)$

By using the grand canonical ensemble, we can scale the pair function itself as[1]

$$g_m(r, \beta, \Omega) = g(\lambda r, A\beta, \lambda^3 \Omega, A^{3/2} z) \qquad (3.8.11)$$

The above results already yield a useful insight into two-component mixtures. In particular, the density–density correlation in a mixture, $g_{NN}(r)$, can be obtained by reinterpreting $A$ and $\lambda$.

The reader is referred to the work of Parrinello et al.[34] for detailed derivations. We shall conclude this brief discussion of the structure of a conformal solution by returning to the concentration–concentration structure factor, which has the form

$$S_{cc}(k) = c(1 - c) + d(\widetilde{g\phi}) \qquad (3.8.12)$$

All the $k$-dependence resides in $\widetilde{g\phi}$, which is the Fourier transform of the product of the radial distribution function $g(r)$ and the pair potential $\phi(r)$ of the reference liquid.

Johnson et al.[35] report numerical results for $\widetilde{g\phi}$, the deviation $\Delta S_{cc}(k)$ from an ideal solution as a function of $k$ for various choices of the reference liquid. They also point out that, in the random-phase approximation, $\Delta S_{cc}(k)$ is simply proportional to the direct correlation function $\tilde{c}(k)$ of the reference liquid.

## 3.9. Elementary Introduction to the Dynamical Structure Factor

In Section 3.6, we introduced the dynamical structure factor of a pure liquid and set up a simple model based on independent density fluctuations.

Here, in connection with two-component systems, we shall consider the simplest classical mixture with nontrivial dynamics, namely a mixture of isotopes! In this case the forces between the components are all the same, as are the static structure factors.

Being an "ideal" solution, the static concentration fluctuations are independent of $k$, the static structure factor $S_{cc}(k)$ being given simply by

$$S_{cc}(k) = c_1 c_2 \qquad (3.9.1)$$

in terms of the concentrations of the two isotopes.

### 3.9.1. Dynamics of an Isotopic Mixture

Dynamic properties of such mixtures have been discussed independently by Parrinello et al.[36] using simple analytic models, and by Ebbsjö et al.[37] with the aid of computer simulation in Lennard-Jones isotopic mixtures.

#### 3.9.1.1. Models of Partial Dynamic Structure Factors

For the ideal solution, which an isotopic mixture constitutes, the dynamic structure factors $S_{NN}(k\omega)$ and $S_{NC}(k\omega)$ have zero-order moments given by

$$\int_{-\infty}^{\infty} \frac{d\omega}{2\pi} S_{NC}(k\omega) = S_{NC}(k) = 0 \tag{3.9.2}$$

and

$$\int_{-\infty}^{\infty} \frac{d\omega}{2\pi} S_{NN}(k\omega) = S(k) \tag{3.9.3}$$

As for liquid metals in Section 3.6.1, the simplest model is again to use the idea of well-defined collective excitations and to find their dispersion relation $\omega_k$ from the second-moment relations, namely

$$\int_{-\infty}^{\infty} \frac{d\omega}{2\pi} \omega^2 S_{\alpha\beta}(k\omega) = \frac{\delta_{\alpha\beta} k_B T k^2}{M_\alpha} \tag{3.9.4}$$

This yields, for the number-concentration dynamic structure factors, the relations

$$\int_{-\infty}^{\infty} \frac{d\omega}{2\pi} \omega^2 S_{NC}(k\omega) = c_1 c_2 k_B T k^2 \left( \frac{1}{M_1} - \frac{1}{M_2} \right) \tag{3.9.5}$$

$$\int_{-\infty}^{\infty} \frac{d\omega}{2\pi} \omega^2 S_{NN}(k\omega) = k_B T k^2 \left( \frac{c_1}{M_1} + \frac{c_2}{M_2} \right) \tag{3.9.6}$$

and

$$\int_{-\infty}^{\infty} \frac{d\omega}{2\pi} \omega^2 S_{cc}(k\omega) = \frac{c_1 c_2 M k_B T k^2}{M_1 M_2} \tag{3.9.7}$$

where $M = c_1 M_1 + c_2 M_2$.

Now except for the lightest isotopes, mass differences are small and the first of these moments in equation (3.9.5) tends to zero as $M_1 - M_2$ tends to zero, the other second moments remaining finite. Thus, in the limit

$M_1 - M_2$ tends to zero, there is no link between the number density $\rho(\mathbf{k}t)$ and the concentration $c(\mathbf{k}t)$ and they can be assumed to oscillate independently, with frequencies $\omega_{10}(k)$ and $\omega_{20}(k)$ say, respectively.

In order to satisfy zeroth and second moments in this limit of small $(M_1 - M_2)$, we therefore adopt the forms

$$S_{NN}(k\omega) \doteq \frac{S(k)}{2}\{\delta[\omega - \omega_{10}(k)] + \delta[\omega + \omega_{10}(k)]\} \qquad (3.9.8)$$

and

$$S_{cc}(k\omega) \cong \tfrac{1}{2}c_1 c_2\{\delta[\omega - \omega_{20}(k)] + \delta[\omega + \omega_{20}(k)]\} \qquad (3.9.9)$$

with $S_{Nc}(k, \omega) \doteq 0$. Then we obtain

$$\omega_{10}^2(k) = \frac{k_B T k^2}{S(k)}\left(\frac{c_1}{M_1} + \frac{c_2}{M_2}\right) \qquad (3.9.10)$$

and

$$\omega_{20}^2(k) = M k_B T k^2 / M_1 M_2 \qquad (3.9.11)$$

Two limitations must be noted:

1. The velocity of sound in an isotopic mixture will only be correct if the ratio of specific heats $\gamma$ is near to unity. (This is the case for simple liquid metals just above freezing point for which $\gamma$ is about 1.1–1.3, but for argon at its triple point with $\gamma = 2.2$, the error would be severe.)
2. The treatment is valid only for $M_1 - M_2$ small.

### 3.9.1.2. Case of Arbitrary Mass Difference

In this case $S_{Nc}(k, \omega)$ is nonzero and the pure modes $\omega_{10}$ and $\omega_{20}$ interact and mix. The theory can be formulated in terms of response functions for which the main results are merely summarized.

The new squared frequencies are given by $(\beta^{-1} = k_B T)$

$$\omega^2(k) = \frac{1}{2}\left\{\omega_{10}^2 + \omega_{20}^2 \pm \left[(\omega_{10} + \omega_{20})^2 - \frac{-k^4}{M_1 M_2 \beta^2 S(k)}\right]^{1/2}\right\} \qquad (3.9.12)$$

When the mass difference is small

$$\frac{4k^2}{M_1 M_2 \beta^2 S(k)} \cong 4\omega_{10}^2\omega_{20}^2 \qquad (3.9.13)$$

giving back the pure-mode frequencies.

Evidently, given $S(k)$, equation (3.9.12) determines the frequencies for arbitrary concentration and masses. Therefore, we will only discuss briefly the limiting case when $M_1/M_2$ becomes large. In this case

$$\omega_1^2 \doteq \frac{c_1}{M_1} \frac{k_B T k^2}{S(k)} \tag{3.9.14}$$

and

$$\omega_2^2 \doteq \frac{c_2}{M_1} k_B T k^2 \tag{3.9.15}$$

When $c_1$ becomes small, namely a light impurity in the host liquid, the two modes are well separated and $\omega_1^2$ becomes very small, although still reflecting the oscillations in $S(k)$ at large $k$. Some discussion of damping of such modes has been given by Parrinello et al. but we shall not go into further detail here.

### 3.9.1.3. Atomic Transport for Mass Difference $(M_1 - M_2)$ Small

Having dealt with models for arbitrary mass differences, we note that the dynamics of isotopic mixtures can be discussed carefully by perturbation theory, the perturbation being the deviation of $M_1$ and $M_2$ from some suitably chosen mass $M$.

The theory can be posed precisely as follows. We take a reference liquid with the same force law as the mixture composed of $N = N_1 + N_2$ identical atoms of mass $M$. Then the difference $H'$ between the Hamiltonian of the mixture and that of the reference liquid is given by

$$H' = -\left[ \Delta_1 \sum_{i=1}^{N_1} p_i^2 + \Delta_2 \sum_{i=1}^{N_2} p_i^2 \right] \Big/ 2M \tag{3.9.16}$$

with $\Delta_i = (M_i - M)/M_i$. With the assumption that $H'$ can be treated as a perturbation, some simple results follow.

1. The interdiffusion constant $D$ is given by

$$D = c_2 D_1 + c_1 D_2 \tag{3.9.17}$$

where $D_1$ and $D_2$ are self-diffusion coefficients for the two isotopes in the mixture. This relation is correct to first order in $M_1 - M_2$.

2. By choosing the "reference mass" $M$ in the form

$$M = \left( \frac{c_1}{M_1} + \frac{c_2}{M_2} \right)^{-1} \tag{3.9.18}$$

the following relations obtain:

$$c_1 D_1 + c_2 D_2 = D_1^0 \left( c_1 + c_2 \frac{M_1}{M_2} \right)^{1/2}$$

$$= D_2^0 \left( c_2 + c_1 \frac{M_2}{M_1} \right)^{1/2} \tag{3.9.19}$$

where $D_2^0$ etc. denote the self-diffusion coefficients in the pure isotopes.

### 3.9.1.4. Velocity Autocorrelation Functions $\langle v(0)v(t)\rangle/\langle v(0)^2\rangle$

Perturbation theory for small mass differences allows one to exhibit explicitly the dependence of $S_{\alpha\beta}(k\omega)$ and $S_{\alpha\alpha}^s(k\omega)$ on concentration and masses.

Molecular dynamics calculations conducted with a velocity autocorrelation function (Ebbsjö et al.[37]) can be compared with the perturbative results. Significant deviations from this theory are observed only for very large mass differences.

## 3.10. Freezing Theory of Simple Liquids and Liquid Mixtures

We referred in the introduction in Section 3.1 to the pioneering work of Kirkwood and Monroe[2] on the statistical mechanical theory of freezing. These workers employed the first equation of the Born–Green–Yvon hierarchy for an inhomogeneous system. This related the singlet density $\rho(\mathbf{r})$, the pair function $g$, and the pair potential $\phi(r)$. They enquired, with some limited success, as to whether this equation had, under suitable circumstances, coexisting solutions of uniform and of periodic singlet density representing, respectively, liquid and crystalline phases.

This topic was reopened by Ramakrishnan and Yussouff.[38] While at first sight their approach seemed to be based on one of the approximate theories of liquid structure (the hypernetted chain theory, see Section 3.5), their work made it quite clear that (1) one should work with the direct correlation function $c(r)$ rather than $g(r)$ and $\phi(r)$, and (2) one should "close" the hierarchy by using experimental data as input information for $c(r)$, i.e., bulk liquid data. Subsequently, it was emphasized independently by Haymet and Oxtoby[39] and March and Tosi[40] that this theory was founded fundamentally in the statistical mechanics of inhomogeneous systems via an equation relating the singlet density and the direct correlation function.[41] Since March and Tosi were interested in their work primarily in the freezing of multicomponent systems, we will outline for this case the statistical mechanical theory following the arguments of Bhatia *et al.*[42]

## 3.10.1. Singlet Density — Direct Correlation Function Equations in Inhomogeneous Systems

The argument given below generalizes that of Lovett *et al.*[41] to multicomponent systems and aims to obtain equations determining the singlet-density profiles in the multicomponent inhomogeneous system.

Let $u_i(r)$ denote the dimensionless one-body potential per particle for species $i$:

$$u_i(r) = \beta[\mu_i - U_i(r)], \qquad \beta = (k_B T)^{-1} \tag{3.10.1}$$

Here $U_i(r)$ is the external potential for species $i$ and $\mu_i$ is its chemical potential.

The system is considered to be open and at constant $V$ and $T$. The single-particle densities are then given by

$$\rho_i(\mathbf{r}) = \langle \hat{\rho}_i(r) \rangle \tag{3.10.2}$$

where $\langle \cdot \cdot \rangle$ denotes the ensemble average as usual.

If all the $u_i(r)$ are given, the various $\rho_j(r)$ are uniquely determined, and vice versa, at given $V$ and $T$. Hence the quantities $u_i(r)$ can be regarded as functionals of the various $\rho_j(r)$, and vice versa. One then has

$$\frac{\delta \rho_i(r)}{\delta u_j(r')} = \langle \hat{\rho}_i(r) \hat{\rho}_j(r') \rangle - \langle \hat{\rho}_i(r) \rangle \langle \hat{\rho}_j(r') \rangle \tag{3.10.3}$$

and

$$\frac{\delta u_i(r)}{\delta \rho_j(r')} = \delta_{ij} \frac{\delta(r - r')}{\rho_i(r)} - c_{ij}(r, r') \equiv K_{ij}(r, r') \tag{3.10.4}$$

$c_{ij}(r, r')$ being the partial direct correlation functions of the mixture. If $u_i(r)$ is written in the form

$$u_i(r) = \ln[\rho_i(r) \Lambda_i^3] - C_i(r) \tag{3.10.5}$$

where $\Lambda_i = h(2\pi m_i k_B T)^{-1/2}$, then

$$c_{ij}(r, r') = \frac{\delta C_i(r)}{\delta \rho_j(r')} \tag{3.10.6}$$

If the functional dependence of the quantity $u_i$ is denoted by

$$u_i(r_1, [\rho_1(r), \rho_2(r) \cdots \rho_\nu(r)]) \equiv u_i(r_1) \tag{3.10.7}$$

say, then translational invariance implies

$$u_i(r_1, [\rho_1(r+\delta) \cdots \rho_\nu(r+\delta)]) \equiv u_i(r_1+\delta) \tag{3.10.8}$$

Hence one finds

$$\nabla u_i(r_1) = \sum_{j=1}^{\nu} \int d\mathbf{r} \, \frac{\delta u_i(r_1)}{\delta \rho_j(r)} \nabla \rho_j(r) \tag{3.10.9}$$

or

$$\nabla u_i(r_1) = \sum_{j=1}^{\nu} \int d\mathbf{r} K_{ij}(r_1, r) \nabla \rho_j(r) \tag{3.10.10}$$

where equation (3.10.4) has been used in equation (3.10.10).

If the external potential is now reduced to zero, i.e., $\nabla u_i(r) = 0$, equation (3.10.10) becomes

$$\sum_{j=1}^{\nu} \int d\mathbf{r} K_{ij}(r_1, r) \nabla \rho_j(r) = 0 \tag{3.10.11}$$

or, using equation (3.10.4),

$$\frac{\nabla \rho_i(r_1)}{\rho_i(r_1)} = \sum_{j=1}^{\nu} \int d\mathbf{r} c_{ij}(r_1, r) \nabla \rho_j(r) \tag{3.10.12}$$

These are the basic equations determining the density profiles in terms of the partial direct correlation functions.

### 3.10.2. Coexistence of Uniform and Periodic Solutions for Singlet Densities

We will present the argument below for a pure liquid, though it has been given also for mixtures by March and Tosi.[43] If equation (3.10.12) is written for a pure liquid, and we apply the equation for freezing in the spirit of the above discussion, the resulting equation can be integrated to yield

$$\ln \rho(\mathbf{r}_1) = \int d\mathbf{r}_2 c(r_{12}) \rho(\mathbf{r}_2) + \text{constant} \tag{3.10.13}$$

This is the central tool employed by March and Tosi. By directly solving equation (3.10.13) they demonstrate that, for a given (experimentally determined, say) liquid direct correlation function, equation (3.10.13) admits not only a solution for which the singlet density is uniform, with value $\rho(\mathbf{r}) = \rho_l$, but also a coexisting periodic solution $\rho_p(\mathbf{r})$.

To proceed with the above program, we apply equation (3.10.13) to both the singlet densities $\rho_l$ and $\rho_p(\mathbf{r})$ and then subtract to yield

$$\ln \left( \frac{\rho_p(\mathbf{r}_1)}{\rho_l} \right) - \int d\mathbf{r}_2 c(|\mathbf{r}_1 - \mathbf{r}_2|)[\rho_p(\mathbf{r}_2) - \rho_l] = 0 \qquad (3.10.14)$$

By assertion $\rho_p$ is periodic, so we can expand in a Fourier series using the reciprocal lattice vectors $\mathbf{G}$ to give

$$\rho_p(\mathbf{r}) = \rho_0 + V^{-1} \sum_{G \neq 0} \rho_G \exp(i\mathbf{G} \cdot \mathbf{r}) \qquad (3.10.15)$$

where $V$ is the total volume. If equation (3.10.15) is substituted in (3.10.14) and the resulting equation integrated over $\mathbf{r}_2$, we obtain

$$\ln \left( \frac{\rho_p(\mathbf{r})}{\rho_l} \right) = \frac{(\rho_0 - \rho_l)}{\rho_l} \tilde{c}(0) + (\rho_l V)^{-1} \sum_{G \neq 0} \rho_G \cdot \tilde{c}(\mathbf{G}) \exp(i\mathbf{G} \cdot \mathbf{r}) \qquad (3.10.16)$$

where, as usual, $\tilde{c}(k)$ is the Fourier transform of $c(r)$. For a given set of Fourier components of the liquid direct correlation function, equation (3.10.16) must be solved for the Fourier components $\rho_G$ of the singlet density.

Next, it will be useful to regard equation (3.10.14) as the Euler equation of a minimum free-energy principle. Of course, the thermodynamic requirement for the two phases to be in equilibrium is that this free-energy difference is zero. Following March and Tosi,[40] it is actually convenient to work with the thermodynamic potential $\Omega$, related to the Helmholtz free energy $F$ and chemical potential $\mu$ by

$$\Omega = F - N\mu \qquad (3.10.17)$$

The Helmholtz free energy can be conveniently divided into two parts, one corresponding to free particles and the other taking acount of the interparticle interactions via the direct correlation function $c(r)$. The first part is well known for uniform density (see, e.g., March and Tosi[40]) and we merely take the free-energy density over into the local density $\rho(\mathbf{r})$. The second part is also available, for example, in the work of Triezenberg and

Zwanzig,[44] and thus we can write

$$\frac{\Delta\Omega}{k_B T} = \int d\mathbf{r} \left\{ \rho_p(r) \ln\left(\frac{\rho_p(\mathbf{r})}{\rho_l}\right) - [\rho_p(\mathbf{r}) - \rho_l] \right\}$$

$$- \frac{1}{2} \iint d\mathbf{r}_1 d\mathbf{r}_2 [\rho_p(\mathbf{r}_1) - \rho_l] c(|\mathbf{r}_1 - \mathbf{r}_2|)[\rho_p(\mathbf{r}_2) - \rho_l] \qquad (3.10.18)$$

Variation of $\Delta\Omega$ with respect to $\rho_p(\mathbf{r})$ is readily shown to lead back to equation (3.10.14).

At this stage, we insert the Fourier expansions (3.10.15) and (3.10.16) into equation (3.10.18) to find, with $N = \rho_l V$,

$$\frac{\Delta\Omega}{k_B T} = \frac{1}{2N} \sum_G \tilde{c}(\mathbf{G}) |\rho_G|^2 - \frac{N(\rho_0 - \rho_l)}{\rho_l} + \frac{1}{2} N\tilde{c}(0)\frac{(\rho_0^2 - \rho_l^2)}{\rho_l^2} \qquad (3.10.19)$$

### 3.10.3. Volume Change on Freezing Related to Structural Properties

Equation (3.10.19) is the desired expression for the free-energy difference in terms of the Fourier components $\rho_G$ of the periodic density, and the volume change reflected in the difference between $\rho_l$ and $\rho_0$.

The possibility of coexistence of homogeneous liquid and periodic phases is clear from equation (3.10.19) because of the balance between positive contributions from the first and third terms on the right-hand side and the negative term from the volume change, provided the periodic phase has the higher density. That these terms are strongly coupled is clear from the highly nonlinear nature of the Euler equation (3.10.16). The actual coexistence point is evidently determined by the properties of $\tilde{c}(\mathbf{G})$, including $\mathbf{G} = 0$, linking the liquid structure intimately with the appearance of the periodic phase.

Essentially, the above equations have been solved numerically for liquid argon by Ramakrishnan and Yussouff[38] and we refer the reader to their paper for the numerical results, which are in good agreement with experiment.

### 3.10.4. Some Results for Freezing of Two-Component Systems

We conclude with a brief application of the above theory to freezing of the two-component systems (March and Tosi[43,45]). In the first of these papers, the theory was given for a two-component liquid alloy as well as in a form appropriate to the freezing of an ionic melt, say NaCl and RbCl. For these two materials, structural data of the type discussed in Chapter 1 has been used in conjunction with equation (3.10.12) above to give a quantitative theory of freezing.[46,47] Also discussed by D'Aguanno *et al.*[47] are fully nonlinear solutions for the problem of the freezing of $BaCl_2$ and $SrCl_2$ into a fast ion conducting phase in which, after freezing, the cations form an almost rigid sublattice. The anions in this sublattice behave as a "liquid in a periodic potential" or a "lattice liquid." Further applications of the theory of freezing given above are in progress.[48]

## Appendix 3.1. Distribution Functions and Fluctuations

The probability that a system in the grand canonical ensemble contains $N$ atoms is given by

$$P_N = \frac{\exp(N\mu/k_B T)Z(N, \Omega, T)}{Z_G} \tag{A3.1.1}$$

where $Z$ is the partition function in the canonical ensemble and $Z_G$ the grand partition function.

Let $\rho^{(n)}(\mathbf{r}_1, \ldots, \mathbf{r}_n)d\mathbf{r}_1 \cdots d\mathbf{r}_n$ be the probability of observing a molecule in $d\mathbf{r}_1 \cdots d\mathbf{r}_n$ at $\mathbf{r}_1, \ldots, \mathbf{r}_n$, irrespective of $N$. Then clearly

$$\rho^{(n)} = \sum_{N \geq n} \rho_N^{(n)} P_N \tag{A3.1.2}$$

where $\rho_N^{(n)}$ are the usual distribution functions in the canonical ensemble. Also we must have

$$\int \cdots \int \rho^{(n)}(\mathbf{r}_1, \ldots, \mathbf{r}_n)d\mathbf{r}_1 \cdots d\mathbf{r}_n = \sum_{N \geq n} P_n \frac{N!}{(N-n)!} \tag{A3.1.3}$$

If averages are always defined with respect to $P_N$, we may write

$$\int \cdots \int \rho^{(n)}(\mathbf{r}_1, \ldots, \mathbf{r}_n)d\mathbf{r}_1 \cdots d\mathbf{r}_n = \left\langle \frac{N!}{(N-n)!} \right\rangle \tag{A3.1.4}$$

Therefore as special cases we have

$$\int \rho^{(1)}(\mathbf{r}_1)d\mathbf{r}_1 = \langle N \rangle = \bar{N}, \text{ say}, \qquad n = 1 \tag{A3.1.5}$$

and hence

$$\iint \rho^{(1)}(\mathbf{r}_1)\rho^{(1)}(\mathbf{r}_2)d\mathbf{r}_1 d\mathbf{r}_2 = (\bar{N})^2 \tag{A3.1.6}$$

For $n = 2$,

$$\int \rho^{(2)}(\mathbf{r}_1\mathbf{r}_2)d\mathbf{r}_1 d\mathbf{r}_2 = \langle N(N-1)\rangle = \bar{N}^2 - \bar{N} \qquad (A3.1.7)$$

Thus remembering the fluid properties that $\rho^{(1)}(\mathbf{r}) = $ constant density $\rho$ and $\rho^{(2)} = \rho^{(2)}(|\mathbf{r}_1 - \mathbf{r}_2|)$, we find by subtraction

$$\Omega \int [\rho^{(2)}(r) - \rho^2]d\mathbf{r} = \bar{N}^2 - (\bar{N})^2 - \bar{N} \qquad (A3.1.8)$$

So we can express $\int [g(r)-1]d\mathbf{r}$ in terms of the difference between $\bar{N}^2$ and $(\bar{N})^2$. The long-wavelength limit of the structure factor $S(k)$ is immediately related to the above integral by

$$S(0) = 1 + \rho \int [g(r)-1]d\mathbf{r}$$

$$= 1 + \frac{\bar{N}^2 - (\bar{N})^2 - \bar{N}}{\rho\Omega}$$

$$= \frac{\bar{N}^2 - (\bar{N})^2}{\rho\Omega} \qquad (A3.1.9)$$

The final step in order to obtain a useful expression for $S(0)$ is to relate $\bar{N}^2 - (\bar{N})^2$ to the isothermal compressibility by noting that

$$\bar{N}Z_G = \sum_N N \exp(N_\mu/k_BT)Z(N, \Omega, T) \qquad (A3.1.10)$$

If we now differentiate with respect to the chemical potential $\mu$, at constant $\Omega$ and $T$, we obtain

$$\left(\frac{\partial\bar{N}}{\partial\mu}\right)_{\Omega,T} Z_G + \bar{N}\sum_N \frac{N}{k_BT}\exp(N_\mu/k_BT)Z(N, \Omega, T)$$

$$= \sum_N N \frac{N}{k_BT}\exp(N_\mu/k_BT)Z(N, \Omega, T) \qquad (A3.1.11)$$

or

$$\left(\frac{\partial\bar{N}}{\partial\mu}\right)_{\Omega,T} = \frac{1}{k_BT}[\bar{N}^2 - (\bar{N})^2] \qquad (A3.1.12)$$

Combining equations (A3.1.9) and (A3.1.12) leads to the important relation $S(0) = \rho k_B T K_T$, using $\partial N/\partial\mu = \Omega\rho(\partial\rho/\partial p)_T$.

## Appendix 3.2. Summary of Some Important Results on the Thermodynamics and Structure of Liquid Mixtures

In this appendix we merely collect together some of the most useful results on liquid mixtures. Most of these results are readily obtained either from the material in the text or from an extension of the methods set out there:

$$S_{cc}(0) = N\langle(\Delta c)^2\rangle = Nk_B T \Big/ \left(\frac{\partial^2 G}{\partial c^2}\right)_{T,p,N} \tag{A3.2.1}$$

$$S_{NN}(0) = (N/\Omega)k_B T K_T + \delta^2 S_{cc}(0) \tag{A3.2.2}$$

where $K_T$ is the mixture compressibility while $\delta$ is the size difference factor:

$$S_{Nc}(0) = -\delta S_{cc}(0) \tag{A3.2.3}$$

In addition we have the "sum" rules

$$\frac{1}{(2\pi)^3} \int [S_{NN}(q) - 1]d\mathbf{q} = -N/\Omega \tag{A3.2.4}$$

$$\int [S_{cc}(q) - c(1-c)]d\mathbf{q} = 0 \tag{A3.2.5}$$

$$\int S_{Nc}(q)d\mathbf{q} = 0 \tag{A3.2.6}$$

For the special case of a conformal solution

$$S_{cc}(0) = \frac{c(1-c)}{1 - (2w/RT)c(1-c)} \tag{A3.2.7}$$

where the interchange energy $w$ is a function of the thermodynamic state. This formula is useful when size differences are not too great, e.g., for Na–K but not for Na–Cs, where $v_2/v_1$ is approximately 3.

Structure factors at $k = 0$ are given in terms of

$$N\langle(\Delta c)^2\rangle \equiv S_{cc} = Nk_B T \Big/ \frac{\partial^2 G}{\partial c^2}\Big|_{T,p,N}$$

size and compressibility, through equations (A3.2.2) and (A3.2.3).

It is often useful to have explicit expressions for partial structure factors $a_{\alpha\beta}(k)$ at $k = 0$, defined by

$$a_{\alpha\beta}(k) = 1 + 4\pi\rho \int_0^\infty [g_{\alpha\beta}(r) - 1] \frac{\sin kr}{kr} r^2 dr \qquad \text{(A3.2.8)}$$

where $\rho$ is the total number of atoms $N$ per unit volume. Then, for arbitrary concentration

$$a_{11} = \theta + \left[ \frac{1}{(1-c)^2} - \frac{2\delta}{1-c} + \delta^2 \right] S_{cc} - \frac{c}{1-c} \qquad \text{(A3.2.9)}$$

$$a_{22} = \theta + \left[ \frac{1}{c^2} + \frac{2\delta}{c} + \delta^2 \right] S_{cc} - \frac{(1-c)}{c} \qquad \text{(A3.2.10)}$$

$$a_{12} = \theta + \left[ \delta^2 - \frac{(2c-1)\delta}{c(1-c)} - \frac{1}{c(1-c)} \right] S_{cc} + 1 \qquad \text{(A3.2.11)}$$

where $\theta = \rho k_B T K_T$ and

$$\delta = -\frac{1}{\Omega} \left( \frac{\partial\Omega}{\partial c} \right)_{T,p,N}$$

These equations were used by Bhatia *et al.*[23] in conjunction with conformal-solution theory, to calculate the theory curves in Figure 3.3.

# References

1. See, for example, N. H. March and M. P. Tosi, *Atomic Dynamics in Liquids*, Macmillan, London (1976).
2. J. G. Kirkwood and E. Monroe, *J. Chem. Phys.* **9**, 514 (1941).
3. N. Kumar, N. H. March, and A. Wasserman, *Phys. Chem. Liq.* **11**, 271 (1982).
4. J. L. Yarnell, M. J. Katz, R. G. Wenzel, and S. H. Koenig, *Phys. Rev. A* **7**, 2130 (1973).
5. G. S. Rushbrooke, *Physica* **26**, 259 (1960).
6. T. Gaskell, *Proc. Phys. Soc.* **86**, 693 (1965).
7. M. D. Johnson and N. H. March, *Phys. Lett.* **3**, 313 (1963); M. D. Johnson, P. Hutchinson, and N. H. March, *Proc. R. Soc. London, Ser. A* **282**, 283 (1964).
8. I. Ebbsjö, G. G. Robinson, and N. H. March, *Phys. Chem. Liq.* **13**, 65 (1983).
9. P. Schofield, *Proc. Phys. Soc.* **88**, 149 (1966).
10. P. Bratby, T. Gaskell, and N. H. March, *Phys. Chem. Liq.* **2**, 53 (1970).
11. P. A. Egelstaff, D. I. Page, and C. R. T. Heard, *J. Phys. C* **4**, 1453 (1971).
12. U. de Angelis and N. H. March, *Phys. Chem. Liq.* **6**, 225 (1977).
13. D. Pines, *Elementary Excitations in Solids*, Benjamin Inc., New York (1963).
14. J. R. D. Copley and J. M. Rowe, *Phys. Rev. A* **9**, 1656 (1974).
15. C. C. Matthai and N. H. March, *Phys. Chem. Liq.* **11** 207 (1982).
16. J. E. Enderby, T. Gaskell, and N. H. March, *Proc. Phys. Soc.* **85**, 217 (1965).
17. R. Eisenshitz and M. J. Wilford, *Proc. Phys. Soc.* **80**, 1078 (1962).

18. O. J. Kleppa, *J. Chem. Phys.* **56**, 2034 (1972).
19. J. S. Rowlinson, *Liquids and Liquid Mixtures*, Cambridge University Press (1969).
20. A. B. Bhatia and D. E. Thornton, *Phys. Rev. B* **2**, 3004 (1970).
21. R. C. Tolman, *The Principles of Statistical Mechanics*, Oxford University Press (1938).
22. T. L. Hill, *Statistical Mechanics*, McGraw-Hill Book Co., New York (1956).
23. A. B. Bhatia, W. H. Hargrove and N. H. March, *J. Phys. C* **6**, 621 (1973).
24. H. C. Longuet-Higgins, *Proc. R. Soc. London, Ser. A* **205**, 247 (1951).
25. S. P. McAlister and R. Turner, *J. Phys. F* **2**, L51 (1972).
26. A. B. Bhatia, Proc. Liquid Metals Conf., Adam Hilger: IOP (1976).
27. A. B. Bhatia and N. H. March, *Phys. Lett. A* **41**, 397 (1972).
28. P. J. Flory, *J. Chem. Phys.* **10**, 51 (1942).
29. A. B. Bhatia and N. H. March, *J. Phys. F* **5**, 1100 (1975).
30. K. Ichikawa and J. C. Thompson, *J. Phys. F* **4**, L 9 (1974).
31. A. R. Miedema, P. F. de Chatel, and F. R. de Boer, *Physica B* **100**, 1 (1980).
32. J. A. Alonso and N. H. March, *Phys. Chem. Liq.* **11**, 135 (1981).
33. D. I. Page, U. de Angelis, and N. H. March, *Phys. Chem. Liq.* **12**, 53 (1982).
34. M. Parrinello, M. P. Tosi, and N. H. March, *Proc. R. Soc. London, Ser. A* **341**, 91 (1974).
35. M. W. Johnson, N. H. March, D. I. Page, M. Parrinello, and M. P. Tosi, *J. Phys. C* **8**, 751 (1975).
36. M. Parrinello, M. P. Tosi, and N. H. March, *J. Phys. C* **7**, 2577 (1974).
37. I. Ebbsjö, P. Schofield, K. Sköld, and I. Waller, *J. Phys. C* **7**, 3891 (1974).
38. T. V. Ramakrishnan and M. Yussouff, *Solid State Commun.* **21**, 389 (1977); *Phys. Rev. B* **19**, 2775 (1979).
39. A. D. J. Haymet and D. W. Oxtoby, *J. Chem. Phys.* **74**, 2559 (1981).
40. N. H. March and M. P. Tosi, *Phys. Chem. Liq.* **11**, 129 (1981).
41. See, for example, R. Lovett, C. Y. Mou, and F. P. Buff, *J. Chem. Phys.* **65**, 570 (1976).
42. A. B. Bhatia, N. H. March, and M. P. Tosi, *Phys. Chem. Liq.* **9**, 229 (1980).
43. N. H. March and M. P. Tosi, *Phys. Chem. Liq.* **11**, 79 (1981).
44. D. G. Triezenberg and R. Zwanzig, *Phys. Rev. Lett.* **28**, 1183 (1972).
45. N. H. March and M. P. Tosi, *Phys. Chem. Liq.* **11**, 89 (1981).
46. M. Rovere, M. P. Tosi, and N. H. March, *Phys. Chem. Liq.* **12**, 177 (1982).
47. B. D'Aguanno, M. Rovere, M. P. Tosi, and N. H. March, *Phys. Chem. Liq.* **13**, 113 (1983).
48. S. A. Rice et al., *J. Chem. Phys.* **81**, 1406 (1984) and other references given there.

# Pair Potentials in Metals and Alloys: Order, Stability, and Dynamics

## J. Hafner

## 4.1. Introduction

A full understanding of the physics of metallic bonding, including the energy, structure, and dynamics of perfect crystals, of defects such as substitutional impurities, vacancies, etc., of liquids and liquid mixtures, and finally of metastable phases such as glasses, requires the solution of the Schrödinger equation for $10^{23}$ electrons, all interacting with one another and with the positive ions. For a perfect crystal, Bloch's theorem allows us to solve the equation only in one unit cell, and the electron–electron interaction problem may be greatly simplified by introducing the local-density approximation assuming that the exchange correlation potential is determined by the local electron density.

Self-consistent calculations on this basis reproduce the observed cohesive energies, lattice constants, and bulk moduli of the pure crystalline metals rather well, and first steps have been made toward calculating the heat of formation of alloys. However, the treatment of alloys with more complex crystal structures is still a severe test of the capability of even

**J. Hafner** · Institut für Theoretische Physik, Technische Universität Wien, Karlsplatz 13, A 1040 Wien, Austria; and Laboratoire de Thermodynamique et Physico-Chimie Métallurgiques, Domaine Universitaire, B. P. 75, F 38402 Saint Martin d'Hères, France.

present-day computers. More important, the band-structure scheme must be dismissed when translational symmetry breaks down and Bloch's theorem no longer applies.

This chapter presents an alternative approach, which starts from a perturbation calculation of the response of the electronic system to the external perturbation introduced by the ions, the justification for using a perturbative approach being derived from pseudopotential theory. At the level of a second-order approximation, this allows the cohesive energy to be divided into a large volume energy and a smaller contribution that can be represented as a sum of effective pairwise interactions between the ions; as such this second term is structure-sensitive. We wish to show that this approach enjoys considerable success in treatments of crystalline, liquid, and amorphous simple metals and their alloys.

## 4.2. Interatomic Potentials and Response Functions

In principle, any metal is a two-component system and its Hamiltonian $H$ consists of an ionic, an electronic, and an electron–ion interaction contribution:

$$H = H_{i-i} + H_{e-e} + H_{i-e} \tag{4.2.1}$$

A first simplication takes place if we realize that two widely different time scales are involved in the problem. As a result the electronic degrees of freedom can be removed by assuming that at any moment the electrons can be considered to be in their ground state corresponding to the instantaneous arrangement of the ions. This is the adiabatic approximation of Born and Oppenheimer[1] that reformulates the problem in terms of an effective Hamiltonian for the ions,

$$H = H_{i-i} + E_e\left(\{\mathbf{R}_l\}\right) \tag{4.2.2}$$

$E_e(\{\mathbf{R}_l\})$ being the exact ground-state energy of the electrons with the ions fixed in their positions $\mathbf{R}_l$. The price to pay for introducing the adiabatic approximation is that the effective ion–ion interaction consists now of a direct (essentially Coulombic) part and an indirect part mediated by the electrons.

Any information on the indirect ion–electron–ion interaction must be deduced from a calculation of the electronic ground-state energy in the field of fixed ions, the electronic Hamiltonian being given by

$$H_e = H_{e-e} + H_{e-i}\left(\{\mathbf{R}_l\}\right) \tag{4.2.3}$$

   We shall seek a transformation of this many-body Hamiltonian into an effective one-particle form, and also a perturbation expansion of the ground-state energy in terms of the electron–ion interaction. This perturbation expansion will allow us to expand the indirect ion–electron–ion interaction expressed by $E_e(\{\mathbf{R}_l\})$ in terms of a volume force and two-, three-, and many-ion potentials.

   The transformation to a one-particle Hamiltonian can be achieved by adopting the "local-density" approximation (LDA) for the exchange and correlation potential of the electrons,[2] the justification for using a perturbative approach deriving from pseudopotential theory.[3,4] The general form of the perturbation expansion for the ground-state energy in powers of the electron–ion interaction is given by[5]

$$E_e = E_0 + E_1 + E_2 + E_3 + \cdots \tag{4.2.4}$$

where

$$E_n = \Omega \sum_{\mathbf{q}_1,\ldots,\mathbf{q}_n} \Gamma^{(n)}(\mathbf{q}_1,\ldots,\mathbf{q}_n) W(\mathbf{q}_1) \cdots W(\mathbf{q}_n)\delta(\mathbf{q}_1 + \cdots + \mathbf{q}_n) \tag{4.2.5}$$

The $\delta$-function expresses the conservation of momentum and $W(\mathbf{q})$ is the matrix element of the pseudopotential (for simplicity, we consider it to be local). The response functions or multipole functions $\Gamma^{(n)}$ are universal characteristics; they depend only on the electron–electron interaction and are independent of the positions and properties of the ions.

   The pseudopotential matrix element is conveniently factorized into a form factor $w(\mathbf{q})$ and the static structure factor $S(\mathbf{q})$:

$$W(\mathbf{q}) = w(\mathbf{q})N^{-1} \sum_l e^{i\mathbf{q}\mathbf{R}_l} = w(\mathbf{q})S(\mathbf{q}) \tag{4.2.6}$$

   The condition of electrical neutrality must be taken into account in an explicit way. In the Fourier representation this leads to a vanishing of the $q = 0$ components of all three terms of the complete Hamiltonian. In spite of this mutual cancellation of the Coulombic interactions at $q = 0$, the pseudopotential matrix element retains a nonzero $q = 0$ component representing the volume-averaged non-Coulombic interactions:

$$\lim_{q \to 0} w(q) = b/\Omega_a \tag{4.2.7}$$

   The multipole functions may be constructed by diagrammatic techniques. The diagrams for the first two terms are given in Figure 4.1. $\Gamma^{(1)}$ is given by a single diagram, where the full line represents the electron Green's

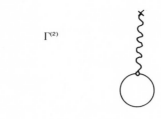

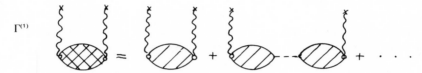

Figure 4.1. Lowest-order diagrams defining the mutipole functions in the series expansion given by equations (4.2.4) and (4.2.5); see text.

function and the wavy line the electron–ion interaction. In the local approximation we immediately have the result[5,6]

$$\Gamma^{(1)}(0) = n_0 \tag{4.2.8}$$

where $n_0$ is the electron density. In the graphical expansion for $\Gamma^{(2)}$, the dashed lines represent the electron–electron interaction and $\chi$ is the irreducible block with respect to these lines, i.e., the polarization operator. The summation may be carried out directly and yields

$$\Gamma^{(2)}(q, -q) = \frac{1}{2} \frac{\chi(q)}{\varepsilon(q)} \tag{4.2.9}$$

where

$$\varepsilon(q) = 1 - \frac{4\pi e^2}{q^2} \chi(q) \tag{4.2.10}$$

is the static dielectric function of the homogeneous electron gas with polarizability $\chi(q)$. Equations (4.2.9) and (4.2.10) are calculated in the random-phase approximation (RPA), which means that the irreducible block is just an empty loop of electron propagators. In this case $\chi(q)$ is just the Lindhard function.[7] Including exchange and correlation among the electrons leads to an expression for the polarizability of the form[4]

$$\bar{\chi}(q) = \frac{\chi(q)}{1 + (4\pi e^2/q^2)G(q)\chi(q)} \tag{4.2.11}$$

where $G(q)$ is the local-field factor determining an effective electron–electron interaction (including exchange and correlation) of the form

$$v_{\text{eff}}(q) = [1 - G(q)] \frac{4\pi e^2}{q^2} \qquad (4.2.12)$$

It is easy to see that this is equivalent to maintaining the form of $\Gamma^{(2)}$ given by equation (4.2.9), but replacing the RPA dielectric function by

$$\tilde{\varepsilon}(q) = 1 - \frac{4\pi e^2}{q^2} [1 - G(q)] \chi(q) \qquad (4.2.13)$$

A great number of different theories have been proposed for computing the local-field correction $G(q)$. The many alternatives in the literature will be discussed here only from a pragmatic point of view. Any dielectric function must satisfy a number of conditions: (1) it must yield an accurate correlation energy; (2) it must satisfy the compressibility sum rule, which requires the equality of the static and dynamic compressibilities; and (3) it must produce a realistic electron–electron pair correlation function even for short electron–electron distances. The most important condition is the compressibility sum rule, which probes the low-$q$ limit of $G(q)$:

$$G(q) \to \gamma^2 q^2 / K_F^2 \qquad (4.2.14)$$

where $\gamma$ is a coefficient related to the second volume derivative of the correlation energy.[8] Condition (3) relates the large-$q$ behavior of $G(q)$ to the electron pair correlation function $g(r)$,[9]

$$\lim_{q \to \infty} G(q) = 1 - g(0) \qquad (4.2.15)$$

and condition (1) tests an integral over the dielectric function. If we eliminate all approximations which violate one of these conditions, only a small number of realistic $G(q)$ survive. Among the most widely used are those proposed by Geldart and Taylor[10] and Vashishta and Singwi.[11] A very effective function has been proposed by Ichimaru and Utsumi.[12]

In the same approximation, the higher-order multiple functions have the general structure[5,6]

$$\Gamma^{(n)}(\mathbf{q}_1, \ldots, \mathbf{q}_n) = \Lambda^{(n)}(\mathbf{q}_1, \ldots, \mathbf{q}_n) / \tilde{\varepsilon}(\mathbf{q}_1) \cdots \tilde{\varepsilon}(\mathbf{q}_n) \qquad (4.2.16)$$

On this basis it is easy to obtain explicit expressions for the various terms of the expansion (4.2.4), energies being given per ion whose valence is $Z$:

$$E_0 = ZE_{eg}$$

$$E_1 = bZ/\Omega_a$$

$$E_2 = \tfrac{1}{2}\Omega_a \sum_{q \neq 0} \frac{\chi(q)}{\tilde{\varepsilon}(q)} |w(q)|^2 |S(\mathbf{q})|^2 \tag{4.2.17}$$

$$E_n = \Omega_a \sum_{\mathbf{q}_1 \cdots \mathbf{q}_n \neq 0} \frac{\Lambda^{(n)}(\mathbf{q}_1, \ldots, \mathbf{q}_n)}{\tilde{\varepsilon}(q_1) \cdots \tilde{\varepsilon}(q_n)} w(q_1) \cdots w(q_n) S(\mathbf{q}_1) \cdots S(\mathbf{q}_n) \delta(\mathbf{q}_1 + \cdots + \mathbf{q}_n)$$

The zeroth-order term is just the ground-state energy of the homogeneous electron gas. By employing the definition of the static structure factor (4.2.6), we find that equation (4.2.17) leads directly to the desired formulation of the effective Hamiltonian $\tilde{H}_i$:

$$\tilde{H}_i = T_i + E_\Omega + \frac{1}{2!\,N} \sum_{i \neq j} V^{(2)}(|\mathbf{R}_i - \mathbf{R}_j|, \Omega_a)$$

$$+ \frac{1}{3!\,N} \sum_{i \neq j \neq k} V^{(3)}(|\mathbf{R}_i - \mathbf{R}_j|, |\mathbf{R}_j - \mathbf{R}_k|, |\mathbf{R}_k - \mathbf{R}_i|, \Omega_a) + \cdots \tag{4.2.18}$$

where $T_i$ is the operator of the kinetic energy of the ions, $E_\Omega$ is a large volume energy, and the $V^{(n)}$ are the potentials describing an $n$-body interaction. For example, consider the second-order contribution in the expansion given by (4.2.17). If we define

$$F(q) = \tfrac{1}{2}\Omega_a \frac{\chi(q)}{\tilde{\varepsilon}(q)} |w(q)|^2 \tag{4.2.19}$$

the energy wave-number characteristic, then

$$E_2 = \sum_{\mathbf{q} \neq 0} |S(\mathbf{q})|^2 F(q)$$

$$= \frac{1}{N} \sum_{i<j} \underbrace{\left( \frac{2!}{N} \sum_{\mathbf{q}} e^{i\mathbf{q}(\mathbf{R}_i - \mathbf{R}_j)} F(q) \right)}_{\equiv V_{ind}^{(2)}(|\mathbf{R}_i - \mathbf{R}_j|)} + \frac{1}{N} \sum_{\mathbf{q}} F(q) - F(0) \tag{4.2.20}$$

We see that $F(q)$ is precisely the Fourier transform of an indirect ion–electron–ion interaction, while the second and third terms in equation (4.2.20) contribute to the volume energy. The direct Coulomb interaction between the ions may be treated in a similar way and combined with the indirect interaction to form an effective ion–ion potential $V^{(2)}$, the $q = 0$ term contributing again to the volume energy. An approximation to second order in the pseudopotential yields

$$E = E_0 + E_1 + E_2 = E_\Omega + E_p \tag{4.2.21}$$

with a volume energy $E_\Omega$ given by

$$E_\Omega = ZE_{eg} + bZ/\Omega_a + \frac{1}{N} \sum_q F(q) + \lim_{q \to 0} \left[ \frac{4\pi Z^2 e^2}{\Omega_a q^2} + F(q) \right] \tag{4.2.22}$$

and a pair energy

$$E_p = \frac{1}{N} \sum_{i<j} V_2^{(2)}(|\mathbf{R}_i - \mathbf{R}_j|, \Omega_a) \tag{4.2.23}$$

defined by the central pair potential

$$V_2^{(2)}(R, \Omega) = \frac{Z^2 e^2}{R} + \frac{2!}{N} \sum_q e^{i\mathbf{q}\mathbf{R}} F(q) \tag{4.2.24}$$

If we use the fact that $N^{-1} \sum_q F(q) = \frac{1}{2} V_{ind}^{(2)}(0)$ and calculate the limit in equation (4.2.22), then for a local potential the volume energy simplifies to[7]

$$E_\Omega = Z(E_{eg} - \frac{1}{2}\Omega_a B_{eg}) + \frac{1}{2} V_{ind}^{(2)}(0) \tag{4.2.25}$$

where both terms have a clear physical meaning: the first is the electron gas term ($B_{eg}$ is the bulk modulus of the electron gas) and the second describes the intra-atomic interaction between the ion core and its own screening cloud.

Similarly, it can be shown that the third-order term contains a three-body interaction and contributions to the two-body potential and the volume energy. In general, we find that the volume energy contains contributions from all orders of the expansion (4.2.17), and the $n$-body potentials $V^{(n)}$ describing the simultaneous indirect interaction between $n$ different ions may be represented in the form of a series in powers $k \geq n$ in the electron–ion interaction

$$V^{(n)}(\{\mathbf{R}_i\}) = \sum_{k \geq n} V_k^{(n)}(\{\mathbf{R}_i\}) \tag{4.2.26}$$

but this is not an expansion in terms of a small parameter. For small momentum transfers the electron–ion pseudopotential is generally not weak, and in cases were small $q$-values are significant the convergence turns out to be slow. In a metal at equilibrium, however, cancellation of the electronic scattering occurs due to interference, for terms of fixed $n$. For a crystal, this interference is complete for wave vectors not equal to vectors of the reciprocal lattice. For a liquid or a glass there is no complete cancellation, but the small values of the static structure factor near $q = 0$ indicate that the importance of the low-$q$ matrix elements is still small. This shows that for terms of a given order in $w(q)/\bar{\varepsilon}(q)$, the separation of this term in pairwise and nonpairwise interactions is considerably less effective than the actual determination of this term. Hence it is not useful to attempt to improve the pair potentials by going to the higher-order terms in (4.2.26), without at the same time including the corresponding many-body interactions.

The particular problem of the low-$q$ contributions becomes very evident if we consider, for example, the relation between the dynamic compressibility derived from the velocity of sound and the static compressibility defined as the second volume derivative of the free energy with respect to the volume. In performing this differentiation we must remember that, in addition to its explicit volume dependence, the free energy depends on the volume through the interatomic distances (or momentum transfers) and through the dependence of the response functions on the electron density. Therefore even though the energy is described by second-order terms only, if the volume changes, not only the distance between the atoms changes but also their interaction potential, due to the dependence on the electronic density. This introduces higher-order interactions between the ions, although they arise from $E^{(2)}$ in the static approach. In the dynamic approach we consider the propagation of sound waves at constant volume, and consequently to second order only pair forces contribute to the dynamic compressibility. To obtain a result consistent with the static approach the third- and fourth-order contributions to $V^{(2)}$ must be included. The necessity for this step stems from a set of identities valid for the multipole functions of any system of fermions.[13] Put most simply, these Ward identities tell us[5,6] that whenever one of the arguments of $\Gamma^{(n)}(\mathbf{q}_{el}, \ldots, \mathbf{q}_n)$ goes to zero, the order of the corresponding contribution to the energy is reduced by one. For our purpose this means that certain contributions to the volume energy and to the pair potential that are formally of third or fourth order must be included in the calculation of the dynamic compressibility in order to make it consistent with the second-order result for the static compressibility. This is a particular aspect of interatomic potentials in metals that is important in the calculation of any property involving volume changes, such as internal pressure, and relaxation around vacancies or defects.[14]

The general form of the pair potential described by equation (4.2.24) is well known, and consists of a strongly repulsive core plus an oscillatory potential at intermediate and long distances. The repulsive form of the potential is well known from Thomas–Fermi theory, where it takes the form of a screened Coulomb potential. The asymptotic (large-$R$) behavior of $V_2^{(2)}(R)$ is given by the Friedel form[4]

$$V_2^{(2)}(R) \propto w(2k_F)^2 \frac{\cos(2k_F R)}{(2k_F R)^3} \tag{4.2.27}$$

which is a consequence of the analytic properties of $\chi(q)$. The oscillations appear not only at large distances, but extend under the repulsive core and may be exposed by a variation of its radius. However, around the nearest-neighbor distance, $V^{(2)}(R)$ does not follow equation (4.2.27): the pseudopotential not only sets the amplitude of the oscillations, but also determines a large-$R$-dependent phase shift.[15,16]

## 4.3. Interatomic Potentials and Pseudopotentials

Pseudopotentials have been introduced to simplify electronic-structure calculations by eliminating the need to include the tightly bound core states and the strong attractive potential necessary for binding them. The first step in this direction was made when Hellmann[3] noted that the additional kinetic energy conferred as a result of the Pauli principle on a conduction electron penetrating the region of the ion core may be treated as a repulsive pseudopotential cancelling the largest part of the strong attraction. Progress has since been achieved by following two characteristically different lines of development, which will be sketched briefly below.

### 4.3.1. The Operator Approach

The first approach starts from the requirement that the conduction orbitals are orthogonal to the core orbitals. This setup results in the familiar nodal behavior of the conduction orbitals and the need to use rather large basis sets to describe them. Hence one proceeds by replacing the valence orbitals $\psi$ by pseudo-orbitals $\varphi$ comprising linear combinations of $\psi$ with the set of core orbitals $\psi_c$:

$$|\varphi\rangle = |\psi\rangle + \sum_c a_c |\psi_c\rangle \tag{4.3.1}$$

The cofficients of this linear combination are chosen so that the pseudo-orbitals possess some desired features, e.g., they are nodeless and close to the exact orbital, or they have minimum kinetic energy. Then one replaces the true Hamiltonian by a pseudo-Hamiltonian, requiring that the pseudo-orbitals are eigenfunctions of this new Hamiltonian and that it reproduces the exact conduction electron spectrum.[4,17] In any case, the true orbital is just the projection of the pseudo-orbital onto the subspace orthogonal to the core states, namely

$$| \psi \rangle = \left( 1 - \sum_c | \psi_c \rangle \langle \psi_c | \right) | \varphi \rangle \equiv (1 - P) | \varphi \rangle \tag{4.3.2}$$

where $P$ is the projection operator onto the subspace spanned by the core states. Equation (4.3.2) clarifies that we can add any linear combination of $\psi_c$ to $\varphi$ and still obtain the same $\psi$, so we are completely free when choosing the coefficients $a_c$. Cohen and Heine[18] have shown that the criterion of the minimum kinetic energy, i.e.,

$$\langle \varphi | T | \varphi \rangle / \langle \varphi | \varphi \rangle \sim \min \tag{4.3.3}$$

corresponds at the same time to selecting the smoothest possible pseudo-orbital and to maximum compensation between the attractive electron–ion potential and the repulsive part of the pseudopotential. Hence we can expect it to yield an optimum convergence for a perturbation series of the type (4.2.4) and (4.2.17). In terms of the interatomic interactions, this means that by optimizing the pseudopotentials we ensure that the largest part of the pair interactions is already contained in a second-order calculation. Calculations using this type of pseudopotential indeed yield accurate results for the cohesive, structural, and dynamic properties of simple metals.[19]

### 4.3.2. The Scattering Approach

The second approach starts by constructing pseudopotentials and pseudowave functions of imposed smoothness in the form of simple analytic expressions. The open parameters of the pseudopotential or orbital are determined by the requirement of reproducing the valence-electron spectrum of a full all-electron calculation for a simple reference configuration (usually the free atom or ion). The prototype of these pseudopotentials is the one proposed by Heine and Abarenkov,[20]

$$w^{HA}(r) = \begin{cases} - A_l(E), & r < r_c \\ - Ze/r, & r \geq r_c \end{cases} \tag{4.3.4}$$

The approach is flexible enough to allow the pseudopotential to meet other requirements, which are thought to improve its effectiveness. Rasolt and co-workers[21] showed that an improved pseudopotential can be obtained by imposing the additional condition that, outside the core, the calculated pseudocharge density must be identical to the charge density induced by a single ion immersed in a homogeneous electron gas, as calculated by a nonlinear iterative technique. The idea (though less clearly developed) is again that, by fitting the nonlinear charge density, one effectively adds some of the higher-order contributions to the pair potential. That this is indeed successful is proved again by a large number of successful applications.

Very recently, this line of development has been further developed by Hamann and co-workers,[22] who constructed pseudopotentials with the following properties: (1) real and pseudovalence orbitals agree for a chosen atomic configuration, (2) real and pseudovalence orbitals agree beyond a core radius $r_c$, (3) the pseudocharge contained inside $r_c$ is identical to the true charge, and (4) the logarithmic derivatives of the real and pseudovalence orbital and their energy derivatives agree for $r = r_c$. Conditions (1) to (4) are imposed on the orbitals and the pseudopotential determined by inverting the wave equation. This new family of pseudopotentials possesses many attractive features and has greatly contributed to recent progress in pseudopotential bound-structure calculations, but it has not yet been applied in perturbation calculations.

### 4.3.3. Model Potentials

The pseudopotentials described above may legitimately be regarded as being derived from first principles, since no experimental data are used to adjust their parameters. If adjustable parameters are introduced, we approach the vast field of empirical pseudopotentials or model potentials of which a very large number have been proposed in the literature. The chief virtue of a model potential is its extreme simplicity, therefore only the very simplest are really useful. A prototype of such a potential is the "empty core" model potential of Ashcroft,[23]

$$w(r) = \begin{cases} 0, & r < r_c \\ -Ze/r, & r > r_c \end{cases} \tag{4.3.5}$$

which is very useful for analyzing physical trends in a simple manner.

### 4.3.4. Multicomponent Systems

If the above theory is to be extended to alloys, the many-body part of the theory presents no difficulty: instead of a single kind of ion, we now introduce a mixture of two different types of ion (represented by their respective pseudopotentials) into the electron gas and screen the ions with the screening functions evaluated at the average electron density. This means that in the perturbation expansion (4.2.5) the multipole functions $\Gamma^{(n)}$ remain the same but the expression for the matrix element of the pseudopotential (4.2.6) must be replaced by

$$W(q) = w_A(q)S_A(q) + w_B(q)S_B(q) \tag{4.3.6}$$

where $w_A$ and $w_B$ are the form factors of the individual pseudopotentials in the alloy and the partial structure factors, $S_A$ and $S_B$, describe the spatial arrangement of the ions A and B. This leads naturally to the introduction of a set of three partial-energy wave-number characteristics $F_{ij}$, $i, j = A, B$ and to the construction of the three related interatomic pair potentials $V_{ij}$,[24] where

$$F_{ij}(q) = \tfrac{1}{2}\Omega_a \frac{\chi(q)}{\bar\varepsilon(q)}\, w_i(q)w_j(q) \tag{4.3.7}$$

and

$$V_{ij}(R, \Omega_a) = \frac{Z_i Z_j e^2}{R} + \frac{2}{N} \sum_q e^{iq^R} F_{ij}(q) \tag{4.3.8}$$

The most important question now is that of the transferability of the pseudopotential: are we entitled to use the same pseudopotential in the pure metal A and also for describing the component A in an A–B alloy? Let us, for example, consider the construction of a pseudopotential in the operator approach.[24] The Pauli principle requires the valence orbital to be orthogonal to the core orbitals of type A and B, i.e., a general pseudo-orbital $\varphi$ will now be given by an expression of the type

$$|\varphi\rangle = |\psi\rangle + \sum_{c(A)} a_c^A |\psi_c^A\rangle + \sum_{c(B)} a_c^B |\psi_c^B\rangle \tag{4.3.9}$$

the coefficients $a_c^A$ and $a_c^B$ being determined by some appropriate criterion, such as that of the minimum kinetic energy of the pseudovalence orbital (4.3.2). Now it is evident that the degree of admixture of core orbitals $\psi_c^A$ will vary with the concentration and with the type of the second component, hence the pseudopotential $w_A$ of an A ion will be different in the pure metal and in an A–B alloy.[24] This reflects the fact that, in general, the pseudopotential is not a purely atomic but a collective property: it describes the scattering of electrons by an ion in a given surrounding medium. If this effective medium changes (e.g., by alloying) the pseudopotential changes too. Within the scattering approach, the chemical effects on the pseudopotential depend on the way the parameters are determined. If this is done by reference to a free atom or free-ion configuration, the chemical effects are reduced to the energy dependence of the pseudopotential. Hence condition (4) for the Hamann–Schluter pseudopotential, namely that the logarithmic derivatives of the exact and the pseudo-orbital match not only for a given energy but for a whole range of neighboring energies, is thought to improve the transferability. If a jellium reference is used, such as in the Rasolt–Taylor–Dagens (RTD) scheme,[21] the resulting pseudopotentials depend strongly on the mean electron density,[25] but not on other characteristics of the second component. For the pure metals one finds that the pair potentials derived from different types of pseudopotential are reasonably well consistent. This is demonstrated in Figure 4.2 for the case of Al. For the alloy case this might be different. Figure 4.3 compares two sets of pair potentials for equiatomic Li–Mg alloys, one derived on the basis of the RTD scheme,[26] the other calculated on the basis of equations (4.3.2) and (4.3.8).[24] They give a rather different picture of the chemistry of the alloy. The RTD potential is strongly nonadditive,

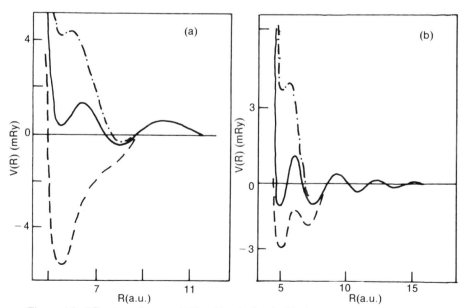

Figure 4.2. Effective pair potential for Al, calculated with the pseudopotentials of Hafner[19] (a) and Dagens et al.[21] (b) and different screening functions: —— Vashista–Singwi, ---- Hubbard–Sham, –·–·–·– Toigo–Woodruff. Note that only the Vashishta–Singwi function satisfies all the conditions for a good dielectric function.

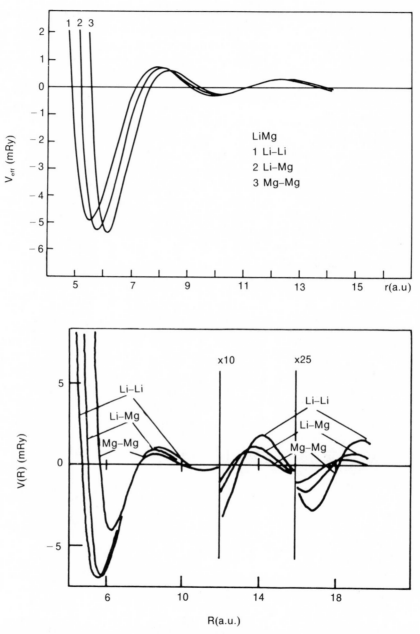

Figure 4.3. Interatomic pair potentials in equiatomic Li–Mg alloys, as calculated using the optimized pseudopotential of Hafner[24] and the DRT potential of Beauchamp et al.[26]; see text.

with strong attractive interactions between the unlike atoms. The ground state belonging to these potentials is a complete CsCl-type ordering of the bcc phase. The other type of potential is only weakly additive and yields a very reasonable ordering energy compatible with the observed short-range ordering.

## 4.4. Pair Potentials and Phase Stability in Crystalline Metals and Alloys

The main topic to be treated in this chapter is the application of pseudopotential-derived pair potentials to the description of the structure and properties of metals and alloys in the liquid and amorphous state. Nonetheless it is still very useful to review briefly the contribution of pair potentials to our understanding of the crystal structures.

### 4.4.1. Pure Metals

Within the framework of our second-order approximation, the ground-state energy is given by the sum of a volume energy $E_\Omega$ and a pair interatomic energy $E_p$. The equilibrium interatomic distance is determined by the zero-pressure condition $p = \partial E / \partial \Omega = 0$, which is satisfied due to the contributions of opposite sign arising from $E_\Omega$ and $E_p$. Which crystal structure will be preferred depends on the form of the pair potential and the nearest-neighbor distance $d_1$, determined by $E_\Omega$ and $E_p$. If $d_1$ for a close-peaked structure falls within the first attractive minimum, then clearly such a structure will be stable. The stacking variant to be preferred is determined by the interference of the next-nearest and more-distant neighbors with the long-range oscillations of the pair potential.

The more interesting case, however, is where the nearest-neighbor distance for close packing $d_{cp}$ falls near a peak in the pair potential. If this happens the crystal can lower its energy by distorting the close-packed structure (or by introducing an entirely new arrangement), which causes some of the atoms to move closer to each other and others to move further away (Figure 4.4).

Very recently, Hafner and Heine[15] have shown that the main trends in the crystal structures of the elements may be interpreted in terms of variations in the pair potential $V(r)$ with respect to electron density and pseudopotential. They found that the position of the main attractive minimum is determined by the core radius $r_c$ of the model potential, and the strength of the minimum is related to the expansion of the repulsive core (and hence to the screening length) and to the strength of the pseudopotential at $q = 2k_F$ determining the amplitude of the oscillations.

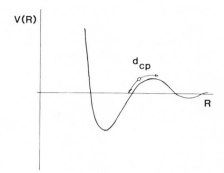

Figure 4.4. The pair potential $V(R)$ for rearrangement of the atoms at constant volume. If the nearest neighbors in a close-packed structure falls close to the peak, as shown, the crystal can lower its energy by a rearrangement, which brings some of them closer and others further away, as indicated by the arrows.

At low electron densities and large core radii, we find pair potentials with a minimum at the nearest-neighbor distance $d_{cp}$, resulting in stable close-packed structures for the A-group elements and Be, Mg, and Al. As the electron density increases and/or the core radius decreases, the minimum is shifted relative to $d_{cp}$ and flattened. This yields distorted structures for Zn, Cd, Ag, Ga, In, Si, Ge, Sn, and the group-V elements at first, but later a return to close-packed structures for Tl and Pb after the last trace of an oscillation of $V(r)$ at $d_{cp}$ has disappeared (Figure 4.5). One of the most important successes of pseudopotential theory is that first-principles calculations of the lattice dynamics of simple metals are possible. Calculated and measured frequencies usually agree within a few percent (Figure 4.6; see elswhere[19] for other simple metals) and this is certainly an important confirmation of the concept of low-order pseudopotential perturbation calculations (for a recent review see, for example, Grimvall[27]).

### 4.4.2. Binary Alloys

Trends in the alloying behavior of simple metals may be discussed very successfully in terms of pseudopotentials and pairwise interactions. Pettifor[28] has shown that the contribution of the two electron-gas terms in equation (4.2.25) to the heat of formation may be written as

$$\Delta H_{eg} = -\frac{5}{9}\bar{Z}(2.21 - 0.641 r_s - 0.023 r_s^2)\left[\Delta\left(\frac{1}{r_s}\right)\right]^2 \qquad (4.4.1)$$

where $r_s$ is the usual electron-density parameter and the bar characterizes concentration-average quantities. Note that $1/r_s$ is proportional to $n^{1/3}$, so that $\Delta H_{eg}$ is proportional to $\{\Delta n^{1/3}\}$ with a prefactor, which depends on the average electron density. The three terms in equation (4.4.1) are the kinetic, exchange, and correlation energy contributions. It follows from

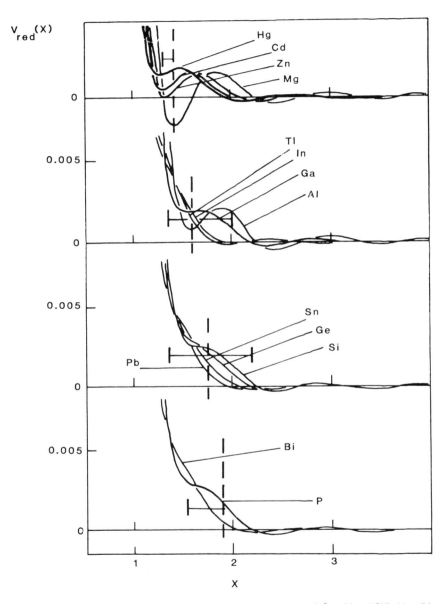

Figure 4.5. Trends in the reduced interatomic potential $V_{red}(X)$ [$V(R) = Z^2/R_a V_{red}(R)$, $X = 2kR/2\pi$] for the B-group elements. Dashed lines mark the nearest-neighbor distance for close packing $d_{cp}$ and arrows show the extent of the rearrangement of the nearest-neighbor shell for the structure with lowest energy in each (after Hafner and Heine[15]).

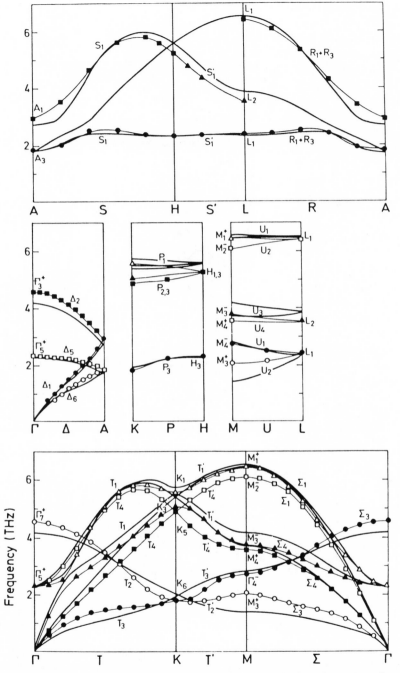

Figure 4.6. Phonon dispersion relations for Zn. Symbols and thin lines denote experimental results and thick lines denote pseudopotential calculation (after Hafner, unpublished).

this equation that the relaxation of the electron-gas kinetic energy provides an attractive contribution to the heat of formation, while the loss of exchange and correlation energy on alloying yields a repulsive contribution, which dominates at low electron densities. If the predictions of equation (4.4.1) are compared with experimental values of the heat of formation, we find that for solid solutions $\Delta H_{eg}$ is usually too low. This means that the pair interactions give a repulsive contribution to $\Delta H$ essentially due to the size misfit necessarily present in disordered solid alloys of ions of different size. If the mismatch exceeds a few percent difference in the atomic radii, the system will try to lower its internal energy by static lattice distortions, ordering, or the formation of an ordered intermetallic compound.

### 4.4.2.1. Static Lattice Distortions in Substitutional Alloys

As a very simple example let us consider the case of K–Rb alloys, which form a continuous series of bcc solid solutions. X-ray diffraction experiments show a pronounced line broadening at low temperatures, indicating the presence of static lattice distortions. The relaxation of the internal strains arising from the size mismatch can easily be calculated by a novel variant of the cluster relaxation technique, which takes into account the volume dependence of the pair potentials.[29] The results are shown in Figure 4.7. It can be seen that the structure is distorted in such a way that the mean interatomic distances are better adapted to the attractive pair potentials. It is also found that the lattice distortion makes a nonnegligible contribution to the heat of formation.

### 4.4.2.2. Ordering

The ordering energy $\Delta E_{ord}$ relative to a substitutional solid solution may be expressed in terms of the alloying potential $\Delta V(R)$ defined by

$$\Delta V(R) = V_{AA}(R) + V_{BB}(R) - 2V_{AB}(R) \qquad (4.4.2)$$

For short-range ordering, $\Delta E_{ord}$ is given by $\Delta V$ and the Warren–Cowley short-range order parameters $\alpha(i)$:

$$\Delta E_{ord} = \sum_i N_i \alpha(i) \Delta V(R_i) \qquad (4.4.3)$$

where summation is over the coordination shells with radius $R_i$ and coordination number $N_i$. For long-range ordering, a similar expression involving the Bethe–Williams long-range order parameter $\eta$ may be

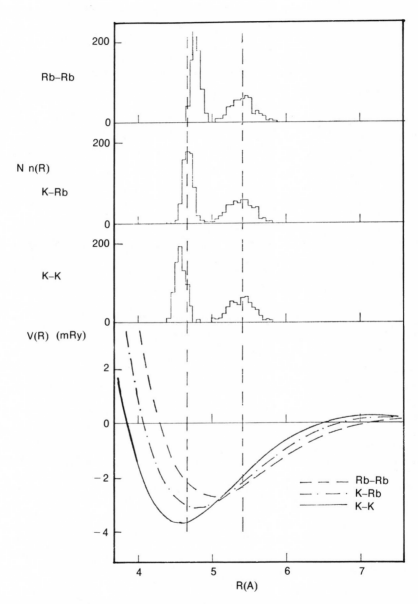

Figure 4.7. Partial radial distribution functions $n_{\alpha\beta}(R)$ and effective interatomic pair potentials in $K_{0.5}Rb_{0.5}$ alloy. Only the peaks representing the nearest- and next-nearest neighbor shells are shown. The dashed vertical lines indicate the interatomic distances in an undistorted bcc lattice.[29]

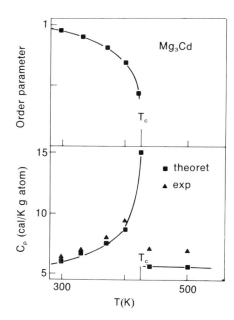

Figure 4.8. Order parameter and specific heat in a Cd₃Mg alloy; see text.[30]

derived.[30] In both cases the equilibrium value of the order parameter is determined by a variational condition,

$$\frac{\partial \Delta E_{\mathrm{ord}}}{\partial \eta} = 0 \quad \text{or} \quad \frac{\partial \Delta E_{\mathrm{ord}}}{\partial \alpha(i)} = 0, \qquad i = 1, 2, \ldots \qquad (4.4.4)$$

An example of such a calculation for $Mg_3Cd$ is shown in Figure 4.8. The order parameter and specific heat show the behavior expected for a second-order phase transition but, of course, the critical exponents are those of mean-field theory.

### 4.4.2.3. Intermetallic Compounds

The formation of disordered or partially disordered solid solutions is the exception rather than the rule. In most cases the formation of an intermetallic compound with an entirely new lattice type will be energetically much more favorable. There have so far been only a very few attempts to examine the structure and stability of intermetallic compounds in terms of pseudopotentials. Hafner[31,32] has shown that the Laves-type structures of $AB_2$ alloys, like $Na_2K$, $K_2Cs$, $CaLi_2$, $CaMg_2$, and $CaAl_2$, are stable because, as shown in Figure 4.9, the nearest neighbors fall right in

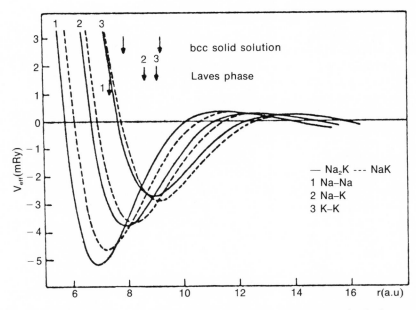

Figure 4.9. Pair potentials for the alloy $Na_2K$. The interatomic distances in the hexagonal (MgZn$_2$-type) Laves phase and in a bcc solid solution are indicated.[31]

the minima of the pair potentials. The Hume–Rothery rules relating the alloy crystal structure to the electron/atom ratio can be explained in terms of the oscillatory part of the pair interactions.[4,24,33]

## 4.5. Structure and Thermodynamics of Liquid Metals and Alloys

Recently, thermodynamic perturbational methods, such as the Gibbs–Bogoljubov variational technique,[34,35] the Weeks–Chandler–Andersen "blip function" expansion,[36] and the optimized cluster theory,[37] have greatly contributed to our understanding of the thermodynamic and structural properties of liquid metals and alloys. The physical picture emerging from these theories is that the liquid's volume is determined by the volume energy and the attractive part of the interatomic potential, but that once the volume has been determined the liquid may be considered as a hard-sphere fluid confined within that volume. Although this picture is generally correct for liquid metals near their melting point, there are situations where the attractive part of the interatomic potential becomes more important:

1. In expanded metals, long-wavelength density fluctuations are observed. These fluctuations cannot be explained in any model by considering only repulsive forces and must be attributed to the attractive interactions.
2. In cases where the mean interatomic distance in the liquid does not fall in a minimum of the pair potential but close to a local maximum, the static structure factor will differ from the hard-sphere type, the distortion being determined by the attractive (long-range) part of the potential.
3. In liquid mixtures nonadditive interactions cause chemical ordering. If the alloying potential $\Delta V(R)$ is strongly repulsive at the nearest-neighbor distance, heterocoordination will be preferred; for an attractive $\Delta V(R)$, the system will exhibit long-wavelength concentration fluctuations with a tendency toward phase separation.

### 4.5.1. Liquid Metals

The simplest form of thermodynamic perturbation theory is a variational method based on the Gibbs–Bogoljubov (GB) inequality

$$F \leq F_0 + \langle V \rangle_0 \tag{4.5.1}$$

which states that the free energy of any system will always be lower than that of a reference system plus the expectation value of the perturbation. If the right-hand side of equation (4.5.1) is minimized with respect to the parameters of a reference system, it constitutes an upper bound to the exact free energy. For liquid metals, the hard-sphere system is an excellent reference system. Only for systems with very soft repulsive interactions might a one-component plasma be more appropriate.[38] For a hard-sphere reference system, the optimal hard-sphere diameter satisfies very well the simple rule[35]

$$V(\sigma) - V_{\min} = \tfrac{3}{2} K_B T \tag{4.5.2}$$

which shows that the physical significance of $\sigma$ is that of a mean collision distance. The GB variational technique describes the structure and thermodynamics of liquid metals quite well.[34,35]

To progress beyond a hard-sphere description of the liquid structure, one must correct (1) for the softness of the repulsive part of the pair potential and (2) for the long-range attractive interactions. The corrections for the softness of the repulsive potential may be performed in the form of a "blip function" expansion, i.e., a Taylor expansion of the free energy in

terms of the difference of the Mayer functions $f(R) = \exp[-\beta V(R)] - 1$ of
the exact repulsive potential $V_0(R)$ and the hard-sphere reference potential $V_\sigma$,

$$\Delta f(R) = f_0(R) - f_\sigma(R) = e^{\beta V_0(R)} - e^{-\beta V_\sigma} \tag{4.5.3}$$

$\sigma$ being determined by the condition that the first-order term of the
functional Taylor series of the free energy in terms of $\Delta f(R)$ vanishes.[36]
The simplest way to account for the attractive interactions $V_1(R)$
$[V(R) = V_0(R) + V_1(R)]$ is via the random-phase approximation (RPA)
for the direct correlation function $C(R)$ given by

$$C(R) = C_0(R) - \beta V_1(R) \tag{4.5.4}$$

$(\beta = 1/K_B T)$, where $C_0(R)$ is the direct correlation function for the
repulsive interactions. The RPA yields nonphysical features for short-
range correlations that may be corrected by imposing the condition

$$g(R) = 0, \qquad R < \sigma \tag{4.5.5}$$

for the pair correlation function. The closure conditions (4.5.4) and (4.5.5)
together with the Ornstein–Zernike relation define the optimized random-
phase approximation (ORPA), which is the simplest form of an optimized
cluster theory.[37] A simple solution of equations (4.5.4) and (4.5.5) may be
found by a variational technique.[39]

Successive improvements of the static structure factor of liquid Rb
over the hard-sphere approximation achieved by the "blip function"
expansion, the RPA, and the ORPA are illustrated in Figure 4.10. The
ORPA can be used to describe very accurately the structure of liquid Rb
over a wide range of densities and temperatures up to the metal-insulator
transition (Figure 4.10[40]).

Regnault et al.[41] have shown that the ORPA can be employed to
calculate the structure of more complicated liquids, such as Ga. Hafner and
Kahl[42] have demonstrated that the arguments used to interpret the crystal
structures of the elements also explain the trends in the static structure
factors of the liquid metals in terms of variations in the interatomic
potentials.

### 4.5.2. Liquid Alloys

The theory for liquid alloys is still at the level of the hard-sphere
variational calculation. For simple liquid alloys such as Na–K or Ca–Mg,
very good agreement is obtained between the calculated static structure
factor[35] and both the results of X-ray and neutron-diffraction experiments

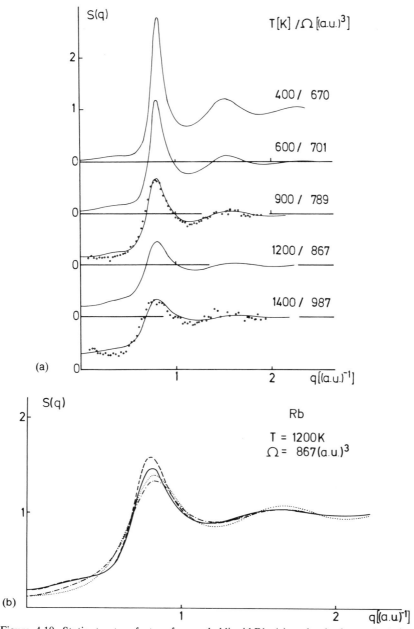

Figure 4.10. Static structure factor of expanded liquid Rb: (a) $\cdots$ hard sphere calculation, -·- including WCA correction for soft repulsive potential, --- random-phase approximation (RPA), —— optimized random phase approximation (ORPA). (b) Comparison of the ORPA (lines) with experiment (circles).[40]

**Table 4.1. Heats of Formation of Equiatomic Liquid Alloys
of Simple Metals near Their Melting Point (in kcal/g.atom)**

|       | $Z$ | $\Delta H_{eg}{}^{a}$ | $\Delta H^{b}$ | |
|-------|-----|-----------------------|----------------|----------------|
|       |     |                       | Calculated     | Experimental   |
| NaK   | 1   | 310                   | 380            | 175            |
| KRb   | 1   | 50                    | 160            | 30             |
| KCs   | 1   | 240                   | 200            | 28             |
| RbCs  | 1   | 75                    | − 39           | − 32           |
| MgCa  | 2   | − 200                 | − 1100         | − 1400         |
| CaSr  | 2   | 60                    |                | − 1800         |
| CaBa  | 2   | 120                   |                | − 2000         |
| AlGa  | 3   | − 270                 |                | 160            |
| AlIn  | 3   | − 1620                |                | 1150           |
| GaIn  | 3   | − 540                 |                | 265            |

<sup>a</sup> As computed using equation (4.4.1).
<sup>b</sup> As compiled by Hafner and Sommer.[44]

and the result of a computer simulation,[43] as well as accurate thermodynamic excess functions.[32,35]

Again, it is very instructive to consider the relative contributions of the electron-gas terms and the pair interaction terms to the heat of formation. Consider the homovalent alloys of the alkali, the alkaline earth, and the group-III metals as an example (Table 4.1). The quantity $\Delta H_{eg}$ is positive for the alkali alloys, very small for the alkaline earth alloys, and strongly attractive for the group-III alloys. The full (calculated or experimental) $\Delta H$ is negative for the alkaline earth alloys, positive for the group-III alloys, and generally weakly positive for the alkali alloys. Thus we find that the contribution of the pair interaction to the heat of formation is strongly attractive for $Z = 2$, strongly repulsive for $Z = 3$, and very weak for $Z = 1$. The explanation of this rather striking circumstance is to be found in the relation between the partial pair correlation functions $g_{ij}(R)$ and the pair potentials $V_{ij}(R)$, shown in Figure 4.11 for Ca–Mg alloys. In this case, we find that the maxima of the pair correlation functions fall into the minima of the pair potential, and evidently this will confer a very low configurational energy on this system. We know that the long-range oscillations of these functions are described by $V_{ij}(R) \rightarrow A\cos(2k_F R)/R^3$ and $g_{ij}(R) \rightarrow 1 + B\sin(Q_{p,ij}R)/R$, where $Q_{p,ij}$ is the wave vector for which the partial structure factor $S_{ij}(q)$ has its first maximum. Hence a necessary condition for the constructive interference shown in Figure 4.10 is

$$Q_{p,ij} \simeq 2k_F \tag{4.5.6}$$

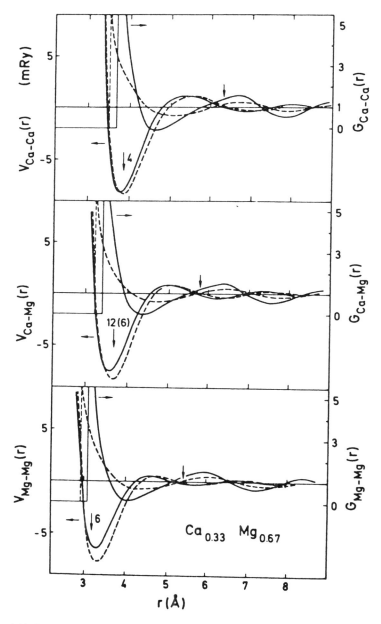

Figure 4.11. Interatomic pair potentials $V_{ij}(R)$ and partial pair correlation functions $g_{ij}(r)$ for a Ca–Mg alloy. Full lines refer to supercooled liquid at $T = 25$ °C and dashed lines to liquid at $T = 900$ °C.[23]

and it is well known that, for not too large size ratios, this condition is best obeyed for an electron/atom ratio close to $Z = 2$. Evidently, equation (4.5.6) is a Hume–Rothery-type criterion for the stability of a liquid alloy. In the next section similar criteria will be shown to determine the stability of amorphous simple-metal alloys.

One of the most interesting topics of current research on liquid alloys is chemical ordering in alloys with at least partially ionic character.[45]

According to pseudopotential calculations, the strength of the interaction between pairs of unlike atoms increases drastically with increasing difference in the electron density of the components,[46] e.g., in the series Li–Na, Li–Mg, Li–Al, and Li–Pb. Although this might be the clue to the problem, there are still same drawbacks: (1) the justification for using perturbation theory for alloys with large differences in valence is questionable, and (2) the calculation of the function $g_{ij}(R)$ from $V_{ij}(R)$ is extremely difficult for realistic potentials. Encouraging results have been obtained using simple hard-sphere models with Yukawa potentials.[47]

## 4.6. Structure Stability and Dynamics of Amorphous Alloys

Amorphous alloys produced by rapid quenching from the vapor or melt exhibit properties very similar to those of the corresponding liquid alloys. Quite generally, the formation of such a highly disordered solid state just before crystallization is expected from Ostwald's step rule.[48] However, not all alloys may be forced into an amorphous state even at the highest quenching rates. Whether that is possible will depend on kinetic factors and on thermodynamic factors,[49] the most important thermodynamic factor being a low value of the driving thermodynamic potential for crystallization, i.e., a small difference in the free enthalpies of the supercooled liquid alloy and the possible crystalline alloys.

### 4.6.1. Structure and Stability

Starting from the interatomic pair potentials, a structural model for an amorphous alloy such as $Mg_{70}Zn_{30}$[32,50,51] or $Ca_{70}Mg_{20}$[52] may be developed using a cluster relaxation technique, starting from a dense random packing of a mixture of hard spheres. The result of such a calculation for $Mg_{70}Zn_{30}$ is shown in Figure 4.12, where the partial pair correlation function $g_{ij}(R)$ is compared with the pair potential $V_{ij}(R)$. This result is remarkable in two respects: (1) the structure of the amorphous solid is found to be identical to that of the supercooled liquid, and (2) again the interference of $g_{ij}(R)$ and $V_{ij}(R)$ contributes to lowering the configurational energy. These findings are confirmed for $Ca_{70}\,Mg_{30}$ alloys.[52] For this system Hafner was able to

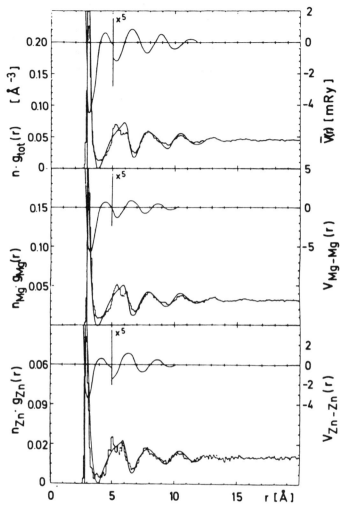

Figure 4.12. Total and partial distribution functions $g_{ij}(R)$ for the amorphous alloy $Mg_{0.7}Zn_{0.3}$. The histogram refers to the cluster relaxation calculation and the full line to the supercooled liquid. The average and partial pair potentials are shown in the inset (cf. text).[32,51]

reconstruct the equilibrium phase and to calculate certain features of the metastable phase diagram, which allows for a complete assessment of the glass-forming ability.[53] Results on amorphous alloys of noble metals and polyvalent simple metals suggest that for these systems the Hume–Rothery-type relation (4.5.6) may again be one of the important factors determining the amorphizability.[54,55] In some sense the amorphous simple metal alloys may thus be regarded as disordered Hume–Rothery phases.

The cluster relaxation model reproduces the neutron-weighted static structure factor[50,56] but fails to produce the small prepeak recently detected in an X-ray diffraction experiment.[57] For a proper description of the weak chemical ordering expressed by this prepeak, it will be necessary to improve both the construction of the starting structure and the relaxation algorithm.

### 4.6.2. Dynamic Properties of Amorphous Alloys

Intensive investigations have very recently been devoted to the problem of atomic vibrations in amorphous systems. Given the pair interaction and a structural model, the vibrational density of states and the dynamic structure factor may be calculated using the recursion method or the equation-of-motion method.[58] The total and partial dynamic structure factors calculated for the metallic glass $Mg_{70}Zn_{30}$[50] are shown in Figure 4.13. Collective excitations are well defined for long wavelengths, but

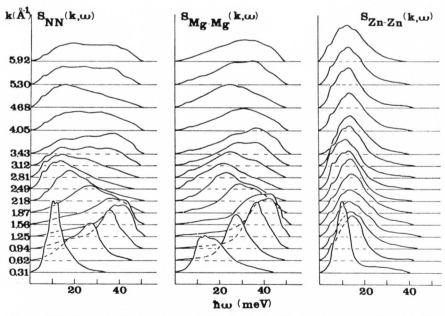

Figure 4.13. Dynamic structure factors for longitudinal excitations in an amorphous $Mg_7Zn_3$ alloy calculated using effective pair potentials and the recursion method.[50]

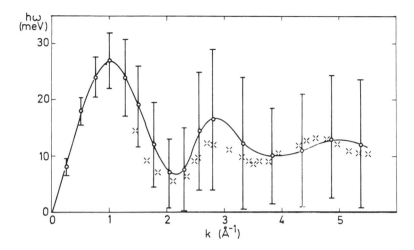

Figure 4.14. Dispersion law for propagating collective excitations in an amorphous $Ca_7Mg_3$ alloy. Circles denote calculation, bars indicate the width of $S(q, \omega)$ at half maximum, and crosses denote experiment.[52]

progressively broadened as the momentum transfer increases. The total $S(q, \omega)$ and that of the majority component show a characteristic dispersion with a minimum near $q = Q_p$. This dispersion minimum has been shown to arise from a process best characterized as "diffuse Umklapp scattering": the sharp maximum of the static structure factor acts like a smeared-out reciprocal lattice vector.[59] The vibrations of the minority atoms, on the other hand, are strongly reminiscent of the incoherent vibrations of substitutional impurities in a crystal. The theoretical predictions are well confirmed by the results of neutron inelastic scattering experiments,[60,61] which show a well-defined Umklapp minimum in the dispersion relation (Figure 4.14).

## 4.7. Discussion and Summary

The approach to metallic cohesion presented in this chapter is based on pseudopotential theory, which gives rise in a second-order perturbation treatment to pair and volume forces. However, one should emphasize that the pseudopotential and hence the pair potential, are not unique, as there are many different pseudopotentials that yield the same valence electron spectrum. All pseudopotentials that scatter the electrons correctly must yield the same answer if the calculation is carried through all orders of perturbation theory, but if we truncate the series at second order, as in traditional pair-potential calculations, one pseudopotential may well be the best. Most of the applications presented here are based on pseudopotentials designed to optimize the convergence of the perturbation series, and we think that the degree of success achieved in these calculations clearly demonstrates that this is a indeed a very fruitful scheme. However, considerable computational complexity is the price to pay for getting an optimized pair potential. For analyzing trends, model potentials whose chief virtue is their simplicity may still be very useful.

ACKNOWLEDGMENTS. The ideas presented in this chapter have taken shape during many discussions with friends and colleagues, but especially Prof. V. Heine. It is a pleasure to thank the Laboratoire de Thermodynamique et Physico-Chimie Métallurgiques at Grenoble, where the final draft of this chapter was written, for their kind hospitality.

# References

1. For a detailed discussion of the Born–Oppenheimer approximation see, e.g., G. V. Chester, *Adv. Phys.* **10**, 357 (1961).
2. W. Kohn and L. Sham, *Phys. Rev. A* **140**, 1133 (1965); for a discussion of the connection between the local density approximation and pseudopotential perturbation theory, see Hafner[19] and Hafner and Eschrig.[19]
3. The first to invent the pseudopotential technique appears to be H. Hellmann, *Acta Physico-Chimica URSS* **1**, 913 (1935); **4**, 225 (1936).
4. V. Heine, D. Weaire, and M. H. Cohen, in: *Solid State Physics*, Vol. 24, Academic Press, New York (1971).
5. For a general review see E. G. Brovman and Yu. M. Kagan, in: *Dynamical Properties of Solids* (A. A. Maradudin and G. K. Horton, eds.), Vol. 1, North-Holland, Amsterdam (1974).
6. The second-order expansion was first given by M. H. Cohen, *Phys. Rev.* **130**, 1301 (1963), an explicit form of the third-order expansion has been presented by P. Lloyd and C. A. Sholl, *J. Phys. C* **1**, 1620 (1968), and an approximate form of the fourth-order term has been given by J. Hammerberg and N. W. Ashcroft, *Phys. Rev. B* **9**, 409 (1974).
7. M. W. Finnis, *J. Phys. F* **4**, 1465 (1974).
8. D. Pines and Ph. Nozières, *Elementary Excitations in Solids*, Benjamin, New York (1962).
9. J. C. Kimball, *Phys. Rev. A* **7**, 1648 (1973).
10. D. J. W. Geldart and R. Taylor, *Can J. Phys.* **48**, 167 (1970).
11. P. Vashishta and K. S. Singwi, *Phys. Rev. B* **6**, 875 (1972).
12. S. Ichimaru and K. Utsumi, *Phys. Rev. B* **24**, 7385. S. Ichimaru, *Rev. Mod. Phys.* **54**, 1017 (1982).
13. J. M. Luttinger and J. C. Ward, *Phys. Rev.* **118**, 1417 (1960).
14. G. Solt, *Phys. Rev. B* **18**, 720 (1978).
15. J. Hafner and V. Heine, *J. Phys. F* **13**, 2479 (1983).
16. D. Pettifor, *Physica Scripta* **T1**, 26 (1982).
17. W. A. Harrison, *Pseudopotentials in the Theory of Metals*, Benjamin, New York (1966).

18. M. H. Cohen and V. Heine, *Phys. Rev.* **122**, 1821 (1961).
19. J. Hafner, *J. Phys. B* **22**, 351 (1975); **24**, 41 (1976), J. Hafner and H. Eschrig, *Phys. Status Solidi B* **72**, 179 (1976).
20. I. V. Abarenkov and V. Heine, *Phil. Mag.* **12**, 529 (1965).
21. L. Dagens, M. Rasolt, and R. Taylor, *Phys. Rev. B* **11**, 2726 (1975). M. Rasolt and R. Taylor, *Phys. Rev. B* **11**, 2717 (1975).
22. D. R. Hamann, M. Schlüter, and C. Chiang, *Phys. Rev. Lett.* **43**, 1494 (1979). G. B. Bachelet, D. R. Hamann, and M. Schlüter, *Phys. Rev. B* **26**, 4199 (1982).
23. N. W. Ashcroft, *Phys. Rev. Lett.* **23**, 48 (1966).
24. J. Hafner, *J. Phys. F* **6**, 1243 (1976).
25. N. Q. Lam, N. van Doan, L. Dagens, and Y. Adda, *J. Phys. F* **11**, 2231 (1981).
26. P. Beauchamp, R. Taylor, and V. Vitek, *J. Phys. F* **4**, 2017 (1975).
27. G. Grimvall, in: *Ab Initio Calculations of Phonon Dispersion Relations* (J. T. Devreese, ed.), Plenum Press, New York (1982).
28. D. Pettifor, in: *Atomistics of Fracture*, Proc. NATO–ASI, Corsica (1981).
29. J. Hafner and G. Punz, *J. Phys. F* **13**, 1393 (1983).
30. C. L. Leung, *J. Phys. F* **9**, 179 (1979).
31. J. Hafner, *Phys. Rev. B* **15**, 617 (1977); **19**, 5094 (1979).
32. J. Hafner, *Phys. Rev. B* **21**, 406 (1980).
33. A. Blandin, in: *Phase Stability in Metals and Alloys* (P. S. Rudman, J. Stringer, and R. I. Jaffee, eds.), McGraw-Hill Book Co., New York (1967).
34. W. H. Young, in: *Liquid Metals 76* (R. Evans and D. A. Greenwood, eds.), p. 1, The Institute of Physics, Bristol (1977).
35. J. Hafner, *Phys. Rev. A* **16**, 351 (1977).
36. H. C. Andersen, D. Chandler, and J. D. Weeks, *Adv. Chem. Phys.* **34**, 105 (1976).
37. H. C. Andersen and D. Chandler, *J. Chem. Phys.* **53**, 547 (1970); ibid. **57**, 1918 (1972).
38. M. Ross, *Phys. Rev. B* **21**, 3140 (1980).
39. G. Kahl and J. Hafner, *Phys. Chem. Liq.* **12**, 109 (1982).
40. G. Kahl and J. Hafner, *Phys. Rev. A* **29**, 3310 (1984).
41. C. Regnault, J. P. Badiali, and M. Dupont, *J. Phys. C* **8**, 603 (1980); *Phys. Lett. A* **74**, 245 (1979).
42. J. Hafner and G. Kahl, *J. Phys. F* **14**, 2259 (1984).
43. G. Jacucci and R. Taylor, *J. Phys. F* **11**, 787 (1981).
44. J. Hafner and F. Sommer, *CALPHAD* **1**, 351 (1977).
45. P. Chieux and H. Ruppersberg, *J. Physique* (Paris), **41**, C8-145 (1980).
46. J. Hafner, in: *Liquid Metals 76* (R. Evans and D. A. Greenwood, eds.), p. 107, The Institute of Physics, Bristol (1977).
47. R. Evans, A. Copestake, H. Ruppersberg, and W. Schirmacher, *J. Phys. F* **13**, 1993 (1983); J. Hafner, A. Pasturel, and P. Hicter, *J. Phys. F* **14**, 1137 (1984); ibid **14**, 2279 (1984).
48. W. Ostwald, *Z. Phys. Chem.* **22**, 289 (1897).
49. J. Hafner, in: *Nuclear Methods in the Investigation of Metallic Glasses* (U. Gonser, ed.), *At. Energy Rev.*, Suppl. 1, 27 (1981).
50. J. Hafner, *J. Phys. C* **16**, 5773 (1983).
51. J. Hafner and L. von Heimendahl, *Phys. Rev. Lett.* **42**, 386 (1979).
52. J. Hafner, *Phys. Rev. B* **26**, 678 (1983).
53. J. Hafner, *Phys. Rev. B* **28**, 1734 (1983).
54. P. Häussler, W. H. Müller, and F. Baumann, *Z. Phys. B* **35**, 67 (1979). P. Häussler and F. Baumann, *Z. Phys. B* **49**, 303 (1983).
55. U. Mizutani, in: *Proc. 4th Int. Conf. On Rapidly Quenched Metals* (T. Masumoto and K. Suzuki, eds.), p. 1279, The Japan Institute of Metals, Sendai (1982).

56. T. Mizoguchi, H. Narumi, N. Akutsu, N. Watanabe, N. Shiotani and M. Ito, *J. Non-Cryst. Solids* **61 + 62**, 285 (1984).
57. E. Nassif, P. Lamparter, W. Sperl, and S. Steeb, *Z. Naturforsch.*, **38a**, 142 (1983).
58. J. Hafner, in: *Ab Initio Calculations of Phonon Dispersion Relations* (J. T. Devreese, ed.), p. 151, Plenum Press, New York (1982).
59. J. Hafner, *J. Phys. C* **14**, L287 (1981).
60. J. B. Suck, H. Rudin, H. J. Güntherodt, and H. Beck, *J. Phys. C* **14**, 2305 (1981); *J. Phys. F* **11**, 1375 (1981); *Phys. Rev. Lett.* **50**, 49 (1983).
61. J. B. Suck and H. Rudin, in: *Glassy Metals II* (J. J. Güntherodt and H. Beck, eds.), Topics in Applied Physics, Vol. 53, Springer-Verlag, Berlin–Heidelberg (1983), p. 217.

## Notes Added in Proof

Since the completion of this manuscript, important progress has been achieved in several areas covered in this chapter. A brief list of pertinent references follows:

*Secs. 4.2 and 4.4.1:* Pettifor and Ward[62] and Hafner and Heine[63] have elaborated highly accurate analytic expressions for the interatomic pair potentials. Their analysis substantiates and extends the physical trends in the interatomic interactions derived from the earlier numerical studies.

*Sec. 4.4.2.1:* The influence of local fluctuations in the interatomic spacings on the phonon spectrum of disordered solid solutions has been studied by Hafner and Punz.[64,65]

*Sec. 4.4.2.3:* The charge density distribution in simple-metal Laves phases has been studied both experimentally and theoretically,[66,67] confirming the analysis presented here. The chemical bonding in Zintl phases has been studied with particular reference to covalent effects.[68,69]

*Sec. 4.5.2:* The problem of the microscopic origin of CSRO has been investigated in many papers: molecular dynamics calculations based on model pair potentials have been performed for $Li_4Pb$.[70] A thermodynamic variational method based on a hard-sphere-Yukawa reference has been developed for describing CSRO.[71] The WCA-blip-function expansion and the optimized random phase approximation have been generalized to binary alloys.[72]

*Sec. 4.6:* The origin of chemical and topological short range order in amorphous alloys and their interrelation to the precise form of the interatomic potentials has been discussed by Hafner.[73,74]

### References

62. D. G. Pettifor and M. A. Ward, *Sol. State Comm.* **49**, 291 (1984).
63. J. Hafner and V. Heine, *J. Phys. F*, in press.
64. J. Hafner and G. Punz, *Phys. Rev. B* **30**, 7336 (1984).
65. G. Punz and J. Hafner, *Z. Phys. B*, in press.
66. T. Ohba, Y. Kitano, and Y. Komura, *Acta Cryst. C* **40**, 1 (1984).
67. J. Hafner, *J. Phys. F*, in press.
68. N. E. Christensen, *Phys. Rev. B*, in press.
69. J. Hafner and W. Weber, *Sol. State Comm.*, in press.
70. G. Jacucci, M. Ronchetti, and W. Schirmacher, in *Neutron Physics Today and Tomorrow* (Oxford, 1984).
71. A. Pasturel and J. Hafner, *Phys. Rev. B*, in press.
72. G. Kahl and J. Hafner, *J. Phys. F*, in press.
73. J. Hafner, in *Amorphous Metals and Nonequilibrium Processing*, ed. by M. von Allmen (Les Editions de physique, Paris 1984), p. 219.
74. J. Hafner, *Proc. 5th Internat. Conference on Rapidly Quenched Metals*, Würzburg 1984, *J. Noncryst. Sol.* **69** (in press).

# 5

# Liquid Surfaces and Solid–Liquid Interfaces

## M. P. Tosi

## 5.1. Introduction

The study of capillarity in terms of interatomic forces originates from the work of Young, Laplace, and Gauss and was based essentially on mechanical considerations, while the use of thermodynamics combined with the notion of a continuous density profile at the interface between a liquid and its vapor is mainly due to van der Waals. The formal statistical mechanics of inhomogeneous fluids is now well developed and is finding a number of applications to liquid–vapor interfaces. These applications have been stimulated by the progress that has taken place over the last two decades in understanding the bulk thermodynamics and structure of liquids (atomic and molecular liquids, liquid metals and alloys, molten salts and other ionic liquids) but have been mostly limited so far to simple fluids. Much remains to be done for a microscopic understanding of important phenomena such as the segregation of the components in the surface region of a liquid alloy. Even less is known at the microscopic level about interfaces between liquids and solids, in spite of the technical relevance of such phenomena as are involved in wetting, nucleation, and electrodics. The aim of this chapter is to provide a short introduction to this vast and important field.

**M. P. Tosi** · Istituto di Fisica Teorica dell'Università and International Centre for Theoretical Physics, Trieste, Italy.

## 5.2. Thermodynamics and Phenomenology

### 5.2.1. Liquid–Vapor Interface

The case of a monatomic liquid in equilibrium with its vapor will serve to introduce some basic concepts. The system, composed of a macroscopic number $N$ of atoms in a volume $V$, is viewed as consisting of the two bulk phases that adjoin through an interface, assumed to be planar (perpendicular to the $z$-axis, say). Thermodynamic equilibrium requires that the temperature $T$ and chemical potential $\mu$ be constant, while mechanical equilibrium requires that the normal pressure $P$ be constant.

The atomic *density profile* $n(z)$ must evidently vary continuously across the interface from the value $n_l$ of the bulk liquid to the value $n_v$ of the bulk vapor, in some manner as sketched in Figure 5.1 (top); this variation may be expected to occur over a distance of a few interatomic spacings, at least if the system is far from its critical point. The anisotropy of the profile implies a net attraction to the liquid phase for an atom in the transition region (Figure 5.1, middle): one must do work to bring an atom from the bulk of the liquid to the surface, i.e., an excess of free energy is associated with the creation of the interface (*surface free energy*). It also implies that the tangential pressure, defined as the force per unit area transmitted perpendicularly across an area element in the $y$–$z$ or $x$–$z$ plane, is a function $P_t(z)$ of position in the transition region. The difference between the components $P_t(z)$ and $P_n = P$ of the stress tensor in the transition region (Figure 5.1, bottom) is negative, i.e., it has the character of a tension (*surface tension*).

The surface free energy is defined by comparing the free energy $F$ of the real system with the sum of the free energies of suitably chosen amounts of *homogeneous* liquid and vapor phases. If $f_l$ and $f_v$ are the free energies per unit volume of the homogeneous phases, we can write the surface free energy $\gamma$ per unit area (or, more precisely, the liquid–vapor interfacial free energy $\gamma_{lv}$) as

$$\gamma = \frac{1}{A} (F - f_l V_l - f_v V_v) \qquad (5.2.1)$$

where $A$ is the surface area, and $V_l$ and $V_v$ are the volumes of the homogeneous phases. These are fixed by

$$V_l + V_v = V, \qquad n_l V_l + n_v V_v = N \qquad (5.2.2)$$

(i.e., no excess matter is attributed to the interface). These conditions can be rewritten as

$$\int_{-\infty}^{z_G} dz\,[n_l - n(z)] + \int_{z_G}^{\infty} dz\,[n_v - n(z)] = 0 \qquad (5.2.3)$$

which fixes the location $z_G$ of the *Gibbs surface* dividing the two hypothetical homogeneous fluids (see Figure 5.1, where $z_G$ has been taken for convenience as the origin of the $z$-axis). For a discussion of multicomponent systems, see Cahn and Hilliard.[1]

The thermodynamic definition (5.2.1) has the following consequences: (1) $\gamma dA$ is the work needed to increase the surface area by the amount $dA$ in *any* isothermal reversible process, and (2) the excess surface entropy is given by $s = -d\gamma/dT$ and hence the excess surface energy is $u = \gamma - Td\gamma/dT$. With regard to (1), we stress in particular that the work done against the surface tension in expanding the surface area by stretching is equal to the surface free energy of the same area of new surface, i.e., the surface tension and the surface free energy (commonly expressed in dyn $cm^{-1}$ and in erg $cm^{-2}$, respectively) are numerically equal for an interface

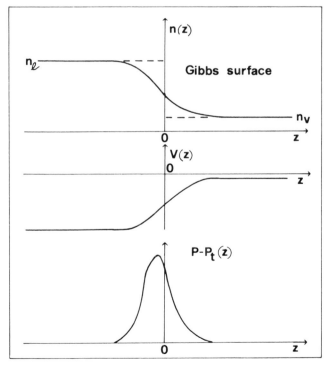

Figure 5.1. Sketches of the density profile $n(z)$, the average potential $V(z)$, and the deficit of tangential pressure $P - P_t(z)$ across a liquid–vapor interface.

between two fluids. This argument can be used to express $\gamma$ in terms of the integrated deficit of tangential pressure as

$$\gamma = \int_{-\infty}^{\infty} dz \, [P - P_t(z)] \qquad (5.2.4)$$

(see, e.g., Croxton[2] for a derivation of this result). The argument assumes that alternative processes lead to the same (equilibrium) surface structure, as will be the case when diffusion can occur.

On the phenomenological side, the values of $\gamma$ for liquid metals near freezing correlate well with the latent heat of vaporization (see, e.g., Faber[3]), in accord with the intuitive notion that the work done in bringing an atom from the bulk liquid up to the surface involves breaking a fraction of its interatomic bonds. The corresponding surface entropy, obtained as noted above from the measured temperature dependence of $\gamma$, is of order $k_B$ per surface atom; such an amount is estimated theoretically[3] by considering the effect of replacing an appropriate number of bulk sound waves by surface capillary waves. Some remarkable exceptions, such as Zn, where $d\gamma/dT$ is positive over a limited range of temperature above the triple point, should however be noted. One expects that $\gamma(T)$ should decrease with increasing temperature and vanish at the critical point, where the distinction between liquid and vapor is lost.

A "law of corresponding states" is known[4] for the surface tension of simple atomic and molecular liquids; the quantity $\gamma(T)v_c^{2/3}/T_c$, where $v_c$ and $T_c$ are the critical volume and temperature, is approximately a universal function of $T/T_c$. The data are described by the empirical relation $\gamma(T) \propto (1 - T/T_c)^{1.27\pm0.02}$.

Values of $\gamma(T)$ for molten salts can be found in the book by Janz.[5] These materials have a large cohesive energy relative to a free-ion state but comparatively modest values of $\gamma$ (see, e.g., Table 5.1), an indication that ionic bonding is largely preserved in the interface.

### 5.2.2. Density Profile and Surface Tension: Cahn–Hilliard Theory

An empirical relation between $\gamma$ and the isothermal compressibility $K$ is known to exist for a large variety of liquids, quite independently of their type of bonding; this is illustrated in Table 5.1, abridged from Egelstaff and Widom.[6] The relation can be written in the form

$$\gamma K = L \qquad (5.2.5)$$

where $L$ has the dimensions of a length and is roughly constant (variations are by a factor of 2–3 only, while $\gamma$ and $K$ vary individually by factors of order 100).

**Table 5.1. Relation between $\gamma$ and $K$ for Liquids near the Triple Point**

|      | $\gamma$ (dyn cm$^{-1}$) | $K$ ($10^{-12}$ cm$^2$ dyn$^{-1}$) | $\gamma K$ (Å) |
|------|------|------|------|
| Na   | 194  | 21   | 0.40 |
| Cs   | 71   | 67   | 0.47 |
| Cd   | 666  | 3.2  | 0.21 |
| Fe   | 1790 | 1.43 | 0.25 |
| NaCl | 116  | 29   | 0.34 |
| KI   | 78   | 50   | 0.39 |
| Ar   | 13.1 | 212  | 0.28 |
| N$_2$ | 11.8 | 211  | 0.25 |
| Br$_2$ | 41.5 | 63   | 0.26 |

The interpretation of the relation (5.2.5) given by Cahn and Hilliard[1] (see also Widom[7]) is important for an introduction to modern statistical–mechanical theories of the liquid surface to be discussed in Section 5.3. The main point is that one can still use a "bulk-like" theory in the presence of a surface provided one keeps account of the gradient of the density profile. Cahn and Hilliard consider two contributions to $\gamma$. For the first they adopt the expression for the free-energy change associated with a density fluctuation $\delta n$ in a volume $v$ of a bulk liquid,

$$F_1 = c_1 v \left(\frac{\delta n}{n_l}\right)^2 \Big/ K$$

taking $\delta n$ as the density change $n_l - n_v$ between liquid and vapor ($c_1$ is a numerical factor which equals $\frac{1}{2}$ in elasticity theory). They then add to $F_1$ a contribution $F_2$ specifically associated with the density gradient,

$$F_2 = c_2 v (\delta n / L)^2$$

where $L$ is a length measuring the gradient of $n(z)$ in the interface and $c_2$ is an unknown parameter. Taking $v = LA$ and minimizing $(F_1 + F_2)/A$ with respect to $L$ one obtains for the surface tension

$$\gamma = \frac{2 c_1 L}{K} \left(\frac{n_l - n_v}{n_l}\right)^2 \simeq \frac{L}{K} \tag{5.2.6}$$

near the triple point where $n_l \gg n_v$. Thus the product $\gamma K$ is a measure of the "thickness of the interface" $L$.

An extension of this line of reasoning to the dependence of $\gamma$ on concentration in liquid alloys has been given by Bhatia and March.[8] The sign of $d\gamma/dc$ is related to the phenomenon of surface segregation, since the solute tends to segregate in the surface region if $\gamma$ decreases with increasing concentration $c$. In the limit $c \to 0$ they find

$$\gamma \simeq \frac{L}{K}\left(1 + \frac{c\delta^2}{n_l k_B TK}\right)^{-1} \tag{5.2.7}$$

Here the quantity $\delta \equiv (\partial \ln V/\partial c)_{T,P,N}$ measures the difference in atomic sizes between solute and solvent, an effect which obviously favors surface segregation irrespective of the sign of $\delta$. From equation (5.2.7) one sees that this effect should be strengthened if $\partial \ln K/\partial c > 0$ (compressibility of the solute larger than that of the solvent). The original work should be consulted for a discussion of experimental data.

### 5.2.3. Interatomic Forces and Cleavage Work: Kohn–Yaniv Theory

While the Cahn–Hilliard argument stresses the importance of the gradient of the density profile across the interface, the role of interatomic forces is illustrated by an equally simple and apparently "universal" argument due to Kohn and Yaniv.[9] This was originally applied to the surface energy of crystalline metals but has subsequently been extended[10] to ionic crystals and to electron–hole liquid drops in semiconductors.

The surface free energy is viewed as the reversible work needed to separate two half-crystals to infinite distance from each other and expressed as

$$\gamma = \frac{1}{2}\int_0^\infty dx F(x) \tag{5.2.8}$$

where $F(x)$ is the force per unit area needed to displace by $x$ the two half-crystals. Using elasticity theory for small $x$ and the Lifshitz theory of van der Waals interactions at large $x$ one has

$$F(x) = \begin{cases} Ax & \text{for } x \ll a \\ \\ Cx^{-3} & \text{for } x \gg a \end{cases} \tag{5.2.9}$$

where $A$ is a suitable elastic constant, $C$ is a van der Waals constant, and $a$ is the near-neighbor distance. To interpolate between the above known limits Kohn and Yaniv introduce a hypothesis of universality, $F(x) = Bf(x/x_0)$, where $f(\xi)$ is assumed to be, at least for similar materials, a universal function of a scaled displacement $\xi = x/x_0$. Contact with the limits (5.2.9) yields $B = (A^3 C)^{1/4}$ and $x_0 = (C/A)^{1/4}$ and hence from equation (5.2.8)

$$\gamma = \alpha (AC)^{1/2} \qquad (5.2.10)$$

where $\alpha = \frac{1}{2} \int_0^\infty d\xi f(\xi)$ should be a universal constant, at least for similar materials.

For a variety of metals, when determining $A$ from phonon dispersion curves and $C$ from the electron plasma frequency, the available values of the surface energy are found[9] to obey this relation with $\alpha = 0.476$. A somewhat lower value $\alpha \simeq 0.35$ is indicated by a similar analysis for ionic crystals[10] ($C$ describes in these materials the van der Waals interactions between the closed electronic shells of the ions and is determined in essence by the optical band gap and the high-frequency dielectric constant).

Clearly, an important quantity in the above simple description of the cleavage force is the scaling length $x_0$, which should roughly measure the separation at which $F(x)$ passes through its maximum in changing over from its initial linear increase to its final $x^{-3}$ decay. At the most elementary level one may extrapolate the limiting forms (5.2.9) until they meet at $x = x_0$: this yields $\alpha = \frac{1}{2}$, in fair accord with the empirical determinations mentioned above, and $\gamma = \frac{1}{2} A x_0^2$. Taking for a liquid $A \simeq (aK)^{-1}$, we finally get

$$\gamma K \simeq \frac{1}{2} \frac{x_0^2}{a} \qquad (5.2.11)$$

which should be compared with equation (5.2.5). It is, of course, not surprising that a relation should exist between the surface thickness of the Cahn–Hilliard theory and the location of the maximum in the cleavage force.

## 5.2.4. Solid–Liquid Interfacial Free Energy

An interface between a solid and its own liquid is usually a less severe perturbation of the single-phase state than those which we have considered above, because the densities and local coordinations of these two phases are often quite similar. The excess free energy $\gamma_{sl}$ associated with this interface plays an important role in the process of solidification and indeed its measurement for metals and many other materials is based on the method of determining the critical supercooling for homogeneous nucleation of the solid inside the melt. The Gibbs free energies $G_s$ and $G_l$ of the two bulk phases cross at the melting point $T_m$ with a difference in slopes that is related to the latent heat of fusion $H_f$ by

$$\left(\frac{\partial G_s}{\partial T}\right)_P - \left(\frac{\partial G_l}{\partial T}\right)_P = \frac{H_f}{T_m} \qquad (5.2.12)$$

Although the bulk solid phase is the absolutely stable one below $T_m$, the liquid can be supercooled because a finite amount of interfacial free energy is to be provided for the formation of the interface between a solid nucleus and the surrounding liquid.

Let us consider a spherical solid particle of radius $r$ in a bulk liquid at a temperature $T$ below $T_m$. If the slopes of $G$ in equation (5.2.12) do not change appreciably over a moderate temperature range, we can write the excess free energy of the solid nucleus as a function of its radius in the form

$$\Delta G = -\frac{4\pi}{3} r^3 \frac{H_f}{T_m} \Delta T + 4\pi r^2 \gamma_{sl} \qquad (5.2.13)$$

where $\Delta T = T_m - T$. This increases initially with $r$ to reach a maximum for $r = r_m$, where

$$r_m = \frac{2\gamma_{sl} T_m}{H_f \Delta T} \qquad (5.2.14)$$

The radius $r_m$ for given $\Delta T$ is the critical size beyond which the solid nucleus will grow at the expense of the liquid phase with a release of free energy. Conversely, in the absence of heterogeneous nucleation processes at impurities or external surfaces, the liquid can be supercooled until it reaches a temperature at which the corresponding critical nucleus is such that there is an appreciable probability that such a nucleus will form out of random fluctuations in the liquid.

The "nucleation transition" is in fact observed to be sharp and measurements of the maximum supercooling to which a liquid can be subjected before homogeneous nucleation occurs allow[11] an estimation of $\gamma_{sl}$ (or, more precisely, of some unspecified average of this quantity over various crystalline planes). This quantity for metals is about 10% of the corresponding liquid–vapor interfacial free energy and correlates rather well with the latent heat $H_f$.

### 5.2.5. Interface Intersections

One of the most general methods of measuring interfacial tensions involves the study of the configuration of intersections between interfaces. Consider for example in Figure 5.2 the case of a sessile liquid drop on a flat

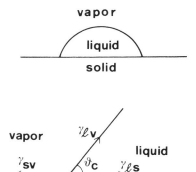

Figure 5.2. A sessile liquid drop on a flat solid: interfacial tensions and contact angle.

solid, whose enlarged view introduces the *contact angle* $\vartheta_c$, measured through the liquid perpendicularly to the three-phase contact line. The equilibrium of the horizontal components of the surface-tension forces yields

$$\gamma_{lv} \cos \vartheta_c = \gamma_{sv} - \gamma_{ls} \tag{5.2.15}$$

which allows one to determine the relative values of the interfacial tensions from a measurement of $\vartheta_c$. Note that the vertical component $\gamma_{lv} \sin \vartheta_c$ is not balanced in the drawing: a local distortion or "puckering" of the solid is supposed to take place near the three-phase contact line to allow a balance of vertical forces, in the absence of a full equilibrium situation that would presumably involve a lenticular shape for the liquid drop.

The case illustrated here ($\vartheta_c < \pi/2$) is conventionally called a "wetting" situation, to be contrasted with the nonwetting case ($\vartheta_c > \pi/2$) and with the "spreading" case where, instead, the liquid covers the whole solid surface. It is convenient to introduce the "work of adhesion" $W_a$ as

$$W_a = \gamma_{sv} + \gamma_{lv} - \gamma_{ls} \tag{5.2.16}$$

This is a positive quantity that measures the work per unit area to pull the liquid away from the solid. We can then rewrite equation (5.2.15) as

$$\cos^2(\tfrac{1}{2}\vartheta_c) = W_a / 2\gamma_{lv} \tag{5.2.17}$$

This shows that no contact angle exists if $W_a > 2\gamma_{lv}$ (spreading situation), while we have a wetting situation ($0 < \vartheta_c < \pi/2$) if $1 < W_a/\gamma_{lv} < 2$ and a nonwetting situation ($\pi/2 < \vartheta_c < \pi$) if $0 < W_a/\gamma_{lv} < 1$. The order of magnitude of the ratio $W_a/\gamma_{lv}$ is the ratio of the strength of a liquid–solid interatomic bond and that of a liquid–liquid bond.

A general treatment of intersections between interfaces has been given by Herring.[12] In applying the method of interface intersections to solid surfaces (e.g., in the study of the equilibrium shapes of grooves formed where grain boundaries intersect a solid surface) one must keep account of the dependence of the surface tension on the crystal plane. The surface tensions are acting in such a way as to change the orientation of the interfaces, which in turn changes the surface tensions of these interfaces: this orientation effect can be described by forces acting on each boundary and tending to orientate it to a low-$\gamma$ orientation. The general form of the equilibrium condition is[12]

$$\sum_i \left( \gamma_i \mathbf{n}_i + \frac{\partial \gamma_i}{\partial \mathbf{n}_i} \right) = 0 \qquad (5.2.18)$$

where the interfaces have surface tensions $\gamma_i$ acting in directions $\mathbf{n}_i$.

The method of studying the equilibrium structure of a solid surface where a grain boundary emerges has been used extensively in evaluating the relative magnitudes of grain boundary and surface free energies. Applications to solid–liquid interfacial free energies have also been made. For a discussion see, e.g., the book of Woodruff.[13]

### 5.2.6. Electrode–Electrolyte Interface

In a system containing mobile charged particles (such as a metal or a molten salt) the anisotropy in the interface results in a redistribution of charge and thus in the creation of an electric double layer and a potential drop across the interface.

A specific example of great practical importance is the case of a liquid electrolyte in contact with a metallic electrode.[14] Orientation of solvent dipoles and redistribution of positive and negative ion densities occur in the electrolytic solution, and redistribution of electrons relative to ions occurs in the metal. The electrode can carry a net charge if a corresponding opposite charge is induced in the electrolyte. An electric double layer, characterized macroscopically by its capacitance, thus exists at the interface. Although the potential drop is not large (typically of the order of one volt) the field strength in the interface, with thickness of order 10 Å, is enormous.

The simplest experimental apparatus is schematized in Figure 5.3 and consists of two electrodes ($M_1$ and $M_2$, say) immersed in an electroytic solution, with a potential difference $\phi$ applied externally across them. Taking account of all the interfaces and applying Kirchhoff's second law,

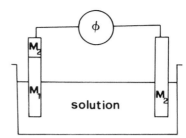

Figure 5.3. A schematized electrolytic cell.

we have

$$\phi = \phi_{M_2/M_1} + \phi_{M_1/S} + \phi_{S/M_2} \qquad (5.2.19)$$

where $\phi_{M_1/S}$ is the potential drop between $M_1$ and the solution S, etc. A liquid-mercury electrode is a very common choice for $M_1$, while the second electrode allows essentially free exchange of one of the ionic species in the solution. We discuss below the case of a 1–1 electrolyte. For a general discussion, see the article of Parsons in the book edited by Bockris *et al.*[14]

The interfacial tension $\gamma$ of the $M_1/S$ interface can be measured as a function of the external potential $\phi$ by electrocapillary techniques (for a discussion see, e.g., Bockris and Reddy[15]). From the differential form of the first law of thermodynamics for the surface excesses at this interface, one can derive the Gibbs adsorption equation in the form

$$d\gamma = -sdT + vdP - \sum_{\alpha=1}^{2} n_\alpha d\mu_\alpha - qd\phi_{M_1/S} \qquad (5.2.20)$$

The quantities $s$, $v$, and $n_\alpha$ are interfacial excesses of entropy, volume, and particle numbers per unit area, while $q$ is the charge on electrode $M_1$ per unit area. Overall neutrality of the interface implies

$$q = -\sum_{\alpha=1}^{2} Z_\alpha e n_\alpha \qquad (5.2.21)$$

where $Z_\alpha$ are the ionic valences.

From equation (5.2.19) we have $d\phi_{M_1/S} = d\phi - d\phi_{S/M_2}$, since the contact potential $\phi_{M_2/M_1}$ remains constant on changing the external potential $\phi$. Furthermore, if the second electrode is reversible to one of the ionic species (species 2, say) we have $d(\mu_2 + Z_2 e\phi_{S/M_2}) = 0$. Hence equation (5.2.20) becomes

$$d\gamma = -sdT + vdP - qd\phi - n_1 d\mu \qquad (5.2.22)$$

where $\mu = \mu_1 + \mu_2$ is the chemical potential of the solute. This is the desired thermodynamic relation between changes in the interfacial tension at the blocking electrode and changes in the "external fields," in particular changes in the potential drop at the measuring instrument and changes in the chemical potential (i.e., the concentration, in essence) of the solute.

Equation (5.2.22) at fixed temperature and pressure yields

$$q = -\left(\frac{\partial \gamma}{\partial \phi}\right)_\mu \tag{5.2.23}$$

and hence a thermodynamic definition for the interfacial capacitance $C$, namely

$$C \equiv \left(\frac{\partial q}{\partial \phi}\right)_\mu = -\left(\frac{\partial^2 \gamma}{\partial \phi^2}\right)_\mu \tag{5.2.24}$$

Furthermore, the surface excess of the "nonreversible" species is given by

$$n_1 = -\left(\frac{\partial \gamma}{\partial \mu}\right)_\phi \tag{5.2.25}$$

The surface excess of the other species follows from equations (5.2.23) and (5.2.25) through the neutrality condition.

The value of $\gamma$ as a function of $\phi$ is found to have a maximum, which defines a fundamental reference potential for the electrified interface, the *potential of zero charge* [see equation (5.2.23)]. At this point the potential drop across the interface is determined purely by polarization processes.

## 5.3. Statistical Mechanics of the Liquid Surface

Statistical mechanical theories of the liquid–vapor interface that go back to the early work of van der Waals[16] aim at calculating the density profile $n(z)$ and the surface tension $\gamma$ from an assumed interatomic force law. Two main lines of approach may be distinguished: the approach of Kirkwood and Buff[17] focuses on the surface stress tensor for the case of pairwise interatomic forces, and the approach systematized by Triezenberg and Zwanzig[18] uses fluctuation-theory results for an inhomogeneous fluid (for a recent review, see Evans[19]). The main role in the two approaches is played by the distribution function $n_2(\mathbf{r}, \mathbf{r}')$ of pairs of atoms and by the corresponding direct correlation function $c(\mathbf{r}, \mathbf{r}')$, respectively, and it will be useful to recall briefly the main properties of these functions for a homogeneous fluid before tackling the interface problem.

### 5.3.1. Pair-Distribution and Direct-Correlation Functions

The pair-distribution function $n_2(\mathbf{r}, \mathbf{r}')$ is the average density of pairs of atoms such that one atom is located at $\mathbf{r}$ and the second at $\mathbf{r}'$. In a homogeneous fluid it depends only on $R \equiv |\mathbf{r} - \mathbf{r}'|$ and is conventionally written as $n_2(R) = n^2 g(R)$. The corresponding direct-correlation function $c(R)$ is defined by the Ornstein–Zernike integral equation

$$g(R) - 1 = c(R) + n \int d\mathbf{R}' c(|\mathbf{R} - \mathbf{R}'|)[g(R') - 1] \qquad (5.3.1)$$

In Fourier transforms we write

$$g(R) = 1 + \frac{1}{(2\pi)^3 n} \int d\mathbf{k} e^{-i\mathbf{k}\cdot\mathbf{R}}[S(k) - 1]$$

$$(5.3.2)$$

$$c(R) = \frac{1}{(2\pi)^3 n} \int d\mathbf{k} e^{-i\mathbf{k}\cdot\mathbf{R}} \hat{c}(k)$$

and hence equation (5.3.1) becomes

$$1 - \hat{c}(k) = 1/S(k) \qquad (5.3.3)$$

where $S(k)$ is the structure factor measured in a diffraction experiment.

For our future discussions it is important to stress the meaning of the two sides of equation (5.3.3) in terms of the response of the fluid to an external perturbation, in the case where classical statistical mechanics applies. We first recall that $S(k)$ is the integral of the dynamic-structure factor $S(k, \omega)$ for coherent inelastic scattering over all energy transfers:

$$S(k) = \int_{-\infty}^{\infty} \frac{d\omega}{2\pi} S(k, \omega) \qquad (5.3.4)$$

We next introduce the density-response function $\chi(k, \omega)$, which gives in Fourier transforms the density change $\delta n(\mathbf{k}, \omega)$ of the fluid in response to a weak external potential $V_{\text{ext}}(\mathbf{k}, \omega)$ as

$$\delta n(\mathbf{k}, \omega) = \chi(k, \omega) V_{\text{ext}}(\mathbf{k}, \omega) \qquad (5.3.5)$$

Recall that the fluctuation-dissipation theorem relates $\text{Im}\,\chi(k, \omega)$ to the dissipation of energy and momentum into the fluid and hence to the inelastic scattering cross section:

$$\text{Im}\,\chi(k,\omega) = \frac{n}{2\hbar}\left[1 - \exp(-\hbar\omega/k_B T)\right]S(k,\omega)$$

$$\xrightarrow[\substack{\text{classical} \\ \text{limit}}]{} \frac{n\omega}{2k_B T}\,S(k,\omega) \tag{5.3.6}$$

The last step is to use the Kramers–Kronig relation to evaluate the static-response function $\chi(k) \equiv \chi(k,\omega = 0)$:

$$\chi(k) = \int_{-\infty}^{\infty} \frac{d\omega}{\pi}\,\frac{\text{Im}\,\chi(k,\omega)}{\omega} \underset{\substack{\text{classical} \\ \text{limit}}}{=} \frac{n}{k_B T}\int_{-\infty}^{\infty}\frac{d\omega}{2\pi}\,S(k,\omega) = \frac{n}{k_B T}\,S(k) \tag{5.3.7}$$

Hence $S(k)$, being simply proportional to $\chi(k)$, measures the "generalized softness" of the fluid to a density perturbation of wave number $k$. Equation (5.3.7) is in fact the generalization to arbitrary wave numbers of the famous Ornstein–Zernike relation for a monatomic classical fluid,

$$\lim_{k\to 0} S(k) = nk_B TK \tag{5.3.8}$$

where the isothermal compressibility $K$ measures the "softness" of the fluid to a uniform squeeze. We then see from equation (5.3.3) that the quantity $1 - \hat{c}(k)$ represents the "generalized stiffness" of the fluid.

Moving now to an inhomogeneous (monatomic) classical fluid, we define the direct-correlation "function" $c(\mathbf{r}, \mathbf{r}')$, which is in fact a matrix in $\mathbf{r}$ and $\mathbf{r}'$, in such a manner that the matrix

$$K(\mathbf{r}, \mathbf{r}') \equiv \frac{\delta(\mathbf{r} - \mathbf{r}')}{n(\mathbf{r})} - c(\mathbf{r}, \mathbf{r}') \tag{5.3.9}$$

is the inverse of the matrix

$$K^{-1}(\mathbf{r}, \mathbf{r}') \equiv n_2(\mathbf{r}, \mathbf{r}') - n(\mathbf{r})n(\mathbf{r}') \tag{5.3.10}$$

$K^{-1}(\mathbf{r}, \mathbf{r}')$ measures the "softness" of the inhomogeneous fluid to an external potential, and $K(\mathbf{r}, \mathbf{r}')$ measures its "stiffness." It is this interpretation of $K(\mathbf{r}, \mathbf{r}')$ that is basic in the fluctuation approach to the liquid–vapor interface.

### 5.3.2. Surface Stress and Surface Tension for Pairwise Interatomic Forces

The approach of Kirkwood and Buff[17] involves the pair-distribution function $n_2(\mathbf{r}, \mathbf{r}')$, which in the case of a planar interface is a function of the vector $\mathbf{R} = \mathbf{r} - \mathbf{r}'$ and of the coordinate $z$ of the first atom. An interatomic potential of the form $\varphi(R)$, depending only on the relative distance $R$ of a pair of atoms, is assumed. The tangential pressure $P_t(z)$ can then be expressed as

$$P_t(z) = k_B Tn(z) - \frac{1}{2} \int d\mathbf{R} \, \frac{X^2}{R} \, \varphi'(R) n_2(\mathbf{R}, z) \qquad (5.3.11)$$

where we write the components of $\mathbf{R}$ as $(X, Y, Z)$ and $\varphi'(R) = d\varphi(R)/dR$. Far away from the interface the $z$-dependence disappears and equation (5.3.11) reduces to the well-known expression for the bulk pressure,

$$P = k_B Tn - \frac{1}{6} \int d\mathbf{R} R \varphi'(R) n_2(R) \qquad (5.3.12)$$

An expression similar to equation (5.3.11) holds for the normal pressure $P_n$, and by imposing $P_n(z) = P$ for hydrostatic equilibrium one derives an equation for the equilibrium density profile,

$$k_B T \frac{dn(z)}{dz} = \int d\mathbf{R} \, \frac{Z}{R} \, \varphi'(R) n_2(\mathbf{R}, z) \qquad (5.3.13)$$

Finally, from equation (5.2.4) one finds

$$\gamma = \frac{1}{2} \int_{-\infty}^{\infty} dz \int d\mathbf{R} \, \frac{X^2 - Z^2}{R} \, \varphi'(R) n_2(\mathbf{R}, z) \qquad (5.3.14)$$

Needless to say, it is very difficult to calculate $n_2(\mathbf{R}, z)$ and therefore the work based on this method has had to relate $n_2$ to the bulk pair function and to $n(z)$. The most drastic simplification is to take the liquid as homogeneous up to the Gibbs surface and to assume that the vapor has negligible density, whereby one recovers from equation (5.3.14) an expression for $\gamma$, due to Fowler,[20]

$$\gamma = \frac{\pi}{8} n_l^2 \int_0^{\infty} dR \varphi'(R) R^4 g(R) \qquad (5.3.15)$$

A delicate point in these calculations is the consistency between $\varphi(R)$ and $g(R)$. Using computer-simulation data as input, McDonald and Freeman[21] find for liquid neon reasonable results on the basis of equation (5.3.15) and an improvement of the results when they allow a smooth density profile of finite thickness. For a more detailed discussion of this approach the article of Berry[22] may be consulted.

### 5.3.3. Theory of Inhomogeneous Fluid and
### Fluctuation Approach to Surface Tension

The Helmholtz free energy $F$ of an inhomogeneous fluid can be written in terms of a position-dependent free energy density $f(\mathbf{r})$ as

$$F = \int d\mathbf{r} f(\mathbf{r}) \tag{5.3.16}$$

The basis of the theory is the theorem of Hohenberg, Kohn, and Mermin (HKM)[23], which shows that $f(\mathbf{r})$ is a unique *functional* of the density profile $n(\mathbf{r}')$, namely $f(\mathbf{r})$ is uniquely determined (in principle, of course!) if the density $n(\mathbf{r}')$ is known at *all* points $\mathbf{r}'$ in the system.

Let us consider the fluid under an external potential $V_{\text{ext}}(\mathbf{r})$, which determines a density profile $n(\mathbf{r})$, and define

$$u(\mathbf{r}) \equiv \beta[\mu - V_{\text{ext}}(\mathbf{r})] \qquad \left(\beta = \frac{1}{k_B T}, \mu = \text{chemical potential}\right) \tag{5.3.17}$$

The HKM theorem ensures that there is a unique functional relationship between $u(\mathbf{r})$ and $n(\mathbf{r}')$, written symbolically as

$$u(\mathbf{r}) = u\{\mathbf{r}; [n(\mathbf{r}')]\} \tag{5.3.18}$$

If we now consider a rigid translation of the whole system by an amount $\boldsymbol{\delta}$, we have

$$u(\mathbf{r} + \boldsymbol{\delta}) = u\{\mathbf{r}; [n(\mathbf{r}' + \boldsymbol{\delta})]\} \tag{5.3.19}$$

In the limit $\boldsymbol{\delta} \to 0$ this yields by a Taylor's expansion

$$\nabla u(\mathbf{r}) = \int d\mathbf{r}' K(\mathbf{r}, \mathbf{r}') \nabla n(\mathbf{r}') \tag{5.3.20}$$

where

$$K(\mathbf{r}, \mathbf{r}') \equiv \delta u(\mathbf{r})/\delta n(\mathbf{r}') \tag{5.3.21}$$

(the symbol $\delta u(\mathbf{r})/\delta n(\mathbf{r}')$ denotes the functional derivative of $u(\mathbf{r})$ with respect to $n(\mathbf{r}')$, i.e., the change in the value of $u$ at $\mathbf{r}$ due to a change in $n$ at $\mathbf{r}'$). From definition (5.3.21) it is clear that $K(\mathbf{r}, \mathbf{r}')$ has the meaning of an inverse density-response matrix and in fact coincides with definition (5.3.9) for a classical fluid. Using that expression in equation (5.3.20) and taking the limit $u \to 0$, we find

$$\frac{\nabla n(\mathbf{r})}{n(\mathbf{r})} = \int d\mathbf{r}' c(\mathbf{r}, \mathbf{r}') \nabla n(\mathbf{r}') \tag{5.3.22}$$

as the equation that determines the equilibrium density profile in an inhomogeneous fluid.

For the case of a planar liquid–vapor interface the matrix $c(\mathbf{r}, \mathbf{r}')$ depends only on $z$, $z'$, and $s = [(x - x')^2 + (y - y')^2]^{1/2}$. Taking the two-dimensional Fourier transform

$$\hat{c}(k; z, z') = \int d^2s\, e^{i\mathbf{k}\cdot\mathbf{s}} c(s; z, z') \qquad (5.3.23)$$

we therefore find

$$\frac{dn(z)}{dz} = n(z) \int_{-\infty}^{\infty} dz'\, c_0(z, z') \frac{dn(z')}{dz'} \qquad (5.3.24)$$

where $c_0(z, z') \equiv \hat{c}(k = 0; z, z') = \int d^2s\, c(s; z, z')$. This equation (which should be contrasted with equation (5.3.13) for the Kirkwood–Buff approach) is the equilibrium condition in the fluctuation approach of Triezenberg and Zwanzig.[18] Their derivation involves examining the effect of a long-wavelength fluctuation of the Gibbs surface, with wave vector along the surface: to lowest order in the wave number of the fluctuation, this corresponds to a rigid translation of the system and equation (5.3.24) follows from the fact that neither a free-energy nor a surface-area change accompanies such a translation.

Such changes arise at the next order in the wave number of the fluctuation and their evaluation leads to an expression for the surface tension. In particular, the free-energy change for a *small* fluctuation $\Delta n(\mathbf{r})$ in the density profile can be written in the form

$$\Delta F = \tfrac{1}{2} k_B T \iint d\mathbf{r}\, d\mathbf{r}'\, \Delta n(\mathbf{r}) K(\mathbf{r}, \mathbf{r}') \Delta n(\mathbf{r}') \qquad (5.3.25)$$

this being an obvious generalization of the usual elasticity-theory expression through the use of the "stiffness" matrix $K(\mathbf{r}, \mathbf{r}')$. A detailed calculation, for which we refer to the original paper,[18] yields

$$\gamma = k_B T \iint_{-\infty}^{\infty} dz\, dz'\, \frac{dn(z)}{dz} c_2(z, z') \frac{dn(z')}{dz'} \qquad (5.3.26)$$

where

$$c_2(z, z') \equiv \tfrac{1}{4} \int d^2s\, s^2 c(s; z, z') = -\tfrac{1}{4}[d^2\hat{c}(k; z, z')/dk^2]_{k\to 0}$$

Equations (5.3.24) and (5.3.26) are formally exact and independent of the detailed nature of the interatomic forces. Of course, the functions $c_0$ and $c_2$ for the inhomogeneous fluid are not known. The theory becomes practicable when these are related to properties of the homogeneous fluid, as one does in practice through the "square density gradient" approximation that we proceed to discuss below. For a general discussion of the free-energy functional the work of Saam and Ebner[24] may be consulted.

### 5.3.4. Square-Gradient Approximation

We now return to equation (5.3.16) and let us be guided by the Cahn–Hilliard argument (Section 5.2.2) in the search for a useful but approximate expression for the free-energy density $f(z)$ as a functional of the density profile $n(z)$ for the planar interface. If $n(z)$ were slowly varying with $z$, it would be reasonable to represent $f(z)$ by a local term $f(n(z))$ [the free-energy density of a *homogeneous* fluid at the local atomic density $n(z)$] plus an infinite series involving derivatives of $n(z)$ and their powers. The Cahn–Hilliard argument suggests that it may be already useful to include only the leading correction in $n'(z) \equiv dn(z)/dz$, which by symmetry is the square power $[n'(z)]^2$. We thus write

$$f(z) \simeq f(n(z)) + \tfrac{1}{2}a(n(z))[n'(z)]^2 \tag{5.3.27}$$

The function $a(n(z))$ is again a property of the homogeneous fluid, and one can show[23,25] that

$$a = \tfrac{1}{6}k_B T \int d\mathbf{r} r^2 c(r) \tag{5.3.28}$$

$c(r)$ being the direct correlation function of the homogeneous fluid at density $n(z)$. Note that, since $n(z)$ changes continuously from $n_l$ to $n_v$ across the interface, equation (5.3.27) requires that we know $f$ and $a$ for the homogeneous fluid also in the density region where it is unstable against two-phase separation.

Let us examine[25] in some detail the consequences of equation (5.3.27). The equilibrium condition can be found by minimizing $F$ at constant $N$, i.e., by minimizing the integral $\int_{-\infty}^{\infty} dz [f(z) - \mu n(z)]$, where the "Lagrange multiplier" $\mu$ is the chemical potential and hence $\omega(z) \equiv \mu n(z) - f(z)$ is the density of thermodynamic potential $\Omega$. The Euler–Lagrange equation for such a variational problem, involving both $n(z)$ and $n'(z)$, is

$$\frac{\partial \omega(z)}{\partial n(z)} - \frac{d}{dz} \frac{\partial \omega(z)}{\partial n'(z)} = 0$$

and hence

$$\mu = \mu(n(z)) - a(n(z))n''(z) - \tfrac{1}{2}a'(n(z))[n'(z)]^2 \qquad (5.3.29)$$

where

$$\mu(n(z)) \equiv \frac{\partial f(n(z))}{\partial n(z)} \quad \text{and} \quad a' = \frac{\partial a(n)}{\partial n}$$

This equilibrium condition determines the density profile $n(z)$.

If equation (5.3.29) is multiplied by $n'(z)$ and then integrated, we get

$$\mu n(z) - f(n(z)) + \tfrac{1}{2}a(n(z))[n'(z)]^2 = P \qquad (5.3.30)$$

where we have identified the constant of integration with the pressure $P$ by taking the limit $z \to \pm\infty$ of the left-hand side. Equation (5.3.30) expresses the condition of mechanical equilibrium and can be shown to be equivalent to the well-known Maxwell construction.

Finally, we can relate the surface tension to the density profile by using equation (5.2.1) and noting that $V_l f(n_l) + V_v f(n_v) = \mu N - PV$:

$$\gamma = \int_{-\infty}^{\infty} dz\{f(n(z)) + \tfrac{1}{2}a(n(z))[n'(z)]^2 - \mu n(z) + P\}$$

$$= \int_{-\infty}^{\infty} dz\, a(n(z))[n'(z)]^2 \qquad (5.3.31)$$

where we have used equation (5.3.30). To see the connection[26] with the Cahn–Hilliard argument, we expand $f(n(z))$ around the liquid density $n_l$,

$$f(n(z)) = f(n_l) + [n(z) - n_l] \frac{\partial f}{\partial n}\Bigg|_{n=n_l} + \frac{1}{2}[n(z) - n_l]^2 \frac{\partial^2 f}{\partial n^2}\Bigg|_{n=n_l} + \cdots \qquad (5.3.32)$$

and use

$$\frac{\partial f}{\partial n}\Bigg|_{n_l} = \mu(n_l) = \mu, \qquad n_l^2 \frac{\partial^2 f}{\partial n^2}\Bigg|_{n_l} = \frac{1}{K}$$

Using again equation (5.3.30) to eliminate $a$, we find

$$\gamma = 2\int_{-\infty}^{\infty} dz\,[f(n(z)) - \mu n(z) + P] \simeq \frac{1}{K}\int_{-\infty}^{\infty} dz \left[\frac{n(z) - n_l}{n_l}\right]^2 \qquad (5.3.33)$$

displaying again the relation between the product $\gamma K$ and the interfacial thickness.

This simple treatment can easily be extended[25] to the case where the fluid is in an external potential $U(z)$, such as a gravitational potential. Nonplanar interfaces can also be treated — for instance, the case of a spherical liquid drop inside its vapor, for which one recovers the Young–Laplace formula. A number of numerical applications to specific systems can be found in the current literature. For extensions to multicomponent systems and to fluids of charged particles, see elsewhere.[27]

### 5.3.5. Inhomogeneous Electron Gas at a Metallic Surface

Simple theories for the suface tension of liquid metals focus attention on the behavior of the conduction electrons. At the most elementary level one constructs[28] the analog of equation (5.3.27) as the sum of a local energy term $\varepsilon(n)$ (the kinetic, exchange, and correlation energy of a homogeneous electron gas at density $n$) and of an inhomogeneity term in $(\nabla n)^2$ associated with kinetic effects:

$$E = \int d\mathbf{r} \left[ \varepsilon(n) + \frac{\lambda \hbar^2}{8m} \frac{(\nabla n)^2}{n} \right] \tag{5.3.34}$$

where $\lambda$ is a numerical factor (the theory of the inhomogeneous Fermi gas yields[29] $\lambda = 1/9$). The variational principle $\delta(E - \mu N)/\delta n = 0$ yields

$$-\frac{\lambda \hbar^2}{4m} \left[ \frac{n''}{n} - \frac{1}{2} \left( \frac{n'}{n} \right)^2 \right] + \frac{d\varepsilon}{dn} = \mu \tag{5.3.35}$$

for the determination of the equilibrium density $n(z)$. Setting $n = \psi^2$, this is formally equivalent to a Schrödinger equation of the form

$$-\frac{\lambda \hbar^2}{2m} \frac{d^2\psi}{dz^2} + \left( \frac{d\varepsilon}{dn} - \mu \right) \psi = 0 \tag{5.3.36}$$

The surface tension, defined as the energy difference between the inhomogeneous electron gas and a homogeneous electron gas, per unit surface area, is given in analogy with equation (5.3.31) by

$$\gamma = \frac{\lambda \hbar^2}{m} \int_{-\infty}^{\infty} dz \, [\psi'(z)]^2 \tag{5.3.37}$$

This elementary theory has the advantage that it can be carried through analytically for certain simple forms of $\varepsilon(n)$, and in particular for the case

$$\varepsilon(n) = n\varepsilon_0 [(n/n_0)^{2/3} - 2(n/n_0)^{1/3}] \tag{5.3.38}$$

which corresponds to the sum of a kinetic energy term $\sim n^{5/3}$ and an exchange term $\sim n^{4/3}$. Clearly $n_0$ is the equilibrium density at zero pressure, and the corresponding electron-gas compressibility is given by $K^{-1} = \frac{2}{9}n_0\varepsilon_0$. The solution of equation (5.3.35) is then

$$n(z)/n_0 = [1 + B \exp(z/l)]^{-3} \tag{5.3.39}$$

where $B$ is a constant and the length $l$, which clearly measures the surface thickness, is given by

$$l = (9\lambda\hbar^2/8m\varepsilon_0)^{1/2} \tag{5.3.40}$$

From equation (5.3.37) we finally obtain

$$\gamma K = \tfrac{3}{4}l \tag{5.3.41}$$

An important advance in the more refined theories of the electronic surface density profile[30,31] is to avoid a density-gradient treatment for the electronic kinetic energy. The equilibrium condition for the profile can still be formulated as a Schrödinger equation in which, however, the Laplacian term accounts for the full single-electron kinetic energy while the potential term includes exchange and correlation contributions. A Hartree-like term for the electron–ion interaction also arises when the ionic and electronic density profiles are allowed to differ. For specific work on liquid metals, including some account of the transition from metallic to localized electron states in the transition region from liquid to vapor, see Rice et al.[32]

## 5.4. Liquid–Solid Transition and Structure of Liquid–Solid Interfaces

### 5.4.1. Statistical Mechanical Theory of Freezing

For an introduction to the discussion of the equilibrium between liquid and solid at the freezing point, it is useful to stress in Figure 5.4 the qualitative differences in the structure factor $S(k)$ for a real gas and a liquid near freezing. Bearing in mind the meaning of $S(k)$ as a "generalized softness function" that we stressed in Section 5.3.1, it is evident that the liquid is not only much harder to compress [cf. equation (5.3.8)] but also considerably "softer" against a density-wave deformation with wave number lying in the region of the main peak of $S(k)$.

Hansen and Verlet[33] have emphasized an empirical relation between the height of the main peak in $S(k)$ for a monatomic liquid and its freezing.

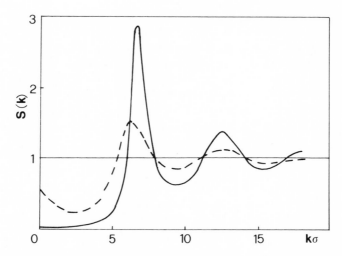

Figure 5.4. Structure factor $S(k)$ for a model fluid in the thermodynamic state of a real gas (broken curve) and of a liquid near freezing (full curve), from computer-simulation work of L. Verlet, *Phys. Rev.* **165**, 201 (1968).

An illustration for the alkali metals is given in Table 5.2. This table reports also the case of the one-component classsical plasma on a uniform neutralizing background (OCP), which can be regarded as a prototype model for these metals.[34] Rough criteria for freezing can clearly be formulated from these data: freezing occurs in these systems when the "plasma parameter" $\Gamma \equiv e^2/ak_B T$ (which measures the ratio of the ionic potential energy to the thermal energy) reaches[35] a value of about 180–200, or alternatively when the main peak in $S(k)$ reaches[33] a value near 3. The last row of Table 5.2 shows that the latter freezing criterion is not unrelated to the well-known Lindemann criterion for melting: it reports the Debye–Waller factor of the solid phase at melting, calculated by harmonic theory at the first reciprocal lattice vector $G_1$, and this quantity clearly measures both the intensity of the corresponding Bragg spots and the ratio of the mean-square displacement to the lattice constant.

**Table 5.2. Freezing Parameters for the Alkali Metals and the Classical Plasma**

|  | Li | Na | K | Rb | Cs | OCP |
|---|---|---|---|---|---|---|
| $\Gamma_f = \dfrac{e^2}{ak_B T_f}$ | 211 | 208 | 186 | 188 | 181 | 177 |
| $S_{\text{peak}}$ at $T_f$ | — | 2.8–3.1 | 2.7–3.1 | ~3 | ~3 | ~3 |
| $\exp(-2W_{G_1})$ | 0.64 | 0.66 | 0.65 | 0.67 | 0.63 | 0.62 |

These considerations lie behind the theory of freezing of Ramakrishnan and Yussouff.[36] We start from equation (5.3.22) for the equilibrium density profile of an inhomogeneous system and, after noting that it is satisfied trivially by the homogeneous liquid phase [where $c(\mathbf{r}, \mathbf{r}') = c_l(|\mathbf{r} - \mathbf{r}'|)$ and $n(\mathbf{r}) = n_l$], we ask whether it will admit at some point a second solution in equilibrium with the homogeneous-liquid solution. We require that this second solution should have the form of a periodic density,

$$n_p(\mathbf{r}) = n_s + \frac{1}{V} \sum_{\mathbf{G} \neq 0} n_\mathbf{G} e^{i\mathbf{G}\cdot\mathbf{r}} \quad (\mathbf{G} = \text{reciprocal lattice vectors}) \quad (5.4.1)$$

and that it should arise as a bifurcation of the solution of the equilibrium equation as we approach the coexistence point from the homogeneous-liquid phase, i.e., that it should satisfy equation (5.3.22) with $c(\mathbf{r}, \mathbf{r}')$ replaced by $c_l(|\mathbf{r} - \mathbf{r}'|)$. We shall next have to impose also that the free energies of the two solutions are equal at coexistence.

If equation (5.3.22) is therefore written as

$$\nabla n(\mathbf{r})/n(\mathbf{r}) = \int d\mathbf{r}' c_l(|\mathbf{r} - \mathbf{r}'|)\nabla n(\mathbf{r}') \quad [n(\mathbf{r}) = n_p(\mathbf{r}) \text{ or } n_l] \quad (5.3.2)$$

integration of the right-hand side by parts and integration over $\mathbf{r}$ yields at the coexistence point

$$\ln\left(\frac{n_p(\mathbf{r})}{n_l}\right) = \int d\mathbf{r}' c_l(|\mathbf{r} - \mathbf{r}'|)[n_p(\mathbf{r}') - n_l] \quad (5.4.3)$$

Our next step is to construct the difference $\Delta\Omega$ in the thermodynamic potential between the two phases in such a manner that equation (5.4.3) follows from it as the Euler equation of a minimum-free-energy principle:

$$\frac{\Delta\Omega}{k_B T} = \int d\mathbf{r} \left\{ n_p(\mathbf{r}) \ln\left[\frac{n_p(\mathbf{r})}{n_l}\right] - [n_p(\mathbf{r}) - n_l] \right\}$$
$$- \frac{1}{2} \int\int d\mathbf{r} d\mathbf{r}'[n_p(\mathbf{r}) - n_l] c_l(|\mathbf{r} - \mathbf{r}'|)[n_p(\mathbf{r}') - n_l] \quad (5.4.4)$$

The first term represents a free-particle contribution and the second is in essence available in equation (5.3.25) for the Triezenberg–Zwanzig theory. By introducing expression (5.4.3) into equation (5.4.4) we obtain at the coexistence point, to within nonlinear terms in $n_s - n_l$, the relation

$$\frac{\Delta\Omega}{k_B T} = \frac{1}{2N} \sum_{\mathbf{G} \neq 0} \hat{c}(\mathbf{G})|n_\mathbf{G}|^2 - \frac{1}{2} N[1 - \hat{c}(k = 0)]\frac{n_s^2 - n_l^2}{n_l^2} \quad (5.4.5)$$

and this must be zero at coexistence. Note that $\hat{c}(G) = 1 - 1/S(G)$ and that $1 - \hat{c}(0)$ is in essence the inverse compressibility of the liquid.

Equations (5.4.3) and (5.4.5) must be solved self-consistently to find the periodic density profile $n_p(\mathbf{r})$ in equilibrium with the homogeneous liquid at density $n_l$. That a solution with $\Delta\Omega = 0$ can exist, is clear from the signs of the two terms in equation (5.4.5): free energy must be expended in modulating the liquid density at wave vectors $\mathbf{k} = \mathbf{G} \neq 0$, but free energy can be gained from the density-change term provided $n_s > n_l$. Freezing occurs in this theory as a first-order phase transition in which the $\mathbf{G} \neq 0$ components and the $\mathbf{k} = 0$ component of the density are strongly coupled by the nonlinear nature of equation (5.4.3).

Freezing of simple liquids is already obtained in this theory if one allows $n_G \neq 0$ only for $\mathbf{G} = \mathbf{G}_1$ (the first star of reciprocal lattice vectors) in equations (5.4.1) and (5.4.5), but this requires an excessive value for $S(G_1)$, i.e., for the main peak in $S(k)$, relative to the data reported in Table 5.2. The inclusion of the first two stars of reciprocal lattice vectors yields very reasonable results.[36] Extensions of these arguments to ionic systems are dealt with elsewhere.[37]

### 5.4.2. Dislocation Theory of Melting

An important feature of the Ramakrishnan–Yussouff theory of *freezing* of a simple liquid is that it accounts correctly for the first-order character of the liquid–solid transition: the $\mathbf{G} \neq 0$ Fourier components of the particle density appear suddenly with a finite value at $T_m$. Many authors in the past have tried to devise a theory of *melting* by viewing the liquid as a crystal which, in thermal equilibrium, contains a high concentration of some particular type of crystal defect (vacancies, dislocations, grain boundaries; for a discussion see, e.g., the book of Nabarro[38]). Work by Edwards and Warner[39] successfully predicts a first-order melting transition on the basis of a dislocation model in which a crucial effect arises from mutual "screening" in a dense assembly of dislocations.

The main contribution to the free energy of a dislocation arises from the long-range elastic distortions induced in the material, but when dislocations of opposite Burgers vectors are brought together most of this distortion is eliminated — in fact, a dense assembly of dislocations can arrange itself so as to maximize this reduction. Edwards and Warner find that the free energy of the dislocation assembly can be expressed schematically as

$$F_d = \rho F_1 + \rho^2 F_2 - \rho^{3/2} F_{3/2} \tag{5.4.6}$$

where $\rho$ is the length of dislocation lines per unit volume and the $F_i$ are

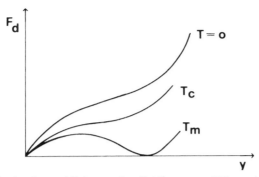

Figure 5.5. Qualitative shape of $F_d$ in equation (5.4.6) vs $y \propto \rho/T$ for various values of $T$; $T_m$ is the melting temperature and $T_c$ the lowest temperature for supercooling of the liquid (after Edwards and Warner[39]).

temperature-dependent coefficients. The successive terms in this expression are in essence (1) the core energy and the entropy of the dislocations, (2) the energy associated with the dilation induced by the dislocations, and (3) a *negative* term arising from many-dislocation effects. This latter term is evaluated by writing the energy of interaction of two dislocation lines as

$$E_{12} \propto \int (\mathbf{b}_1 \cdot d\mathbf{l}_1)(\mathbf{b}_2 \cdot d\mathbf{l}_2)/r_{12}$$

where $\mathbf{b}_i$ are the Burgers vectors, and by calculating the corresponding contribution to the partition function with techniques taken from the theory of networks of random walks. The calculated free energy, as a function of a variable $y \propto \rho/T$, has the form indicated in Figure 5.5 for various values of $T$.

The melting temperature $T_m$ is identified with the temperature at which the equilibrium condition $\partial F_d/\partial y = 0$ yields a finite concentration of dislocations with no excess free energy over the undislocated crystal ($F_d = 0$). The first-order character of the predicted transition is apparent from the fact that such a finite concentration of dislocations at equilibrium arises suddenly in the theory. Over a finite range of temperatures below $T_m$, one still has a local minimum of $F_d$ at finite $y$, but corresponding to a metastable situation ($F_d > 0$), which may be useful to describe a supercooled liquid or a glass. The original work should be consulted for further details.

### 5.4.3. Outline of a Molecular Theory of the Solid–Liquid Interface

Haymet and Oxtoby[40] have shown how the Ramakrishnan–Yussouff theory of freezing may be extended to develop a molecular theory of the solid–liquid interface. For a monatomic system the interface region is assumed to have a density profile that again has the form of equation (5.4.1), but now the density difference $n_s - n_l$ as well as the various Fourier components $n_G$ become functions of position across the interface. More precisely, Haymet and Oxtoby set

$$n(\mathbf{r}) = n_l [1 + \eta(\mathbf{r})] + n_l \sum_{G \neq 0} \mu_G(\mathbf{r}) e^{i\mathbf{G} \cdot \mathbf{r}} \qquad (5.4.7)$$

in the interface region, with limiting values for $\eta(\mathbf{r})$ and $\mu_G(\mathbf{r})$, which are obviously given by

$$\eta(\mathbf{r}) \to (n_s - n_l)/n_l \quad \text{and} \quad \mu_G(\mathbf{r}) \to n_G/n_l V \qquad \text{in the solid}$$

$$\eta(r) \to 0 \quad \text{and} \quad \mu_G(\mathbf{r}) \to 0 \qquad \text{in the liquid} \qquad (5.4.8)$$

If $\eta(\mathbf{r})$ and $\mu_G(\mathbf{r})$ are assumed to be slowly varying functions of position, so that only their square gradients need to be included, one can develop an approximate theory for the liquid–solid interface that is the analog of the theory for the liquid–vapor interface described in Section 5.3.4.

In the particular case of a planar liquid–solid interface, the excess thermodynamic potential per unit area can be written in this square-gradient approximation as

$$\frac{\Delta\Omega}{k_B T} = \frac{\Delta\Omega_u}{k_B T} + \frac{1}{4} \hat{c}''(k = 0) \int_{-\infty}^{\infty} dz \left[ \frac{d\eta(z)}{dz} \right]^2$$

$$+ \frac{1}{4} \sum_{G \neq 0} \hat{c}''(\mathbf{G}) G_z^2 \int_{-\infty}^{\infty} dz \left[ \frac{d\mu_G(z)}{dz} \right]^2 \qquad (5.4.9)$$

This is supplemented by suitable equilibrium conditions for the profiles, which enter the determination of $\Delta\Omega_u$, the excess thermodynamic potential for a "uniform phase" characterized by given values of $\eta$ and $\mu_G$, and given basically by equation (5.4.5).

The authors briefly discuss also the case of a spherical solid particle in its own liquid and point out the relevance of such a model to calculations of the rate of solid phase nucleation in liquids (see Section 5.2.4).

### 5.4.4. Structure of the Solid–Liquid Interface and Crystal-Growth Mechanisms

The question of the structure of the solid–liquid interface as a function of the material and of thermodynamic conditions is of great importance in the discussion of crystal-growth mechanisms. It is clear that an interface with an atomically smooth equilibrium configuration will advance into the liquid by successive processes of two-dimensional nucleation of new atomic layers (or through screw-dislocation mechanisms[38]), while a rough interface, with several atomic layers involved in the transition zone, has a large number of suitable sites at which new atoms can be continuously accommodated. The equilibrium structure of the interface will thus influence the kinetics of the growth process, at least if the rate of motion of the atoms in the interface is sufficiently fast relative to the rate of growth that such equilibrium structure is maintained during growth.

Following the early work on the free solid surface by Burton, Cabrera, and Frank,[41] a simple treatment of the solid–liquid interface has been given by Jackson.[42] Consider the free-energy change $\Delta G$ associated with randomly adding $N_A$ atoms from the liquid phase to the solid surface at the equilibrium temperature $T_m$ for the phase change, on the assumption that the product $P\Delta V$ is negligible. An energy change $\Delta E$ arises from interactions of the adatoms with the substrate and with each other,

$$\Delta E = - N_A \left( 2H_f \frac{n_0}{n} + H_f \frac{n_1}{n} \frac{N_A}{N} \right) \qquad (5.4.10)$$

where $H_f$ is the latent heat of fusion, $N$ is the number of atoms in a complete adlayer, $n_0$ and $n_1$ give the numbers of first neighbors to the adatom in the substrate and in the adlayer, and $n = 2n_0 + n_1$ is the coordination number in the bulk solid. The Bragg–Williams approximation has been adopted in the second term. An entropy change arises both from the transfer of the adatoms from liquid to solid and from configurational disorder in the adlayer,

$$\Delta S = - \frac{H_f}{T_m} N_A + k_B \ln \left[ \frac{N!}{N_A!(N-N_A)!} \right] \qquad (5.4.11)$$

Thus

$$\Delta G = Nk_B T_m \left( \alpha N_A \frac{N-N_A}{N^2} - \ln \frac{N}{N-N_A} - \frac{N_A}{N} \ln \frac{N-N_A}{N_A} \right) \qquad (5.4.12)$$

where $\alpha = n_1 H_f/(nk_B T_m)$. On minimizing $\Delta G$ with respect to $N_A/N$ to find the equilibrium structure of the adlayer, which in this simple theory represents the transition region between liquid and solid, one finds a sharp transition from a rough interface for $\alpha < 2$ to a smooth interface for $\alpha > 2$. More precisely, for $\alpha < 2$ the minimum of $\Delta G$ lies at $N_A/N = \frac{1}{2}$, while for $\alpha > 2$ one finds a maximum at $N_A/N = \frac{1}{2}$ and minima at $N_A/N \simeq 0$ and $N_A/N \simeq 1$. There is good correlation between these predictions and observation, in that high-$\alpha$ materials grow with a faceted habit while low-$\alpha$ materials (typically but not exclusively metals) tend to show dendritic growth — although intermediate cases are known of materials (such as Bi) that can show both modes of growth. The parameter $\alpha$ should in essence measure the solid–liquid interfacial free energy $\gamma_{sl}$ (see also Section 5.2.4).

A sharp transition from rough to smooth interfaces disappears in more refined models, such as that due to Temkin,[43] which allows for a multiplicity of layers in the transition region. In the equilibrium situation the mean position of the interface is nevertheless fixed relative to the lattice periodicity, so that one can envisage a roughly sinusoidal succession of free-energy barriers, which have to be surmounted as the mean position of the interface advances into the liquid. In the supercooled liquid, a driving force for surmounting these barriers is provided by the supercooling acting through a chemical potential difference $\Delta\mu \sim H_f \Delta T / T_m$. The model then indicates two different crystal-growth behaviors as a function of $\alpha$ and $\Delta\mu$. An equilibrium solution can still be found if $\alpha$ is sufficiently large and/or the supercooling is still moderate, implying that a free-energy barrier still exists for crystal growth. No equilibrium solution is found if $\Delta\mu$ is large or $\alpha$ is small; a continuous, barrierless growth is then indicated.

### 5.4.5. The Gouy–Chapman–Stern Model of Electrified Interfaces

A simple model for an electrode–electrolyte interface was introduced long ago by Gouy[44] and Chapman.[45] The model considers a fluid of point ions in a uniform dielectric medium, which builds up a diffuse layer of screening charge in front of a charged hard wall. We shall examine this model in some detail because it is the prototype of self-consistent theories of charged-particle fluids, self-consistency being imposed by the long-range nature of the Coulomb interactions. The best-known simple example of such theories is the Debye–Hückel theory for bulk electrolytes, which was actually preceded in time by the Gouy–Chapman development.

In the Gouy–Chapman theory, the approximation is made that the particle densities $n_\alpha(z)$ are simply related to the electric potential energy $\phi(z)$ by a Boltzmann-distribution expression,

$$n_\alpha(z) = n_\alpha^0 \exp[-Z_\alpha \phi(z)/k_B T] \qquad (5.4.13)$$

as for a system of noninteracting particles in an external potential; here $n_\alpha^0$ are the particle densities in the bulk electrolyte, where we have chosen $\phi(\infty) = 0$. Self-consistency is introduced by requiring that $\phi(z)$ be determined both by the charge on the wall and by the charge density $n_Q = \Sigma_\alpha Z_\alpha e n_\alpha(z)$ induced in the electrolyte, the whole being screened by the dielectric constant $\varepsilon$ of the uniform medium, which represents the solvent.

Thus $\phi(z)$ satisfies the Poisson equation

$$\frac{d^2\phi(z)}{dz^2} = -\frac{4\pi e}{\varepsilon}\, n_0(z)$$

$$= -\frac{4\pi e^2}{\varepsilon}\sum_\alpha Z_\alpha n_\alpha^0 \exp[-Z_\alpha\phi(z)/k_B T] \qquad (z > 0) \qquad (5.4.14)$$

with the boundary condition (following from Gauss' theorem)

$$q = -\frac{\varepsilon}{4\pi e}\left[\frac{d\phi(z)}{dz}\right]_{z=0} \qquad (5.4.15)$$

For a 1–1 electrolyte (where $|Z_\alpha| = 1$ and $n_1^0 = n_2^0 = n_0$, say), the solution of equation (5.4.14) is

$$\phi(z) = 4k_B T\,\mathrm{tgh}^{-1}(se^{-\kappa z}) \qquad (5.4.16)$$

where $s = \mathrm{tgh}[\phi(0)/4k_B T]$ and $\kappa = (8\pi n_0 e^2/\varepsilon k_B T)^{1/2}$ is the Debye–Hückel inverse screening length. This result corresponds to an approximately exponential decay of $\phi(z)$, as can be seen in the case where $s$ is small, $\phi(z) \simeq \phi(0)e^{-\kappa z}$.

The surface charge density on the wall is related to $\phi(0)$ by equation (5.4.15),

$$q = \left(\frac{2k_B T\varepsilon n_0}{\pi}\right)^{1/2}\sinh\frac{\phi(0)}{2k_B T} \qquad (5.4.17)$$

and hence the capacitance $C_d$ of the diffuse layer is given by

$$C_d = e\left(\frac{\partial q}{\partial\phi(0)}\right)_{composition} = \left(\frac{\varepsilon n_0 e^2}{2\pi k_B T}\right)^{1/2}\cosh\frac{\phi(0)}{2k_B T} \qquad (5.4.18)$$

This yields $C_d = (\varepsilon/4\pi)\kappa$ at the point of zero charge, followed by an initially parabolic rise with voltage. The surface excess of particles of either species can also be calculated:

$$n_\alpha = \int_0^\infty dz[n_\alpha(z) - n_\alpha^0]$$

$$= 2n_0 Z_\alpha \int_0^\infty dz\,\exp[-Z_\alpha\phi(z)/2k_B T]\frac{d\phi(z)}{dz}\left(\frac{32\pi k_B T n_0 e^2}{\varepsilon}\right)^{1/2}$$

$$= \left(\frac{\varepsilon k_B T n_0}{2\pi e^2}\right)^{1/2}\{\exp[-Z_\alpha\phi(0)/2k_B T] - 1\} \qquad (5.4.19)$$

the last step involving a change from an integral over $z$ to an integral over $\phi$. Finally the interfacial tension, relative to its value at the potential of zero charge, is easily evaluated by integration of equations (5.4.17) and (5.4.19) with the thermodynamic relations of Section 5.2.6:

$$\gamma = -2k_{\mathrm{B}}T\left(\frac{2k_{\mathrm{B}}T\varepsilon n_0}{\pi e^2}\right)^{1/2}\left[\cosh\left(\frac{\phi(0)}{2k_{\mathrm{B}}T}\right) - 1\right] \qquad (5.4.20)$$

In applying the results of the Gouy–Chapman diffuse-layer theory to an analysis of experimental data on electrode–electolyte interfaces, it has become customary, following Stern,[46] to recognize that in addition to the diffuse layer contribution $C_{\mathrm{d}}$, a second so-called "inner layer" contribution $C_{\mathrm{i}}$ is present. From the simplest viewpoint, this accounts for finite ionic sizes by allowing for a nonvanishing value of the distance of closest approach of the ions to the wall. The interface is then viewed as a series of two capacitors,

$$C = \left(\frac{1}{C_{\mathrm{i}}} + \frac{1}{C_{\mathrm{d}}}\right)^{-1} \qquad (5.4.21)$$

Much effort has been devoted to modeling the inner layer, the Gouy–Chapman theory often being used to isolate the corresponding capacitance on the assumption that equation (5.4.21) holds. We simply mention here some of the problems that have been discussed, with some recent references:

1. Models of the dipolar orientation contribution in the inner layer and effects of dielectric saturation (see, e.g., the article of Reeves in Bockris *et al.*[14] and the paper of Liu.[47]).
2. Preferential adsorption of ions and solvent molecules at the electrode, implying a multiplicity of Stern layers (see, e.g., the article of Habib and Bockris in Bockris *et al.*[14]).
3. Nonlocal effects in dielectric screening and spilling out of the electron distribution from the electrode (see, e.g., the work of Kornyshev and Vorotyntsev[48]).
4. Improved theories of the ionic screening cloud (see, e.g., papers by Blum and others[49] and Grimson and Rickayzen,[50] the latter especially in connection with the problem of solvation forces and the behavior of colloidal dispersions[51]).

# References

1. J. W. Cahn and J. E. Hilliard, *J. Chem. Phys.* **28**, 258 (1958).
2. C. A. Croxton, *Introduction to Liquid State Physics*, Wiley, New York (1975).

3. T. E. Faber, *An Introduction to the Theory of Liquid Metals*, Cambridge University Press (1973).
4. F. P. Buff and R. A. Lovett, in: *Simple Dense Fluids*, Academic Press, New York (1968).
5. G. J. Janz, *Molten Salts Handbook*, Academic Press, New York (1967).
6. P. A. Egelstaff and B. Widom, *J. Chem. Phys.* **53**, 2667 (1970).
7. B. Widom, *J. Chem. Phys.* **43**, 3892 (1965).
8. A. B. Bhatia and N. H. March, *J. Chem.. Phys.* **68**, 4651 (1978).
9. W. Kohn and A. Yaniv, *Phys. Rev. B* **20**, 4948 (1979).
10. E. Roman, G. Senatore, and M. P. Tosi, *J. Phys. Chem. Solids* **43**, 1093 (1982); K. Singwi and M. P. Tosi, *Phys. Rev. B* **23**, 1640 (1981).
11. D. Turnbull, *J. Chem. Phys.* **18**, 768 (1950); *J. Appl. Phys.* **21**, 1022 (1950); J. H. Perepezko, *Rapid Solidification Processing — Principles and Technologies* (M. Cohon, B. H. Kear, and R. Mehrabian, eds.), in press.
12. C. Herring, in: *The Physics of Powder Metallurgy*, McGraw-Hill, New York (1951).
13. D. P. Woodruff, *The Solid–Liquid Interface*, Cambridge University Press (1973).
14. J. O'M. Bockris, B. E. Conway, and E. Yeager (eds.), *Comprehensive Treatise of Electrochemistry*, Vol. 1, Plenum Press, New York (1980).
15. J. O'M. Bockris and A. K. N. Reddy, *Modern Electrochemistry*, Plenum Press, New York (1977).
16. The original paper of van der Waals, published in 1893, has been translated into English by J. S. Rowlinson, *J. Stat. Phys.* **20**, 197 (1979).
17. J. G. Kirkwood and F. Buff, *J. Chem. Phys.* **17**, 338 (1949).
18. D. G. Triezenberg and R. Zwanzig, *Phys. Rev. Lett.* **28**, 1183 (1972).
19. R. Evans, *Adv. Phys.* **28**, 143 (1979).
20. R. H. Fowler, *Proc. R. Soc. London, Ser. A* **159**, 229 (1937).
21. I. R. McDonald and K. S. C. Freeman, *Mol. Phys.* **26**, 529 (1973).
22. M. V. Berry, in: *Surface Science*, Vol. 1, International Atomic Energy Agency, Vienna (1975).
23. P. Hohenberg and W. Kohn, *Phys. Rev.* **136**, B864 (1964); N. D. Mermin, *Phys. Rev.* **137**, A1441 (1965); W. Kohn and L. J. Sham, *Phys. Rev.* **140**, A1133 (1965).
24. W. F. Saam and C. Ebner, *Phys. Rev. A* **15**, 2566 (1977).
25. A. J. M. Yang, P. D. Fleming, and J. H. Gibbs, *J. Chem.. Phys.* **64**, 3732 (1976).
26. A. B. Bhatia and N. H. March, *J. Chem. Phys.* **68**, 1999 (1978).
27. P. D. Fleming, A. J. M. Yang, and J. H. Gibbs, *J. Chem. Phys.* **65**, 7 (1976); A. B. Bhatia, N. H. March, and M. P. Tosi, *Phys. Chem. Liq.* **9**, 229 (1980); G. Senatore and M. P. Tosi, *Nuovo Cimento* **56B**, 169 (1980).
28. R. C. Brown and N. H. March, *J. Phys. C* **6**, L363 (1973).
29. C. F. von Weizsäcker, *Z. Phys.* **96**, 431 (1935); D. A. Kirznits, *Sov. Phys. JETP* **5**, 64 (1957).
30. N. D. Lang and W. Kohn, *Phys. Rev. B* **1**, 4555 (1970); *B* **3**, 1215 (1971); N. D. Lang, *Solid State Phys.* **28**, 225 (1973).
31. J. A. Appelbaum and D. R. Hamann, *Rev. Mod. Phys.* **48**, 479 (1976).
32. M. P. D'Evelyn and S. A. Rice, *Phys. Rev. Lett.* **47**, 1844 (1981); D. Sluis, M. P. D'Evelyn, and S. A. Rice, *J. Chem. Phys.* **78**, 1611 (1983).
33. J. P. Hansen and L. Verlet, *Phys. Rev.* **184**, 150 (1969).
34. D. K. Chaturvedi, G. Senatore, and M. P. Tosi, *Lett. Nuovo Cimento* **30**, 47 (1981); D. K. Chaturvedi, M. Rovere, G. Senatore, and M. P. Tosi, *Physica B* **111**, 11 (1981).
35. A. Ferraz and N. H. March, *Solid State Commun.* **36**, 977 (1980).
36. T. V. Ramakrishnan and M. Yussouff, *Solid State Commun.* **21**, 389 (1977); *Phys. Rev. B* **19**, 2775 (1979).
37. N. H. March and M. P. Tosi, *Phys. Chem. Liq.* **10**, 185 (1980); **11**, 79 and 89 (1981); M. Rovere, M. P. Tosi, and N. H. March, *Phys. Chem. Liq.* **12**, 177 (1982).

38. F. R. N. Nabarro, *Theory of Crystal Dislocations*, Clarendon Press, Oxford (1967); see also D. Kuhlmann-Wilsdorf, *Phys. Rev.* **140**, A1599 (1965).

39. S. F. Edwards and M. Warner, *Phil. Mag.* **A40**, 257 (1979).

40. A. D. J. Haymet and D. W. Oxtoby, *J. Chem. Phys.* **74**, 2559 (1981).

41. W. K. Burton, N. Cabrera, and F. C. Frank, *Philos. Trans. R. Soc. London, Ser. A* **243**, 299 (1951).

42. K. A. Jackson, *Liquid Metals and Solidification*, ASM, Cleveland (1959).

43. D. E. Temkin, *Crystallization Processes*, Consultants Bureau, New York (1966).

44. G. Gouy, *J. Chim. Phys.* **29**, 145 (1903); *J. Phys. (Paris)* **9**, 457 (1910).

45. D. L. Chapman, *Phil. Mag.* **25**, 475 (1913).

46. O. Stern, *Z. Elektrochem.* **30**, 508 (1924).

47. S. H. Liu, *Surf. Sci.* **101**, 49 (1980).

48. A. A. Kornyshev and M. A. Vorotyntsev, *Surf. Sci.* **101**, 23 (1980).

49. L. Blum, *J. Phys. Chem.* **81**, 136 (1977); S. Levine and C. W. Outhwaite, *J. Chem. Soc., Faraday Trans. 2* **74**, 1670 (1978); S. Levine, C. W. Outhwaite, and L. B. Bhuiyan, *J. Electroanal. Chem.* **123**, 105 (1981).

50. M. J. Grimson and G. Rickayzen, *Mol. Phys.* **42**, 767 (1981); **44**, 817 (1981); **45**, 221 (1982).

51. See Chapter 6 by G. Rickayzen, in this book.

# Solvation Forces and the Electric Double Layer

## G. Rickayzen

## 6.1. Introduction

In this chapter, some aspects of the role of liquid structure in determining the forces between solids immersed in liquids will be surveyed. This is a problem of particular importance for understanding the stability of colloids, especially sols. Sols comprise solid particles dispersed through a liquid medium. Although the particles are large compared with the liquid molecules, they are still microscopic and their surfaces play an important part in their behavior.

Seen from a distance these particles are usually neutral and the dominant long-range force between them is the van der Waals attraction. This force arises from the polarizability of the particles and of the medium in between. Because of this polarizability and of the long range of the interaction of the dipoles, the polarization of the whole system has normal modes whose frequencies depend on the separation of the particles. The zero-point energies of these oscillations contribute to the total energy of the system and since the zero-point energies are lower the closer the particles, this gives rise to an attraction between the particles. For separations $r$ large compared with their size (but not so large that retardation effects become important), the potential energy of interaction is of the form

$$V = - C/r^6 \qquad (6.1.1)$$

**G. Rickayzen** · The Physics Laboratory, University of Canterbury, Kent CT2 7NR, England.

where $C$ is a quantity which depends on the dynamic polarizabilities of the particles and of the liquid. On the other hand, for separations small compared with their size the particles can be treated as thick flat plates (see below). In this case the interaction energy is given by

$$V = -A/12\pi r^2 \tag{6.1.2}$$

where $A$ depends on the polarizabilities and is known as the Hamaker constant for the system. There is an extensive theory of Hamaker constants, which we shall not explore here.

Since this force is attractive, it will tend to bring the suspended particles together into a coagulate. However, many sols have very long lives and appear to be stable. Hence this attractive force must be opposed by a shorter-ranged repulsive force, which prevents coagulation. In electrolytes this force is believed to be an electrostatic force due to electric double layers surrounding the colloidal particles. The solid particles acquire a net charge, which the surrounding ions in the liquid will tend to neutralize. However, because of their finite size and kinetic energy there will be some separation of the net charge on the particle from the net charge in the fluid. This is the electric double layer. When two such layers approach each other, like charges in the liquid will come closest together and this will cause a repulsion that opposes coagulation. This chapter is concerned with the calculation of this repulsion according to the classical DLVO[1] theory and to recent modifications of this work.

Throughout this chapter we shall consider, for simplicity, only the forces between two thick parallel flat plates separated by a distance $h$. Derjaguin[2] has shown that the force between two large spherical particles of radius $R$ and separation $h$ ($\ll R$) can be obtained from that between two plates. In fact, if $V(h)$ is the potential between the plates, the force between the spheres is

$$F(h) = \pi R V(h) \tag{6.1.3}$$

Thus the problem of flat plates is of practical importance. One important consequence of the classical theory (dealt with below) is the existence of a critical concentration of electrolyte, $\rho_{0c}$, below which the colloid is stable. In the theory $\rho_{0c}$ satisfies the Schulze–Hardy rule[3]

$$\rho_{0c} \propto T^5/Z^6 \tag{6.1.4}$$

where $T$ is the temperature and $Z$ the valency of the ions. This rule appears to be in reasonable agreement with experiment, although it has not been tested very severely.

## 6.2. Gouy–Chapman Theory of the Electric Double Layer

For simplicity we consider the case of equally charged surfaces (Figure 6.1). In the theory of Gouy and Chapman (GC),[4] one treats the solution as a medium of dielectric constant $\varepsilon_r$ in which the ions are situated. If there are different kinds of ion of charge $e_i$ and number density $\rho_i(x)$ at position $x$, there is an electrostatic potential $V(x)$ that satisfies the Poisson equation

$$\nabla^2 V(x) = -\sum_i e_i \rho_i(x)/\varepsilon_0 \varepsilon_r \qquad (6.2.1)$$

Now at low density the ions can be treated as independent and in a potential $V(x)$ each ion has energy $e_i V(x)$. The number of ions at a particular point is then given by the Boltzmann distribution

$$\rho_i(x) = \rho_{0i} \exp[-\beta e_i V(x)] \qquad (6.2.2)$$

where $\rho_{0i}$ is the number density in the bulk $[V(x)=0]$ of species $i$.

For a given surface-charge density $Q$ on each plate the potential is fixed by the condition of overall charge neutrality

$$\sum_i e_i \int \rho_i(x)\,dx = -2Q \qquad (6.2.3)$$

Given the densities $\rho_{0i}$ equations (6.2.1) to (6.2.3) are sufficient to determine $V(x)$ completely. Since the bulk is neutral these densities satisfy

$$\sum_i e_i \rho_{0i} = 0 \qquad (6.2.4)$$

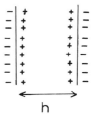

Figure 6.1. Scheme of the charges on two plates and in the liquid.

We consider two special cases*:

*Case 1.* $\beta e_i V(x) \ll 1$, *all i*
This is the dilute limit and leads to Debye–Hückel theory.[5]
If equation (6.2.2) is used in equation (6.2.1) we have

$$\frac{d^2 V(x)}{dx^2} = -\sum_i \frac{e_i \rho_{0i} \exp[-\beta e_i V(x)]}{\varepsilon_0 \varepsilon_r} \tag{6.2.5}$$

Due to the assumption of a weak potential, the exponents can be expanded and we retain only first-order terms in $V(x)$. The zeroth-order terms vanish because of equation (6.2.4). Thus

$$\frac{d^2 V(x)}{dx^2} = \sum_i \frac{e_i^2 \rho_{0i} \beta V(x)}{\varepsilon_0 \varepsilon_r} = \frac{V(x)}{\lambda^2} \tag{6.2.6}$$

where $\lambda$ is the Debye length defined by

$$\lambda^{-2} \equiv \kappa^2 = \beta \sum_i e_i^2 \rho_{0i} / \varepsilon_0 \varepsilon_r \tag{6.2.7}$$

If the plates are placed at $x = \pm h/2$ so that their separation is $h$, we require the solution of (6.2.6) that is symmetric about the origin, namely

$$V(x) = A \cosh(x/\lambda) \tag{6.2.8}$$

The field at any point in the fluid is given by

$$E(x) = -\frac{dV}{dz} = -\frac{A}{\lambda} \sinh\left(\frac{x}{\lambda}\right) \tag{6.2.9}$$

However, just outside the left-hand plate the field is

$$E(-h/2) = \frac{Q}{\varepsilon_0 \varepsilon_r} = \frac{A}{\lambda} \sinh\left(\frac{h}{2\lambda}\right) \tag{6.2.10}$$

whence

$$A = \frac{\lambda Q}{\varepsilon_0 \varepsilon_r \sinh(h/2\lambda)} \tag{6.2.11}$$

$$V(x) = \frac{\lambda Q \cosh(x/\lambda)}{\varepsilon_0 \varepsilon_r \sinh(h/2\lambda)} \tag{6.2.12}$$

This solution automatically conserves charge.

* The case of 1–1 electrolytes is discussed rather fully in Chapter 5.

The Debye length is the parameter which characterizes the distance over which the potential (and hence the ion densities) tend to vary.

The energy in the system $\Omega$ is the energy required for the plates to acquire charge $Q$ from zero and is given by

$$\Omega(Q) = 2 \int_0^O V(Q')dQ' \qquad (6.2.13)$$

where $V(Q')$ is the potential of a plate when the charge is $Q'$. Since $V(Q)$ is proportional to $Q$, this becomes

$$\Omega(Q) = V(Q)Q = (\lambda Q^2/\varepsilon_0\varepsilon_r)\coth(h/2\lambda)$$

which is the potential energy from which the force can be obtained at constant charge on the plates. For the more common case of constant potential on the plates, one must use the function $\mathscr{F}(V_0)$, related to $\Omega(Q)$ by a Legendre transformation,[6]

$$\mathscr{F}(V_0) = \Omega(Q) - 2V_0Q = -(\varepsilon_0\varepsilon_r V_0^2/\lambda)\tanh(h/2\lambda) \qquad (6.2.14)$$

where $V_0$ is the potential on a plate such that

$$V_0 = (\lambda Q/\varepsilon_0\varepsilon_r)\coth(h/2\lambda)$$

Measured relative to the potential energy when the separation is infinite,

$$\mathscr{F}(V_0) = \frac{\varepsilon_0\varepsilon_r V_0^2}{\lambda}\left[1 - \tanh\left(\frac{h}{2\lambda}\right)\right] \qquad (6.2.15)$$

This is positive, so the plates tend to repel each other — not surprising, as we are trying to bring like charges close together. The force of repulsion when $V_0$ is held fixed is

$$f_V(h) = -\frac{\partial\mathscr{F}}{\partial h} = \frac{\varepsilon_0\varepsilon_r V_0^2}{2\lambda^2}\operatorname{sech}^2\left(\frac{h}{2\lambda}\right) \qquad (6.2.16)$$

As $h \to \infty$ this becomes $(2\varepsilon_0\varepsilon_r V_0^2/\lambda^2)\exp(-h/\lambda)$ and decays exponentially with decay length $\lambda$.

A useful estimate of the size of $\lambda$ is that, when $T = 298$ K, $\varepsilon_r = 78$, and we have a 1–1 electrolyte of 0.5 M, then $\lambda = 4$ Å. Note that $\lambda \propto \rho^{-1/2}$. For $\lambda \sim 4$ Å, one would certainly expect to take account of solvent structure. We discuss this case further after considering the Gouy–Chapman theory.

*Case 2. z–z electrolyte*

In this case there are two ions with $e_1 = -e_2 = Ze$, where $e$ is the charge on an electron and $\rho_{01} = \rho_{02} = \rho_0/2$. Equations (6.2.1) and (6.2.2) then yield

$$\frac{d^2V}{dx^2} = -\frac{Ze\rho_0}{2\varepsilon_0\varepsilon_r}\left[\exp(-\beta ZeV) - \exp(\beta ZeV)\right]$$

$$= \frac{Ze\rho_0}{\varepsilon_0\varepsilon_r}\sinh(\beta ZeV) \tag{6.2.17}$$

Hence

$$\frac{1}{2}\frac{d}{dV}\left(\frac{dV}{dx}\right)^2 = \frac{Ze\rho_0}{\varepsilon_0\varepsilon_r}\sinh(\beta ZeV)$$

which can be integrated to yield

$$\left(\frac{dV}{dx}\right)^2 = \frac{2\rho_0}{\beta\varepsilon_0\varepsilon_r}\left[\cosh(\beta ZeV) - \cosh(\beta ZeV_m)\right] \tag{6.2.18}$$

where $V = V_m$ when $dV/dx = 0$, at the midpoint between the plates. Equation (6.2.18) can be integrated further in terms of elliptic integrals. This is useful for computational purposes, but it does not alter the qualitative results.

For infinite separation of the plates, $V_m$ is zero and equation (6.2.18) can be integrated using elementary functions. In fact the equation can then be written as

$$\left(\frac{dV}{dx}\right)^2 = \frac{4\rho_0}{\beta\varepsilon_0\varepsilon_r}\sinh^2\left(\frac{\beta ZeV}{2}\right) \tag{6.2.19}$$

Hence

$$\frac{dV}{dx} = \pm\sqrt{\frac{4\rho_0}{\beta\varepsilon_0\varepsilon_r}}\sinh\left(\frac{\beta ZeV}{2}\right) \tag{6.2.20}$$

If we choose the origin so that the plate is at $x = 0$, the integral of this equation is given by

$$\tanh\left(\frac{\beta ZeV}{4}\right) = \tan\left(\frac{\beta ZeV_0}{4}\right)e^{-\kappa x} \tag{6.2.21}$$

where $V_0$ is the potential on the plate. When $V_0$ is so small that

$$\beta Ze V_0 \ll 1$$

this reduces to the Debye–Hückel result,

$$V = V_0 \exp(-\kappa x) \tag{6.2.22}$$

## 6.3. Conditions for Coagulation

The interaction between two plates now comprises two parts: a repulsion, which arises from the Coulomb interaction of the charges, which decays exponentially at large separations, and tends to a finite limit as the plates approach; and a van der Waals attraction, which depends on the separation according to an inverse power law. The different contributions to the potential energy of the plates are shown schematically in Figure 6.2. At both large and very close separations the van der Waals forces are dominant. Between them the repulsion is comparable, and the total potential rises to a maximum $W_m$. If $W_m < 0$ there is no impediment to the plates coming together and coagulation will take place rapidly. On the other hand, if $W_m > 0$ there is a macroscopic barrier to be overcome and coagulation will be slowed down. Thus a rough criterion for stability is $W_m > 0$. Hence there is a critical ion concentration $\rho_{0c}$ given by

$$\frac{dW(h)}{dh} = 0, \qquad W(h) = 0 \tag{6.3.1}$$

at which the colloid becomes unstable. For $\rho_0 < \rho_c$ the colloid is stable, while for $\rho_0 > \rho_c$ it is unstable.

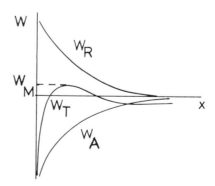

Figure 6.2. Schematic plot of the potential energy between two plates as a function of their separation. The curve labeled $W_A$ represents the van der Waals attraction, $W_R$ the repulsion due to the double layers, and $W_T$ is their sum.

The total potential energy between the walls, including the van der Waals contribution, is $(Z\beta e V_0 \leq 1)$

$$W(h) = \frac{\varepsilon_0 \varepsilon_r V_0^2}{\lambda} \left[ 1 - \tan\left(\frac{h}{2\lambda}\right) \right] - \frac{A}{12\pi h^2} \qquad (6.3.2)$$

where $A$ is the Hamaker constant. For $\exp(h/\lambda) \gg 1$ this has the form

$$W(h) = \frac{2\varepsilon_0 \varepsilon_r V_0^2}{\lambda} e^{-h/\lambda} - \frac{A}{12\pi h^2} \qquad (6.3.3)$$

From equation (6.3.1) this leads to the conditions

$$h/\lambda = 2 \qquad (6.3.4)$$

and

$$2\varepsilon_0 \varepsilon_r V_0^2 e^{-2}/\lambda = A/12\pi h^2 \qquad (6.3.5)$$

Since $\lambda$ depends on $\rho_0$ this is an equation for the critical concentration $\rho_c$. It has been derived assuming that $e^{h/\lambda} \gg 1$ and $Z\beta e V_0/2 \ll 1$ (i.e., $V_0 \ll 100/z$ mV). Neither of these approximations is entirely satisfactory. Improving on the first changes the value of $h/\lambda$ slightly and changes the numbers in equation (6.3.5), but it makes little qualitative difference. However, it seems that in most practical cases $Z\beta e V_0/2 \gg 1$ and the second approximation must be improved. In turn, this means that the Gouy–Chapman equation (6.2.18) must be solved. The result is that $Z\beta e V_0/4$ is replaced by $\tanh Z\beta e V_0/4 \sim 1$.

The new condition becomes, for a $z-z$ electrolyte,

$$\rho_c = \frac{(192)^2 (4\pi\varepsilon_0\varepsilon_r)^3 (kT)^5 \exp(-4)}{\pi z^6 e^6 A^2}$$

For water at 25 °C

$$\rho_c = 17.30 \times 10^{-39}/Z^6 A^6 \qquad (6.3.6)$$

a result known as the Schulze–Hardy rule.[3] Some data are given in Table 6.1. Agreement with experiment here is good, but the value of $A$ is overestimated.

## 6.3.1. Criticisms

Four unjustified assumptions of the theory are:

1. Additivity of electrostatic and van der Waals forces.

**Table 6.1. Critical Coagulation Concentration Values in Millimoles per Liter (after Verwey and Overbeek,[7] with Permission)**

| $Z$ for counter-ions | Sol of $As_2S_2$ (negatively charged) | | Sol of Au (negatively charged) | | Sol of $Fe_2O_3$ (positively charged) | | Sol of $Al_2O_3$ (positively charged) | |
|---|---|---|---|---|---|---|---|---|
| 1 | LiCl | 58 | | | | | | |
| | NaCl | 51 | NaCl | 24 | NaCl | 9.25 | NaCl | 43.5 |
| | KCl | 49.5 | | | KCl | 9.0 | KCl | 46 |
| | | | $KNO_3$ | 25 | $KNO_3$ | 12 | $KNO_3$ | 60 |
| 2 | $MgCl_2$ | 0.72 | | | | | | |
| | $MgSO_4$ | 0.81 | | | $MgSO_4$ | 0.22 | | |
| | $BaCl_2$ | 0.69 | $BaCl_2$ | 0.35 | | | | |
| | $CaCl_2$ | 0.65 | $CaCl_2$ | 0.41 | | | | |
| | | | | | $K_2SO_4$ | 0.205 | $K_2SO_4$ | 0.30 |
| | | | | | $K_2Cr_2O_7$ | 0.195 | $K_2Cr_2O_7$ | 0.63 |
| 3 | $AlCl_3$ | 0.093 | | | | | | |
| | $\frac{1}{2}Al_2(SO_4)_3$ | 0.096 | $\frac{1}{2}Al_2(SO_4)_3$ | 0.009 | | | | |
| | | | | | | | $K_2Fe(CN)_6$ | 0.080 |

2. The potential in which an ion sits is assumed to be the average potential.
3. Uniform solvent, unperturbed by colloid, with a uniform dielectric constant.
4. Point charges.

One would expect (3) and (4) to be unsatisfactory when $\lambda \sim \sigma$, a molecular diameter. For water with $\sigma \sim 4$ Å this happens at a concentration of 0.5 M, which is high for a colloid, but not for all applications of the double layer. Further, the theory leads to unrealistic values for the concentration of ions near the surface when $V_0 \gtrsim 100$ mV, which is the surface potential assumed in deriving the Schulze–Hardy rule. The results are symmetric with repect to the sign of the potential, but this is not true in practice. Finally, the rule and the law of force have not yet been subject to very stringent tests. For all these reasons attempts have been made to take some account of liquid structure. There are also other reasons connected with understanding the dynamic properties of the system that require some extension of the theory as well.

### 6.3.2. Stern Layer

This is a phenomenological attempt to introduce liquid structure, especially to improve our understanding of a single double layer. It is simply assumed that the ions can only approach to within some distance $\sigma_0$ of the wall, known as the Outer Helmholtz Plane. Then one must solve the electrostatic problem separately in the inner region and the outer one, which is again treated as a diffuse layer. The two solutions are then joined. Models of the inner layer can be quite sophisticated and have led to a greater understanding of double-layer properties. They are often expressed in terms of the capacity $C$ of the double layer, defined by

$$C = dQ/dV_0 \qquad\qquad (6.3.7)$$

Then, for a simple model with the diffuse layer treated by Debye–Hückel theory (i.e., dilute electrolyte) and a charge-free inner layer with dielectric constant $\varepsilon_s$,

$$C^{-1} = \frac{\lambda}{\varepsilon_0 \varepsilon_r} + \frac{\sigma_s}{\varepsilon_0 \varepsilon_s} \qquad\qquad (6.3.8)$$

the sum of capacitors in series. The theory involves the new constant $\sigma_s/\varepsilon_s$.

However, the split of a liquid into two regions is completely artificial and, in any case, when several molecules of different diameters are present, as is usually the case, it is not at all clear what values to use for $\sigma_s$ and $\varepsilon_s$. Modern understanding of bulk liquid structure has therefore led to attempts to apply that knowledge to the behavior of liquids near walls and, most recently, to the double layer. The main methods used are computer-simulation and integral-equation methods. Given sufficient time and money the former will presumably be able to solve all problems. However, we are a long way from achieving this for the double layer because of the enormous number of particles one would have to include in the simulation to get reasonable statistics. In fact, present computer simulations cope with about 1024 particles. However, even in a concentrated electrolyte of about 0.5 M only about 1 in 30 molecules is an ion. Therefore, to get reasonable statistics for the electric potential one would need many more particles. For dilute electrolytes the situation is worse, although this is the area where analytic theories are probably best. Thus analytic approaches are useful. Although the approximations that are made are usually uncontrolled, they can usefully be monitored by computer simulation for cases where both are applicable. This is then a subject where computer simulation and analytic techniques supplement each other.

Several different techniques have been applied to these problems, most notably the BYG equations,[8] generalizations of the Boltzmann equation,[9] and density functional[10] or equivalent methods. All have strengths and weaknesses. The density functional method will be examined below. The initial developments were undertaken for neutral fluids, but since some of the effects are important in charged fluids as well and it is in any case useful pedagogically, the case of neutral fluids will be treated first.

## 6.4. Density-Functional Approach to Neutral Fluids*

We start from the fact that if a fluid has an average local density $\rho(\mathbf{r})$ imposed on it, there exists a thermodynamic potential $\Omega[\rho]$, which depends on $\rho(\mathbf{r})$ and which is a minimum in the equilibrium state. This was a result proved mathematically by Mermin,[11] but it makes good physical sense. In general, $\Omega$ for a fluid in an external potential $U(\mathbf{r})$ is defined by[12]

$$\exp(-\beta\Omega) = \sum_N \frac{z^N}{N!} \int \exp\left[-\beta V(\mathbf{r}_1, \mathbf{r}_2, \ldots, \mathbf{r}_N) - \beta \sum_i U(\mathbf{r}_i)\right] d^3r_1 \cdots d^3r_N$$

(6.4.1)

where $V(\mathbf{r}_1, \ldots, \mathbf{r}_N)$ is the interparticle potential, and $z$ depends on the chemical potential and originates from the kinetic degrees of freedom. The average local density is

$$\rho(\mathbf{r}) = e^{\beta\Omega} \sum_N \frac{z^N}{N!} \int \exp\left[-\beta V - \beta \sum_i U(\mathbf{r}_i)\right] \sum_j \delta(\mathbf{r} - \mathbf{r}_j) d^3r_1 \cdots d^3r_N$$

(6.4.2)

However, if we change $U(\mathbf{r})$ by a small amount $\delta U(\mathbf{r})$, $\Omega$ will change by a small amount $\delta\Omega$ such that

$$-\beta\delta\Omega = -\beta e^{\beta\Omega} \sum_N \frac{z^N}{N!} \int \exp(-\beta F) \sum_i \delta U(\mathbf{r}_i) d^3r_1 \cdots d^3r_N$$

$$= -\beta e^{\beta\Omega} \sum_N \frac{z^N}{N!} \int \exp(-\beta F) \sum_i \delta(\mathbf{r} - \mathbf{r}_i) \delta U(\mathbf{r}) d^3r\, d^3r_1 \cdots d^3r_N$$

$$= -\beta \int \rho(\mathbf{r}) \delta U(\mathbf{r}) d^3r$$

(6.4.3)

where

$$F = V(\mathbf{r}_1, \ldots, \mathbf{r}_N) + \sum_i U(\mathbf{r}_i)$$

Hence $\Omega[\rho]$ will have the form

$$\Omega[\rho] = \Omega_0[\rho] + \int d^3r \rho(\mathbf{r}) U(\mathbf{r})$$

(6.4.4)

* See also the related discussion in Chapter 5.

where $\Omega_0$ is independent of $U(\mathbf{r})$; $\Omega$ and $\Omega_0$ depend on $\rho(\mathbf{r})$ at all points of the fluid and are called functionals of $\rho$.

If we know $\Omega_0[\rho]$, then we have a well-defined problem for a given potential $U(\mathbf{r})$ such as that due to walls, and $\rho(\mathbf{r})$ must then be obtained by minimizing (6.4.4), a problem in the calculus of variations. However, we do not know $\Omega_0[\rho]$ and a first step in the application of the method is to make an educated guess for $\Omega_0[\rho]$.

In some circumstances, one can make reliable guesses as to the form of $\Omega_0[\rho]$. For example, this is the case when $\rho$ is believed to be varying slowly in space. Then $\Omega_0[\rho]$ can be expanded in powers of the spatial derivatives of $\rho$. This approach has been used successfully in studies of the liquid–vapor interface.[13] However, near walls one expects short-range order with layers of atoms nearby and a corresponding oscillation in $\rho(\mathbf{r})$ with period equal to a molecular diameter, a behavior similar to $g(\mathbf{r})$ for a bulk liquid. Thus a gradient expansion does not seem to be appropriate.

### 6.4.1. Linear Theory

If the potential is weak then a perturbation theory is applicable. In fact, the discontinuous effect of a wall is not weak. Nevertheless, this approach is helpful qualitatively and not bad quantitatively. Let us suppose that when $U$ is zero the equilibrium density is uniformly $\rho_0$. A weak potential $U(\mathbf{r})$ will change this to $\rho_0 + \delta\rho(\mathbf{r})$. Since $\rho_0$ minimizes $\Omega[\rho]$, an expansion of $\Omega_0$ in powers of $\delta\rho(\mathbf{r})$ will contain no linear terms. To second order we have

$$\Omega[\rho] - \Omega[\rho_0] = A \int d^3r \delta^2\rho(\mathbf{r}) + \frac{1}{2} \int d^3r d^3r' \delta\rho(\mathbf{r})K(\mathbf{r},\mathbf{r}')\delta\rho(\mathbf{r}')$$

$$+ \int d^3r[\rho_0 + \delta\rho(\mathbf{r})]U(\mathbf{r}) \tag{6.4.5}$$

If we minimize with respect to $\delta\rho(\mathbf{r})$ we obtain the equation

$$A\delta\rho(\mathbf{r}) + \int d^3r' K(\mathbf{r},\mathbf{r}')\delta\rho(\mathbf{r}') + U(\mathbf{r}) = 0 \tag{6.4.6}$$

Now $A$ and $K(\mathbf{r},\mathbf{r}')$ depend on $\rho_0$ and so are properties of the bulk fluid. In fact, it is shown in standard fluid theory (e.g., on p. 98 of Hansen and MacDonald[12]), from equation (6.4.2), that

$$\delta\rho(\mathbf{r}) = -\beta\rho_0 U(\mathbf{r}) - \beta\rho_0^2 \int h(\mathbf{r},\mathbf{r}')U(\mathbf{r}')d^3r' \tag{6.4.7}$$

where $h(\mathbf{r}, \mathbf{r}')$ is the indirect correlation function related to the pair distribution function $g(\mathbf{r}, \mathbf{r}')$ by

$$g(\mathbf{r}, \mathbf{r}') = 1 + h(\mathbf{r}, \mathbf{r}') \tag{6.4.8}$$

To compare equations (6.4.6) and (6.4.7) we need to invert, and this can be done using the Ornstein–Zernike (OZ) equation

$$h(\mathbf{r}, \mathbf{r}') = c(\mathbf{r}, \mathbf{r}') + \int d^3r'' c(\mathbf{r}, \mathbf{r}'')\rho_0 h(\mathbf{r}'', \mathbf{r}') \tag{6.4.9}$$

where this equation provides one definition of the direct correlation function $c(\mathbf{r}, \mathbf{r}')$. With this function equation (6.4.7) can be rewritten in the form

$$\frac{\delta\rho(\mathbf{r})}{\beta\rho_0} - \frac{1}{\beta}\int c(\mathbf{r}, \mathbf{r}')\delta\rho(\mathbf{r}')d^3r' + U(\mathbf{r}) = 0 \tag{6.4.10}$$

Comparison of equations (6.4.6) and (6.4.10) yields

$$A = (\beta\rho_0)^{-1}, \qquad K(\mathbf{r}, \mathbf{r}') = -\beta^{-1}c(\mathbf{r}, \mathbf{r}') \tag{6.4.11}$$

Hence the approximate form of $\Omega$ is

$$\Omega[\rho] - \Omega[\rho_0] = (2\beta\rho_0)^{-1}\int d^3r\,\delta^2\rho(\mathbf{r}) - (2\beta)^{-1}\int d^3r\,d^3r'\,\delta\rho(\mathbf{r})c(\mathbf{r}, \mathbf{r}')\delta\rho(\mathbf{r}')$$

$$+ \int d^3r[\rho_0 + \delta\rho(\mathbf{r})]U(\mathbf{r}) \tag{6.4.12}$$

If the direct correlation function for the bulk liquid is then known exactly or approximately, this functional can be used to determine $\delta\rho$ in an inhomogeneous fluid. For a fluid trapped between two walls at $x = 0$ and $x = h$, we set $\rho(\mathbf{r}) = 0$ for $x < 0$ and $x > h$. Hence

$$\delta\rho(\mathbf{r}) = -\rho_0 \qquad \text{for } x < 0 \text{ and } x > h \tag{6.4.13}$$

For the density between the walls, we find that

$$\frac{\rho(x) - \rho_0}{\beta\rho_0} - \frac{1}{\beta}\int_0^h dx \int dy'dz'\,c(\mathbf{r}, \mathbf{r}')\rho(x') + \frac{\rho_0}{\beta}\int dv'\,c(\mathbf{r}, \mathbf{r}') + U(x) = 0 \tag{6.4.14}$$

where $U(x)$ is the potential (which may be zero) acting on the liquid between the walls. In some special cases, this equation can be solved analytically but generally it must be solved by computer.

Equation (6.4.14) is a linear theory for $\rho(x)$ and is in that sense analogous to Debye–Hückel theory. In fact, as we shall see when ions are introduced and their sizes allowed to tend to zero, it reduces to Debye–Hückel theory. It is the equation obtained when one applies the Percus–Yevick approximation[12] to a bulk liquid comprising two species of different radius and density. If one allows the radius of the second species to tend to infinity while letting its density tend to zero, the pair distribution function of the first liquid becomes $\delta\rho(\mathbf{r})$ and satisfies equation (6.4.14) for the case of a single wall. Therefore this approach is also called the P–Y approximation.

The force on the plates at separation $h$ is given by

$$f(h) = -\frac{\partial\Omega}{\partial h} \qquad (6.4.15)$$

From equations (6.4.12) and (6.4.14), one finds that (if $U = 0$) this force is

$$f(h) = \frac{\rho_s^2(h)}{2\rho_0\beta} \qquad (6.4.16)$$

where $\rho_s(h)$ is the liquid density at one plate when the separation is $h$. This, however, gives the force due to the liquid between the plates. There is also a force due to the liquid on the other side of the plates. Since there is no plate facing this side, the situation corresponds to infinite separation. Hence the net force is

$$F(h) = \frac{\rho_s(h)^2 - \rho_s(\infty)^2}{2\rho_0\beta} \qquad (6.4.17)$$

We discuss the results obtained from this theory after examining a second approximate density functional.

### 6.4.2. Nonlinear Theory

One obvious defect of the previous approximation for the density functional is that, if the fluid comprises independent particles, it does not lead to the Bolzmann factor for the dependence of density on $U(\mathbf{r})$,

$$\rho(\mathbf{r}) = \rho_0 \exp[-\beta U(\mathbf{r})] \qquad (6.4.18)$$

Let us see what form of $\Omega$ does lead to this result.

To make $\Omega[\rho]$ stationary note that if $\rho(\mathbf{r})$ increases by $\delta\rho$, $\Omega$ increases by $\delta\Omega$, and $\Omega_0$ by $\delta\Omega_0$ where $\delta\Omega_0$ depends linearly on $\delta\rho$ and can be written as

$$\delta\Omega_0 = \int d^3r \frac{\delta\Omega_0}{\delta\rho(\mathbf{r})} \delta\rho(\mathbf{r}) \tag{6.4.19}$$

Here $\delta\Omega_0/\delta\rho$ is defined by this equation and written in this suggestive form by analogy with the behavior of functions of several variables. Thus for $\Omega$ to be a minimum $\rho(\mathbf{r})$ must satisfy

$$\int d^3r \left[ \frac{\delta\Omega_0}{\delta\rho(\mathbf{r})} + U(\mathbf{r}) \right] \delta\rho(\mathbf{r}) = 0$$

for all functions $\delta\rho(\mathbf{r})$. This requires that

$$\frac{\delta\Omega_0}{\delta\rho(\mathbf{r})} + U(\mathbf{r}) = 0 \tag{6.4.20}$$

However, for the solution (6.4.18),

$$\frac{1}{\beta} \ln \frac{\rho(\mathbf{r})}{\rho_0} + U(\mathbf{r}) = 0 \tag{6.4.21}$$

and for the density functional to yield this equation we must have

$$\frac{\delta\Omega_0}{\delta\rho(\mathbf{r})} = \frac{1}{\beta} \ln \frac{\rho(\mathbf{r})}{\rho_0} \tag{6.4.22}$$

As $\rho$ at only one point is involved, this equation can be integrated to yield, for independent particles,

$$\Omega_{\mathrm{I_0}}[\rho] = \frac{1}{\beta} \int \rho(\mathbf{r}) \left\{ \ln \left[ \frac{\rho(\mathbf{r})}{\rho_0} \right] - 1 \right\} d^3r \tag{6.4.23}$$

Relationship (6.4.23) can also be obtained directly from equations (6.4.1) and (6.4.2). This contribution is evidently the entropy of the fluid. In the absence of interaction we want $\Omega_0[\rho]$ to reduce to $\Omega_{\mathrm{I_0}}[\rho]$.

We can relate the difference to the inhomogeneous direct correlation function as follows. Suppose we change $U(\mathbf{r})$ by a small amount. Then $\delta\rho(\mathbf{r})$ changes by a small amount and the two changes are corrected by the change in equation (6.4.20). Thus

$$\int \frac{\delta}{\delta\rho(\mathbf{r}')} \left[ \frac{\delta\Omega_0}{\delta\rho(\mathbf{r})} \right] \delta\rho(\mathbf{r}')d^3r' + \delta U(\mathbf{r}) = 0 \tag{6.4.24}$$

But from the general theory we have

$$\delta\rho(\mathbf{r}) = -\beta\rho_0(\mathbf{r})\delta U(\mathbf{r}) - \beta\rho(\mathbf{r})\int h(\mathbf{r},\mathbf{r}')\rho(\mathbf{r}')\delta U(\mathbf{r}')d^3r' \quad (6.4.25)$$

This differs from equation (6.4.7) because we are now dealing with a small change in an inhomogeneous fluid. Nevertheless, we have a generalized OZ equation

$$h(\mathbf{r},\mathbf{r}') = c(\mathbf{r},\mathbf{r}') + \int d^3r'' c(\mathbf{r},\mathbf{r}'')\rho(\mathbf{r}'')h(\mathbf{r}'',\mathbf{r}') \quad (6.4.26)$$

and this can be used to invert equation (6.4.25) to yield

$$\frac{\delta\rho(\mathbf{r})}{\beta\rho(\mathbf{r})} - \frac{1}{\beta}\int d^3r' c(\mathbf{r},\mathbf{r}';\rho)\delta\rho(\mathbf{r}') + \delta U(\mathbf{r}) = 0 \quad (6.4.27)$$

where $C(\mathbf{r},\mathbf{r}',\rho)$ now depends on $\rho$ at all points. Comparison of equations (6.4.24) and (6.4.27) shows that

$$\frac{\delta^2\Omega_0}{\delta\rho(\mathbf{r}')\delta\rho(\mathbf{r})} = \frac{\delta(\mathbf{r}-\mathbf{r}')}{\beta\rho(\mathbf{r})} - \frac{1}{\beta}c(\mathbf{r},\mathbf{r}';\rho) \quad (6.4.28)$$

The first term comes from $\Omega_{I_0}$, and the second term results from the effect of interactions between the particles.

To find $\Omega_0$, we must now "integrate" with respect to $\rho$. Unfortunately, $c$ is a completely unknown function for general $\rho$ and this cannot be done. Formally we have not advanced at all. However, the direct correlation function is a short-range function and it seems that comparatively simple approximations for it yield reasonable results. In the remainder of this chapter we shall make the drastic approximation that we can replace this function by the direct correlation function of the homogeneous fluid. By comparison with machine calculations where possible, we shall be able to judge how well it works. Thus we use

$$c(\mathbf{r},\mathbf{r}';\rho) \approx c(\mathbf{r},\mathbf{r}',\rho_0) \equiv c(\mathbf{r},\mathbf{r}') \quad (6.4.29)$$

In this case equation (6.4.28) can be integrated to yield

$$\frac{\delta\Omega_0}{\delta\rho(\mathbf{r})} = \frac{1}{\beta}\ln\frac{\rho(\mathbf{r})}{\rho_0} - \frac{1}{\beta}\int d^3r' c(\mathbf{r},\mathbf{r}')[\rho(\mathbf{r}') - \rho_0] \quad (6.4.30)$$

and

$$\Omega_0 = \frac{1}{\beta} \int d^3r \rho(\mathbf{r}) \left\{ \ln \left[ \frac{\rho(\mathbf{r})}{\rho_0} \right] - 1 \right\}$$

$$- \frac{1}{2\beta} \int d^3r d^3r' [\rho(\mathbf{r}) - \rho_0] c(\mathbf{r}, \mathbf{r}') [\rho(\mathbf{r})' - \rho_0] \tag{6.4.31}$$

Equation (6.4.30) is connected with the HNC approximation in the same way as the linear theory is connected with the P–Y approximation. The approximation (6.4.29) is not the only one to have been used. However, the others are much more complex and have not yet shown themselves to be sufficiently superior to justify the complexity. From equations (6.4.15), (6.4.30), and (6.4.31) we find that, in this approximation, the force between two plates immersed in a hard-sphere fluid is

$$F(h) = [\rho_s(h) - \rho_s(\infty)]/\beta \tag{6.4.32}$$

This can be shown to be the exact result if $\rho_s(h)$ is exact.[14] For weak perturbations, where $\rho_s(h) - \rho_0$ is small, equations (6.4.17) and (6.4.32) agree. The first term of equation (6.4.31) is the entropy of the fluid while the second is the interaction energy.

For the neutral hard-sphere fluid the bulk dcf can be obtained in analytic form in the P–Y approximation.* The result is

$$c_{hs}(r) = \begin{cases} -\lambda_1 - 6\eta\lambda_2 \left( \dfrac{r}{\sigma} \right) - \dfrac{1}{2}\eta\lambda_1 \left( \dfrac{r}{\sigma} \right)^3, & r < \sigma \\ 0, & r > \sigma \end{cases} \tag{6.4.33}$$

where

$$\lambda_1 = \frac{(1 + 2\eta)^2}{(1 - \eta)^4}, \qquad \lambda_2 = \frac{(1 + \frac{1}{2}\eta)^2}{(1 - \eta)^4}, \qquad \eta = \frac{\pi\rho_0\sigma^3}{6} \tag{6.4.34}$$

### 6.4.3. Results for Neutral Fluids

Figures 6.3 to 6.8 present a small sample of results for neutral fluids obtained by a variety of methods. Figure 6.3 shows the density profile for a hard-sphere fluid contained between hard walls obtained by Monte-Carlo simulation.[15] The characteristic oscillations of the density profile with peak-to-peak distances of the order of a molecular diameter are evident. Similar results are obtained for the two-dimensional system of a fluid of hard disks confined between hard walls. In this case random walks of

* See also Appendix 3.1 of Chapter 3.

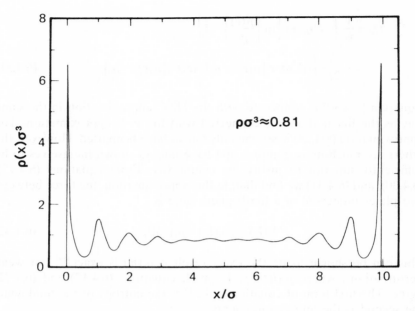

Figure 6.3. Monte-Carlo density profile for a hard-sphere fluid between hard walls (after Snook and Henderson[15]).

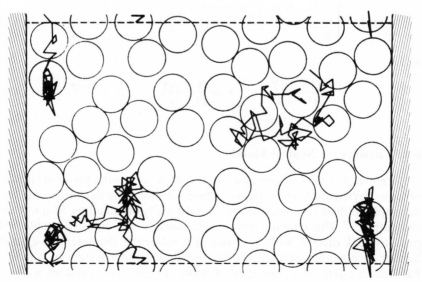

Figure 6.4. Hard-disk random walks for four molecules of a hard-disk fluid between hard walls (after Snook and Henderson[15]).

individual molecules have been examined in detail. Figure 6.4 shows the random paths of four such disks. Disks close to a wall tend to be confined to this position by collisions with neighboring disks, while disks well within the fluid tend to make wider excursions. This accounts for the high average density found near the wall.

In Figure 6.5 the results of the P–Y theory for the density of a hard-sphere fluid near one wall[16] (equivalent to density-functional theory) are compared with computer simulation. Except within a molecular radius

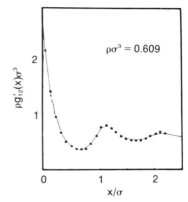

Figure 6.5. Density profile for hard spheres near a hard wall. The points give the Monte-Carlo values and the solid and broken curves give the generalized MSA and P–Y results, respectively; $\rho_0 = 0.609\sigma^{-3}$ (after Waisman *et al.*[16]).

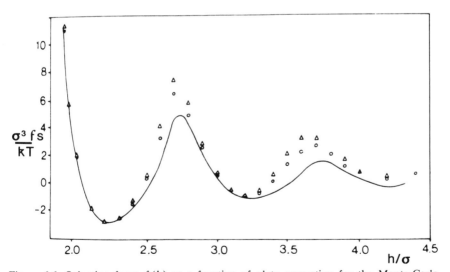

Figure 6.6. Solvation force $f_s(h)$ as a function of plate separation for the Monte-Carlo simulation of a Lennard-Jones fluid between two parallel Lennard-Jones surfaces as a function of their separation and for different values of the chemical potential $\mu^* : \mu^* = -2.477$ (solid line), $\mu^* = -1.786$ (○), $\mu^* = -1.237$ (△) (after van Megen and Snook[17]).

of the closest distance of approach, agreement between the two is very good indeed. Unfortunately, as we can see from equations (6.4.17) and (6.4.32), the force between two plates depends on the density of the fluid at the closest distance of approach where the theory is worst. To improve the theory, Waisman *et al.* suggest a phenomenological way, known as the generalized MSA, of getting the correct density at the wall. With this improvement agreement between the integral-equation approach and computer simulation is very good indeed.

In Figures 6.6 and 6.7 results are shown for the solvation force between two walls. Figure 6.6 is obtained from computer simulation[17] on a Lennard-Jones liquid (LJ) with an LJ interaction with the walls. Figure 6.7 is the result for a hard-sphere liquid using density-functional theory.[10] It can be seen that the two figures display the same characteristics and, in particular, the oscillations in force that arise from the difficulty of squeez-

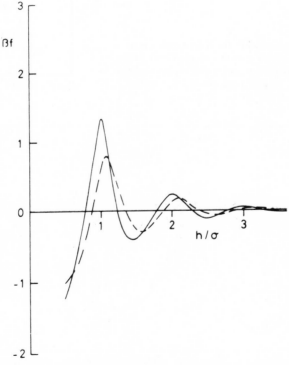

Figure 6.7. Reduced solvation force $\beta f(h)$ between two walls as a function of their separation $h$, when they are separated by a hard-sphere fluid of molecules of diameter $\sigma$. The approximations are those of the density-functional approach of Grimson and Rickayzen,[10] the solid curve corresponding to the nonlinear theory and the dashed curve to the linear theory (after Grimson and Rickayzen[10]).

ing out different layers of the liquid as the walls approach. Since the two figures are obtained for different models it is not possible to compare them quantitatively. However, it seems that even though the density-functional theory does not give the correct wall density of the fluid, it may give the correct solvation force. This could be because the force depends on a difference of wall densities rather than on the densities themselves.

The oscillations in density have been observed in a beautiful experiment of Horn and Israelachvili.[18] Their results for the liquid OMCTS ($[(CH_3)SiO]_4$) are shown in Figure 6.8. The peak-to-peak distances are of the order of the molecular diameter, in agreement with theory. In practice, the oscillations are difficult to observe and are easily wiped out by small concentrations of impurity. Theoretical calculations[19] on fluids comprising spheres of different size show that, under these conditions, the oscillations are lost.

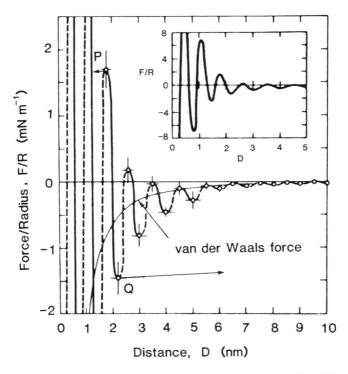

Figure 6.8. Force between two crossed mica cylinders immersed in OMCTS, $[(CH_3)_2SiO]_4$, as a function of separation. Distance measurements are accurate to $\pm 0.1$–$0.2$ mm, forces to $\pm 5 \times 10^{-8}$ N (after Horn and Israelachvili[18]).

## 6.5. The Electric Double Layer

In this case we have to include at least three kinds of molecule, the two kinds of ion and the solvent molecules, which either carry dipoles, or are polarizable, or both. We discuss in some detail the simplest case which is the primitive model (PM) electrolyte. This is a fluid comprising two kinds of hard spheres with equal diameters that carry equal and opposite charges at their centers (see Figure 6.9). The solvent is still taken to be a structureless medium of dielectric constant $\varepsilon_r$.

The potentials between the molecules are then given by

$$V_{\lambda\nu}(r) = \begin{cases} \infty, & r < \sigma \\ \dfrac{e_\lambda e_\nu}{4\pi\varepsilon_0\varepsilon_r r}, & r > \sigma,\; e_1 = -e_2 \end{cases} \tag{6.5.1}$$

In this case, in order to specify the state of the fluid we must give the average densities of the two components $\rho_1(\mathbf{r})$ and $\rho_2(\mathbf{r})$, and $\Omega_0$ will be a function of both of these. Perturbation theory and the nonlinear approach lead, respectively, to the thermodynamic potentials

$$\Omega_0[\rho] = \Omega_0[\rho_0] + \frac{1}{2\beta} \sum_{\lambda,\nu} \int d^3r\, d^3r'\, \delta\rho(\mathbf{r}) \left[ \frac{\delta_{\mu\nu}\delta(\mathbf{r}-\mathbf{r}')}{\rho_{0\lambda}} - c_{\lambda\nu}(\mathbf{r}-\mathbf{r}') \right] \delta\rho_\nu(\mathbf{r}') \tag{6.5.2}$$

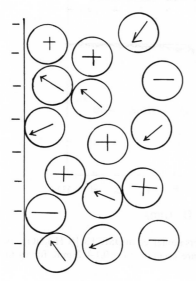

Figure 6.9. Schematic diagram of oppositely charged ions and dipolar molecules near a charged wall.

and

$$\Omega_0[\rho] = \Omega_0[\rho_0] + \frac{1}{\beta} \sum_\lambda \int d^3r \rho_\lambda(\mathbf{r}) \left[ \ln \frac{\rho_\lambda(\mathbf{r})}{\rho_{0\lambda}} - \rho_\lambda(\mathbf{r}) + \rho_{0\lambda} \right]$$

$$- \frac{1}{2\beta} \sum_{\lambda,\nu} \int d^3r d^3r' [\rho_\lambda(\mathbf{r}) - \rho_{0\lambda}] c_{\lambda\nu}(\mathbf{r}, \mathbf{r}') [\rho_\nu(\mathbf{r}) - \rho_{0\nu}] \qquad (6.5.3)$$

There now exist four direct correlation functions, which must be known before we can apply the functional. These are not known exactly, so they have been obtained using the Mean Spherical Approximation[20] in the bulk liquid. This approximates

$$c_{\lambda\nu}(\mathbf{r}) = - \beta V_{\lambda\nu}(\mathbf{r}), \qquad r > \sigma \qquad (6.5.4)$$

which is correct only as $r \to \infty$, and uses the exact result

$$g_{\lambda\nu}(\mathbf{r}) = 1 + h_{\lambda\nu}(\mathbf{r}) = 0, \qquad r < \sigma \qquad (6.5.5)$$

Then, in the region $r < \sigma$, the dcf can be found by solving the OZ equation, which is valid in this case:

$$h_{\lambda\nu}(\mathbf{r}) = c_{\lambda\nu}(\mathbf{r}) + \sum_\mu \int d^3r' c_{\lambda\mu}(\mathbf{r}') \rho_{0\mu} h_{\mu\nu}(\mathbf{r}' - \mathbf{r}) \qquad (6.5.6)$$

Hence

$$c_{\lambda\nu}(\mathbf{r}) = c_{hs}(\mathbf{r}) - \frac{\beta e_\lambda e_\nu}{4\pi\varepsilon_0\varepsilon_r\sigma} \left( 2B - \frac{B^2r}{\sigma} \right), \qquad r < \sigma \qquad (6.5.7)$$

where $c_{hs}(r)$ is the dcf for the hard-sphere liquid given in equation (6.4.33),

$$B = \frac{\lambda}{\sigma} \left[ 1 + \frac{\sigma}{\lambda} - \left( 1 + \frac{2\sigma}{\lambda} \right)^{1/2} \right] \qquad (6.5.8)$$

and $\lambda$ is the Debye length for this electrolyte. A measure of the strength of the Coulomb interaction is given by the dimensionless parameter

$$\beta^* = \frac{\beta e_\lambda^2}{4\pi\varepsilon_0\varepsilon_r\sigma} \qquad (6.5.9)$$

For $\varepsilon_r = 78.54$, $T = 298$ K, $\sigma = 4.25$ Å, and monovalent ions, we obtain $\beta^* = 1.68$, and it increases in size as the square of the valency.

The total free energy includes the interaction of the walls with the charges and, if we are to derive the total force, we should include the interaction of the walls with each other. If our attention is confined to the linear case, then the total thermodynamic potential becomes

$$\Omega[\rho] = \Omega_0[\rho] + \sum_\lambda \int d^3r\rho(r) \frac{e_\lambda\sigma(r')}{4\pi\varepsilon_0\varepsilon_r(\mathbf{r}-\mathbf{r}')} d^3r'$$

$$+ \frac{1}{2} \int d^3rd^3r' \frac{\sigma(\mathbf{r})\sigma(\mathbf{r}')}{4\pi\varepsilon_0\varepsilon_r(\mathbf{r}-\mathbf{r}')} \tag{6.5.10}$$

where $\sigma(\mathbf{r})$ is the charge density on the walls. Minimization of $\Omega[\rho]$ with respect to $\rho_1(\mathbf{r})$ and $\rho_2(\mathbf{r})$ leads to equations for these densities in terms of the charge on the walls.

For the PM electrolyte, the simple form (6.5.7) for $c_{\lambda\nu}(r)$ allows a useful division of the thermodynamic potential into two mutually independent parts, one which depends on the total density and one on the charge density. To make this explicit, we replace $\rho_1$ and $\rho_2$ by

$$\rho(\mathbf{r}) = \rho_1(\mathbf{r}) + \rho_2(\mathbf{r}) \tag{6.5.11}$$

and

$$q(\mathbf{r}) = \sum_\lambda e_\lambda\rho_\lambda(\mathbf{r}) = e_1[\rho_1(\mathbf{r}) - \rho_2(\mathbf{r})] \tag{6.5.12}$$

Then

$$\Omega[\rho_\lambda] = \Omega_\rho[\rho] + \Omega_q[q] \tag{6.5.13}$$

where $\Omega_\rho$ is the thermodynamic potential (6.4.12) and

$$\Omega_q = \frac{1}{2\beta\rho_0 e^2} \int d^3rq^2(\mathbf{r}) + \frac{1}{8\pi\varepsilon_0\varepsilon_r} \int d^3rd^3r'q(\mathbf{r})c'(\mathbf{r}-\mathbf{r}')q(\mathbf{r}')$$

$$+ \int \frac{q(\mathbf{r})\sigma(\mathbf{r}')}{4\pi\varepsilon_0\varepsilon_r|\mathbf{r}-\mathbf{r}'|} d^3rd^3r' + \int d^3rd^3r' \frac{\sigma(\mathbf{r})\sigma(\mathbf{r}')}{8\pi\varepsilon_0\varepsilon_r|\mathbf{r}-\mathbf{r}'|} \tag{6.5.14}$$

where

$$c'(\mathbf{r}) = \begin{cases} \frac{1}{\sigma}\left(2B - \frac{B^2r}{\sigma}\right), & r < \sigma \\ 1/r, & r > \sigma \end{cases} \tag{6.5.15}$$

Thus $\rho(\mathbf{r})$ is the same as for the corresponding neutral fluid and gives rise to the same force, while the charge density is to be determined by minimizing expression (6.4.14). Hence the charge satisfies the integral equation

$$\frac{q(\mathbf{r})}{\beta\rho_0e^2} + \frac{1}{4\pi\varepsilon_0\varepsilon_r} \int d^3r' \left[C'(\mathbf{r}-\mathbf{r}')q(\mathbf{r}') + \frac{\sigma(\mathbf{r}')}{|\mathbf{r}-\mathbf{r}'|}\right] = 0 \tag{6.5.16}$$

In the limit when the spheres tend to point charges, $c'(r)$ becomes $r^{-1}$ and the equation becomes

$$\frac{q(\mathbf{r})}{\beta \rho_0 e^2} + \frac{1}{4\pi \varepsilon_0 \varepsilon_r} \int d^3r' \frac{q(\mathbf{r'}) + \sigma(\mathbf{r'})}{|\mathbf{r} - \mathbf{r'}|} = 0 \qquad (6.5.17)$$

This is the integrated form of the Debye–Hückel equation for this problem, as can be seen by operating with $\nabla^2$ on the equation to yield

$$\frac{\nabla^2 q(\mathbf{r})}{\beta \rho_0 e^2} = \frac{q(\mathbf{r}) + \sigma(\mathbf{r})}{\varepsilon_0 \varepsilon_r} \qquad (6.5.18)$$

But equation (6.5.17) can also be written as

$$\frac{q(\mathbf{r})}{\beta \rho_0 e^2} + V(\mathbf{r}) = 0 \qquad (6.5.19)$$

where $V(\mathbf{r})$ is the electric potential. Hence inside the fluid where $\sigma(\mathbf{r}) = 0$,

$$\nabla^2 V(\mathbf{r}) = \frac{\beta \rho_0 e^2}{\varepsilon_0 \varepsilon_r} V(\mathbf{r}) = \frac{V(\mathbf{r})}{\lambda^2} \qquad (6.5.20)$$

Therefore as $\sigma$, the diameter of the molecules, tends to zero we obtain Debye–Hückel theory. Equation (6.5.16) is a generalization, which takes into account the finite size of the molecules. In the same way the nonlinear theory reduces to the Gouy–Chapman theory as $\sigma \to 0$.

If the equations are specialized to the use of electrolyte between two parallel plates, there are a couple of points to be noted. The first is that, because the plates extend to infinity, the electric field does not necessarily tend to zero at infinity. Hence charge is not necessarily conserved. This must be taken care of explictly by introducing a Lagrange multiplier and adding a term to $\Omega$, namely

$$\theta \int [q(\mathbf{r}) + \sigma(\mathbf{r})] d^3r \qquad (6.5.21)$$

where $\theta$ is chosen to impose

$$\int [q(\mathbf{r}) + \sigma(\mathbf{r})] d^3r = 0$$

Further, $\rho$ and $\sigma$ become functions of $x$, the coordinate normal to the plates, and the correlation function required is

$$C'(x) = \int c(r)\,dy\,dz \tag{6.5.22}$$

Now because of the Coulombic dependence of $c$ for large $r$ [see equations (6.5.4) and (6.5.1)], this integral diverges. However, if the system is overall neutral this divergence makes no contribution to the total energy and $C'(x)$ is the solution of the one-dimensional Poisson equation

$$\frac{d^2 C'(x)}{dx^2} = \frac{\beta}{\varepsilon_0 \varepsilon_r}\,\delta(x) \tag{6.5.23}$$

given by

$$C'(x) = -\frac{2\pi\beta}{\varepsilon_0 \varepsilon_r}|x| \tag{6.5.24}$$

The force depends on whether the surface charge or surface potential is being held fixed. In the latter case and with the one-dimensional geometry for a PM electrolyte between hard plates, the force is

$$F_V(h) = \frac{1}{2\beta\rho_0 e^2}[q_s^2(h) - q_s^2(\infty)] - \frac{1}{2\varepsilon_0\varepsilon_r}[Q^2(h) - Q^2(\infty)] \tag{6.5.25}$$

where $Q(h)$ is the charge on the plate required to keep the potential at $V$ for separation $h$ and $q_s(h)$ is the surface charge on the fluid for the same separation. If the charge is held fixed, $Q(h) = Q(\infty)$ and

$$F_Q(h) = \frac{1}{2\beta\rho_0 e^2}[q_s^2(h) - q_s^2(\infty)] \tag{6.5.26}$$

### 6.5.1. Results for Model Electrolytes

A sample of the results obtained for the restricted primitive model (RPM) are shown in Figures 6.10–6.13. Torrie and Valleau[21] have carried out Monte-Carlo computer simulations on the RPM and compared their results with those given by Gouy-Chapman theory with an appropriate Stern layer. Figure 6.10 shows a comparison of the electric potentials for a concentration of 0.1 M, 1–1 electrolyte, and a reduced surface charge $Q^*$ of 0.3. The equation used is

$$Q^* = Q\sigma^2/e$$

where $Q$ is the actual surface charge. In making the comparison there is

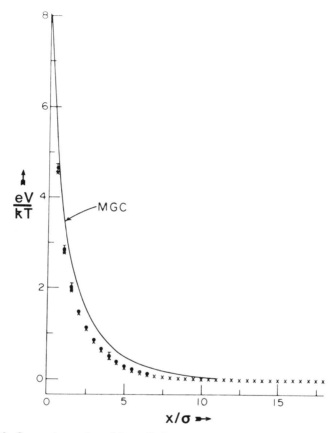

Figure 6.10. Comparison of a Monte-Carlo computer simulation of the RPM with Gouy–Chapman theory with a Stern layer (MGC) (solid line) at a concentration of 0.1 M, 1–1 electrolyte. The reduced surface charge is 0.3. The separate points are the results of the simulation for the electric potential (after Torrie and Valleau[21]).

ambiguity about the value of $\varepsilon_s$ to be used. Torrie and Valleau take $\varepsilon_s = \varepsilon_r$. At this concentration the MGC theory is quite close to the computed results. Figure 6.11 shows the concentrations of the two ions at the higher concentration of 1 $M$ and with the greater reduced surface charge of 0.7. At this concentration there are significant differences between theory and simulation. In particular, according to the simulation there is a peak in the counterion density at a distance of $1.5\sigma$ from the wall. This work therefore shows that at high concentrations of electrolyte, MGC theory is not satisfactory. However, the concentration at which this discrepancy is important is much greater than that commonly used in colloids.

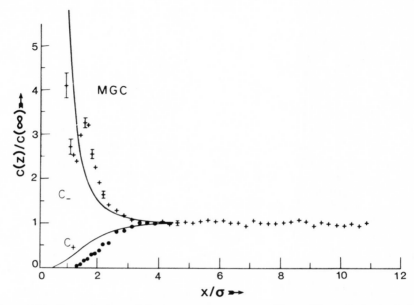

Figure 6.11. Concentrations of the two components according to Monte-Carlo simulation and MGC theory at a concentration of 1 M, 1–1 electrolyte. The reduced surface charge is 0.7 (after Torrie and Valleau[21]).

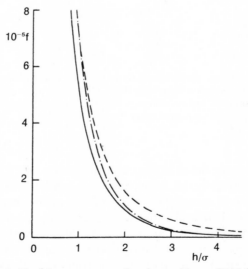

Figure 6.12. Force in Nm$^{-2}$ between two walls separated by an RPM fluid as a function of separation $R$ at 0.5 M. The parameters are $\sigma = 4.25$ Å, $\beta^* = 1.680$, $Q = -1.75 \times 10^{-2}$ cm$^{-2}$. Curves refer to nonlinear density-functional theory (——), linear theory (–·–·), and MGC theory (– – –) (after Grimson and Rickayzen[22]).

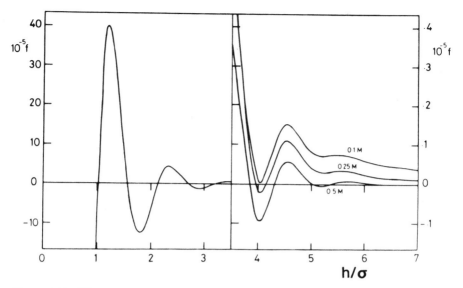

Figure 6.13. Effect of increasing density of neutral fluid on the force between walls. Linear theory with three-component fluid, $\rho_0\sigma^3 = 0.4$, $\beta^* = 1.680$, $V_0 = -10$ mV. Note the change of vertical scale at $3.5\sigma$ (after Grimson and Rickayzen[23]).

Figure 6.12 shows a comparison of the results of density-functional theory and of MGC theory for the force between two walls separated by an RPM fluid.[22] Although there are differences, they again only become evident at very high concentrations of electrolyte and at high surface charge.

Of course, the RPM takes into account only the molecular structure of the ions but not that of the solvent. To remedy this to some extent, Grimson and Rickayzen[23] have included a third neutral component. Some of their results are shown in Figure 6.13 for concentrations of electrolyte between 0.1 M and 0.5 M. Note that even at 0.1 M and with $h \sim 5\sigma$ the curve does not resemble the monotonic Debye–Hückel result. In fact these results approach Debye–Hückel theory only for $h > 6\sigma$.

In more recent work, the uniform solvent has been replaced by a fluid of hard spheres with dipoles at their centers. This has been called the civilized model by Carnie and Chan.[24] The problem of a single charged wall immersed in this fluid with electrolyte has, in the P–Y approximation, been solved analytically by Carnie and Chan[24] and by Henderson and Blum.[25] For very low concentrations of electrolyte they find that the capacity $C$ of the double layer can be written in the form

$$C^{-1} = \frac{\lambda}{\varepsilon_0 \varepsilon_r} \left( 1 + \frac{\varepsilon_r \sigma_s}{\varepsilon_s \lambda} \right) \tag{6.5.27}$$

with

$$\frac{\sigma_s}{\varepsilon_s} = \frac{\sigma}{\varepsilon_r} \left( 1 + \frac{\varepsilon_r - 1}{\gamma} \right), \qquad \gamma^2 (1 + \gamma)^4 = 16 \varepsilon_r, \qquad \gamma \geqslant 1 \tag{6.5.28}$$

For this purpose then, the fluid behaves as if it possesses a Stern layer with parameters given by equation (6.5.28). However, it is doubtful whether the behavior of the fluid with regard to other properties can also be analyzed in terms of a Stern layer with the same parameters.

This work has been extended by Augousti and Rickayzen[26] to the case of the fluid between two walls. They conclude that for this model structural effects are important when $\varepsilon_r \approx 78$ even at a concentration of 0.005 M. This may be important for understanding the behavior of colloids.

## References

1. E. J. W. Verwey and J. Th. G. Overbeek, *Theory of Stability of Lyophobic Colloids*, Elsevier, Amsterdam (1948).
2. B. V. Derjaguin, *Trans. Faraday Soc.* **36**, 203 (1940).
3. H. Schulze, *J. Prakt. Chem.* **25**, 431 (1882); **27**, 320 (1883); W. P. Hardy, *Proc. R. Soc. London* **66**, 110 (1900).
4. G. Gouy, *J. Phys.* **9**, 457 (1910); *Ann. Phys.* **7**, 129 (1917); D. L. Chapman, *Phil. Mag.* **25**, 475 (1913).
5. P. Debye and E. Hückel, *Phys. Z.* **24**, 185 (1923); P. Debye, *Phys. Z.* **25**, 93 (1924).
6. M. J. Grimson and G. Rickayzen, *Mol. Phys.* **44**, 817 (1981).
7. E. J. W. Verwey and J. Th. G. Overbeek, *Theory of Stability of Lyophobic Colloids*, pp. 8–9, Elsevier, Amsterdam (1948).
8. J. P. Hansen and I. R. McDonald, *Theory of Simple Liquids*, p. 43, Academic Press, London (1976).
9. S. Levine and C. W. Outhwaite, *J. Chem. Soc., Faraday Trans. 2* **74**, 1670 (1978). This paper contains references to earlier work by this group.
10. W. F. Saam and C. Ebner, *Phys. Rev. A* **15**, 2566 (1977); D. Y. C. Chan, B. A. Pailthorpe, J. S. McCaskill, D. J. Mitchell, and B. W. Ninham, *J. Coll. Interface Sci.* **72**, 27 (1979); M. J. Grimson and G. Rickayzen, *Mol. Phys.* **42**, 767 (1981).
11. N. D. Mermin, *Phys. Rev. A* **137**, 1441 (1965).
12. J. P. Hansen and I. R. McDonald, *Theory of Simple Liquids*, pp. 30–31, Academic Press, London (1976).
13. R. J. Evans, *Adv. Phys.* **28**, 143 (1979). F. F. Abraham, *Phys. Rep.* **53**, 93 (1979).
14. I. Z. Fisher, *Statistical Theory of Liquids*, p. 109, University of Chicago Press (1964).
15. I. K. Snook and D. Henderson, *J. Chem. Phys.* **68**, 2134 (1978).
16. E. Waisman, D. Henderson, and J. L. Lebowitz, *Mol. Phys.* **32**, 1373 (1976).
17. W. van Megen and I. K. Snook, *J. Chem. Soc., Faraday Trans. 2* **75**, 1095 (1979); *J. Chem. Phys.* **74**, 1409 (1981).

18. R. G. Horn and J. N. Israelachvili, *Chem. Phys. Lett.* **71**, 192 (1980); *J. Chem. Phys.* **75**, 1400 (1981).
19. M. J. Grimson, *J. Chem. Soc., Faraday Trans. 2* **78**, 1905 (1982).
20. E. Waisman and J. L. Lebowitz, *J. Chem. Phys.* **56**, 3086 (1972).
21. G. M. Torrie and J. P. Valleau, *J. Chem. Phys.* **73**, 5807 (1980).
22. M. J. Grimson and G. Rickayzen, *Mol. Phys.* **45**, 221 (1982).
23. M. J. Grimson and G. Rickayzen, *Chem. Phys. Lett.* **86**, 71 (1982).
24. S. L. Carnie and D. Y. C. Chan, *Chem. Phys.* **73**, 2949 (1980); *J. Chem. Soc., Faraday Trans. 2* **78**, 695 (1982).
25. L. Blum and D. Henderson, *J. Chem. Phys.* **74**, 1902 (1981); D. Henderson and L. Blum, *J. Chem. Soc., Faraday Symp.* **16**, 151 (1981).
26. A. Augousti and G. Rickayzen, *J. Chem. Soc., Faraday Trans. 2* **80**, 141 (1984).

# II

# Dynamic Structure and Transport

# Self-Motion in Liquids

## M. Gerl

### 7.1. Introduction

The aim of this chapter is to outline the principal methods used to describe atomic motions in monatomic liquids like rare-element liquids, or liquid metals within the adiabatic approximation. We shall be mainly concerned with the self-motion of a tagged particle immersed in a fluid, but the general methods presented can on the whole be applied when studying the dynamics of the fluid itself.

In the first section we recall the basic definitions of the diffusion constant $D$, of the velocity autocorrelation function $\psi$, which retains many more details of the dynamics of the marked atom, and we show what kind of information can be gained from neutron-scattering and molecular-dynamics experiments.

In Brownian-motion theory, valid for a large and heavy particle immersed in a fluid of light particles, the diffusion constant is expressed in terms of the friction coefficient. We shall examine in some detail the basic approximations made in this theory and show how the diffusion constant can be estimated using simple models. Memory effects are introduced phenomenologically in the generalized Langevin equation and the Mori–Zwanzig formalism is presented.

The last section is devoted to kinetic theories of the self-transport in fluids. The Fokker–Planck equation is useful when the particle undergoes frequent but soft collisions, as in Brownian-motion theory. On the other

**M. Gerl** · Laboratoire de Physique du Solide (L.A. au C.N.R.S. n° 155), Faculté des Sciences, B.P. 239, 54506 Vandoeuvre Les Nancy Cedex, France.

hand, the Boltzmann and Enskog equations, although valid when the tagged particle experiences hard but rare collisions, as in a gas, provide a good estimation of $D$ at low and moderate densities. Finally, the problem of describing the dynamic correlations between collisions is briefly presented in the last section of this chapter.

## 7.2. The Self-Diffusion Constant of a Fluid

It is well known that the flow of marked particles (such as radioactive atoms) in an isotropic fluid can be written in the form

$$\mathbf{J} = -D\nabla c$$

where $D$ is the diffusion constant and $c$ is the concentration of marked atoms. This, together with the equation of conservation of solute atoms, leads to the diffusion equation

$$\frac{\partial c}{\partial t} = D\nabla^2 c \tag{7.2.1}$$

which should also be obeyed by the self-correlation function $G_s(r, t)$ at large times, so that

$$\tau \gg \tau_c: \qquad G_s(r, t) = (4\pi Dt)^{-3/2} \exp(-r^2/4Dt) \tag{7.2.2}$$

According to this equation, it is easy to show that the mean-square displacement $\langle r^2(t) \rangle$ of a given marked atom varies linearly with time ($\langle r^2 \rangle = 6Dt$). The diffusion constant then measures the rate of variation of $\langle r^2 \rangle$ with $t$.

It is instructive to correlate $D$ with the velocity autocorrelation function (VAF) $Z(t)$ defined by[1–3]

$$Z(t) = \langle u_x(t)u_x(0) \rangle \tag{7.2.3}$$

where $u_x$ is the $x$-component of the velocity of a given tagged particle. Given that

$$D = \lim_{t \to \infty} \frac{1}{2t} \langle x^2(t) \rangle$$

and

$$x(t) = \int_0^t u_x(s)\,ds$$

it is easy to show that $D$ is the time integral of $Z(t)$,[1,2] i.e.,

$$D = \int_0^\infty Z(t)dt \qquad (7.2.4)$$

Indeed $D$ is governed by the behavior of $Z(t)$ at all times and it is of interest to study $Z(t)$.

At very short times, when a marked atom has not yet collided with neighbors, it behaves as in an ideal gas. The leading term in the time expansion of $Z(t)$ is $\langle u_x(0)u_x(0) \rangle = \langle u_x^2 \rangle$. In order to obtain the full short-time expansion of $Z(t)$, we note first that, in a stationary system, any correlation function $C(t) = \langle A(t)B(0) \rangle$ must be independent of the origin of time,

$$C(t) = \langle A(t)B(0) \rangle = \langle A(t+s)B(s) \rangle$$

from which it follows that

$$\left. \frac{dC(t)}{ds} \right|_{s=0} = \langle \dot{A}B \rangle + \langle A\dot{B} \rangle = 0 \qquad (7.2.5)$$

If we now expand $u_x(t)$ in powers of $t$ in equation (7.2.3) and make use of equation (7.2.5), we easily obtain

$$Z(t) = \langle u_x^2 \rangle \left( 1 - \frac{t^2}{2!} \frac{\langle aa \rangle}{\langle u_x^2 \rangle} + \frac{t^4}{4!} \frac{\langle \dot{a}\dot{a} \rangle}{\langle u_x^2 \rangle} - \cdots \right) \qquad (7.2.6)$$

where $a = a_x(0)$ is the $x$-component of the acceleration $(1/m)F_x$ of the particle subjected to the force $F_x$. It is easy to calculate $\langle aa \rangle$ in terms of the interatomic potential $v(r)$:

$$\langle aa \rangle = \frac{1}{m^2} \langle F_x^2 \rangle = \frac{1}{m^2} \left\langle \frac{\partial U}{\partial x} \frac{\partial U}{\partial x} \right\rangle \qquad (7.2.7)$$

where $U$ is the potential energy of the system.

If we express relationship (7.2.7) as a configurational average and integrate by parts, we obtain

$$\frac{\langle aa \rangle}{\langle u_x^2 \rangle} = \Omega_0^2 = \frac{n}{3m} \int d\mathbf{r}\, g(\mathbf{r}) \nabla^2 v(\mathbf{r}) \qquad (7.2.8)$$

where $n$ is the number density of the fluid and $g(\mathbf{r})$ is the radial distribution function; $\Omega_0$ would represent the vibration frequency of the tagged particle if all the atoms in the liquid were maintained at their average equilibrium position. The average restoring force $-m\Omega_0^2 \mathbf{r}$ on a given tagged particle accounts for the harmonic-oscillator form of $Z(t)$ at very short times:

$$Z(t) = \frac{k_B T}{m}\left(1 - \Omega_0^2 \frac{t^2}{2!} + \cdots\right) \tag{7.2.9}$$

It is also of interest to study the short-time expansion of the intermediate scattering function $F_s(\mathbf{k}, t)$ given by

$$F_s(\mathbf{k}, t) = \langle e^{-ikx(t)} e^{ikx(0)} \rangle \tag{7.2.10}$$

where $x$ is the component of $\mathbf{r}$ along $\mathbf{k}$. By differentiating this expression with respect to $t$ and using (7.2.5), we obtain

$$-\lim_{k\to 0}\frac{1}{k^2}\ddot{F}(\mathbf{k}, t) = Z(t) \tag{7.2.11}$$

so that we can expect to obtain $Z(t)$ from measurements of $F_s(\mathbf{k}, t)$ or $S_s(\mathbf{k}, \omega)$ at very small momentum transfers. Moreover, a direct expansion of equation (7.2.10) leads to

$$F_s(\mathbf{k}, t) = 1 - \omega_0^2 \frac{t^2}{2!} + \omega_0^2(3\omega_0^2 + \Omega_0^2)\frac{t^4}{4!} - \cdots \tag{7.2.12}$$

where

$$\omega_0^2 = \frac{k_B T}{m} k^2$$

The structure of the liquid enters the third term of this expression, through the frequency $\Omega_0$. A convenient way of using expansion (7.2.12) is to calculate the frequency moments $\langle \omega^n \rangle$ of $S_s(k, \omega)$, defined by

$$\langle \omega^n \rangle = \int_{-\infty}^{+\infty} d\omega\, \omega^n S_s(k, \omega) \tag{7.2.13}$$

From the relation between $F_s(\mathbf{k}, t)$ and $S_s(\mathbf{k}, \omega)$,

$$F_s(\mathbf{k}, t) = \int_{-\infty}^{+\infty} d\omega\, e^{-i\omega t} S_s(\mathbf{k}, \omega)$$

it is easy to derive

$$\langle \omega \rangle = 1, \qquad \langle \omega^2 \rangle = \omega_0^2, \qquad \langle \omega^4 \rangle = \omega_0^2(3\omega_0^2 + \Omega_0^2) \tag{7.2.14}$$

This provides useful consistency equations, which must be obeyed by any experimentally or theoretically determined $S_s(\mathbf{k}, \omega)$.

Values of function $S_s(\mathbf{k}, \omega)$ can be measured experimentally by inelastic scattering of neutrons from the liquid, where $S_s(\mathbf{k}, \omega)$ is actually the incoherent part of $S(\mathbf{k}, \omega)$, which can be obtained through scattering experiments on mixtures of suitable isotopes.[4,5] Figure 7.1 shows, for instance, the experimental values of $\bar{S}_{inc}(\mathbf{k}, \omega)$ obtained by Sköld et al.[4] in argon. In order to interpret the data and explain the overall shape of $\bar{S}_{inc}(\mathbf{k}, \omega)$ it is convenient to consider two limits:

    1. In an ideal gas,

$$F_s(\mathbf{k}, t) = \langle e^{-iku_x t} \rangle = \exp\left(-\frac{\omega_0^2}{2} t^2\right)$$

so that

$$S_s(\mathbf{k}, \omega) = (2\pi\omega_0^2)^{-3/2} \exp(-\omega^2/2\omega_0^2)$$

(7.2.15)

In this limit (large $\mathbf{k}$, $\omega$), $S_s(\mathbf{k}, \omega)$ is a Gaussian function of $\omega$ with HWHM

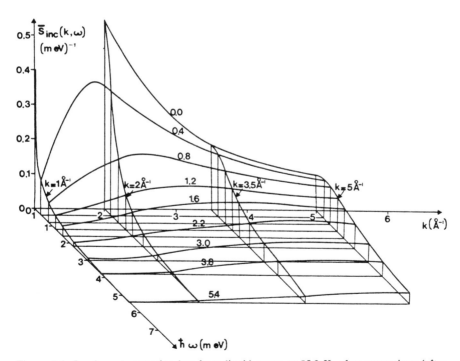

Figure 7.1. Incoherent scattering law from liquid argon at 85.2 K, after corrections (after Sköld et al.[4]).

$W(k)$ proportional to $k$. In practice, this ideal gas limit is useful for $k > 10$ $\text{Å}^{-1}$.

2. When the tagged particle has undergone a large number of collisions (limit of small $\mathbf{k}$, $\omega$), the diffusion approximation becomes valid and

$$F_s(\mathbf{k}, t) = \exp(-k^2 Dt)$$
$$\left.\begin{array}{l} \\ S_s(\mathbf{k}, \omega) = \frac{1}{\pi} \frac{Dk^2}{\omega^2 + (Dk^2)^2} \end{array}\right\} \qquad (7.2.16)$$

Therefore $W(k)$ varies parabolically with $k$. The range of validity of this approximation can be appreciated in Figure 7.2.

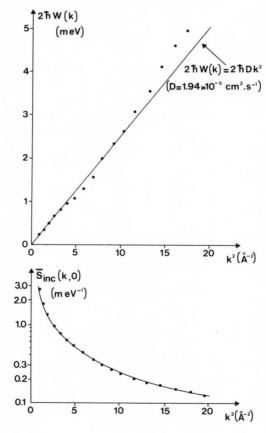

Figure 7.2. (a) Full width at half maximum $2\hbar W(k)$ of $\overline{S}_{\text{inc}}(k, \omega)$; ●, experimental points after correction (after Sköld et al.[4]). (b) —, simple diffusion result; ●, experimental data after the resolution correction.[4]

3. In both limits of large and small $k$, $F_s(\mathbf{k}, t)$ exhibits a Gaussian dependence on $k$. Hence it is reasonable to assume that, in the whole range of variation of $k$,

$$F_s(\mathbf{k}, t) = \exp[-\tfrac{1}{2}\alpha(t)k^2] \tag{7.2.17}$$

where $\alpha(t) = \langle u_x^2 \rangle t^2$ in the short-time limit and $\alpha(t) = 2Dt$ within the diffusion approximation. Egelstaff and Schofield[6] noted that the width function

$$\alpha(t) = 2D[(t^2 + c^2)^{1/2} - c] \tag{7.2.18}$$

with $c = mD/k_B T$ has the desired behavior at short and long times, and represents fairly well the molecular-dynamics results of Levesque and Verlet.[7] The only concern with the ansatz (7.2.18) is that it does not satisfy the fourth-moment requirement (7.2.14). An improvement over relationship (7.2.18), due to Lovesey,[8,9] is given by the equation

$$S_s(\mathbf{k}, \omega) = \frac{\pi^{-1}\tau\omega_0^2(\omega_s^2 - \omega_0^2)}{[\omega\tau(\omega^2 - \omega_s^2)]^2 + (\omega^2 - \omega_0^2)^2} \tag{7.2.19}$$

where

$$\omega_s^2 = 3\omega_0^2 + \Omega_0^2 \quad \text{and} \quad \tau^{-1} = \frac{mD}{k_B T}\Omega_0(\omega_s^2 - \omega_0^2)^{1/2}$$

This $S_s(\mathbf{k}, \omega)$ satisfies the fourth-moment condition. The results given by equations (7.2.18) and (7.2.19) are compared in Figure 7.3.

The Gaussian approximation to $F_s(\mathbf{k}, t)$ has been studied in detail by Nijboer and Rahman.[10] In particular, it is easy to show that the short-time expansion of $\alpha(t)$ is

$$\tfrac{1}{2}\alpha(t) = \frac{k_B T}{m}\left(\frac{t^2}{2!} - \Omega_0^2\frac{t^4}{4!} + \cdots\right) \tag{7.2.20}$$

Figure 7.3. Approximations to $F_s(k,t)$ in argon: ——, Fourier transform of equation (7.2.19) with $\Omega_0^2 = 0.58 \times 10^{26}$ s$^{-2}$, $(mD/k_B T)\Omega_0 = 0.81$[8,9]; ●, Egelstaff and Schofield model [equation (7.2.18)] with $c = mD/k_B T = 0.11 \times 10^{-12}$ s[6]; ---, molecular-dynamics result of Levesque and Verlet.[7]

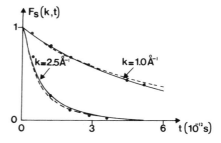

which leads to

$$S_s(\mathbf{k}, \omega) = (2\pi\omega_0^2)^{-3/2} \exp\left(-\frac{\omega^2}{2\omega_0^2}\right)\left[1 + \frac{1}{4!}\frac{\Omega_0^2}{\omega_0^2} H_4\left(\frac{\omega}{\omega_0}\right) + \cdots\right] \qquad (7.2.21)$$

where $H_4$ is the fourth Hermite polynomial.

However, the Gaussian form (7.2.17) of $F_s(\mathbf{k}, t)$ remains an approximation. The next term in the exponent is $\gamma k^4$ and Schofield[11] has shown that the time expansion of $\gamma(t)$ starts with $t^8$. According to Nijboer and Rahman,[10] the expansion of $F_s(\mathbf{k}, t)$ is actually of the form

$$F_s(\mathbf{k}, t) = \exp[-\tfrac{1}{2}\alpha(t)k^2]\left[1 + \alpha_2(t)\frac{(\alpha k^2/2)^2}{2!} - ak^6 + bk^8 + \cdots\right] \qquad (7.2.22)$$

where

$$\alpha_2(t) = \frac{3\langle r^4\rangle}{5\langle r^2\rangle^2} - 1$$

This expansion has been checked experimentally by Sköld et al.[4], and numerically by Levesque and Verlet[12] and Rahman.[13] Figure 7.4 shows that, in the Gaussian approximation, the crossover between the ideal-gas and the diffusion regimes occurs at about $5 \times 10^{-12}$ s and that the maximum departure from Gaussian behavior is in the vicinity of $10^{-12}$ s. More accurately, Levesque and Verlet showed that a good fit to the molecular-dynamics data can be obtained with the function

$$\alpha_2(t) = C\left(\frac{t_c}{t}\right)\exp\left[-\left(\frac{t_c}{t} - 1\right)\right]$$

where $C = 0.1186$ and $t_c = 1.835 \times 10^{-12}$ s in liquid argon at 85.2 K.

The velocity autocorrelation function has been studied in detail by molecular dynamics. After the pioneering work of Alder et al.[14] in hard-sphere fluids, and Rahman[13] and Levesque and Verlet[12] in Lennard-Jones liquids, the VAF varies as follows (Figure 7.5):

1. At intermediate density ($V/V_0 \sim 3$), where $V_0$ is the close-packing volume, the VAF shows an initial decay [equation (7.2.9)] followed by a plateau and a long-time $t^{-3/2}$ tail, which can be understood in the following manner: If the tagged particle is given an initial momentum, it may excite long-lasting hydrodynamic modes that may assist the tagged particle in its forward motion. The $t^{-3/2}$ law can be qualitatively understood by employing the argument that the momentum given to the particle is shared by a volume of fluid whose size increases at $t^{3/2}$. These qualitative features have been demonstrated by Alder and Wainwright by integration of the Navier–Stokes equations.

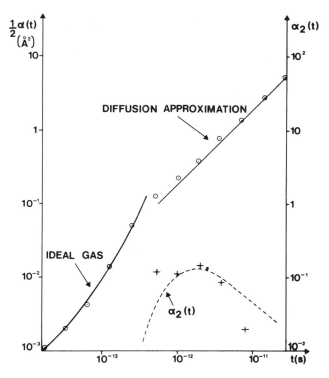

Figure 7.4. The width function $\frac{1}{2}\alpha(t)$ as defined in equation (7.2.17). Open circles and crosses are experimental points obtained by Sköld *et al.*[4] Solid lines represent the limit behavior of $\frac{1}{2}\alpha(t)$ and the dotted line is the result of computer calculation by Levesque and Verlet.[12] The full circle represents the result obtained by Rahman.[13]

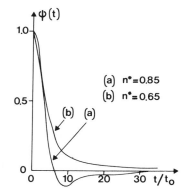

Figure 7.5. Normalized velocity autocorrelation function $\psi(t)$ in a Lennard-Jones fluid at two different number densities.[2]

2. At larger densities ($V/V_0 \sim 1.6$) the VAF has a tendency to oscillate: it becomes negative, exhibits a negative plateau, and decreases slowly at longer times. This backscattering effect leads to a diffusion coefficient that is smaller than expected if the collisions were assumed independent.

We mention further that molecular-dynamics studies have also been made in systems with a tagged particle of mass and size different from those of the matrix,[14,15] in ionic systems,[16] in mixtures,[17] and in isotopic mixtures[18,19]

## 7.3. The Langevin Equation

Though not directly applicable to the self-diffusion problem, much can be learned by studying the dynamics of a large and heavy particle immersed in a bath of small and light particles. This is the Langevin problem.

### 7.3.1. The Simple Langevin Equation[20-23]

#### 7.3.1.1. General Method[23]

A large particle in a fluid of small particles does not respond to individual impacts of bath atoms but to fluctuations of the total rate of impacts over its surface. The motion of the particle therefore exhibits an average behavior, on which is superimposed an erratic motion due to the fluctuations of the force experienced by the particle. The equation of motion of the particle can be written in the form

$$m\dot{\mathbf{u}} = \mathbf{F}(t, \mathbf{u}[t']) \tag{7.3.1}$$

where the total force $\mathbf{F}$ at time $t$ depends on the velocity $\mathbf{u}(t')$ at earlier times $t'$. The following approximations are introduced in order to obtain the Langevin equation:

1. The characteristic time $\tau_c$ of the fluctuations of $\mathbf{F}$ is assumed to be small with respect to $\tau_R$, the decay time of the velocity of the particle, so that $\mathbf{F}$ depends only on the velocity $\mathbf{u}(t)$ at time $t$.
2. $\mathbf{F}$ can be split into its average value $\langle \mathbf{F} \rangle_u$ (given that the velocity of the particle is $\mathbf{u}$) and a stochastic part $\mathbf{R}(t)$.
3. The systematic force $\langle \mathbf{F} \rangle_u$, called the friction force, is assumed to be

proportional to **u**, namely

$$\mathbf{F}(t) = -\zeta\mathbf{u} + \mathbf{R}(t) \tag{7.3.2}$$

where $\zeta = \beta m$ is the friction coefficient.

4. On the time scale of the motion of the particle, $\mathbf{R}(t)$ and $\mathbf{R}(t+\tau)$ are correlated over a very short time,

$$\langle X(t+\tau)X(t)\rangle = G\varphi(\tau) \sim G\delta(\tau) \tag{7.3.3}$$

where $X$ is the $x$-component of **R**; the function $\varphi(\tau)$ is sharply peaked at $\tau = 0$.

According to these assumptions the Langevin equation becomes

$$\frac{d\mathbf{u}}{dt} = -\beta\mathbf{u} + \frac{1}{m}\mathbf{R}(t) \tag{7.3.4}$$

This is not an ordinary differential equation because $\mathbf{R}(t)$ has only statistically defined properties. We can, however, solve equation (7.3.4) directly:

$$\mathbf{A}(t) = \mathbf{u}(t) - \mathbf{u}_0 e^{-\beta t} = \int_0^t e^{\beta(s-t)} \frac{\mathbf{R}(s)}{m} \, ds \tag{7.3.5}$$

After a large number of collisions ($t \gg \tau_c$) the probability distribution $W(\mathbf{A})$ of **A** becomes Maxwellian, because **A** is the sum of a large number of random contributions. The distribution $W(\mathbf{A})$ can be determined, from which the probability distributions of **u** and **r** are obtained. In practice, we are more interested in the average $\langle \cdots \rangle_1$ of a physical quantity, given that at time $t = 0$ the Brownian particle was at the point $\mathbf{r}_0$ with velocity $\mathbf{u}_0$. The calculation shows that:

(i) $$\langle \mathbf{u}(t)\rangle_1 = \mathbf{u}_0 \exp(-\beta t) \tag{7.3.6}$$

from which it follows that the VAF $Z(t) = \langle u_x(t)u_x(0)\rangle$ is given by

$$Z(t) = \langle u_x^2 \rangle \exp(-\beta t) \tag{7.3.7}$$

This form is clearly incorrect at small $t$ as $Z(t)$ must be an even function of $t$.

(ii) $$\langle \mathbf{r}(t) - \mathbf{r}_0\rangle_1 = \beta^{-1}\mathbf{u}_0[1 - \exp(-\beta t)] \tag{7.3.8}$$

This shows that the center of gravity of the possible positions of the Brownian particle at time $t \gg \tau_R$ is displaced from $\mathbf{r}_0$ by the so-called "diffusive step" $\beta^{-1}\mathbf{u}_0$. [Note that $(2\beta)^{-1}\mathbf{u}_0$ is the distance the particle would travel to rest if it were subjected to the constant friction force $-\zeta\mathbf{u}_0$.]

(iii)     $\langle u^2(t) \rangle_1 = u_0^2 \exp(-2\beta t) + 3\dfrac{k_B T}{m}[1 - \exp(-2\beta t)]$

$\left.\begin{array}{l} \\ \langle (\mathbf{r} - \mathbf{r}_0)^2 \rangle_1 = \beta^{-2}u_0^2[1 - \exp(-\beta t)]^2 \\[2mm] \qquad + 3\dfrac{k_B T}{m\beta^2}(2\beta t - 3 + 4e^{-\beta t} - e^{-2\beta t}) \end{array}\right\}$     (7.3.9)

This last equation is especially interesting in the case where the initial velocity $\mathbf{u}_0$ is not fixed but has a Maxwell distribution. Averaging over $\mathbf{u}_0$ yields

$$\langle (\Delta\mathbf{r})^2 \rangle = \frac{6k_B T}{m\beta^2}(\beta t - 1 + e^{-\beta t})$$     (7.3.10)

which shows that, for large times,

$$\langle (\Delta r)^2 \rangle = 6Dt \qquad \text{with } D = \frac{k_B T}{\zeta}$$

(the Einstein formula).

### 7.3.1.2. Calculation of the Friction Coefficient

The friction coefficient $\zeta$ is actually related to the autocorrelation of the total force acting on the particle. From equation (7.3.5) we have

(i)                    $\langle (\Delta\mathbf{u})^2 \rangle_1 = \dfrac{3G}{2\beta m^2}(1 - e^{-2\beta t})$     (7.3.11)

and, by comparison with equations (7.3.9), we see that $G = 2\beta m k_B T$, which emphasizes the connection between $G$ (fluctuations) and $\beta$ (friction).

(ii)                    $\langle \mathbf{R}(0) \cdot \mathbf{u}(t) \rangle_1 = \dfrac{3G}{m}\theta(t)e^{-\beta t}$     (7.3.12)

where $\theta$ is the step function.

(iii)                    $\langle \mathbf{u}(0) \cdot \mathbf{u}(t) \rangle_1 = \dfrac{3k_B T}{m}e^{-\beta|t|}$     (7.3.13)

From these equations it is easy to calculate the autocorrelation function of the total force $\mathbf{F}(t)$, namely

$$\mathscr{G}(t) = \langle \mathbf{F}(0) \cdot \mathbf{F}(t) \rangle = 3G[\delta(t) - \tfrac{1}{2}\beta e^{-\beta t}] \qquad (7.3.14)$$

The behavior of $\mathscr{G}(t)$ is shown in Figure 7.6. The integral $I(t) = \int_0^t \mathscr{G}(s)\,ds$ increases rapidly until the plateau time $t_1$ is reached, where $t_1$ is the characteristic time of the fluctuations of $\mathbf{R}$. After $t_1$, $I(t)$ decreases slowly (time scale $\beta^{-1}$) and the plateau value of $I(t)$ is given by

$$I(t_1) = \int_0^{t_1} \langle \mathbf{F}(0) \cdot \mathbf{F}(t) \rangle\,dt = \frac{3G}{2} = 3k_{\mathrm{B}}T\zeta \qquad (7.3.15)$$

This equation enables one to easily calculate $\zeta$, for instance, using molecular dynamics. In order to obtain an order of magnitude of $\zeta$, note that, within time $t_1$, $\mathbf{F}(t)$ essentially maintains the value it had at $t = 0$. Therefore

$$\zeta \sim \frac{1}{k_{\mathrm{B}}T} \langle F_x^2 \rangle t_1 \qquad (7.3.16)$$

where $\langle F_x^2 \rangle = k_{\mathrm{B}}Tm\Omega_0^2$. Thus

$$\zeta \sim m\Omega_0^2 t_1 \qquad \text{and} \qquad D = \frac{k_{\mathrm{B}}T/m}{\Omega_0^2 t_1} \qquad (7.3.17)$$

Expression (7.3.15) has been used extensively to calculate diffusion coefficients in liquids and its validity has been demonstrated by Lebowitz and Rubin[24] and Résibois and Davis.[25] According to this equation, the particle is being accelerated in the time interval 0 to $t_1$, during which $\mathbf{F}$ and the position of the particle do not vary much. After time $t_1$ the direct

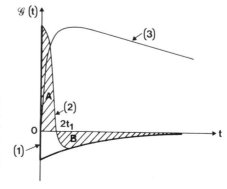

Figure 7.6. Autocorrelation function $\mathscr{G}(t)$ of the total force $\mathbf{F}$ acting on the Brownian particle: (1) from equation (7.3.14), (2) replacing $\delta(t)$ by the more realistic function $\varphi(t) \sim (2t_1)^{-1}$ over the time interval $[0, t_1]$, (3) integral of $\mathscr{G}(t) = \langle \mathbf{F}(0) \cdot \mathbf{F}(t) \rangle$. Note that areas $A$ and $B$ are equal.[23]

correlation between the fluctuating forces disappears and a new effect dominates, i.e., a friction force that tends to decelerate the particle.

The main point of equation (7.3.15) is that it involves the total force on the particle. There is therefore no need to know the friction mechanism in order to calculate $\zeta$.[26] Many calculations of $\zeta$ using equation (7.3.15) have been made[27-31] along the principles decided carefully by Helfand.[32]

1. In the case of small forces, it is possible to evaluate $\langle \mathbf{F}(0) \cdot \mathbf{F}(t) \rangle$ using the linear trajectory approximation.[28,32] It is assumed that, in the time interval $(0, t_1)$, the trajectories of all particles are linear and that they do not experience any acceleration. In the case of self-diffusion this approximation leads to

$$\zeta^S = -\frac{1}{3} \left( \frac{\pi m}{k_B T} \right)^{1/2} \frac{n}{(2\pi)^2} \int_0^\infty dk \, k^3 \tilde{V}^S(k) \tilde{G}(k) \qquad (7.3.18)$$

where $n$ is the number density, and $\tilde{V}^S(k)$ and $\tilde{G}(k)$ are, respectively, the Fourier transforms of $V^S(\mathbf{r})$, the soft potential, and of $[g(r) - 1]$.

2. In the case of hard-sphere interactions Kirkwood[26] proposed the same technique for calculating $\zeta$. Since $t_1 \ll \beta^{-1}$, equation (7.3.11) yields

$$\langle (\Delta \mathbf{p})^2 \rangle_1 = 3Gt_1 = 6k_B T \zeta t_1 \qquad (7.3.19)$$

so that

$$\zeta = \frac{1}{6k_B T t_1} \langle (\Delta \mathbf{p}_{t_1})^2 \rangle$$

where $\Delta \mathbf{p}_{t_1}$ is the increment of the momentum of the Brownian particle in the time interval $t_1$. We are therefore led to a simple kinetic problem: to count how many collisions occur in time $t_1$, calculate the momentum transfer to the tagged particle, and average. For a system of hard spheres of diameter $\sigma$, this method leads to the well-known result

$$\zeta^H = \tfrac{8}{3} ng(\sigma) \sigma^2 (\pi m k_B T)^{1/2} \qquad (7.3.20)$$

Enskog,[33] Longuet-Higgins and Pople,[34] and Rice et al.[35] also obtained this expression using similar arguments. The reader is referred elsewhere[36] for a comprehensive review of binary-collision techniques for calculating $\zeta$ in hard-sphere systems.

3. In real systems we can split the two-body potential into a hard-core repulsion $V^H(r)$ and a soft part $V^S(r)$ and accordingly decompose the autocorrelation function of the total force. The friction coefficient is therefore given by

$$\zeta = \zeta^S + \zeta^H + \zeta^{HS}$$

where $\zeta^{HS}$ arises from cross-correlations between $F^H$ and $F^S$. This technique has been used by Davis and Palyvos[28,29] and Rice and Gray.[22] It is shown qualitatively that $\zeta^S$ predominates at low temperature and high density and $\zeta^H$ at high temperature and low density. In a Lennard-Jones fluid $\zeta^S$ accounts for 80% of $\zeta$ at the melting temperature.

## 7.3.2. Hydrodynamic Calculations of the Friction Coefficient

By solving the Navier–Stokes equations for the surrounding fluid, Stokes[37] calculated $\zeta$ for the steady motion at small velocity $\mathbf{u}$ of a large sphere in a viscous, incompressible fluid. According to the boundary conditions at the surface of the sphere, the friction coefficient is $\zeta = 3\pi\eta\sigma$ ("stick" condition) or $\zeta = 2\pi\eta\sigma$ ("slip" condition), where $\eta$ is the viscosity of the fluid. These results lead to diffusion coefficients of remarkable approximate validity.

However, Zwanzig and Bixon[38] noted that the characteristic decay time of the velocity ($\beta^{-1} = m/\zeta$) is of the order of $10^{-13}$ s, which is of the same order of magnitude as the time necessary for a sound wave to propagate over an interatomic distance. Consequently, the approximation of fluid incompressibility is certainly not safe. They therefore removed the approximation of steady motion (which automatically leads to the incompressibility assumption) and conducted a Fourier analysis in time of the Navier–Stokes equations. Moreover, at high frequencies the fluid behaves elastically while at low frequencies it can be considered as only viscous. In order to cover the whole range of frequencies they took account of this viscoelastic behavior by defining a complex viscosity

$$\tau(\omega) = \frac{\eta_0}{1 - i\omega\tau}$$

where $\tau$ ($\sim 10^{-13}$ s) is the relaxation time for the viscosity. This generalized hydrodynamic calculation[39] leads to a frequency-dependent friction coefficient $\zeta(\omega)$,

$$\mathbf{F}(\omega) = -\zeta(\omega)\mathbf{u}(\omega) \qquad (7.3.21)$$

where $\mathbf{u}(\omega)$ is the $\omega$ Fourier component of the particle velocity. When $\zeta$ is assumed to be frequency-independent, one can write the normalized VAF $\psi(t)$ as

$$\psi(t) = \frac{Z(t)}{\langle u_x^2 \rangle} = e^{-\beta|t|} = \frac{2}{\pi} \operatorname{Re} \int_0^\infty d\omega \cos \omega t \frac{1}{-i\omega + \beta} \qquad (7.3.22)$$

Zwanzig and Bixon replaced $\beta$ by $\beta(\omega) = \zeta(\omega)/m$, obtained through the solution of the Navier–Stokes equations, and derived the VAF shown in Figure 7.7, together with its Fourier transform. The limits of large and small $\omega$ are of interest.

1. When $\omega \to \infty$, $\psi(\omega) \sim \omega^{-2}$. Hence at short times the VAF has the same dependence as

$$\int_\alpha^\infty \frac{d\omega}{\omega^2} \cos \omega t$$

where $\alpha$ is a cutoff frequency. Therefore

$$\psi(t) \underset{t \to 0}{\sim} a - bt$$

Consequently $\psi(t)$ exhibits a nonphysical cusp at $t = 0$ arising from the fact that the actual interatomic interactions have been described by boundary conditions at the surface of the sphere.

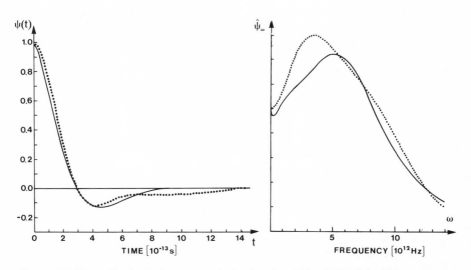

Figure 7.7. Normalized velocity autocorrelation function $\psi(t) = Z(t)\langle u_x^2 \rangle^{-1}$ and its spectrum $\psi(\omega)$ in argon: ..., computer simulation (after Rahman[39]); ——, result of Zwanzig and Bixon.[38]

2. When $\omega \to 0$, the generalized hydrodynamic calculation leads to

$$\psi(\omega) \underset{\omega \to 0}{\sim} a' - b'\omega^{1/2}$$

so that

$$\psi(t) \underset{t \to 0}{\sim} t^{-3/2}$$

A more careful analysis of the Navier–Stokes equations led Dorfman and Cohen[40] and Ernst *et al.*[41] to the following asymptotic expression for $\psi(t)$:

$$\psi(t) \underset{t \to \infty}{\sim} \frac{1}{d(d-1)} \frac{1}{n} \frac{1}{[4\pi(D+\nu)t]^{3/2}} \qquad (7.3.23)$$

where $d$ is the dimensionality of the system, $n$ the number density, and $\nu = \eta/nm$ the kinematic viscosity. This long-time behavior of $\psi(t)$ has been confirmed using molecular dynamics by Levesque and Ashurst[42] in Lennard-Jones fluids and by Alder and Wainwright[14] in hard-sphere fluids.

### 7.3.3. The Generalized Langevin Equation and the Memory-Function Formalism

The simple Langevin equation predicts an incorrect exponential decrease of $\psi(t)$ [equation (7.3.7)], and retardation effects have been neglected. At times $t' < t$, the tagged particle in fact creates a perturbation in the fluid that propagates in time and influences the force on the particle at time $t$. This effect can be described by the generalized Langevin equation

$$\frac{du_x}{dt} + \int_0^t K(t-\tau)u_x(\tau)d\tau = \frac{R_x(t)}{m} = a_x(t) \qquad (7.3.24)$$

where $R_x$ and $a_x$ are, respectively, the fluctuating force and acceleration of the tagged particle, and $K(t)$ is called the generalized friction coefficient or memory function. As in the simple Langevin equation [where $K(t) = 2\beta\delta(t)$], $K(t)$ is equal to the autocorrelation function of the fluctuating acceleration $a_x(t)$. This can be demonstrated as follows. Since $\langle a_x(t)u_x(0)\rangle = 0$, we obtain from equation (7.3.24)

$$\dot{\psi}(t) + \int_0^t K(t-\tau)\psi(\tau)d\tau = 0 \qquad (7.3.25)$$

from which it follows that $\dot{\psi}(0) = 0$, as it must be. After differentiation and integration by parts, this equation can also be written in the form

$$\ddot{\psi}(t) + K(t)\psi(0) + \int_0^t K(t - \tau)\dot{\psi}(\tau)d\tau = 0 \qquad (7.3.26)$$

which shows that $K(0) = -\ddot{\psi}(0)/\psi(0) = \Omega_0^2$. The autocorrelation function of the fluctuating acceleration is now

$$\langle a_x(t)a_x(0) \rangle = \langle \dot{u}_x(t)\dot{u}_x(0) \rangle + \int_0^t \langle u_x(\tau)\dot{u}_x(0) \rangle K(t - \tau)d\tau$$

$$= \langle u_x^2 \rangle K(t) \qquad (7.3.27)$$

from equation (7.3.26) and the fact that $\langle \dot{u}_x(t)\dot{u}_x(0) \rangle = -\ddot{\psi}(t)$.

This important result may be summarized as follows. The velocity autocorrelation function $\psi(t)$ obeys equation (7.3.25), where the memory function $K(t)$ is the autocorrelation function of the fluctuating acceleration $a_x(t)$, which enters the equation of motion of $u_x(t)$.

We may likewise assume that $a_x(t)$ itself obeys a similar generalized Langevin equation

$$\dot{a}_x(t) + \int_0^t M(t - \tau)a_x(\tau)d\tau = b_x(t) \qquad (7.3.28)$$

with $\langle b_x(t)a_x(0) \rangle = 0$, where $M(t)$, the new memory function, is related to the autocorrelation function $\langle b_x(t)b_x(0) \rangle$ of the new variable $b$. This process can be continued by writing a generalized Langevin equation for $b_x(t)$ and so on. The sequence of equations similar to equation (7.3.25) can be solved by using the Fourier–Laplace transformation. We define

$$\tilde{f}(s) = \int_0^\infty e^{-st}f(t)dt \qquad (s > 0) \qquad (7.3.29)$$

and obtain the continued fraction expansion of $\tilde{\psi}(s)$:

$$\tilde{\psi}(s) = \cfrac{\psi(0)}{s + \cfrac{K(0)}{s + \tilde{M}(s)}} = \cfrac{\psi(0)}{s + \cfrac{K(0)}{s + \cfrac{M(0)}{s + \cdots}}} \qquad (7.3.30)$$

Note that $K(0), M(0), \ldots$ can be obtained from the short-time expansion of

$\psi(t)$,

$$\psi(t) = 1 - \mu_2 \frac{t^2}{2!} + \mu_4 \frac{t^4}{4!} - \cdots$$

from which it follows that

$$\tilde{\psi}(s) = \frac{1}{s} - \mu_2 \frac{1}{s^3} + \mu_4 \frac{1}{s^5} - \cdots$$

Comparison with the expansion of equation (7.3.30) therefore yields

$$K(0) = \mu_2 = \Omega_0^2$$

$$M(0) = \frac{\mu_4 - \mu_2^2}{\mu_2}$$

where $\mu_n$ are the frequency moments of $\psi(\omega)$.

In order to obtain an approximate VAF we must truncate (7.3.30) at some step; this procedure guarantees that a number of moments of $\psi(\omega)$ will be reproduced correctly. Let us consider some examples.

In the simple Langevin theory $K(t) = 2\beta\delta(t)$, $\tilde{K}(s) = \beta$, and $\tilde{\psi}(s) = (s + \beta)^{-1}$, so

$$\psi(\omega) = \frac{1}{\pi} \operatorname{Re} \psi(-i\omega) = \frac{1}{\pi} \operatorname{Re} \frac{1}{-i\omega + \beta}$$

which is the Fourier transform of $\exp(-\beta|t|)$.

The next step is called the relaxation-time approximation. Here $K(t)$ is assumed to decay exponentially with time,[43,44]

$$K(t) = \Omega_0^2 \exp(-t/\tau)$$

in which case

$$\tilde{\psi}(s) = \frac{1}{s + \Omega_0^2/(s + 1/\tau)} \tag{7.3.31}$$

Within this approximation, the diffusion coefficient is given by

$$D = \frac{k_B T}{m} \tilde{\psi}(0) = \frac{k_B T}{m \Omega_0^2 \tau} \tag{7.3.32}$$

an expression similar to equation (7.3.17).

If $\tau$ is infinite, equation (7.3.31) shows that $\psi(t)$ oscillates with natural frequency $\Omega_0$. Therefore, as $\tau$ increases from zero to infinity, $\psi(t)$ evolves from a pure exponential to an oscillatory form. More accurately, it is easy to calculate the Laplace inverse of (7.3.31),

$$\psi(t) = \frac{1}{s_+ - s_-} (s_+ e^{-s_- t} - s_- e^{-s_+ t})$$

where $s_\pm$ are the poles of $\tilde{\psi}(s)$:

$$s_\pm = -\frac{1}{2\tau} [1 \mp (1 - 4\Omega_0^2 \tau^2)^{1/2}]$$

Therefore $\psi(t)$ oscillates when $s_\pm$ are complex numbers, i.e., when

$$\frac{Dm\Omega_0}{k_B T} < 2$$

a condition that is easily satisfied in classical fluids.

A number of postulated memory functions have been used in the literature,[43,45–47] and they can be inserted in the integrodifferential equation (7.3.25) to calculate $\psi(t)$. Harp and Berne[44] have checked the corresponding result against molecular-dynamics calculations.

### 7.3.4. The Mori–Zwanzig Formalism

In the preceding section it was assumed that the velocity, acceleration, etc., obey generalized Langevin equations. Mori and Zwanzig[48–50] showed that this procedure is valid and is actually equivalent to solving the Liouville equation.

The argument is as follows. If the tagged particle were only acted on by the friction force, its velocity $u_x(t)$ would be completely determined by $u_x(0)$. The role of the random force is to make $u_x(t)$ explore some states not necessarily correlated with $u_x(0)$. In the language of vectors, we can say that $u_x(t)$ explores a space that in some sense is orthogonal to the subspace spanned by $u_x(0)$.

Following Mori and Zwanzig, we may define the Hilbert space of dynamic variables $A$ with inner product

$$(A \mid B) = \int A^* B f_0 \, dp \, dq = \langle A^* B \rangle_0 \qquad (7.3.33)$$

where, for instance, $f_0$ is the equilibrium distribution function of the whole system. The dynamic variable $B$ has a component along $A$ (i.e., is correlated with $A$) if

$$P \mid B) = \mid A) \frac{(A \mid B)}{(A \mid A)} \text{ is nonzero}$$

In this Hilbert space, the time evolution of $A$ is given by the Liouville equation

$$\frac{dA(t)}{dt} = iLA(t) \tag{7.3.34}$$

with

$$iLA = \sum_{p,q} \left( \frac{\partial \mathcal{H}}{\partial p} \frac{\partial A}{\partial q} - \frac{\partial \mathcal{H}}{\partial q} \frac{\partial A}{\partial p} \right) = \{A, \mathcal{H}\}$$

where $\mathcal{H}(\mathbf{p}, \mathbf{q}, t)$ is the Hamiltonian of the system.

If $L$ does not depend on time the formal solution of (7.3.34) is

$$A(t) = e^{iLt}A$$

and, in general, the Liouville operator $e^{iLt}$ makes the variable $A(t)$ explore regions of the Hilbert space that are orthogonal to $A$. We may define the projector $P$ on $\mid A)$ and the complementary operator $Q = 1 - P$ so that

$$A(t) = PA(t) + QA(t) = A_S(t) + A_R(t) \tag{7.3.35}$$

where $A_S(t)$ remains correlated with $A$ while $A_R(t)$ is orthogonal to $A$.

By projecting out the Liouville equation (7.3.34) on the subspace spanned by $A$ and in the space orthogonal to $A$, it is easy to show that[48]

$$\frac{dA_S(t)}{dt} = PiLA_S(t) + PiL \int_0^t ds\, e^{iQL(t-s)} iQLA(s) \tag{7.3.36}$$

The scalar product of this equation with $\mid A)$ gives the following equation for the correlation function $C(t) = (A \mid A(t))$:

$$\dot{C}(t) - i\Omega C(t) + \int_0^t ds\, K(t-s)C(s) = 0 \tag{7.3.37}$$

where

$$i\Omega = \frac{(A \mid iLA)}{(A \mid A)}$$

$$K(t) = \frac{(LA \mid e^{iQLt}Q \mid LA)}{(A \mid A)} \tag{7.3.38}$$

Now if the correlation function $C(t)$ obeys the Langevin equation (7.3.37), it is clear that the variable $A(t)$ evolves in time according to the equation

$$\dot{A}(t) - i\Omega A(t) + \int_0^t ds\, K(t-s)A(s) = F(t)$$

provided $F(t)$ is orthogonal to $A$. A simple calculation shows that

$$F(t) = e^{iQLt}QiLA \tag{7.3.39}$$

and that the memory function $K(t)$ is actually the autocorrelation function of the random force $F(t)$,

$$K(t) = \frac{(F(0)\,|\,F(t))}{(A\,|\,A)} \tag{7.3.40}$$

This process can be continued by writing a Langevin equation for $F(t)$ in terms of a new random force $G(t)$ and a new memory function, which is the autocorrelation function of $G$. Such an approach has been used[51-57] for calculating the spectrum of the velocity autocorrelation function of a tagged particle. Equation (7.3.37) is written for $\psi(t)$, where $\Omega = 0$ because $(u_x \,|\, iLu_x) = 0$, and it is shown that

$$K(t) = (Lu_x \,|\, e^{iQLQt} \,|\, Lu_x)(u_x \,|\, u_x)^{-1} \tag{7.3.41}$$

where $L = QLQ$ is the restriction of the Liouville operator to the space orthogonal to $u_x$. In the same way

$$\dot{K}(t) + \int_0^t ds\, M(t-s)K(s) = 0 \tag{7.3.42}$$

with

$$M(t) = (QL^2 u_x \,|\, e^{iQ'QLQQ't} \,|\, QL^2 u_x)(u_x \,|\, L^2 \,|\, u_x)^{-1}$$

where $P'$ is the projector on the random force and $Q' = 1 - P'$.

The Langevin equations are then Laplace transformed to obtain the Fourier spectrum $\psi(\omega)$. The analysis is conducted in terms of two-mode variables $j_0(-\mathbf{k})\rho(\mathbf{k})$ and $j(\mathbf{k})\rho_0(-\mathbf{k})$, where $\rho_0(\mathbf{k}) = e^{-i\mathbf{k}\cdot\mathbf{r}_0}$ is the tagged particle density and $j(\mathbf{k}) = \sum_n \mathbf{u}_n e^{-i\mathbf{k}\cdot\mathbf{r}_n}$ is the current of bath particles. We refer the reader to the original papers for further details. The memory function $M(t)$ is split into $M_L(t) + 2M_T(t)$, where $M_L$ represents the coupling of the tagged particle to longitudinal modes and $M_T$ arises from

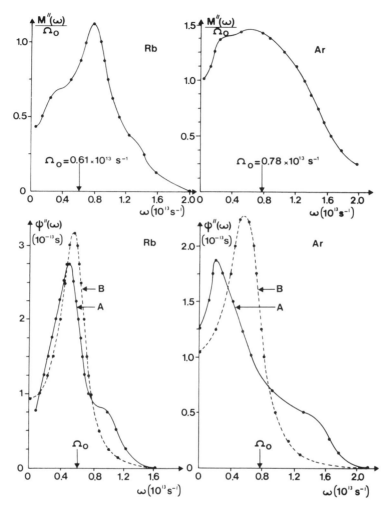

Figure 7.8. Fourier spectra of memory function $M(t)$ and velocity autocorrelation function $\psi(t)$. Curves A correspond to mode–mode coupling theory[54] and curves B to relaxation-time approximation.

coupling to shear excitations of the fluid. Figure 7.8 gives the results obtained by Bosse *et al.*[54] in argon and rubidium as compared with the relaxation-time approximation [where $\tilde{M}(s) = 1/\tau$]. Function $M''(\omega)$ shows a peak at $\Omega_0$, and coupling of the tagged particle to coherent bath modes ensures that the hydrodynamic modes are included correctly.

## 7.4. Kinetic Theory of Atomic Transport

If $f_s(1) = f_s(\mathbf{r}_1, \mathbf{v}_1, t_1)$ denotes the phase-space probability density of the tagged particle at point $\mathbf{r}_1$, $\mathbf{v}_1$ at time $t_1$, then the velocity autocorrelation function can be written in the form

$$\langle u_x(t_1)u_x(t_2)\rangle = \int d1\,d2\,v_x(t_1)v_x(t_2)\langle f_s(1)f_s(2)\rangle \qquad (7.4.1)$$

Kinetic theories attempt to calculate $f_s(\mathbf{r}, \mathbf{v}, t)$. This is a complicated many-body problem because at time $t_1$ the tagged particle disturbs the surrounding fluid, the disturbance propagates and reacts back on the tagged particle at time $t_2$. Again we shall satisfy two simple limits:

1. The tagged particle experiences frequent but soft collisions, as in Brownian motion. In this case we obtain the well-known Fokker–Planck equation.
2. The tagged particle undergoes rare but violent collisions. This leads to the Boltzmann and Enskog theories, useful in gases and moderately dense fluids.

The actual situation in liquids lies between these two extreme limits and we shall give some insight on modern kinetic theories of self-motion in liquids.

### 7.4.1. The Fokker–Planck Equation

If we assume a time interval $\tau$ can be defined that is both long compared to the time scale of molecular fluctuations and short with respect to the relaxation time of the velocity, we can write

$$f(v, t + \tau) = \int dv'f(v', t)p(v, v', \tau) \qquad (7.4.2)$$

where $f(v, t)$ is the velocity distribution of the tagged particle in a one-dimensional system. The transition probability $p(v, v', \tau)$ is a function of the velocity $v$ and the velocity increment $w = v - v'$. It can be written formally as[23]

$$p(v, w, \tau) = \left\langle \delta\left[ w - \int_0^\tau \frac{F(t')}{m}\,dt' \right] \right\rangle \qquad (7.4.3)$$

where $F$, the total force on the particle, can be split into the systematic retarding force $-\zeta v$ and the random force $R$. The moments of $p(v, w, \tau)$

are therefore

$$\frac{\overline{w(\tau)}}{\tau} = \frac{1}{\tau} \int_{-\infty}^{+\infty} p(v, w, \tau) w \, dw = \frac{1}{\tau} \int_0^\tau \frac{\langle F(t') \rangle}{m} \, dt' = -\frac{\zeta}{m} v$$

$$\frac{\overline{w^2(\tau)}}{2\tau} = \frac{1}{2\tau} \int_0^\tau \int_0^\tau dt' dt'' \frac{\langle R(t') R(t'') \rangle}{m^2}$$

$$= \int_0^\tau \frac{\langle R(t') R(0) \rangle dt'}{m^2} = k_B T \frac{\zeta}{m^2}$$

(7.4.4)

according to equation (7.3.15).

The velocity transfer is a random process whose distribution becomes centered on $(-\zeta v/m)\tau$ and of width given by $\overline{w^2(\tau)}$. We can now write

$$f(v, t + \tau) = \int_{-\infty}^{+\infty} f(v - w, t) p(v, w, \tau) d\tau$$

and expand $f(v - w, t)$ in a Taylor series to obtain

$$\frac{\partial f(v, t)}{\partial t} = \sum_{n=1}^\infty \frac{(-1)^n}{n!} \frac{\partial^n}{\partial v^n} \left( \frac{\overline{w^n}}{\tau} f \right)$$

When limited to the first two terms, this is the Fokker–Planck equation

$$\frac{\partial f(v, t)}{\partial t} = \frac{\zeta}{m} \frac{\partial}{\partial v} \left( vf + \frac{k_B T}{m} \frac{\partial f}{\partial v} \right)$$

(7.4.5)

which becomes in the general case

$$\frac{\partial f(\mathbf{r}, \mathbf{v}, t)}{\partial t} + \mathbf{v} \cdot \frac{\partial f}{\partial \mathbf{r}} = \frac{\zeta}{m} \frac{\partial}{\partial \mathbf{v}} \cdot \left( \mathbf{v} + \frac{k_B T}{m} \frac{\partial}{\partial \mathbf{v}} \right) f = \hat{A} f$$

(7.4.6)

where $\hat{A}$ is the Fokker–Planck operator.

In principle, it is possible to obtain the solution of this equation compatible with the initial conditions, and to deduce the same kind of information as in the Langevin theory. If we are only interested in the diffusion constant, we must study the long-time behavior of $f(\mathbf{r}, \mathbf{v}, t)$ or its space Fourier transform $\tilde{f}(\mathbf{q}, \mathbf{v}, t)$, because we know that at long times

$$\int \tilde{f}(\mathbf{q}, \mathbf{v}, t) d\mathbf{v} \sim \exp(-q^2 Dt)$$

Fourier transformation of equation (7.4.6) yields

$$\frac{\partial \tilde{f}}{\partial t} = (\hat{A} - i\mathbf{q} \cdot \mathbf{v})\tilde{f} \tag{7.4.7}$$

where $-i\mathbf{q} \cdot \mathbf{v}$ can be treated as a small perturbation in the diffusion regime ($q$ small).

The nonperturbed equation

$$\frac{\partial \tilde{f}}{\partial t} = \hat{A}\tilde{f} \tag{7.4.8}$$

can be solved if the eigenfunctions of $\hat{A}$ are known,[57]

$$\hat{A} \left| \psi_n^0 \right\rangle = A_n^0 \left| \psi_n^0 \right\rangle \tag{7.4.9}$$

The vectors $\left| \psi_n^0 \right\rangle$ belong to the Hilbert space with inner product

$$\langle f \,|\, g \rangle = \int d\mathbf{v} \varphi_0^{-1}(\mathbf{v}) f^*(v) g(v)$$

where $\varphi_0(\mathbf{v})$ is the equilibrium distribution function of the velocity. The general solution of equation (7.4.8) is therefore

$$\tilde{f} = \sum_n a_n \exp(A_n^0 t)$$

where the $a_n$ are determined by the initial conditions. All $A_n^0$ are negative and the smallest in absolute value, which determines the long-time behavior of $\tilde{f}$, is $A_0^0 = 0$ $[\tilde{f} = a_0 = \varphi_0(v)]$.

The leading term in the perturbation expansion of $\tilde{f}$ is thus given by the change in $A_0^0$ when the perturbation is introduced. This can be obtained using standard perturbation theory:

$$A_0^q = A_0^0 + \langle \varphi_0 \left| - iqv_x \right| \varphi_0 \rangle + \sum_{n \neq 0} \frac{\langle \varphi_0 \left| - iqv_x \right| \psi_n^0 \rangle \langle \psi_n^0 \left| - iqv_x \right| \varphi_0 \rangle}{- A_n^0}$$

We may as well introduce $n = 0$ in the sum over $n$ for $\langle \psi_0 \,|\, v_x \,|\, \psi_0 \rangle = 0$, but we must correspondingly add an infinitesimal $\varepsilon$ to the denominator to avoid divergence. Hence

$$D = -\frac{A_0^q}{q^2} = -\lim_{\varepsilon \to 0} \left\langle \varphi_0 \left| v_x \frac{1}{\hat{A} - \varepsilon} v_x \right| \varphi_0 \right\rangle \tag{7.4.10}$$

It is easy to show that $|v_x\varphi_0\rangle$ is an eigenvector of $\hat{A}$ with eigenvalue $-\zeta/m$, in which case

$$D = \lim_{\varepsilon \to 0} \left\langle \varphi_0 v_x \left| \frac{1}{\zeta/m + \varepsilon} \right| v_x\varphi_0 \right\rangle = \frac{m}{\zeta} \langle \varphi_0 | v_x^2 | \varphi_0 \rangle = \frac{k_B T}{\zeta} \quad (7.4.11)$$

This method of calculation will be used for solving the Boltzmann and Enskog equations.[58]

### 7.4.2. The Boltzmann and Enskog Equations

Now consider the case when the tagged particle undergoes rare but large momentum transfers, as in dilute gas. Again we assume that there exists a time $\tau$ long compared with the duration $\tau_c$ of a collision but short with respect to the time $\tau_R$ between successive collisions. We can again use equation (7.4.2) for $f(v, t)$ with[23]

$$p(v, v', \tau) = \lambda \delta(v' - v) + \tau W(v', v) \quad (7.4.12)$$

where the first term expresses the absence of collisions during the time interval $\tau$ and $W(v', v)$ is the transition probability $v' \to v$ per unit time. If, in equation (7.4.2), we expand $f(v, t + \tau)$ in a Taylor series we obtain the Master Equation

$$\frac{\partial f(\mathbf{r}, \mathbf{v}, t)}{\partial t} + \mathbf{v} \cdot \frac{\partial f}{\partial \mathbf{r}} = nCf \quad (7.4.13)$$

where $C$ is the operator defined by

$$nCf(\mathbf{r}, \mathbf{v}, t) = \int d\mathbf{v}'[W(\mathbf{v}', \mathbf{v}, \mathbf{r})f(\mathbf{r}, \mathbf{v}', t) - W(\mathbf{v}, \mathbf{v}', \mathbf{r})f(\mathbf{r}, \mathbf{v}, t)] \quad (7.4.14)$$

assuming that the particle size is negligible. Function $W(\mathbf{v}', \mathbf{v}, \mathbf{r})$ depends on the number of bath particles located at $\mathbf{r}$ with velocity $\mathbf{v}'$ and such that, after the collision, the velocity of the tagged particle becomes $\mathbf{v}$.

The Boltzmann equation can be obtained using the approximation that there is no spatial or velocity correlation between the tagged and bath particles. The number of bath particles at $\mathbf{r}$ and in the range $d\mathbf{v}'$ is then $n\varphi_b(\mathbf{v}')d\mathbf{v}'$ and equation (7.4.13) becomes linear:

$$\frac{\partial f}{\partial t} + \mathbf{v} \cdot \frac{\partial f}{\partial \mathbf{r}} = nC^{(1)}f \quad (7.4.15)$$

By analogy with the solution to the Fokker–Planck equation the diffusion coefficient is obtained in the form

$$D = -\lim_{\varepsilon \to 0} \left\langle \varphi_0 \left| v_x \frac{1}{nC^{(1)} - \varepsilon} v_x \right| \varphi_0 \right\rangle \qquad (7.4.16)$$

The eigenvalues of $C^{(1)}$ depend on the detailed dynamics of the collisions. In a fluid of hard spheres of diameter $\sigma$ it can be shown that[58]

$$D_0 = \frac{3}{8n\sigma^2} \left( \frac{k_B T}{\pi m} \right)^{1/2} \qquad (7.4.17)$$

Enskog[59] much improved this result. He retained the assumption of two-body collisions but took into account spatial correlations. The collision rate is no longer controlled by $n$, the number density, but by $ng(\sigma)$, where $g(\sigma)$ is the radial distribution function at contact. The argument can be made more rigorous[59] and leads to the Enskog diffusion coefficient in a hard-sphere fluid:

$$D_E = \frac{D_0}{g(\sigma)} = \frac{3}{8n\sigma^2 g(\sigma)} \left( \frac{k_B T}{\pi m} \right)^{1/2} \qquad (7.4.18)$$

Although dynamic polarization of the bath by the tagged particle has been neglected, this equation is extremely useful for evaluating $D$ in van der Waals and metallic liquids. An effective hard-sphere diameter $\sigma(T)$ can be defined and allows us to simulate the equation of state of the actual liquid. This procedure was employed by Dymond and Alder[60] in rare-gas liquids, and by Ascarelli and Paskin[61] and Vadovic and Colver[62] in metallic liquids.

The best results are obtained when the Carnahan–Starling equation of state[63,64] is used to determine $g(\sigma)$:

$$Z = \frac{p}{nk_B T} = \frac{1 + \eta + \eta^2 - \eta^3}{(1 - \eta)^3} = 1 + 4\eta g(\sigma) \qquad (7.4.19)$$

where $\eta = \pi n\sigma^3/6$ is the packing fraction, so that $D_E$ is entirely determined by $n$ and the diameter $\sigma(T)$. The effective diameter $\sigma(T)$ depends on the temperature, because the closest distance of approach decreases as the energy of two-body collisions increases. This parametrization of $\sigma(T)$ has been studied in detail by Protopapas et al.,[65] who suggest the following expression for $\sigma(T)$:

$$\sigma(T) = \sigma_0[1 - B(T/T_m)^{1/2}] \qquad (7.4.20)$$

where $B$, the relative vibration amplitude of the atoms around $\sigma_0$ at the melting point, is known to vary little among metallic systems.

### 7.4.3. Full Kinetic Theories

In the spirit of the Boltzmann–Enskog theory the binary collisions are regarded as uncorrelated. This is valid at low density, but as the density increases the tagged particle polarizes dynamically the bath and successive collisions can be highly correlated. The correlation processes are schematized in Figure 7.9, where scheme (b) shows a sequence of two correlated collisions: at the space–time point B the tagged particle collides again with the particle (2) with which it had previously collided and that may have retained the memory of the previous collision. These collisions are of the general type described in scheme (c): after the first interaction between (1) and (2) at A, the tagged particle undergoes a series of uncorrelated collisions with bath particles and collides at point B with a bath particle $(n)$, which (directly or indirectly) has retained the memory of the first encounter at A. The particle $(n)$ may be the same as (2) or different. In this "ring-collision" mechanism, the collisions at A and B are correlated and it is clear that such correlated sequences must be properly accounted for in order to properly describe self-diffusion in dense fluids.

A first step when introducing dynamic correlations consists in the

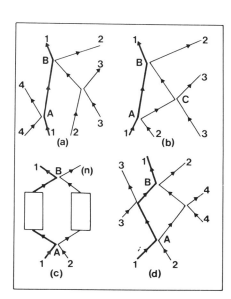

Figure 7.9. Successive collisions of the tagged particle (1) may be uncorrelated (a) or correlated [(b) and (d)]; scheme (c) shows a ring collision.

generalization of equation (7.4.16),

$$D \sim \frac{1}{C}$$

where

$$C = nC^{(1)} + n^2 C^{(2)} + \cdots \tag{7.4.21}$$

The quantity $C^{(2)}$, for example, describes processes involving two collisions of the marked atom with bath particles. According to equation (7.4.21) we could hope to obtain a density expansion of the diffusion constant $D$. Unfortunately some terms in the expansion of $D$ diverge; for instance, the fraction of binary collisions that give rise to a recollision at any later time is infinite (in a two-dimensional system). Physically, these divergences occur because we allow for recollisions that take place a long time after the first encounter. In the actual situation, however, particles (1) and (2) are not free but interact with bath particles, which act as a screen that on the whole prevents long-time recollision processes.[58] This kind of argument leads to a nonanalytic expansion of $D$ in terms of the number density $n$ and hard-sphere diameter $\sigma$:

$$D \sim \sigma \bar{v} [n + K_1 (n\sigma^3)^2 + K_2 (n\sigma^3)^3 \ln(n\sigma^3) + \cdots]^{-1} \tag{7.4.22}$$

The proper way to tackle this divergence problem is to consider rings, which exhaust the most diverging diagrams. This problem has been studied in detail and we refer the reader to the original papers.[66–73] The problem is to calculate the velocity autocorrelation function $\psi(t)$ from first principles. According to equation (7.4.1), the Laplace transform (LT) of $\psi(t)$ is given by

$$\psi(z) = \frac{1}{n_i \Omega} \int d1\, d2\, v_{1x} C_s(12) v_{2x} \tag{7.4.23}$$

where $n_i$ is the number of tagged particles per unit volume, $\Omega$ is the volume of the system, and $C_s(12)$ is the Laplace transform of the tagged-particle density correlation function:

$$C_s(12) = \mathrm{LT}[\langle f_s(1, t_1) f_s(2, t_2) \rangle]$$

It is easy to show that $C_s(12)$ obeys the kinetic equation

$$[z - L_0(1)] C_s(12) - \int d3\, \phi_s(13) C_s(32) = \tilde{C}_s(12) \tag{7.4.24}$$

where $L_0(1)$ is the tagged-particle free-streaming Liouville operator, $\phi_s(13)$ is the LT of the memory function, and $\tilde{C}_s(12)$ is the equal time correlation function. The equation of motion of $\phi_s$ involves higher-order correlation functions, which must be approximated in some way, and $\phi_s$ can be formally written as

$$\phi_s = -V_s \bar{G}_M V_s + \delta\phi_s \qquad (7.4.25)$$

where $\bar{G}_M$ is the mean field value of the four-point correlation function $G_s(1\ \bar{1},\ 2\ \bar{2})$, describing the event where the tagged particle at point 1 interacts with a bath particle at $\bar{1}$ and later with another (or the same) bath particle at point 2. Geometric correlations are included in $G_M$ but all dynamic correlations are neglected, so that the first term in equation (7.4.25) contains the Enskog approximation to $\phi_s$. The quantity $V_s$ in this equation represents the interaction between the tagged particle and bath particles. As a result of the screening and excluded volume effects, the particles interact by the mean force potential $-kT \ln g(r)$ rather than by the bare interatomic potential $V(r)$. If we use equation (7.4.25), neglect the correction term $\delta\phi_s$, and make a simple approximation to $\bar{G}_M$, we recover the Enskog theory where the tagged particle undergoes only uncorrelated collisions with bath particles.

The corrections to the Enskog theory are contained in $\delta\phi_s$ and can be separated into two classes.

*Class 1.* In the *ring approximation*, we can write

$$\delta\phi_s = -\tau G' \tau^T \qquad (7.4.26)$$

where $\tau$ and its transpose $\tau^T$ represent Enskog binary encounters between the tagged particle and bath particles. Equation (7.4.26) represents an initial collision between the tagged particle $\alpha$ and a bath particle $\beta$ (through $\tau^T$), then some complicated intermediate propagation (through $G'$), and a terminating collision ($\tau$) between $\alpha$ and $\beta$ or another bath particle, which has interacted with $\beta$ since the first collision; $\tau^T$ and $\tau$ therefore represent correlated collisions. Using this description, Mehaffey and Cukier[73] calculated $\psi(t)$ and the diffusion constant $D_i$ of hard spheres of diameter $\sigma_i$ and mass $m_i$ in a bath of spheres of diameter $\sigma_s$ and mass $m_s$, in the limit $\sigma_i \gg \sigma_s \gtrsim l$, where $l$ is the mean free path of bath particles. If the Enskog velocity-relaxation frequency $\lambda_E$ is defined by

$$\lambda_E = \frac{4m_s}{3(m_i + m_s)\tau_{is}}$$

where $\tau_{is}^{-1}$ is the tagged particle–bath particle collision frequency, then the

Enskog approximation becomes

$$\psi_E(t) = \frac{k_B T}{m_i} \exp(-\lambda_E t) \tag{7.4.27}$$

and

$$D_i^E = \frac{k_B T}{m_i \lambda_E} \tag{7.4.28}$$

This time dependence of $\psi_E(t)$ is clearly incorrect at either short or long times.

In the ring approximation, equations (7.4.27) and (7.4.28) are replaced by

$$\psi_R(t) \underset{t \to \infty}{\sim} \frac{2}{3} \left(1 - \frac{r_1}{\lambda_E}\right)^2 \frac{k_B T}{n_s m_s} (4\pi \nu_E t)^{-3/2} \tag{7.4.29}$$

and

$$\frac{D_i^R}{D_i^E} = \frac{1}{(1 - r_1/\lambda_E)} \tag{7.4.30}$$

where

$$\frac{r_1}{\lambda_E} \sim n_s \sigma_s^3 \frac{\sigma_{is}}{\sigma_s} \left(\frac{m_i}{m_i + m_s}\right)^{1/2}$$

and $\nu_E$ is the Enskog kinematic shear viscosity.

Equation (7.4.29) exhibits the correct long-time behavior of $\psi(t)$ but, when the tagged particle is large ($\sigma_i \gg \sigma_s$), the coefficient of $t^{-3/2}$ depends on $(\sigma_i/\sigma_s)^2$ while fluctuating hydrodynamics predicts that this coefficient is independent of $\sigma_i$. Moreover, equation (7.4.30) predicts that as $r_1/\lambda_E$ increases, $D_i^R/D_i^E$ increases and may become negative, clearly a nonphysical result.

Note, however, that for tagged particles not too large the ring theory provides a good description of $\psi(t)$ and $D$. For instance, Furtado et al.[73] express the memory function $K(t)$ of the velocity autocorrelation function as follows:

$$K(t) = \lambda_E \delta(t) + \delta K(t)$$

where $\lambda_E \delta(t)$ is the Enskog contribution. The part of $\delta K(t)$ due to the ring collision is calculated in a quasi-hydrodynamic calculation: the contribution of the five hydrodynamic states is calculated exactly and the nonhydrodynamic modes are taken into account approximately. The VAF $\psi(t)$

and ratio $D/D_E$ calculated using this procedure exhibit the same variation with density as the computer-simulation results (Figure 7.10).

In particular, the enhancement of $D/D_E$ at intermediate density is obtained, as well as the long-time tail of $\psi(t)$. These features are also apparent in the calculation of Mazenko,[73] who considers only hydro-dynamic contributions, and in the work of Résibois,[69] who determines the contribution of correlated sequences of two binary collisions between three particles.

*Class 2.* As the size of the tagged particle increases, repeated colli-sions become important because the bath particles repeatedly return to the tagged particle. The *repeated ring approximation* is now necessary in order to remove the nonphysical divergence of $D$ found in the ring theory. Using this description, Mehaffey and Cukier[73] show that

$$\psi(t) \underset{t \to \infty}{\sim} \frac{2}{3}\left(\frac{k_B T}{n_s m_s}\right)(4\pi\nu_E t)^{-3/2} \tag{7.4.31}$$

and

$$D_i^{RR} = D_i^E + \frac{2k_B T}{5\pi\eta_E\sigma_i}$$

where $\eta_E = \nu_E n_s m_s$ is the Enskog value of the fluid shear viscosity.

The coefficient of $t^{-3/2}$ in $\psi(t)$ now agrees well with the results of fluctuating hydrodynamics. For very large particles the second term in $D_i^{RR}$, representing the contribution of successive correlated collisions, predominates and the diffusion constant assumes the Stokes form.

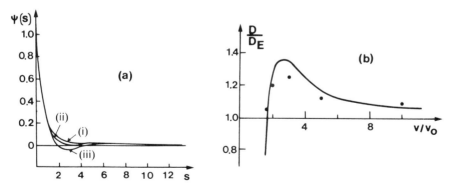

Figure 7.10. (a) VAF $\psi(s)$ for a hard-sphere fluid with different values of the packing fraction $y$: (i) $y = 0.0741$, (ii) $y = 0.2468$, (iii) $y = 0.4628$, $s = \frac{3}{2}\lambda_E t$. (b) Variation of the self-diffusion constant ratio $D/D_E$ with $V/V_0$, where dots denote computer molecular-dynamics results (after Furtado *et al.*[73]).

A generalization of these calculations to systems with continuous-interaction potentials can be found in a series of papers by Sjögren and Sjölander,[72] who show in particular that the Laplace transform of the memory function $K(t)$ of the VAF can be written in the form

$$K(z) = \frac{K^B(z) + R_{00}(z) + K^B(z)R_{01}(z)}{1 - R_{01}(z) - K^B(z)R_{11}(z) - R'_{22}(z)} \qquad (7.4.32)$$

where

$$R'_{22}(z) = [K^B(z) + R_{00}(z) + K^B(z)R_{01}(z)]R_{22}(z)$$

Here $K^B(z)$ represents the memory function due to single binary collisions, $R_{00}$, $R_{01}$, and $R_{11}$ describe the coupling to the density and the longitudinal current, and $R_{22}$ is the contribution arising from coupling to the transverse current. Expression (7.4.32) can be shown to lead to the following memory function $K(t)$:

$$K(t) = A \exp(-at^2) + Bt^4 \exp(-bt)$$

which is valid at short times, as shown by Levesque and Verlet.[12]

In the long-time limit, $K(t)$ exhibits the correct $t^{-3/2}$ behavior. Figure 7.11 shows a typical result obtained using equation (7.4.32) in rubidium and the comparison with molecular-dynamics data.

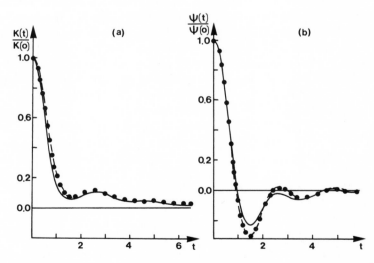

Figure 7.11. Memory function $K(t)$ of (a) the VAF and (b) the VAF $\psi(t)$ in rubidium. The unit of time is $\tau_0 = 3.212 \times 10^{-13}$ s (after Sjögren[72]).

## 7.5. Conclusion

In this chapter we have sketched some general methods used to describe the thermal motion of a tagged particle immersed in a fluid. The computational difficulty arises essentially from the fact that the diffusion behavior takes place after the particle has experienced a minimum number of correlated collisions. Two simple limits are useful to consider:

1. If the marked atom undergoes frequent but soft collisions from an equilibrium bath, the classical Brownian motion theory — through the simple Langevin or the usual Fokker–Planck equations — easily provides the velocity autocorrelation function and the diffusion constant. Although it is only valid for a large tagged particle, this theory sheds some light on the relationship between the friction coefficient and the correlation function of the force acting on the particle.

2. If the collisions are violent but rare — as in a dilute gas — they can be considered independent of one another. The simple kinetic models provided by the Boltzmann–Enskog equations lead to diffusion constants, which are of the right order of magnitude, although dynamic correlations are completely neglected.

As the density increases from that of a gas, one can no longer ignore dynamic correlations between collisions. The tagged particle induces a backflow in the fluid that reacts back on it at later times, leading to memory effects. These effects can be described phenomenologically through the memory-function formalism, but their description from first principles is a complicated many-body problem. Molecular-dynamics calculations have largely contributed to our understanding of some simple effects, such as the backscattering effect at large densities and the persistence of the velocity at intermediate densities. This persistence effect is attributed to coupling of the tagged particle to the hydrodynamic modes, which decay slowly and give a long-time tail to the velocity autocorrelation function. It is comparatively easy to describe either the short-time dynamics or the long-time behavior of a tagged particle. The difficult problem is to bridge the gap in between and to pass from individual collisions to the diffusion behavior of the particle.

## References

1. N. H. March, in: *Theory of Condensed Matter*, p. 93, I. A. E. A., Vienna (1968).
2. J. P. Hansen and I. R. McDonald, *Theory of Simple Liquids*, Academic Press, New York (1976).
3. J. P. Hansen, in: *Microscopic Structure and Dynamics of Liquids* (J. Dupuy and A. J. Dianoux, eds.), Vol. 33, p. 3, Nato Advanced Study Institutes Series. Plenum Press, New York (1978).

4. K. Sköld, J. M. Rowe, G. Ostrowski, and P. D. Randolph, *Phys. Rev. A* **6**, 1107 (1972).
5. J. R. D. Copley and S. W. Lovesey, *Rep. Prog. Phys.* **38**, 461 (1975).
6. P. A. Egelstaff and P. Schofield, *Nucl. Sci. Eng.* **12**, 260 (1962).
7. D. Levesque, L. Verlet, and J. Kurkijärvi, *Phys. Rev. A* **7**, 1690 (1973).
8. S. W. Lovesey, *J. Phys. C* **6**, 1856 (1973).
9. S. W. Lovesey, in: *Dynamics of Solids and Liquids by Neutron Scattering* (S. W. Lovesey and T. Springer, eds.), Springer-Verlag, Berlin (1977).
10. B. R. A. Nijboer and A. Rahman, *Physica* **32**, 415 (1966).
11. P. Schofield, *Proc. I. A. E. A. Symposium*, p. 39, Vienna (1961).
12. D. Levesque and L. Verlet, *Phys. Rev. A* **2**, 2514 (1970).
13. A. Rahman, *Phys. Rev. A* **136**, 405 (1964).
14. B. J. Alder and T. E. Wainwright, *Phys. Rev. Lett.* **18**, 988 (1967). B. J. Alder and T. E. Wainwright, *J. Phys. Soc. Jpn., Suppl.* **26**, 267 (1968). B. J. Alder and T. E. Wainwright, *Phys. Rev. A* **1**, 18 (1970). B. J. Alder, D. M. Gass, and T. E. Wainwright, *J. Chem. Phys.* **53**, 3813 (1970). B. J. Alder, W. E. Alley, and J. H. Dymond, *J. Chem. Phys.* **56**, 987 (1972); *J. Chem. Phys.* **61**, 1415 (1974).
15. K. Toukubo and K. Nakanishi, *J. Chem. Phys.* **65**, 1937 (1976).
16. J. P. Hansen and I. R. McDonald, *Phys. Rev. A* **11**, 2111 (1975).
17. G. Jacucci and I. R. McDonald, *Physica A* **80**, 607 (1975).
18. I. Ebbsjö, P. Schofield, K. Sköld, and I. Waller, *J. Phys. C* **7**, 3891 (1974).
19. M. Parrinello, M. P. Tosi, and N. H. March, *J. Phys. C* **7**, 2577 (1974).
20. A. Einstein, *Ann. Physik* **17**, 549 (1905); **19**, 371 (1906). P. Langevin, *C. R. Acad. Sci. Paris* **146**, 530 (1908). S. Chandrasekhar, *Rev. Mod. Phys.* **15**, 1 (1943).
21. L. S. Ornstein and W. R. van Wijk, *Physica* **1**, 235 (1933).
22. S. A. Rice and P. Gray, *The Statistical Mechanics of Simple Liquids*, Monographs in Statistical Physics, Vol. 3, Interscience Publ., New York (1965).
23. Ph. Noziĕres, Lecture Notes, Grenoble University (unpublished).
24. J. Lebowitz and E. Rubin, *Phys. Rev.* **131**, 2381 (1963).
25. P. Résibois and H. T. Davis, *Physica* **30**, 1077 (1964).
26. J. G. Kirkwood, *J. Chem. Phys.* **14**, 180 (1946).
27. G. Boato, G. Casanova, and A. Levi, *J. Chem. Phys.* **40**, 2419 (1964).
28. J. A. Palyvos and H. T. Davis, *J. Phys. Chem.* **71**, 439 (1967).
29. H. T. Davis and J. A. Palyvos, *J. Chem. Phys.* **46**, 4043 (1967).
30. R. V. Gopala Rao and A. K. Murthy, *Z. Naturforsch.* **30a**, 619 (1975).
31. K. Ichikawa and M. Shimoji, *Phil. Mag.* **20**, 341 (1969).
32. E. Helfand, *Phys. Fluids* **4**, 681 (1961).
33. S. Chapman and T. G. Cowling, *The Mathematical Theory of Non-Uniform Gases*, Cambridge University Press, New York (1958).
34. H. C. Longuet-Higgins and J. A. Pople, *J. Chem. Phys.* **25**, 884 (1956).
35. S. A. Rice, J. G. Kirkwood, J. Ross, and R. W. Zwanzig, *J. Chem. Phys.* **31**, 575 (1959).
36. J. T. O'Toole and J. S. Dahler, *J. Chem. Phys.* **33**, 1496 (1960).
37. G. C. Stokes, *Trans. Camb. Phil. Soc.* **8**, 287 (1845). J. Boussinesq, *J. Math. Pures Appl.* (2) **13**, 377 (1868). Barré de Saint Venant, *C. R. Acad. Sci. Paris* **17**, 1240 (1843).
38. R. Zwanzig and M. Bixon, *Phys. Rev. A* **2**, 2005 (1970).
39. A. Rahman, *J. Chem. Phys.* **45**, 2585 (1966).
40. J. R. Dorfman and E. G. D. Cohen, *Phys. Rev. Lett.* **25**, 1257 (1970); *Phys. Rev. A* **6**, 776 (1972); *Phys. Rev. A* **12**, 292 (1975).
41. M. H. Ernst, E. H. Hauge, and J. M. J. van Leeuwen, *Phys. Rev. Lett.* **25**, 1254 (1970); *Phys. Rev. A* **4**, 2055 (1971).
42. D. Levesque and W. T. Ashurst, *Phys. Rev. Lett.* **33**, 277 (1974).
43. B. J. Berne, J. P. Boon, and S. A. Rice, *J. Chem. Phys.* **45**, 1086 (1966).

44. G. D. Harp and B. J. Berne, *Phys. Rev. A* **2**, 975 (1970).
45. K. Singwi and S. Tosi, *Phys. Rev.* **157**, 153 (1967).
46. P. Martin and S. Yip, *Phys. Rev.* **170**, 151 (1968).
47. R. Desai and S. Yip, *Phys. Rev.* **166**, 129 (1968).
48. R. Zwanzig, in: *Molecular Fluids* (R. Balian and G. Weill, eds.), Gordon and Breach, London (1973).
49. H. Mori, *Prog. Theor. Phys.* **33**, 423 (1965).
50. H. Mori, *Prog. Theor. Phys.* **34**, 399 (1965).
51. J. Bosse, W. Götze, and M. Lücke, *Phys. Rev. A* **17**, 434 (1978); *Phys. Rev. A.* **17**, 447 (1978).
52. A. Zippelius and W. Götze, *Phys. Rev. A* **17**, 414 (1978).
53. J. Bosse, W. Götze, and M. Lücke, *Phys. Rev. A* **18**, 1176 (1978).
54. J. Bosse, W. Götze, and A. Zippelius, *Phys. Rev. A* **18**, 1214 (1978).
55. T. Munakata and I. Igarashi, *Prog. Theor. Phys.* **58**, 1345 (1977); *Prog. Theor. Phys.* **60**, 45 (1978).
56. G. S. Dubey, V. K. Jindal, and K. N. Pathak, *Prog. Theor. Phys.* **64**, 1893 (1980).
57. B. F. McCoy, *Phys. Rev. A* **12**, 1678 (1975).
58. P. Résibois and M. de Leener, *Classical Kinetic Theory of Fluids*, John Wiley and Sons, New York (1977).
59. S. Chapman and T. G. Cowling, *The Mathematical Theory of Non-Uniform Gases*, Cambridge University Press, New York (1970).
60. J. H. Dymond and B. J. Alder, *J. Chem. Phys.* **48**, 343 (1968). J. H. Dymond, *Physica* **75**, 100 (1974). J. H. Dymond, *J. Chem. Phys.* **60**, 969 (1972).
61. P. Ascarelli and A. Paskin, *Phys. Rev. B* **165**, 222 (1968).
62. C. J. Vadovic and C. P. Colver, *Phys. Rev. B* **1**, 4850 (1970).
63. N. F. Carnahan and K. E. Starling, *J. Chem. Phys.* **51**, 635 (1969).
64. L. Verlet and J. J. Weis, *Phys. Rev. A* **5**, 939 (1971).
65. P. Protopapas, H. C. Andersen, and N. A. D. Parlee, *J. Chem. Phys.* **59**, 15 (1973). P. Protopapas and N. A. D. Parlee, *J. Chem. Phys.* **11**, 201 (1975). P. Protopapas and N. A. D. Parlee, *High Temp. Sci.* **7**, 259 (1975).
66. K. S. Singwi and A. Sjölander, *Phys. Rev. A* **167**, 152 (1968). K. S. Singwi, in: *Theory of Condensed Matter*, p. 603, I.A.E.A., Vienna (1968).
67. A. Z. Akcasu and E. Daniels, *Phys. Rev. A* **2**, 962 (1970). A. Z. Akcasu and J. J. Duderstadt, *Phys. Rev.* **188**, 479 (1969).
68. J. L. Lebowitz, J. K. Percus, and J. Sykes, *Phys. Rev.* **188**, 487 (1969).
69. P. Résibois and J. L. Lebowitz, *J. Stat. Phys.* **12**, 483 (1975). P. Résibois, *J. Stat. Phys.* **13**, 393 (1975). Y. Pomeau and P. Résibois, *Phys. Rep. C* **19**, 63 (1975).
70. C. D. Boley, *Ann. Phys.* **86**, 91 (1974); *Phys. Rev. A* **11**, 328 (1975).
71. E. P. Gross, *J. Stat. Phys.* **11**, 503 (1974); **15**, 181 (1976).
72. L. Sjögren and A. Sjölander, *Ann. Phys.* **110**, 122 (1978); **110**, 421 (1978); *J. Phys. C* **12**, 4369 (1979). L. Sjögren, *Ann. Phys.* **110**, 156 (1978); **110**, 173 (1978); **113**, 304 (1978); *J. Phys. C* **11**, 1493 (1978); **13**, 705 (1980). A. Sjölander, in: *Liquid and Amorphous Metals* (E. Lüscher and H. Coufal, eds.), p. 63, Nato Advanced Study Institutes Series, Vol. 36, Sijthoff and Noordhoff International Publishers B.V., Alphen aan den Rijn, The Netherlands (1980).
73. G. F. Mazenko, *Phys. Rev. A* **7**, 209 (1973); **7**, 222 (1973); **9**, 360 (1974). P. M. Furtado, G. F. Mazenko, and S. Yip, *Phys. Rev. A* **12**, 1653 (1975). J. R. Mehaffey, R. C. Desai, and R. Kapral, *J. Chem. Phys.* **66**, 1665 (1977). J. R. Mehaffey and R. I. Cukier, *Phys. Rev. A* **17**, 1181 (1978).

# Structure and Dynamics of Charged Fluids

## J. P. Hansen

### 8.1. Introduction

"Charged fluids" is a generic name for a vast variety of gaseous or liquid systems containing charged particles like positive or negative ions and radicals, charged polymers, and free electrons. Systems of charged particles occur in many fields of physics and chemistry, ranging from astrophysics and plasma physics to electrochemistry and colloid science. The common link between all these widely different systems is the predominance of long-range Coulomb interactions between the charged particles that confer to these systems a certain number of characteristic collective properties not found in fluids of neutral atoms or molecules. The present chapter is devoted to an overview of the essential structural and dynamic properties of some charged fluids of importance in condensed matter and chemical physics. This does exclude the very important field of plasma physics, although contact will be made with concepts as well as simple models borrowed from that field. We shall in fact be essentially concerned with ionic liquids and solutions, but some reference will be made to more "exotic" systems like macromolecular ionic systems and two-dimensional Coulomb fluids.

The properties of these charged fluids will be examined from an essentially microscopic point of view, within the framework of statistical

**J. P. Hansen** · Laboratoire de Physique Théorique des Liquides (Equipe Associée au C.N.R.S.), Université P. et M. Curie, 75230 Paris Cedex 05, France.

mechanics. Whenever possible the results of theoretical analysis and calculations will be confronted with experimental probes of the microscopic structure and dynamics. We shall restrict ourselves to the essential features of the charged fluids under consideration, without going into the details of quantitative determinations of the properties of specific fluids. This means that, in order to gain a qualitative understanding of basic physical mechanisms, we shall give preference to simplified *models* (which still retain the essential features) over very "realistic" calculations which, in any case, always hinge on a precise knowledge of intermolecular forces.

### 8.1.1. Classes of Charged Fluids

Any classification of material systems is in some sense arbitrary, but for the sake of clarity we list here the various Coulomb fluids, which will be considered in greater or lesser detail below.

1. *Liquid metals* will be looked upon as "cold" two-component plasmas, comprising positive ions and degenerate conduction (or valence) electrons. We shall restrict ourselves to metals having simple band structures (essentially alkali metals) and their *alloys*. Electrical and thermal conductivities are electronic in character, with electrical conductivities $\sigma$ of the order of $10^4 \ \Omega^{-1} \ cm^{-1}$. Some liquid alloys made up of two metallic elements become ionic (with conductivities of the order of $1 \ \Omega^{-1} \ cm^{-1}$) at a definite stoichiometric composition; the equimolar Cs–Au alloy is the best known example of such ionic melts.

2. A second class of ionic liquids comprises *molten salts*, made up of two ionic species of opposite charge. Salts are characterized by large Coulomb binding and a correspondingly high melting temperature and by ionic conductivities of the order of $1 \ \Omega^{-1} \ cm^{-1}$ in the liquid phase. At supercritical temperatures these systems exhibit a continuous change from an insulating (molecular) vapor to a dense conducting fluid. Certain crystalline salts exhibit very high conductivities, typical of the molten phase; these are the so-called *superionic conductors* in which, putting it very schematically, one of the sublattices (corresponding generally to the smaller ionic species) melts, while the opposite ionic species maintains crystalline long-range order. The best known example is silver iodide ($\alpha$-AgI) where the $Ag^+$ ions are delocalized in the temperature range $450 < T < 850$ K.

3. *Metal salt solutions* of the form $M - MX$ (with M an alkali metal and X a halide) are very interesting combinations of the two former classes, which generally exhibit a miscibility gap and a rapid transition from an ionic to a metallic behavior with increasing metal concentration.

4. All preceding classes of charged fluids comprise exclusively particles carrying an electric charge (ions and free electrons). In the direction of

increasing complexity we consider next the very broad class of *ionic solutions* consisting of a solvent (made up of polar molecules) and a solute (made up of at least two species of oppositely charge ions). If both species of ions are of similar size and absolute charge, one deals with an *electrolyte solution* (e.g., $Na^+Cl^-$ in water). *Macromolecular ionic solutions* contain macroions (charged polymer chains or coils, micelles, charged colloidal particles, etc.) and microscopic counterions; important members of this class of complex ionic liquids are the polyelectrolytes and the charged colloidal suspensions, which play an important role in biochemistry and in many other fields.

5. *Electron layers* at the surface of liquid helium and in MOS devices are two-dimensional charged fluids, which are being extensively studied in the laboratory; they will be briefly considered in the last section.

### 8.1.2. Some Important Physical Parameters

We consider a charged fluid made up of $\nu$ species of particles of mass $m_\alpha$ $(1 \leqslant \alpha \leqslant \nu)$, carrying an electric charge $Z_\alpha e$ ($e$ being the elementary proton charge). Let $n_\alpha = N_\alpha / \Omega$ be the number density (number of particles per unit volume) of species $\alpha$. Overall electroneutrality requires that

$$\sum_{\alpha=1}^{\nu} n_\nu Z_\nu = 0 \tag{8.1.1}$$

A characteristic feature common to all Coulomb fluids is the phenomenon of *screening*: the electric potential due to any local excess charge is effectively reduced at large distances by a rearrangement of the surrounding charges. For sufficiently low densities and high temperatures (i.e., in the weak-coupling limit), this mechanism is characterized by the Debye screening length $\lambda_D$; with each ionic species one associates a partial Debye wave number given by

$$k_{D\alpha} = (4\pi n_\alpha Z_\alpha^2 e^2 / \varepsilon k_B T)^{1/2} \tag{8.1.2}$$

in terms of which the screening length is given by

$$\lambda_D^2 = \sum_\alpha k_{D\alpha}^{-2} \tag{8.1.3}$$

In equation (8.1.2) $\varepsilon$ is the dielectric constant of the medium in which the charges move (e.g., the solvent). The classical formula (8.1.2) does not apply to the degenerate gas of conduction electrons in metallic liquids.

The Debye length allows a rough distinction between two important classes of ionic liquids. Let $n = \Sigma\, n_\alpha$ be the total number density (irrespective of species) and

$$a = (3/4\pi n)^{1/3} \tag{8.1.4}$$

the mean "ion sphere" radius. A third relevant length on a microscopic scale is the ionic diameter $d$; the corresponding packing fraction is defined as

$$\eta = \tfrac{1}{6}\pi n d^3 \tag{8.1.5}$$

A convenient Coulomb coupling constant is the dimensionless parameter

$$\Gamma = \frac{Z^2 e^2}{a k_B T} \tag{8.1.6}$$

where $Z$ is some mean ionic valence. Under typical molten salt conditions ($T \simeq 10^3$ K, $\varepsilon = 1$) one finds that

$$\lambda_D < a \simeq d, \qquad \eta \simeq 0.4, \qquad \Gamma \gtrsim 50$$

This corresponds to the strong coupling regime, where Debye screening loses its physical significance and the microscopic fluid structure is dominated by short-range order effects.

In a dilute ($10^{-3}$ molar, say) electrolyte on the other hand, the following double inequality holds ($T \simeq 300$ K, $\varepsilon = 80$ in water):

$$d < a < \lambda_D, \qquad \eta \simeq 10^{-3}, \qquad \Gamma \simeq 10^{-1}$$

which corresponds to weak-coupling conditions that are correctly described by the classical Debye–Hückel theory.[1] Concentrated electrolyte solutions (e.g., 1 molar) are somewhat intermediate between the two preceding cases.

In simple metals and alloys, the screening is essentially governed by the degenerate conduction electrons and the relevant screening length is roughly equal to the Thomas–Fermi length of an ideal Fermi gas, i.e.,

$$\lambda_{TF} = \frac{1}{2}\left(\frac{\pi}{3}\right)^{1/6}\left(\frac{\hbar^2}{m_e e^2 n_e^{1/3}}\right)^{1/2} \tag{8.1.7}$$

where the index e refers to electronic properties. Under typical liquid-metal conditions, $\lambda_{TF} \simeq a$ and $\Gamma \simeq 200$, which corresponds to very strong coupling.

### 8.1.3. Simple Models

The simplest, although somewhat artificial model of a Coulomb fluid is the so-called "one-component plasma" (OCP), a system of $N$ point particles of charge $Ze$ and mass $m$ immersed in a neutralizing uniform background of opposite charge density; the total potential energy $V_N$ of a periodic OCP in a volume $\Omega$ includes particle–particle, particle–background, and background–background terms, and can be expressed in terms of the Fourier components of the microscopic particle density

$$\rho_\mathbf{k} = \sum_{i=1}^{N} e^{i\mathbf{k} \cdot \mathbf{r}_i}$$

in the compact form

$$V_N(\Omega) = \frac{1}{\Omega} \sum_\mathbf{k}' \frac{4\pi Z^2 e^2}{k^2} (\rho_\mathbf{k} \rho_\mathbf{k}^* - N) \tag{8.1.8}$$

Omission of the $\mathbf{k} = \mathbf{0}$ Fourier component (denoted by a prime in the summation) accounts for the neutralizing background, while the subtraction of $N$ corresponds to the omission of the infinite self-energy of the point particles.

Under the name of "jellium" the OCP has been widely used as a model for the degenerate electron gas in solids on the assumption that the ionic charge distribution is uniformly "smeared out" over the whole volume. We shall be more concerned with the opposite, high-temperature limit where the classical ions are represented by the discrete point charges, while the uniform background is provided by the degenerate Fermi gas of conduction electrons. This assumption is closest to reality at high densities, since the ratio of the electron screening length (8.1.7) to the ion-sphere radius (8.1.4) can be expressed in the form

$$\frac{\lambda_{\mathrm{TF}}}{a} = \left(\frac{\pi}{12Z}\right)^{1/3} r_s^{-1/2} \tag{8.1.9}$$

where $r_s = a/(Z^{1/3} a_0)$ is the usual electron-density parameter and $a_0 = \hbar^2/m_e e^2$ is the Bohr radius. Equation (8.1.9) shows that in the high-density limit ($r_s \ll 1$), the electron screening length becomes much larger than the interionic spacing; this means that the Pauli principle inhibits the polarization of the electron gas by the ionic charge distribution, so that the former reduces essentially to a rigid uniform background. Such extreme situations are in fact achieved in very dense degenerate stars (such as white dwarfs), but we shall see that the OCP is also in a very useful starting point for the study of liquid metals. The model has the simplifying feature (which it shares with all systems of particles interacting via inverse-power potentials) that all its static and dynamic properties, when properly reduced to a dimensionless form, depend on a single variable, which is conveniently chosen to be the coupling constant (8.1.6). This property is easily verified by inspecting any equilibrium statistical average, and noting that in the Boltzmann factor the ratio of the Coulomb potential to the thermal energy can be written as

$$\frac{v(r)}{k_B T} = \frac{Z^2 e^2}{r k_B T} = \frac{Z^2 e^2}{a k_B T} \frac{a}{r} = \frac{\Gamma}{x} \tag{8.1.10}$$

Two extensions of the OCP model are particularly useful. If finite size effects of the ions are important, a model of charged hard spheres (of diameter $d$) in a uniform background might be more appropriate. The situations where the background penetrates the spheres and where it is excluded, but uniform outside the spheres, can be shown to be equivalent.[2] A thermodynamic state of this model is characterized by two independent variables, which are conveniently chosen to be $\eta$ and $\Gamma$.

The second extension of the OCP model corresponds to the case where there are several ionic species immersed in the neutralizing background. The binary case ("binary ionic mixture" or BIM model) has been extensively studied[4]; the two relevant variables are generally chosen to be $\Gamma$, and one of the concentrations

$$x_\alpha = \frac{n_\alpha}{n}, \qquad \alpha = 1, 2$$

The BIM model can be taken as a starting point for the description of binary alloys.

The simplest model retaining the essential features of ionic liquids (molten salts or electrolytes) is the "primitive model" of oppositely charged hard spheres. Let $Z_\alpha e$ and $d_\alpha$ be the charges and diameters of the two ionic species; the three pair potentials are of the form

$$v_{\alpha\beta}(r) = \begin{cases} \infty, & r < d_{\alpha\beta} = \frac{1}{2}(d_\alpha + d_\beta) \\ \dfrac{Z_\alpha Z_\beta}{\varepsilon r}, & r > d_{\alpha\beta} \end{cases} \tag{8.1.11}$$

The impenetrable hard core accounts for the sharply repulsive forces acting between ions at short distances; the Coulomb interaction is repulsive for ions of the same species and attractive for ions of opposite species. In a molten salt $\varepsilon = 1$, while for electrolytes $\varepsilon > 1$ accounts for the solvent, which reduces the Coulomb interaction between ions. Note that in this model the discrete molecular nature of the solvent is ignored and the latter is modeled by a dielectric continuum.

The "restricted primitive model" (RPM) is a symmetric version of the primitive model in which the ions have equal diameters ($d_1 = d_2 = d$) and opposite charges ($Z_1 = -Z_2$); equation (8.1.1) then implies $n_1 = n_2$. A thermodynamic state is characterized by the two variables $\eta$ and $\Gamma$ or

$$\gamma = \frac{Z^2 e^2}{\varepsilon d k_B T} = \frac{\Gamma}{2\eta^{1/3}} \qquad (8.1.12)$$

The various models introduced in this section must be regarded as starting points for a qualitative understanding of charged fluids, but they are generally too crude for a quantitative description of such fluids; more "realistic" potential models are required for this task.

## 8.2. Microscopic Structure and Thermodynamics

In this section we introduce the fundamental static (equal-time) density–density correlation functions and various related quantities. We recall their relationship with thermodynamics and with the static linear response to external perturbations that will allow us to derive the long-wavelength behavior of the static structure factors. Finally, we sketch some theoretical techniques for the computation of static correlation functions.

### 8.2.1. Distribution Functions and Structure Factors

Consider a fluid made up of $\nu$ particle species, with number densities $n_\alpha$ and concentrations $x_\alpha = n_\alpha/n$ ($1 \leq \alpha \leq \nu$). With each species we associate the microscopic density

$$\rho_\alpha(\mathbf{r}, t) = \sum_{i=1}^{N_\alpha} \delta[\mathbf{r} - \mathbf{r}_{i\alpha}(t)] \qquad (8.2.1)$$

and its Fourier components

$$\rho_{\mathbf{r}\alpha}(t) = \sum_{i=1}^{N_\alpha} e^{i\mathbf{k}\cdot\mathbf{r}_{i\alpha}(t)} \qquad (8.2.2)$$

where $\mathbf{r}_{i\alpha}(t)$ denotes the center-of-mass position of particle $i$ of species $\alpha$ at time $t$. Since we are interested in static (equal-time) correlations in a fluid in thermodynamic equilibrium (which satisfies time-translation invariance), we may ignore at present any explicit time dependence. The one-particle densities are defined as the statistical averages (e.g., in the canonical ensemble) of the corresponding microscopic densities:

$$\rho_\alpha^{(1)}(\mathbf{r}) = \langle \rho_\alpha(\mathbf{r}) \rangle = N_\alpha \langle \delta(\mathbf{r} - \mathbf{r}_{1\alpha}) \rangle \tag{8.2.3}$$

The two-particle density matrix is defined in terms of the density–density correlation matrix by

$$\rho_{\alpha\beta}^{(2)}(\mathbf{r}, \mathbf{r}') = \langle \rho_\alpha(\mathbf{r})\rho_\beta(\mathbf{r}') \rangle - \rho_\alpha^{(1)}(\mathbf{r})\delta_{\alpha\beta}\delta(\mathbf{r} - \mathbf{r}')$$

$$= \sum_i \sum_j{}' \langle \delta(\mathbf{r} - \mathbf{r}_{i\alpha})\delta(\mathbf{r}' - \mathbf{r}_{j\beta}) \rangle \tag{8.2.4}$$

where the prime in the summation means that the "self"-term ($i = j$) is to be omitted if $\alpha = \beta$.

For a rotationally and translationally invariant (i.e., isotropic and homogeneous) fluid,

$$\rho_\alpha^{(1)}(\mathbf{r}) = n_\alpha \tag{8.2.5a}$$

$$\rho_{\alpha\beta}^{(2)}(\mathbf{r}, \mathbf{r}') = n_\alpha n_\beta g_{\alpha\beta}(|\mathbf{r} - \mathbf{r}'|) \tag{8.2.5b}$$

where the $g_{\alpha\beta}$ are the (center-of-mass) pair distribution functions (pdf). Clearly $4\pi n_\beta g_{\alpha\beta}(r)r^2 dr$ is the mean number of particles of species $\beta$ situated at distance $r$ (within $dr$) from a particle of species $\alpha$. For $r \to \infty$, the positions of two particles in a fluid are completely decorrelated (no long-range order!) so that

$$\lim_{r\to\infty} g_{\alpha\beta}(r) = 1 \tag{8.2.6}$$

It is customary to define the pair correlation functions

$$h_{\alpha\beta}(r) = g_{\alpha\beta}(r) - 1 \tag{8.2.7}$$

which vanish in the limit $r \to \infty$.

In the low-density (weak-coupling) limit, only positions of pairs of particles are correlated, hence

$$\lim_{n\to\infty} g_{\alpha\beta}(r) = \exp[-v_{\alpha\beta}(r)/k_B T] \tag{8.2.8}$$

where $v_{\alpha\beta}(r)$ denotes the pair potential between species $\alpha$ and $\beta$ (assumed spherical).

Two related correlation function matrices are

$$S_{\alpha\beta}(r) = \delta_{\alpha\beta}\delta(\mathbf{r}) + n(x_\alpha x_\beta)^{1/2} h_{\alpha\beta}(r) \qquad (8.2.9)$$

and its inverse, defined by the relation

$$\sum_\gamma \int S_{\alpha\gamma}^{-1}(\mathbf{r}-\mathbf{r}'')S_{\gamma\beta}(\mathbf{r}''-\mathbf{r}')d^3r'' = \delta_{\alpha\beta}\delta(\mathbf{r}-\mathbf{r}') \qquad (8.2.10)$$

The inverse defines the *direct correlation functions* $c_{\alpha\beta}(r)$ via

$$S_{\alpha\beta}^{-1}(r) = \delta_{\alpha\beta}\delta(\mathbf{r}) - n(x_\alpha x_\beta)^{1/2} c_{\alpha\beta}(r) \qquad (8.2.11)$$

Inserting equations (8.2.9) and (8.2.11) into (8.2.10) we find that function $h_{\alpha\beta}(r)$ can be expressed in terms of the $c_{\alpha\beta}(r)$ by the Ornstein–Zernike (OZ) relations[3]:

$$h_{\alpha\beta}(r) = c_{\alpha\beta}(r) + n\sum_\gamma x_\gamma c_{\alpha\gamma}\otimes h_{\gamma\beta} \qquad (8.2.12)$$

where $\otimes$ denotes a convolution product. When dimensionless Fourier transforms

$$\hat{f}(k) = n\int e^{i\mathbf{k}\cdot\mathbf{r}} f(r)d^3r \qquad (8.2.13)$$

are introduced, the OZ relations assume a simple form in $k$-space:

$$\hat{h}_{\alpha\beta}(k) = \hat{c}_{\alpha\beta}(k) + \sum_\gamma x_\gamma \hat{c}_{\alpha\gamma}(k)\hat{h}_{\gamma\beta}(k) \qquad (8.2.14)$$

It is easily verified that the partial structure factors

$$S_{\alpha\beta}(k) = \frac{1}{(N_\alpha N_\beta)^{1/2}}\langle \rho_{\mathbf{k}\alpha}\rho_{\mathbf{k}\beta}^*\rangle \qquad (8.2.15)$$

are precisely the Fourier transforms of the $S_{\alpha\beta}(r)$:

$$S_{\alpha\beta}(k) = \delta_{\alpha\beta} + (x_\alpha x_\beta)^{1/2}\hat{h}_{\alpha\beta}(k) \qquad (8.2.16a)$$

$$S_{\alpha\beta}^{-1}(k) = \delta_{\alpha\beta} - (x_\alpha x_\beta)^{1/2}\hat{c}_{\alpha\beta}(k) \qquad (8.2.16b)$$

The $S_{\alpha\beta}(k)$ are directly accessible by radiation-diffraction techniques, through their linear combinations:

$$I(\theta) \sim \sum_{\alpha} \sum_{\beta} f_\alpha f_\beta S_{\alpha\beta}(k) \qquad (8.2.17)$$

where $f_\alpha$ denotes the ($k$-dependent) atomic form factor (X-ray diffraction) or the mean nuclear scattering length (neutron diffraction) of species $\alpha$, and $I(\theta)$ is the scattered intensity per unit volume, with the scattering angle given by

$$k = 2\pi \sin \theta / \lambda$$

$\lambda$ being the radiation wavelength.

### 8.2.2. Number, Concentration, and Charge Structure Factors

In order to illustrate the definitions of the preceding section, we now restrict ourselves to systems containing two ionic species and consider two important classes of such binary fluids: molten salts and metallic binary alloys. If $n_\alpha$ ($\alpha = 1, 2$) denotes the partial densities of the two ionic species, the two classes of fluids differ fundamentally in that the partial densities cannot be varied independently in the molten salt due to the electroneutrality condition (8.1.1), while in the alloy charge neutrality is achieved by the conduction electrons, so that $n_1$ and $n_2$ can be varied freely. An equivalent set of independent parameters in the latter case comprises the total ionic number density $n = n_1 + n_2$ and one of the concentrations, $x_\alpha = n_\alpha / n$ ($x_1 + x_2 = 1$).

If one now examines typical neutron-scattering or computer-simulation data for the three partial structure factors of such binary systems, one is immediately struck by a fundamental difference between molten salts and binary alloys. In the case of molten alkali halides (say $Na^+Cl^-$), the three $S_{\alpha\beta}(k)$ exhibit a first extremum at about the same wave number ($k \simeq 1.7$ Å$^{-1}$ for NaCl[5,6]), but while this corresponds to a sharp maximum for $S_{++}(k)$ and $S_{--}(k)$, the cross factor $S_{+-}(k)$ has a pronounced (negative) minimum at the same wave number. In the case of a liquid alloy, on the other hand, the three partial structure factors exhibit sharp maxima at *different* wave numbers (see, e.g., the classic example of the copper–tin alloy $Cu_6Sn_5$[7]).

In order to analyze this significant qualitative difference, it is physically instructive to consider certain linear combinations of the partial microscopic densities (8.2.1). The total microscopic number density is just the sum

$$\rho_N(\mathbf{r}, t) = \rho_1(\mathbf{r}, t) + \rho_2(\mathbf{r}, t) \qquad (8.2.18)$$

In the case of the molten salt, we define the charge density in the form

$$\rho_Z(\mathbf{r}, t) = Z_1 \rho_1(\mathbf{r}, t) + Z_2 \rho_2(\mathbf{r}, t) \tag{8.2.19}$$

Equations (8.2.18) and (8.2.19) can be used to define the following three structure factors:

$$S_{NN}(k) = \frac{1}{N} \langle \rho_{\mathbf{k}N} \rho_{\mathbf{k}N}^* \rangle = \sum_\alpha \sum_\beta (x_\alpha x_\beta)^{1/2} S_{\alpha\beta}(k) \tag{8.2.20a}$$

$$S_{NZ}(k) = \frac{1}{N} \langle \rho_{\mathbf{k}N} \rho_{\mathbf{k}Z}^* \rangle = \sum_\alpha \sum_\beta (x_\alpha x_\beta)^{1/2} Z_\beta S_{\alpha\beta}(k) \tag{8.2.20b}$$

$$S_{ZZ}(k) = \frac{1}{N} \langle \rho_{\mathbf{k}Z} \rho_{\mathbf{k}Z}^* \rangle = \sum_\alpha \sum_\beta (x_\alpha x_\beta)^{1/2} Z_\alpha Z_\beta S_{\alpha\beta}(k) \tag{8.2.20c}$$

Adams et al.[6] show these structure factors in the case of molten NaCl: $S_{NN}(k)$ turns out to be a relatively structureless function of $k$, due to a considerable amount of cancellation between the $S_{\alpha\beta}(k)$, while the linear combination (8.2.20c) strongly enhances the structure and yields a pronounced peak in $S_{ZZ}(k)$ $[S_{ZZ}(k_{max}) \simeq 4.5!]$. This is a clear manifestation of *charge ordering* in the molten salt, that becomes very apparent if one looks at the partial pair distribution functions $g_{\alpha\beta}(r)$. Computer simulations on molten alkali halides[8,9,6] show that, while $g_{++}(r) \simeq g_{--}(r)$, the oscillations of $g_{+-}(r)$ are exactly out of phase with those in $g_{++}(r)$. Consequently the linear combination $g_{NN}(r)$ is a very "flat" function of $r$ leading to a rather structureless $S_{NN}(k)$, while the combination $g_{ZZ}(r)$ enhances the oscillations, yielding a pronounced main peak in $S_{ZZ}(k)$. Local charge neutrality is most efficiently achieved by a regular alternation of opposite and equal sign neighbor shells, and clearly manifests itself in the peculiar structure of the $g_{\alpha\beta}(r)$. Moreover, the short-range order is not unlike that found in a solid NaCl structure, with an average nearest-neighbor coordination number close to 6.

In the case of a binary alloy, on the other hand, we define, in addition to the microscopic number density (8.2.18), the microscopic deviation from the mean concentration (or concentration "density")[10]:

$$\rho_C(\mathbf{r}, t) = x_2 \rho_1(\mathbf{r}, t) - x_1 \rho_2(\mathbf{r}, t) \tag{8.2.21}$$

Note that in view of the electroneutrality condition (8.1.1), the charge density (8.2.19) of a molten salt is similar to the concentration variable $\rho_C$ of an alloy. The physical interpretation of this new variable is clear: if $\rho_C$ is integrated over a small volume inside the alloy, the result is zero only if the numbers of particles of the two species inside this volume are exactly proportional to the mean (macroscopic) concentrations $x_1$ and $x_2$; $\rho_C(\mathbf{r}, t)$ is hence the natural variable to study concentration fluctuations. The corresponding structure factors are, apart from (8.2.20a),

$$S_{CC}(k) = \frac{1}{N} \langle \rho_{\mathbf{k}C} \rho^*_{\mathbf{k}C} \rangle$$

$$= x_1 x_2 [x_2 S_{11}(k) - 2(x_1 x_2)^{1/2} S_{12}(k) + x_1 S_{22}(k)] \qquad (8.2.22a)$$

$$S_{NC}(k) = \frac{1}{N} \langle \rho_{\mathbf{k}N} \rho^*_{\mathbf{k}C} \rangle$$

$$= x_1 x_2 \left[ S_{11}(k) - S_{22}(k) + \frac{x_2 - x_1}{(x_1 x_2)^{1/2}} S_{12}(k) \right] \qquad (8.2.22b)$$

Note that for an *ideal* mixture (two species of identical, but tagged, ions) it is a trivial exercise to check that $S_{CC}(k) = x_1 x_2$ and $S_{NC}(k) = 0$ for all $k$.

By making the appropriate linear combinations of experimental data for the partial structure factors of a typical liquid alloy,[7,10] one obtains a sharply peaked $S_{NN}(k)$ (not unlike the structure factor of a pure liquid), while $S_{CN}(k)$ and especially $S_{CC}(k)$ turn out to be relatively structureless, indicating the absence of any significant ordering ("random" alloy) in contradistinction to the charge ordering observed in the molten salt.

### 8.2.3. Relation to Thermodynamics

If the interparticle forces are central and pairwise additive, the total potential energy is of the form

$$V_N = \frac{1}{2} \sum_\alpha \sum_\beta \sum_i \sum_j{}' v_{\alpha\beta}(|\mathbf{r}_i - \mathbf{r}_j|) \qquad (8.2.23)$$

Within the framework of classical statistical mechanics, the internal energy of the fluid is then expressible in terms of the pdfs $g_{\alpha\beta}(r)$:

$$\frac{U}{N} = \tfrac{3}{2} k_B T + \frac{1}{N} \langle V_N \rangle$$

$$= \tfrac{3}{2} k_B T + \frac{1}{2} \sum_\alpha \sum_\beta x_\alpha x_\beta n \int v_{\alpha\beta}(r) g_{\alpha\beta}(r) d^3 r \qquad (8.2.24)$$

Similarly, the virial theorem leads immediately to the following expression for the pressure $P$:

$$\frac{P\Omega}{Nk_B T} = 1 - \frac{1}{6} \sum_\alpha \sum_\beta x_\alpha x_\beta n \int r \frac{dv_{\alpha\beta}(r)}{dr} g_{\alpha\beta}(r) d^3 r \qquad (8.2.25)$$

For fluids of charged particles, the $v_{\alpha\beta}(r)$ contain the Coulomb part

$$v_{\alpha\beta}^C(r) = \frac{Z_\alpha Z_\beta e^2}{\varepsilon r} \tag{8.2.26}$$

Use of the charge-neutrality condition (8.1.1) then allows the $g_{\alpha\beta}(r)$ to be replaced by the correlation functions $h_{\alpha\beta}(r)$, so that the integrals appearing in equations (8.2.24) and (8.2.25) remain separately finite.

### 8.2.4. Static Response and Long-Wavelength Behavior

It is well known that the static structure factors, which are a measure of equilibrium fluctuations, are intimately related to the static (linear) response functions (or susceptibilities), which describe the response of the system to a weak external perturbation.

Let $\Phi_\beta(\mathbf{r})$ be the potential of an external force field coupled to the density of particles of species $\beta$. Since we are interested in linear response, we may restrict ourselves to a single Fourier component of the potential:

$$\Phi_\beta(\mathbf{r}) = \phi_\beta(k)e^{-i\mathbf{k}\cdot\mathbf{r}}$$

The Hamiltonian $H$ in the presence of the external field (with $H_0$ the Hamiltonian of the unperturbed system) is given by

$$H = H_0 + \int \Phi_\beta(\mathbf{r})\rho_\beta(\mathbf{r})d^3r$$

$$= H_0 + \phi_\beta(\mathbf{k})\rho_{\mathbf{k}\beta}^* \tag{8.2.27}$$

We are interested in the response of the system as measured by the deviation of the Fourier component of the local density of species $\alpha$ [namely $\rho_\alpha^{(1)}(\mathbf{r})$] from its equilibrium value $n_\alpha$:

$$\delta\rho_\alpha(\mathbf{k}) = \frac{1}{\Omega}\int e^{i\mathbf{k}\cdot\mathbf{r}}\delta\rho_\alpha(\mathbf{r})d^3r$$

$$= \frac{1}{\Omega}(\langle\rho_{\mathbf{k}\alpha}\rangle - \langle\rho_{\mathbf{k}\alpha}\rangle_0)$$

$$= \frac{1}{\Omega}\left[\frac{1}{Q_N}\int \exp\left\{-\frac{1}{k_BT}[H_0 + \phi_\beta(\mathbf{k})\rho_{\mathbf{k}\beta}^*]\right\}\rho_{\mathbf{k}\alpha}d\Gamma_N - \langle\rho_{\mathbf{k}\alpha}\rangle_0\right] \tag{8.2.28}$$

where $d\Gamma_N$ denotes a phase space element, $Q_N$ the partition function, and the index 0 an equilibrium average over an ensemble of unperturbed systems. Linearization of the Boltzmann factor in equation (8.2.28) with respect to the weak external perturbation immediately yields the desired result:

$$\delta\rho_\alpha(\mathbf{k}) = -\frac{1}{\Omega} \langle \rho_{\mathbf{k}\alpha} \rho^*_{\mathbf{k}\beta} \rangle \frac{1}{k_B T} \phi_\beta(\mathbf{k})$$

$$= -\frac{(n_\alpha n_\beta)^{1/2}}{k_B T} S_{\alpha\beta}(k) \phi_\beta(k)$$

By identification with the defining relation for the static response function $\chi_{\alpha\beta}(k)$, namely

$$\delta\rho_\alpha(\mathbf{k}) = \chi_{\alpha\beta}(k) \phi_\beta(\mathbf{k})$$

we derive the static version of the (classical) fluctuation-dissipation theorem[11,3] in the form

$$\chi_{\alpha\beta}(k) = -\frac{(n_\alpha n_\beta)^{1/2}}{k_B T} S_{\alpha\beta}(k) \qquad (8.2.29)$$

We are more specifically interested in the response of a fluid of rigid ions to the electric field of an external charge density having Fourier components $\rho_e(\mathbf{k})$. From Poisson's equation the corresponding electric potential is given by

$$\phi^e_Z(\mathbf{k}) = \frac{4\pi e^2}{k^2} \rho_e(\mathbf{k}) \qquad (8.2.30)$$

which couples to the microscopic charge density $\rho_{\mathbf{k}Z}$ of the fluid. The dielectric behavior is characterized by the charge–charge response function $\chi_{ZZ}(k)$, where

$$\delta\rho_Z(\mathbf{k}) = \chi_{ZZ}(k) \phi^e_Z(\mathbf{k}) \qquad (8.2.31)$$

while the electrostrictive behavior is measured by the response of the particle density,

$$\delta\rho_N(\mathbf{k}) = \chi_{NZ}(k) \phi^e_Z(\mathbf{k}) \qquad (8.2.32)$$

The charge response to the external charge density is also described by the inverse (longitudinal) static dielectric function:

$$\frac{1}{\varepsilon(k)} = \frac{\mathbf{k} \cdot \mathbf{E}(\mathbf{k})}{\mathbf{k} \cdot \mathbf{D}(\mathbf{k})} = \frac{\rho_Z(\mathbf{k})}{\rho_e(\mathbf{k})} = 1 + \frac{\delta\rho_Z(\mathbf{k})}{\rho_e(\mathbf{k})} \qquad (8.2.33)$$

where $\mathbf{E}$ and $\mathbf{D}$ denote the electric field and electric displacement vectors, respectively, $\rho_z = \rho_e + \delta\rho_z$ is the total (i.e., external plus induced) charge density, and use was made of Poisson's equation.

Comparison of equation (8.2.33) with equations (8.2.30) and (8.2.31) yields the standard relation

$$\frac{1}{\varepsilon(k)} = 1 + \frac{4\pi e^2}{k^2} \chi_{ZZ}(k) \tag{8.2.34}$$

while the fluctuation-dissipation theorem applied to the charge response yields

$$\chi_{ZZ}(k) = -\frac{n}{k_B T} S_{ZZ}(k) \tag{8.2.35}$$

From its definition (8.2.20c), the charge structure factor $S_{ZZ}(k)$ is nonnegative, so that equations (8.2.34) and (8.2.35) entail the stability condition

$$\frac{1}{\varepsilon(k)} < 1 \tag{8.2.36}$$

Negative values of the dielectric function are a common feature of strongly coupled charged fluids.[12]

Inside the fluid an external charge is screened due to the polarization of the medium, and each ion experiences a local electric field that is the superposition of the external field and the field due to the induced charge density. The response of the system to the local (rather than external) electric potential is described by the screened response function[15]

$$\delta\rho_z(\mathbf{k}) = \tilde{\chi}_{ZZ}(k)[\phi_z^e(\mathbf{k}) + \delta\phi_z(\mathbf{k})] \tag{8.2.37}$$

where

$$\delta\phi_z(\mathbf{k}) = \frac{4\pi e^2}{k^2} \delta\rho_z(\mathbf{k}) \tag{8.2.38}$$

Comparison of equations (8.2.31) and (8.2.37) yields the relation between external and screened response functions:

$$\chi_{ZZ}(k) = \frac{\tilde{\chi}_{ZZ}(k)}{1 - (4\pi e^2/k^2)\tilde{\chi}_{ZZ}(k)} \tag{8.2.39}$$

while

$$\varepsilon(k) = 1 - (4\pi e^2/k^2)\tilde{\chi}_{ZZ}(k) \tag{8.2.40}$$

Perfect screening in a conducting fluid imposes the condition

$$\lim_{k \to 0} \rho_Z(\mathbf{k}) = \lim_{k \to 0} \left[ \rho_e(\mathbf{k}) + \delta \rho_Z(\mathbf{k}) \right] = 0 \qquad (8.2.41)$$

When combined with equations (8.2.30), (8.2.31), and (8.2.35), condition (8.2.41) determines the long-wavelength behavior of the charge structure factor in the form

$$\lim_{k \to 0} \frac{k_D^2}{k^2} S_{ZZ}(k) = \overline{Z^2} \qquad (8.2.42)$$

where

$$\overline{Z^2} = \sum_{\alpha} x_{\alpha} Z_{\alpha}^2 \qquad (8.2.43)$$

and $k_D^2 = \Sigma_\alpha k_{D\alpha}^2$ is the Debye wave number defined by equations (8.1.2) and (8.1.3).

The charge structure factor vanishes as $k^2$ for long wavelengths, a consequence of the $k^2$ singularity of the Coulomb potential. The partial structure factors, on the other hand, remain finite in the limit $k \to 0$.

For the sake of illustration we now specifically consider the case of a binary ionic liquid. The total charge surrounding an ion of a given species must exactly cancel the charge of that ion, hence we have the electroneutrality constraints

$$\left. \begin{array}{l} n_1 Z_1 \displaystyle\int g_{11}(r) d^3 r + n_2 Z_2 \int g_{12}(r) d^3 r = -Z_1 \\[2ex] n_1 Z_1 \displaystyle\int g_{12}(r) d^3 r + n_2 Z_2 \int g_{22}(r) d^3 r = -Z_2 \end{array} \right\} \qquad (8.2.44)$$

With the aid of equations (8.1.1) and (8.2.16a), in conjunction with the compressibility equation for binary fluids,[13] it is then a straightforward exercise to show that the partial structure factors have the following long-wavelength limit:

$$\lim_{k \to 0} S_{\alpha\beta}(k) = (x_\alpha x_\beta)^{1/2} n k_B T K_T \qquad (8.2.45)$$

where $K_T = -(\delta\Omega/\partial P)_{N,T}/\Omega$ is the isothermal compressibility.

We finally examine the long-wavelength behavior of the dielectric function $\varepsilon(k)$, restricting ourselves now for the sake of simplicity to the one-component plasma (OCP) model for which number and charge density fluctuations coincide. A simple macroscopic calculation will give us the required long-wavelength behavior.[14] In the absence of any flow, the force due to the local electric field must be exactly cancelled by the force due to the pressure gradient:

$$n(Ze)\mathbf{E}(\mathbf{r}) = \nabla P(\mathbf{r}) \tag{8.2.46}$$

Poisson's equation reads

$$\nabla \cdot \mathbf{E}(\mathbf{r}) = 4\pi e[\rho_e(\mathbf{r}) + \delta\rho_z(\mathbf{r})] \tag{8.2.47}$$

Assuming local thermodynamic equilibrium and considering an isothermal process, the pressure fluctuation can be expressed as

$$\delta P(\mathbf{r}) = P(\mathbf{r}) - P_0 = \left(\frac{\partial P}{\partial n}\right)_T \delta n(\mathbf{r}) = \frac{1}{Z}\left(\frac{\partial P}{\partial n}\right)_T \delta\rho_z(\mathbf{r}) \tag{8.2.48}$$

If equations (8.2.46)–(8.2.48) are combined, the following differential equation for $\partial\rho_z$ is obtained:

$$\frac{1}{4\pi Z^2 e^2}\left(\frac{\partial P}{\partial n}\right)_T \nabla^2 \delta\rho_z(\mathbf{r}) - \delta\rho_z(\mathbf{r}) = \rho_e(\mathbf{r}) \tag{8.2.49}$$

This is easily solved in Fourier space:

$$\delta\rho_z(\mathbf{k}) = -\frac{\rho_e(\mathbf{k})}{1 + k^2/k_s^2} \tag{8.2.50}$$

where

$$k_s^2 = 4\pi Z^2 e^2 n^2 K_T \tag{8.2.51}$$

is the square of the screening wave number. The long-wavelength limit of $\varepsilon(k)$ than follows immediately from equation (8.2.33):

$$\varepsilon(k) = 1 + k_s^2/k^2 \tag{8.2.52}$$

Note that the perfect screening condition (8.2.41) is contained in equation (8.2.50). The electrostatic potential around the screened external charge is, according to Poisson's equation,

$$\phi_z(\mathbf{k}) = \frac{4\pi e^2}{k^2}[\rho_e(\mathbf{k}) + \delta\rho_z(\mathbf{k})]$$

$$= \frac{4\pi e^2}{k^2 \varepsilon(k)}\rho_e(\mathbf{k}) = \frac{4\pi e^2}{k^2 + k_s^2}\rho_e(\mathbf{k})$$

If one of the point charges in the fluid is regarded as the external charge $[\rho_e(r) = Ze\delta(\mathbf{r})]$, the preceding result shows that the *effective* electric potential due to that charge decays exponentially at large distances:

$$v_{\text{eff}}(r) = \frac{Z^2 e^2}{r} \exp(-k_s r) \tag{8.2.53}$$

In the weak-coupling limit, $K_T$ goes over to its ideal gas limit $(nk_B T)^{-1}$ and $k_s$ reduces to the Debey wave number $k_D$ [equation (8.1.2)]. In the strong-coupling regime $(\Gamma > 1)$, the compressibility of the OCP becomes negative so that $k_s$ takes on imaginary values,[14] corresponding to an oscillatory behavior of $v_{\text{eff}}(r)$, characteristic of short-range ordering.

### 8.2.5. Approximations for the Pair Distribution Functions

In this section we briefly examine some of the more efficient approximations for the calculation of the pair distribution functions. Again, for the sake of simplicity, we shall first restrict ourselves to the case of the simplest model ionic fluid, the OCP (with $Z = 1$), the pair structure of which is characterized by a single pdf $g(r)$. In the weak-coupling limit $(\Gamma \to 0)$, correlations become negligible and we make the familiar "random-phase" approximation (RPA), which amounts to replacing the screened response function $\tilde{\chi}(k)$ in equation (8.2.39) by the response function of a system of noninteracting particles, $\chi_0(k)$, which reduces to $-n/k_B T$ in the classical limit. The structure factor then follows directly from equation (8.2.35):

$$S_{\text{RPA}}(k) = \frac{k^2}{k^2 + k_D^2} \tag{8.2.54}$$

It is noteworthy that similar reasoning leads to the familiar Lindhard dielectric function of the degenerate electron gas.[15] Function $S_{\text{RPA}}(k)$ obviously satisfies the long-wavelength limit (8.2.42). From the OZ relation (8.2.16b) it is immediately clear that the RPA is equivalent to setting

$$\hat{c}(k) = -\hat{v}(k)/k_B T = -k_D^2/k^2 \tag{8.2.55a}$$

or

$$c(r) = -v(r)/k_B T \tag{8.2.55b}$$

Equation (8.2.55a) is believed to be true asymptotically $(r \to \infty)$ for any pair potential[16]; the RPA replaces the direct correlation function by its asymptotic form, for all distances $r$. Equations (8.2.55) allow a straightforward generalization of the RPA to multicomponent fluids, in conjunction with the OZ relations (8.2.16).

Historically, the RPA was first derived by Debye and Hückel (DH)[1] from the Poisson–Boltzmann equation. The charge density around a central ion is related to the electrostatic potential by the Boltzmann factor:

$$\rho_Z(r) = n[g(r) - 1] = n\{\exp\{-[v(r) + \Phi(r)]/k_B T\} - 1\} \qquad (8.2.56a)$$

where $v(r) = e^2/r$ and $\Phi(r)$ are the potentials due to the central ion and to the polarization cloud, respectively. The latter satisfies Poisson's equation

$$\nabla^2 \Phi(r) = -4\pi e^2 \rho_Z(r) \qquad (8.2.56b)$$

which admits the solution

$$\Phi(r) = \int d^3r' \rho_Z(r') v(|\mathbf{r} - \mathbf{r}'|)$$

$$= n \int d^3r' h(r') v(|\mathbf{r} - \mathbf{r}'|) \qquad (8.2.56c)$$

where the definition (8.2.7) was used. The coupled equations (8.2.56) form a closed, nonlinear set. Upon linearization of equation (8.2.56a) and subsequent Fourier transformation, the RPA (or DH) structure factor (8.2.54) is recovered. The linearization is only justified on the weak-coupling (i.e., low-density or high-temperature) limit. For intermediate couplings, the full, nonlinear Poisson–Boltzmann equation yields reasonable results[17,18] despite the neglect of correlations between particles in the polarization cloud. These correlations are partly included in the so-called "hypernetted-chain" (HNC) approximation,[3,19] which amounts to replacing the bare potential $v(r)$ by the "renormalized" potential $-k_B T \cdot c(r)$ in equation (8.2.56c). Combining equations (8.2.56) and (8.2.12) we express the HNC approximation in the standard form

$$g(r) = \exp[-v(r)/k_B T + h(r) - c(r)] \qquad (8.2.57)$$

The multicomponent generalization is

$$g_{\alpha\beta}(r) = \exp[-v_{\alpha\beta}(r)/k_B T + h_{\alpha\beta}(r) - c_{\alpha\beta}(r)] \qquad (8.2.58)$$

Equations (8.2.58) and (8.2.12) form a closed set, which must be solved numerically by an iterative procedure. Comparison with computer-simulation data shows that the HNC approximation generally yields reasonable pair distribution functions for Coulomb fluids, even under strong-coupling conditions (see, e.g., various papers on the subject[9,19–22]). Some improved versions of the HNC scheme have recently been proposed and lead to nearly perfect agreement with the simulation data.[23,24]

Another powerful approximation scheme, which applies to systems of charged hard spheres, is the so-called "mean-spherical" approximation (MSA), which has the advantage of admitting analytic solutions in some cases. The pair potentials being of the form (8.1.11), the MSA completes the *exact* requirements

$$g_{\alpha\beta}(r) = 0, \qquad r < d_{\alpha\beta} \tag{8.2.59a}$$

by the *approximate* closure

$$c_{\alpha\beta}(r) = -v_{\alpha\beta}(r)/k_B T, \qquad r > d_{\alpha\beta} \tag{8.2.59b}$$

In the limit of vanishing hard-sphere diameters (point ions), the MSA reduces to the RPA [cf. equation (8.2.55b)]; for finite diameters, the MSA includes short-range correlations via equation (8.2.59a). Equations (8.2.59) together with the OZ relations (8.2.12) form a closed set, which has been solved analytically in the case of the restricted primitive model (equal diameters, opposite charges).[25] The extension to unequal diameters and arbitrary numbers of components is also available, but requires the numerical solution of an algebraic equation.[26] The MSA solutions reproduce the main qualitative features of the partial pdfs in the concentrated electrolyte and in the molten salt regime, and have been used to analyze X-ray and neutron-diffraction data for molten alkali halides.[27] In order to achieve quantitative agreement with experimental or simulation data, one must resort to semiempirical modifications of the original MSA scheme, but at the expense of considerable numerical complications.[28,29]

## 8.3. Microscopic Dynamics and Transport

In this section we introduce a certain number of tools, mainly time-dependent correlation functions and their associated memory functions, which serve in a quantitative analysis of single-particle and collective motions in neutral or charged fluids, and which lead to transport coefficients within the framework of linear response theory. Many of the concepts of this section have already been introduced in Chapter 7 and can be found in numerous textbooks or review articles.[3,11,30–32] Consequently we shall be very brief as regards generalities, and concentrate on aspects which are specific for charged fluids.[12,19]

### 8.3.1. Time-Dependent Correlations: A Brief Reminder

A microscopic dynamic variable will be any scalar, vectorial, or tensorial function of the instantaneous positions and (or) momenta (or velocities) of some, or all, particles of a many-particle system. Examples are the partial microscopic densities (8.2.1) or their Fourier components (8.2.2). The time evolution of a dynamic variable $A$ is governed by

$$\dot{A} = \frac{dA}{dt} = iLA \tag{8.3.1}$$

where $L$ denotes the Liouville operator. In quantum statistical mechanics $L$ is the commutator

$$L = \frac{1}{\hbar}[H, \quad] \tag{8.3.2a}$$

while in classical statistical mechanics, to which we shall henceforth restrict ourselves, $L$ is the Poisson bracket:

$$L = i\{H, \quad\} = i \sum_{j=1}^{N} \left( \frac{\partial H}{\partial \mathbf{r}_j} \cdot \frac{\partial}{\partial \mathbf{p}_j} - \frac{\partial H}{\partial \mathbf{p}_j} \cdot \frac{\partial}{\partial \mathbf{r}_j} \right) \tag{8.3.2b}$$

Equation (8.3.1) has the formal solution

$$A(t) = \exp(iLt)A \tag{8.3.3}$$

where $A \equiv A(0)$ denotes the initial value of the dynamic variable.

The equilibrium time correlation function (TCF) of two dynamic variables $A$ and $B$ is defined as

$$\begin{aligned} C_{AB}(t', t'') &= \langle A(t')B^*(t'') \rangle \\ &= \langle A(t' - t'')B^*(0) \rangle \end{aligned} \tag{8.3.4}$$

where the angle brackets denote an ensemble average over the initial phase, and advantage is taken of the stationarity of equilibrium averages. More precisely, if $t = t' - t''$, then

$$C_{AB}(t) = \int d^{3N}r \int d^{3N}p \, (e^{iLt}A)B^* f_0^{(N)}(\mathbf{r}^N, \mathbf{p}^N) \tag{8.3.5}$$

where $f_0^{(N)}$ denotes the equilibrium distribution function in $6N$-dimensional phase space. More generally, if the dynamic variables vary in space [like the partial densities (8.2.1)], and if the fluid is homogeneous (translationally invariant in space), the corresponding TCF will also depend on the spatial variable $\mathbf{r} = \mathbf{r}' - \mathbf{r}''$:

$$C_{AB}(\mathbf{r}, t) = \langle A(\mathbf{r}', t)B^*(\mathbf{r}'', 0)\rangle$$
$$= \langle A(\mathbf{r}, t)B^*(\mathbf{0}, 0)\rangle \qquad (8.3.6)$$

It is clear that spatial homogeneity implies that only Fourier components of opposite wave numbers have nonvanishing correlations, so that

$$C_{AB}(\mathbf{k}, t) = \int e^{i\mathbf{k}\cdot\mathbf{r}}C_{AB}(\mathbf{r}, t)$$
$$= \langle A_{\mathbf{k}}(t)B^*_{\mathbf{k}}(0)\rangle \qquad (8.3.7)$$

We shall only consider dynamic variables of zero mean, $\langle A \rangle = 0$. After a sufficiently long time interval, any dynamic variable will be completely decorrelated from the initial value of the same, or any other, variable, so that

$$\lim_{t\to\infty} C_{AB}(t) = \langle A(t)\rangle\langle B^*(0)\rangle$$
$$= \langle A \rangle\langle B^* \rangle = 0 \qquad (8.3.8)$$

This property allows us to define the spectral function (or power spectrum) of any TCF as its Fourier transform in time:

$$\hat{C}_{AB}(\omega) = \frac{1}{2\pi}\int_{-\infty}^{+\infty} e^{i\omega t}C_{AB}(t)dt \qquad (8.3.9)$$

It is also convenient to introduce the Laplace transform:

$$\tilde{C}_{AB}(z) = \int_0^\infty e^{izt}C_{AB}(t)dt, \qquad \mathrm{Im}\, z > 0 \qquad (8.3.10)$$

which is related to the spectrum by a Hilbert transformation:

$$\tilde{C}_{AB}(z) = i\int_{-\infty}^{+\infty} d\omega \frac{\hat{C}_{AB}(\omega)}{z - \omega} \qquad (8.3.11)$$

Inversely

$$\hat{C}_{AB}(\omega) = \lim_{\varepsilon\to 0} \frac{1}{\pi}\tilde{C}'_{AB}(\omega + i\varepsilon) \qquad (8.3.12)$$

where the prime denotes a real part (a double prime will denote an imaginary part). If $A$ and $B$ are both either even or odd functions of the momenta, their TCF is an even function of time and the spectrum is an even function of frequency; $C_{AB}(t)$ then admits a Taylor expansion in even powers of $t$:

$$C_{AB}(t) = \sum_{n=0}^{\infty} \frac{t^{2n}}{(2n)!} C_{AB}^{(2n)}(t=0)$$

$$= \sum_{n=0}^{\infty} \frac{t^{2n}}{(2n)!} (-1)^n \langle A^{(n)}(0) B^{*(n)}(0) \rangle$$

$$= \sum_{n=0}^{\infty} \frac{t^{2n}}{(2n)!} \langle (L^n A)(L^n B^*) \rangle \qquad (8.3.13)$$

where $(n)$ denotes an $n$th order derivative with respect to time. If the inverse Fourier transform of equation (8.3.9) is differentiated $2n$ times with respect to $t$, then by setting $t = 0$ we derive

$$\Omega_{AB}^{(2n)} = \int_{-\infty}^{+\infty} \omega^{2n} \hat{C}_{AB}(\omega) d\omega$$

$$= (-1)^n C_{AB}^{(2n)}(0) \qquad (8.3.14)$$

Thus the frequency moments of the spectral function are directly related to derivatives of the TCF taken at $t = 0$. The latter are static (equal-time) correlation functions, similar to the static structure factors introduced in the preceding section. The short-time expansions of $C_{AB}(t)$ in equation (8.3.13) implies, via equation (8.3.11), the following high-frequency expansion for the Laplace transform:

$$\tilde{C}_{AB}(z) = \frac{i}{z} \sum_{n=0}^{\infty} \frac{\Omega_{AB}^{(2n)}}{z^{2n}} \qquad (8.3.15)$$

The exact time evolution of any dynamic variable is described by equation (8.3.1). For practical purposes, a more phenomenological approach is widely used. We start from the familiar Langevin equation for the velocity of a Brownian particle. If $u$ denotes one of its Cartesian components, $\xi$ the friction coefficient, and $R(t)$ a component of the stochastic random force, this equation becomes

$$m\dot{u}(t) = -m\xi u(t) + R(t) \qquad (8.3.16)$$

Equipartition of energy, and the assumption that $R(t)$ is orthogonal to $u(0)$

for all $t$, lead to the following relation between $\xi$ and $\mathbf{R}^{(11)}$:

$$\xi = \frac{1}{mk_B T} \int_0^\infty \langle R(t)R(0)\rangle dt \qquad (8.3.17)$$

When retarded effects and the action of an external force field $X(t)$ are taken into account, equation (8.3.16) is generalized to

$$\dot{u}(t) = -\int_0^t \xi(t-s)u(s)ds + \frac{1}{m}R(t) + \frac{1}{m}X(t) \qquad (8.3.18)$$

Considering first the case $X = 0$, projecting both sides of equation (8.3.18) onto the initial velocity $u(0)$, and making the usual assumption that

$$\langle R(t)u(0)\rangle = 0 \qquad \forall t > 0 \qquad (8.3.19)$$

we immediately derive an expression for the Laplace transform (8.3.10) of the velocity autocorrelation function (ACF) $C_{uu}(t)$:

$$\tilde{C}_{uu}(z) = \frac{C_{uu}(t=0)}{-iz + \tilde{\xi}(z)} = \frac{k_B T/m}{-iz + \tilde{\xi}(z)} \qquad (8.3.20)$$

where $\tilde{\xi}(z)$ is given by a generalization of equation (8.3.17):

$$\tilde{\xi}(z) = \frac{1}{mk_B T} \int_0^\infty \langle R(t)R(0)\rangle e^{izt}dt \qquad (8.3.21)$$

The generalized Langevin equation (8.3.20) can be extended to any dynamic variable, and the generalized "random force" can be given a precise statistical-mechanics interpretation in terms of projected time evolution.[33,3] If $A$ denotes a set of dynamic variables, the correlation function matrix obeys the "memory" function equation

$$\dot{C}_{AA}(t) - i\Omega C_{AA}(t) + \int_0^t M(t-s)C_{AA}(s)ds = 0 \qquad (8.3.22)$$

where the frequency matrix $\Omega$ is given by

$$\Omega = \langle \dot{A}A^*\rangle(\langle AA^*\rangle)^{-1}$$

and the memory function matrix $M$ plays the role of the generalized friction coefficient $\xi$. If Laplace transforms are taken, equation (8.3.22) yields

$$\tilde{C}_{AA}(z) = C_{AA}(t=0)[-iz - i\Omega + \hat{M}(z)]^{-1} \qquad (8.3.23)$$

The practical interest of equation (8.3.22) or (8.3.23) lies in the fact that, for a judicious choice of the set of dynamic variables $A$, the memory functions have a simpler structure (in particular, a faster decay) than the corresponding correlation functions, and are hence more easily amenable to simple relaxation-time approximations.

### 8.3.2. Mobilities and Conductivities

If an ensemble of Brownian particles is initially ($t=0$) located at the origin, their mean-square deviation in the $x$-direction is given by Einstein's law

$$\langle x^2(t) \rangle = 2Dt \qquad (8.3.24)$$

where $D$ is the self-diffusion constant given by the time integral of the velocity ACF:

$$D = \int_0^\infty C_{uu}(t)dt = \tilde{C}_{uu}(z=0) \qquad (8.3.25)$$

If the following periodic external force is applied to the Brownian particles [cf. equation (8.3.18)]

$$X(t) = RX_0 e^{-i\omega t}$$

the mean velocity at time $t$ (which measures the response of the system to the external field) is given by

$$\langle u(t) \rangle = R\mu(\omega)X_0 e^{-i\omega t} \qquad (8.3.26)$$

where the expression for the frequency-dependent *mobility* $\mu(\omega)$ is easily derived from equation (8.3.18) in the form

$$\mu(\omega) = \frac{1}{m} \frac{1}{-i\omega + \tilde{\xi}(\omega)} = \frac{1}{k_B T} \tilde{C}_{uu}(\omega) \qquad (8.3.27)$$

In particular, the static mobility $\mu = \mu(\omega = 0)$ and the self-diffusion constant $D$ obey the Einstein relation

$$\mu = \frac{D}{k_B T} \qquad (8.3.28)$$

$\mu(\omega)$ is the simplest example of a linear response function (or dynamic susceptibility), and equation (8.3.27) is a special case of the fluctuation-dissipation theorem.[11]

These results for a Brownian particle are easily generalizable to the case of a multicomponent charged fluid. With each species we associate a velocity ACF:

$$C_\alpha(t) \equiv C_{v_\alpha v_\alpha}(t) = \langle \mathbf{v}_\alpha(t) \cdot \mathbf{v}_\alpha(0) \rangle \tag{8.3.29}$$

where $\mathbf{v}_\alpha$ denotes the velocity of any one of the $N_\alpha$ particles of species $\alpha$. The corresponding self-diffusion constant is given by

$$D_\alpha = \frac{1}{3} \int_0^\infty C_\alpha(t) dt \tag{8.3.30}$$

and the (static) mobility of ions of species $\alpha$ follows from the Einstein relation

$$\mu_\alpha = \frac{Z_\alpha e D}{k_B T} \tag{8.3.31}$$

where the factor $Z_\alpha e$ arises because we adopt the convention that the mobility measures the response of a single ion to an applied electric field rather than to the corresponding force. The collective response to a (local) periodic field is characterized by the (complex) frequency-dependent electric conductivity

$$\sigma(\omega) = \sigma'(\omega) + \sigma''(\omega) = \frac{e^2}{3k_B T} \int_0^\infty e^{i\omega t} C_{zz}(t) dt \tag{8.3.32}$$

where $C_{zz}(t)$ is the ACF of the fluctuating microscopic charge current in the absence of an external electric field:

$$C_{zz}(t) = \frac{1}{N} \langle \mathbf{j}_z(t) \cdot \mathbf{j}_z(0) \rangle \tag{8.3.33a}$$

$$\mathbf{j}_z(t) = \sum_\alpha Z_\alpha \mathbf{j}_\alpha(t) \tag{8.3.33b}$$

$$\mathbf{j}_\alpha(t) = \sum_{i=1}^{N_\alpha} \mathbf{v}_{i\alpha}(t) \tag{8.3.33c}$$

The usual static electric conductivity $\sigma$ can then be identified as

$$\sigma = \lim_{\omega \to 0} \sigma(\omega) = \frac{\omega_P^2}{4\pi} \int_0^\infty J_z(t) dt \tag{8.3.34}$$

where

$$\omega_p^2 = \sum_\alpha \omega_{p\alpha}^2 = \sum_\alpha \frac{4\pi n_\alpha Z_\alpha^2 e^2}{m_\alpha} \tag{8.3.35}$$

is the square of the plasma frequency, and $J_Z(t) = C_{ZZ}(t)/C_{ZZ}(0)$ is the normalized ACF of the fluctuating charge current. Equations (8.3.32) and (8.3.35) easily lead to the following sum rule for the real part of the frequency-dependent conductivity:

$$\int_{-\infty}^{+\infty} \frac{d\omega}{\pi} \sigma'(\omega) = \frac{\omega_p^2}{4\pi} \tag{8.3.36}$$

An approximate relation between the static conductivity and the mobilities of the various ionic species follows from equations (8.3.30), (8.3.31), and (8.3.34) if all cross-correlations between the velocities of different ions (of the same or of different species) are neglected, i.e., if one assumes

$$\langle \mathbf{v}_{i\alpha}(t) \cdot \mathbf{v}_{j\beta}(0) \rangle = 0 \qquad i \neq j$$

In classical statistical mechanics this assumption is exact only for $t = 0$. If it is generalized to all times, it leads to a simple relation between $C_{ZZ}(t)$ and $C_\alpha(t)$ of the form

$$C_{ZZ}(t) = \sum_\alpha x_\alpha Z_\alpha^2 C_\alpha(t)$$

from which the familiar Nernst–Einstein relation follows immediately:

$$\sigma = \frac{e^2}{k_B T} \sum_\alpha n_\alpha Z_\alpha^2 D_\alpha = \sum_\alpha n_\alpha Z_\alpha e \mu_\alpha \tag{8.3.37}$$

This approximate relation is well verified in electrolyte solutions, but in molten alkali halides equation (8.3.37) typically overestimates $\sigma$ by 20%.

The equations of this section are directly applicable to the study of ionic conductivity. In liquid metals and alloys the much larger conductivity is due to the degenerate Fermi gas of conduction electrons. For simple metals, like the alkali, the conductivity is accurately calculated from Ziman's equation in the framework of a Lorentz model, where the mutually noninteracting electrons are individually scattered by the spatially correlated ions; the relatively weak electron–ion interaction is treated in a Born approximation.[34] The resulting equation for the resistivity ($\rho = 1/\sigma$) of a liquid metal is

$$\rho = \frac{m_e^2}{12\pi^3 \hbar^3 e^3 n_e} \int_0^{2k_F} |\hat{v}(k)|^2 S(k) k^3 dk \tag{8.3.38}$$

where $\hat{v}(k)$ is the Fourier transform of the electron–ion pseudopotential, $S(k)$ is the ionic structure factor, and $k_F$ is the Fermi momentum of the electron gas. Excellent results for the temperature variation of the resistivity of liquid alkali have been obtained from equation (8.3.38) on the basis of the OCP model for the static structure factor.[35]

### 8.3.3. Density and Current Correlation Functions

In the study of wave-number-dependent collective modes, the basic dynamic variables are the Fourier components (8.2.2) of the partial microscopic densities, and the associated partial particle currents:

$$\mathbf{j}_{\mathbf{k}\alpha}(t) = \sum_{i=1}^{N_\alpha} \mathbf{v}_{i\alpha}(t) e^{i\mathbf{k}\cdot\mathbf{r}_{i\alpha}(t)} \tag{8.3.39}$$

The two are related by the continuity equations expressing particle conservation:

$$\dot{\rho}_{\mathbf{k}\alpha}(t) = i\mathbf{k}\cdot\mathbf{j}_{\mathbf{k}\alpha}(t) \tag{8.3.40}$$

We define the partial density TCFs in the form

$$F_{\alpha\beta}(k, t) = \frac{1}{(N_\alpha N_\beta)^{1/2}} \langle \rho_{\mathbf{k}\alpha}(t)\rho_{\mathbf{k}\beta}^*(0)\rangle \tag{8.3.41}$$

which depend only on $k = |\mathbf{k}|$, due to the rotational invariance of an isotropic fluid. Their spectra (8.3.9) are the so-called dynamic structure factors,

$$S_{\alpha\beta}(k, \omega) \equiv \hat{F}_{\alpha\beta}(k, \omega)$$

The initial values of the $F_{\alpha\beta}$ are clearly the static structure factors defined in equation (8.2.15):

$$F_{\alpha\beta}(k, t = 0) = S_{\alpha\beta}(k) = \int_{-\infty}^{+\infty} S_{\alpha\beta}(k, \omega)d\omega \tag{8.3.42}$$

From equations (8.3.14) and (8.3.40) it is immediately clear that the dynamic structure factors obey the $f$-sum rules:

$$\Omega_{\alpha\beta}^{(2)}(k) = \int_{-\infty}^{+\infty} \omega^2 S_{\alpha\beta}(k, \omega)d\omega = \frac{k^2}{(N_\alpha N_\beta)^{1/2}} \langle j_{\mathbf{k}\alpha}^{(x)} j_{\mathbf{k}\beta}^{(x)*}\rangle$$

$$= k^2 v_{0\alpha}^2 \delta_{\alpha\beta} \tag{8.3.43}$$

where $v_{0\alpha}$ denotes the thermal velocity $(k_B T/m_\alpha)^{1/2}$. The longitudinal and transverse partial current TCFs are defined in terms of the projections of $\mathbf{j}_{\mathbf{k}\alpha}(t)$ parallel and perpendicular to the wave vector $\mathbf{k}$:

$$C_{\alpha\beta}^{L}(k, t) = \frac{1}{(N_\alpha N_\beta)^{1/2} k^2} \langle \mathbf{k} \cdot \mathbf{j}_{\mathbf{k}\alpha}(t) \mathbf{k} \cdot \mathbf{j}_{\mathbf{k}\beta}(0) \rangle \qquad (8.3.44a)$$

$$C_{\alpha\beta}^{T}(k, t) = \frac{1}{(N_\alpha N_\beta)^{1/2} k^2} \text{Tr}\{\langle [\mathbf{k} \wedge \mathbf{j}_{\mathbf{k}\alpha}(t)][\mathbf{k} \wedge \mathbf{j}_{\mathbf{k}\beta}^{*}(0)]\rangle\} \qquad (8.3.44b)$$

From the continuity equations (8.3.40) we immediately deduce

$$C_{\alpha\beta}^{L}(k, t) = -\frac{1}{k^2} \frac{d^2}{dt^2} F_{\alpha\beta}(k, t) \qquad (8.3.45)$$

so that the $C_{\alpha\beta}^{L}$ do not contain any new information. The $C_{\alpha\beta}^{T}$ are, however, independent functions, since the transverse currents are completely decoupled from the longitudinal currents.

### 8.3.4. Concentration Fluctuations

In many situations, certain linear combinations of the partial densities or currents have a physically more transparent significance, as was already stressed in Section 8.2.2 for static fluctuations. For binary mixtures and alloys we have introduced the variables $\rho_N$ and $\rho_C$ defined by equations (8.2.18) and (8.2.21). The corresponding number–number, number–concentration, and concentration–concentration dynamic structure factors are linear combinations of the $S_{\alpha\beta}(k, \omega)$, obvious generalizations of equations (8.2.22).

The hydrodynamic (i.e., small $k$ and $\omega$) limit of $S_{CC}(k, \omega)$ can be easily inferred from the macroscopic Fick's law:

$$\mathbf{j}_c(\mathbf{r}, t) = -D\nabla\rho_C(\mathbf{r}, t) \qquad (8.3.46)$$

where $\mathbf{j}_c = x_2\mathbf{j}_1 - x_1\mathbf{j}_2$ is the interdiffusion current, $D$ is the (mutual) diffusion constant, and $\rho_C$ is the local concentration variable, which is also related to $\mathbf{j}_c$ by the continuity equation

$$\dot{\rho}_C(\mathbf{r}, t) + \nabla \cdot \mathbf{j}_c(\mathbf{r}, t) = 0 \qquad (8.3.47)$$

It should be noted that on a macroscopic scale, the local variables $\rho_C$ and $\mathbf{j}_c$ are averages of the corresponding microscopic variables over a small (but macroscopic) volume element of the fluid. Eliminating $\mathbf{j}_c$ from equations

(8.3.46) and (8.3.47) yields a closed equation for $\rho_C$:

$$\dot{\rho}_C(\mathbf{r}, t) = D\nabla^2 \rho_C(\mathbf{r}, t) \qquad (8.3.48a)$$

or

$$\dot{\rho}_C(\mathbf{k}, t) = -k^2 D\rho_C(\mathbf{r}, t) \qquad (8.3.48b)$$

In terms of Laplace transforms we obtain

$$\tilde{\rho}_C(\mathbf{k}, z) = \frac{\rho_C(\mathbf{k}, t = 0)}{-iz + Dk^2} \qquad (8.3.49)$$

The long-wavelength, low-frequency concentration correlation function then becomes

$$\tilde{F}_{CC}(k, z) = \frac{1}{N} \langle \tilde{\rho}_C(\mathbf{k}, z)\rho_C^*(\mathbf{k}, t = 0) \rangle$$

$$= \frac{1}{N} \langle |\rho_C(\mathbf{k}, 0)|^2 \rangle [-iz + Dk^2]^{-1} \qquad (8.3.50)$$

In the small-$k$ limit we have[13]

$$S_{CC}(k) = \frac{1}{N} \langle |\rho_C(\mathbf{k}, 0)|^2 \rangle_{k \to 0} = \left[ \frac{\partial^2 g}{\partial x_1^2} \right]_{T,P,N}^{-1} \qquad (8.3.51)$$

where $g = G/Nk_B T$ denotes the reduced Gibbs free energy per ion. In the "hydrodynamic limit" the resulting dynamic structure factor is given by

$$S_{CC}(k, \omega) = \frac{1}{\pi} \tilde{F}'_{CC}(k, z = \omega)$$

$$= \frac{1}{\pi} \left( \frac{\partial^2 g}{\partial x_1^2} \right)^{-1} \frac{Dk^2}{\omega^2 + (Dk^2)^2} \qquad (8.3.52)$$

This equation shows that the mutual diffusion constant can be calculated from the limit

$$D = \pi \left( \frac{\partial^2 g}{\partial x_1^2} \right) \lim_{\omega \to 0} \lim_{k \to 0} \frac{\omega^2}{k^2} S_{CC}(k, \omega) \qquad (8.3.53)$$

With the aid of equation (8.3.44a) this relation can be expressed in a standard Kubo form, similar to equation (8.3.34) for the conductivity:

$$D = \left( \frac{\partial^2 g}{\partial x_1^2} \right) \frac{1}{3N} \int_0^\infty \langle \mathbf{j}_c(t) \cdot \mathbf{j}_c(0) \rangle dt \qquad (8.3.54)$$

where $\mathbf{j}_c(t) = x_2\mathbf{j}_1(t) - x_1\mathbf{j}_2(t)$ is the fluctuating interdiffusion current. Note that for an *ideal* mixture, $\partial^2 g/\partial x_1^2 = (x_1 x_2)^{-1}$; most alloys of simple metals are fairly close to ideality. Finally, by making the same assumption leading to the Nernst–Einstein relation (8.3.37), we arrive at the approximate relation

$$D = x_2 D_1 + x_1 D_2$$

which is fairly well verified for mixtures of neutral fluids[36] but has yet to be checked for alloys.

### 8.3.5. Charge-Density Fluctuations

We now examine the case of binary ionic fluids containing ions of opposite charge. The important dynamic variables are now the Fourier components of the charge density (8.2.19) and of the associated electric current:

$$\mathbf{j}_{kz} = Z_1\mathbf{j}_{k1}(t) + Z_2\mathbf{j}_{k2}(t) \tag{8.3.55}$$

The corresponding correlation functions are

$$F_{zz}(k, t) = \frac{1}{N} \langle \rho_{kz}(t)\rho_{kz}^*(0)\rangle \tag{8.3.56}$$

and the longitudinal and transverse charge current correlation functions $C_{zz}^L(k, t)$ and $C_{zz}^T(k, t)$ defined as in equations (8.3.44).

The response of the fluid to a time-dependent external potential is measured by the dynamic (frequency-dependent) generalization of the static charge response function introduced in equation (8.2.31). By virtue of the fluctuation-dissipation theorem this complex dynamic response function is entirely determined by the equilibrium charge fluctuation spectrum, i.e.,

$$\chi_{zz}''(k, \omega) = -\frac{n\omega}{k_B T} S_{zz}(k, \omega) \tag{8.3.57}$$

while the real part follows from the standard Kramers–Kronig relation

$$\chi_{zz}'(k, \omega) = P\frac{1}{\pi} \int_{-\infty}^{+\infty} \frac{1}{\omega' - \omega} \chi_z''(k, \omega')d\omega' \tag{8.3.58}$$

The longitudinal complex dielectric function is determined by the dynamic generalization of equation (8.2.34):

$$\frac{1}{\varepsilon(k, \omega)} = 1 + \frac{4\pi e^2}{k^2} \chi_{zz}(k, \omega) \tag{8.3.59}$$

The function $\chi_{zz}$ and $\varepsilon^{-1}$ measure the linear response of the plasma to an *external* field. As was already pointed out for the static case, this external field polarizes the fluid and creates a local internal electric field, which is the superposition of the external field (or electric displacement field) and the field due to the induced-charge density. This local (or screened) electric field is, of course, the field experienced by the particles. The response of the fluid to the local electric field is characterized by the screened response function $\chi_{zz}^s(k, \omega)$, which is related to the dielectric function by [cf. equation (8.2.40)]

$$\varepsilon(k, \omega) = 1 - \frac{4\pi e^2}{k^2} \chi_{zz}^s(k, \omega) \tag{8.3.60}$$

The electric-conductivity tensor relates the Fourier components of the induced electric current to the Fourier components of the local electric field:

$$\left. \begin{array}{l} \mathbf{j}_z(\mathbf{k}, t) = \int_0^t \boldsymbol{\sigma}(\mathbf{k}, t - t') \cdot \mathbf{E}(\mathbf{k}, t') \\[2mm] \tilde{\mathbf{j}}_z(\mathbf{k}, \omega) = \boldsymbol{\sigma}(\mathbf{k}, \omega) \cdot \tilde{\mathbf{E}}(\mathbf{k}, \omega) \end{array} \right\} \tag{8.3.61}$$

The decomposition of $\mathbf{E}$ into its longitudinal and transverse parts leads to a similar distinction between longitudinal and transverse conductivities $\sigma^L$ and $\sigma^T$. The former is directly related to the (longitudinal) dielectric function via the familiar expression

$$\varepsilon(k, \omega) = 1 + \frac{4\pi i}{\omega} \sigma^L(k, \omega) \tag{8.3.62}$$

In the long-wavelength limit, spatial isotropy imposes that longitudinal and transverse conductivities become equal, namely

$$\sigma^L(0, \omega) = \sigma^T(0, \omega) \equiv \sigma(\omega) \tag{8.3.63}$$

and, in particular, comparison of equations (8.3.60) and (8.3.62) shows that the static conductivity is given by

$$\sigma = -e^2 \lim_{\omega \to 0} \omega \left[ \lim_{k \to 0} \frac{1}{k^2} \chi_{zz}^{s''}(k, \omega) \right] \tag{8.3.64}$$

Note that $\sigma(\omega)$ is *not* directly related to the $k \to 0$ limit of the charge density fluctuation spectrum [see equation (8.3.57)] or, equivalently, of the longitudinal current fluctuation spectrum [cf. equation (8.3.45)!], but rather to the corresponding spectrum of the *transverse* current correlation function, a fact intimately related to the $k \to 0$ singularity of the Coulomb potential.[37,31,12]

To obtain the $k, \omega \to 0$ limit of the charge-fluctuation spectrum, we proceed as in Section 8.3.4 for the case of the concentration fluctuations. The continuity equation (8.3.40) here assumes the form (with $z = \omega$)

$$- i\omega \tilde{\rho}_Z (\mathbf{k}, \omega) = \rho_Z (\mathbf{k}, t = 0) + i\mathbf{k} \cdot \tilde{\mathbf{j}}_Z (\mathbf{k}, \omega) \qquad (8.3.65)$$

while Poisson's equation is written in the form

$$- i\mathbf{k} \cdot \tilde{\mathbf{E}}(\mathbf{k}, \omega) = 4\pi \tilde{\rho}_Z (\mathbf{k}, \omega) \qquad (8.3.66)$$

These two equations are combined with the longitudinal projection of Ohm's law (8.3.61), yielding

$$\tilde{\rho}_Z (\mathbf{k}, \omega) = \frac{\rho_Z (\mathbf{k}, t = 0)}{- i\omega + 4\pi \sigma^L (\mathbf{k}, \omega)} \qquad (8.3.67)$$

If equation (8.3.67) is multiplied by $\rho_Z^*(\mathbf{k}, 0)$ and thermally averaged, we obtain

$$\tilde{F}_{ZZ} (k, \omega) = \frac{S_{ZZ}(k)}{- i\omega + 4\pi \sigma^L (\mathbf{k}, \omega)} \qquad (8.3.68)$$

which shows that $\sigma^L(k, \omega)$ is the memory function for the charge-density correlation function. In the limit $k \to 0$

$$\lim_{k \to 0} \frac{\tilde{F}_{ZZ} (k, \omega)}{S_{ZZ}(k)} = \frac{1}{- i\omega + 4\pi \sigma(\omega)} \qquad (8.3.69)$$

The corresponding spectrum (8.3.12) takes the form

$$\lim_{k \to 0} \frac{S_{ZZ} (k, \omega)}{S_{ZZ}(k)} = \frac{1}{\pi} \frac{4\pi \sigma'(\omega)}{[\omega - 4\pi \sigma''(\omega)]^2 + [4\pi \sigma'(\omega)]^2} \qquad (8.3.70)$$

which assumes the following low-frequency limit $[\sigma'(\omega) \to \sigma; \sigma''(\omega) \to 0]$:

$$S_{ZZ} (k, \omega) = \frac{1}{\pi} \frac{4\pi \sigma k^2 \overline{Z^2}/k_D^2}{\omega^2 + (4\pi\sigma)^2}, \qquad k, \omega \to 0 \qquad (8.3.71)$$

a result to be contrasted with the hydrodynamic limit (8.3.52) for $S_{CC}(k, \omega)$. The width of the concentration-fluctuation spectrum vanishes with $k$, but it stays finite for the charge-fluctuation spectrum in the limit $k \to 0$. The long-wavelength *high*-frequency dielectric behavior, on the other hand, follows directly from the sum rule (8.3.36) and the high-frequency expansion (8.3.15), which immediately leads to

$$\sigma(\omega) = \frac{i\omega_p^2}{4\pi\omega}\left[1 + O\left(\frac{1}{\omega^2}\right)\right]$$

Using (8.3.62) we find

$$\varepsilon(0, \omega) = 1 - \frac{\omega_p^2}{\omega^2} + O\left(\frac{1}{\omega^4}\right) \tag{8.3.72}$$

which indicates the possible existence of a high-frequency "optic mode" in ionic liquids, similar to the optic mode observed in ionic crystals, or to the familiar plasma oscillations in plasmas.[9]

To conclude this section we stress once more the fact that the long range of the Coulomb interaction leads to different $k \to 0$ limits of the longitudinal and transverse electric-current correlation functions. This is exemplified by the different $k \to 0$ limits of the longitudinal and transverse frequency moments (8.3.14). If the characteristic longitudinal and transverse "optic mode" frequencies are defined by

$$\omega_{L,T}^2(k) = \frac{\Omega_{ZZ}^{(2)L,T}(k)}{\Omega_{ZZ}^{(0)L,T}(k)}$$

a straightforward calculation[6] leads to the result

$$\lim_{k \to 0}\left[\omega_L^2(k) - \omega_T^2(k)\right] = \omega_p^2 \tag{8.3.73}$$

which generalizes the Lyddane–Sachs–Teller sum rule for ionic crystals.

### 8.3.6. Mass-Density Fluctuations

Another, physically important, linear combination of the partial densities is the mass density

$$\rho_{kM}(t) = \sum_\alpha m_\alpha \rho_{k\alpha}(t) \tag{8.3.74}$$

The corresponding mass density correlation function is

$$F_{MM}(k, t) = \frac{1}{N} \langle \rho_{kM}(t) \rho_{kM}^*(0) \rangle$$

$$= \sum_\alpha \sum_\beta m_\alpha m_\beta (x_\alpha x_\beta)^{1/2} F_{\alpha\beta}(k, t) \qquad (8.3.75)$$

The hydrodynamic limit of the associated spectrum can be derived from the Navier–Stokes equations and leads, in particular, to the Kubo limit[38] given by

$$\lim_{\omega \to 0} \lim_{k \to 0} \frac{\omega^4}{k^4} S_{MM}(k, \omega) = 2k_B T(\tfrac{4}{3}\eta + \zeta) \qquad (8.3.76)$$

where $\eta$ and $\zeta$ are the shear and bulk viscosities, respectively. The result (8.3.76) is the same for charged and for neutral fluids.

The coupling between mass and concentration fluctuations (in the case of binary alloys), or between mass and charge fluctuations (for ionic liquids), is characterized by the cross-correlation function $F_{MC}(k, t)$ or $F_{MZ}(k, t)$.

## 8.4. Selected Applications: Part 1

In this and the following section we briefly review some salient features of a few typical Coulomb fluids in the light of the more general framework introduced in the preceding sections. The present section will be mostly devoted to the "one-component plasma" (OCP) model and some of its extensions and applications; the last section will be mainly devoted to two-component ionic liquids. These sections should only be considered as a brief introductory guide to a few aspects of this rich variety of charged fluids, and the reader should consult the (incomplete) list of references for more details.

### 8.4.1. The One-Component Plasma

The OCP model has been defined in Section 8.1.3. Its essential virtue is its simplicity, but the model does exhibit some of the essential characteristics of Coulomb fluids, despite some very unphysical features associated with the rigid uniform background. In particular, mass- and charge-density fluctuations coincide in the OCP, so that the electric conductivity is zero, due to total momentum conservation. The model is thoroughly reviewed elsewhere[19] where, in particular, a rather complete list of original references can be found.

The OCP thermodynamics and static structure are very accurately known from computer simulations[19]; these static properties are reasonably reproduced by HNC theory and its extensions. The isothermal compressibility goes negative for $\Gamma \gtrsim 3$ without any incidence on thermodynamic stability, due to local electric-field fluctuations that inhibit large-scale (small-$k$) density fluctuations.[14] Short-range order [i.e., an oscillatory $g(r)$] appears for $\Gamma > 2$. In the strong-coupling limit ($\Gamma \gg 1$), the excess internal energy is within a few percent of the simple ion-sphere result

$$\frac{U^{\text{ex}}}{Nk_{\text{B}}T} = -\tfrac{9}{10}\Gamma$$

which represents in fact an exact lower bound to the energy. The OCP crystallizes into a BCC lattice at $\Gamma \simeq 170$.[39]

The dynamic properties of the OCP have been extensively studied by "molecular-dynamics" simulations.[40,19] The most striking features are the following:

1. The velocity ACF exhibits marked oscillations at roughly the plasma frequency $\omega_{\text{p}}$ for sufficiently strong couplings ($\Gamma > 10$). These oscillations are more and more pronounced and long-lived as $\Gamma$ increases; they are a clear manifestation of a strong coupling of the single-particle motion to the collective charge-density fluctuations.[41]

2. The shear viscosity first decreases with increasing coupling, passes through a minimum at $\Gamma \simeq 20$, and then increases gradually until crystallization.[42]

3. Conservation of total momentum implies that the high-frequency plasmon mode is undamped in the $k \to 0$ limit. This mode exhibits *negative* dispersion (i.e., $d\omega/dk < 0$) for $\Gamma \gtrsim 10$.

The simulation results are surprisingly well reproduced by a simple memory-function analysis of the density ACF, $F(k, t)$. Equations (8.2.22) and (8.3.42) yield

$$\tilde{F}(k, z) = \frac{S(k)}{-iz + \tilde{M}_1(k, z)} \tag{8.4.1}$$

When the high-frequency expansions (8.3.15) of $\tilde{F}$ and $\tilde{M}_1$ are compared, we find that

$$M_1(k, t = 0) = \frac{\Omega^{(2)}(k)}{\Omega^{(0)}(k)} = \frac{v_0^2 k^2}{S(k)} \equiv \omega_{0l}^2(k) \tag{8.4.2}$$

where $v_0 = (k_B T/m)^{1/2}$ is the thermal velocity. The first-order memory function $M_1$ can itself be expressed in terms of a second-order memory function $M_2$ via equation (8.3.23). For the latter function we introduce a single relaxation-time approximation

$$M_2(k, t) = M_2(k, t = 0)\exp[-t/\tau(k)]$$

$$\tilde{M}_2(k, z) = \frac{M_2(k, t = 0)}{-iz + 1/\tau(k)} \tag{8.4.3}$$

The initial value of $M_2$ is again easily derived from the high-frequency expansion (8.3.15):

$$M_2(k, t = 0) = \omega_{1l}^2(k) - \omega_{0l}^2(k) \tag{8.4.4}$$

where

$$\omega_{1l}^2(k) = \frac{\Omega^{(4)}(k)}{\Omega^{(2)}(k)} = \frac{1}{v_0^2 k^2} \Omega^{(4)}(k) \tag{8.4.5}$$

The static quantity is expressible in terms of the pair-distribution function[40]

$$\omega_{1l}^2(k) = \omega_p^2 \left\{ 1 + \frac{3v_0^2 k^2}{\omega_p^2} + 2 \int_0^\infty \frac{dr}{r} [g(r) - 1] j_2(kr) \right\} \tag{8.4.6}$$

where $j_2$ denotes the second-order spherical Bessel function.

Inserting equations (8.4.2)–(8.4.4) into equation (8.4.1) and taking the real part, we obtain the following expression for the dynamic structure factor:

$$S(k, \omega) = \frac{1}{\pi} \tilde{F}'(k, \omega)$$

$$= \frac{1}{\pi} \frac{\tau(k) v_0^2 k^2 [\omega_{1l}^2(k) - \omega_{0l}^2(k)]}{\{\omega\tau(k)[\omega^2 - \omega_{1l}^2(k)]\}^2 + [\omega^2 - \omega_{0l}^2(k)]^2} \tag{8.4.7}$$

Following Lovesey,[43] we construct the unknown relaxation time $\tau(k)$ from the two characteristic frequencies $\omega_{0l}(k)$ and $\omega_{1l}(k)$ by choosing the combination which ensures that for large $k$ (wavelengths much shorter than the interparticle spacing), $S(k, \omega = 0)$ goes over correctly to its free particle limit:

$$\lim_{k \to 0} S(k, 0) = (2\pi v_0^2 k^2)^{-1/2}$$

This leads immediately to

$$\tau(k) = \left\{ \frac{4}{\pi} [\omega_{1l}^2(k) - \omega_{0l}^2(k)] \right\}^{-1/2} \tag{8.4.8}$$

The dielectric function is then derived from $S(k, \omega)$ via equations (8.3.57)–(8.3.59). The plasmon dispersion curve $\omega = \omega(k)$ is determined by the equation

$$\varepsilon[k, \omega(k)] = 0 \tag{8.4.9}$$

For small wave-numbers, the damping of the plasmon mode is negligibly small and $\omega(k)$ is practically determined by the position of the plasmon peak in the charge-fluctuation spectrum $S(k, \omega)$. The dispersion relation reads

$$\omega(k) = \omega_p \left[ 1 + \gamma \frac{k^2}{k_D^2} + O(k^4) \right] \tag{8.4.10}$$

with $\gamma = \frac{3}{2} + 2(U^{ex}/Nk_B T)/15$. Since the excess internal energy behaves essentially as $-0.9\Gamma$, $\gamma$ changes sign for $\Gamma \simeq 13$, in agreement with the negative dispersion observed in the computer simulations.[40] A more fundamental analysis has recently been given by Carrini and Kalman.[70]

### 8.4.2. Liquid Metals

The OCP model is a reasonable starting point for the description of very dense Coulomb matter occurring in extreme astrophysical situations. For instance, in a white-dwarf star densities are of the order of $10^6$–$10^8$ g/cm$^3$ and temperatures are typically $10^7$–$10^8$ K. Under such conditions matter is metallic and comprises fully stripped ions (such as $C^{6+}$ nuclei) and highly degenerate electrons (Fermi temperature $T_F$ being approximately $10^{10}$ K). The electron screening length (8.1.9) far exceeds the interionic spacing, so the electron gas can, to a good approximation, be regarded as providing a rigid, uniform background in which the classical positive ions move. Under typical white-dwarf conditions the ionic-coupling constant is large ($\Gamma > 10$), and during the cooling process the ionic plasma will finally crystallize ("diamonds" in the sky[44]).

When the density decreases (and hence parameter $r_s$ increases), the electron gas is increasingly polarized by the ionic-charge distribution and electron-screening effects must be taken into account. If the screening remains moderate ($r_s < 1$), these effects can be treated by thermodynamic perturbation theory.[45] Such calculations show that the thermodynamic properties and the pair structure are surprisingly little affected by electron screening, as long as $\lambda_{TF} \gtrsim a$. The OCP is hence a reasonable model for metallic hydrogen under physical conditions occurring in the interior of Jupiter or Saturn.

When the density is further decreased, some of the electrons recombine with the nuclei to form ions having a finite core ($r_c \simeq 1$ Å in the alkali); the nearly free conduction electrons interact with these ions via relatively weak pseudopotentials, which can be treated by perturbation theory. The OCP has been used as a successful starting point for the description of simple liquid metals.[46] The success of this approach is linked to two simple observations. Under triple-point conditions the Coulomb coupling parameter $\Gamma$ for the liquid alkali, as determined from their density and temperature, is typically of the order of 180–200, strikingly close to the OCP melting conditions ($\Gamma \simeq 170$). By a simple density scaling of the wave numbers, the static-structure factors of the alkali just above melting are practically indistinguishable, and are surprisingly well represented by the OCP structure factor taken under the same temperature–density conditions.[35,47] The only disagreement is at long wavelengths ($k \to 0$), where the ionic-structure factor of the alkali tends toward isothermal compressibility [cf. equation (8.2.45)], while the OCP structure factor vanishes as $k^2$ [cf. equation (8.2.42)].

This defect can be overcome by accounting for electron-polarization effects through perturbation theory.[46] A fluctuation in the ion density induces a fluctuation in the electron density. Within the linear response relation (8.2.29) this takes the form

$$\delta\rho_e(\mathbf{k}) = \chi_{ee}(k)\phi_e(\mathbf{k})$$

$$= \chi_{ee}(k)\hat{v}_{ei}(k)\delta\rho_i(\mathbf{k}) \tag{8.4.11}$$

where $\hat{v}_{ei}(k)$ is the (weak) electron–ion potential. The Fourier components of the *effective* potential acting on the ions due to the fluctuation $\delta\rho_i(\mathbf{k})$ is the sum of a direct term and an indirect electron-polarization term:

$$\hat{v}_{eff}(k)\delta\rho_i(\mathbf{k}) = \hat{v}_{ii}(k)\delta\rho_i(\mathbf{k}) + \hat{v}_{ei}(k)\delta\rho_e(\mathbf{k})$$

$$= \{\hat{v}_{ii}(k) + \chi_{ee}(k)[\hat{v}_{ei}(k)]^2\}\delta\rho_i(\mathbf{k}) \tag{8.4.12}$$

Consequently

$$\hat{v}_{eff}(k) = \hat{v}_{ii}(k) + \hat{w}(k)$$

$$\hat{w}(k) = \frac{[\hat{v}_{ei}(k)]^2}{4\pi e^2/k^2}\left[\frac{1}{\varepsilon_e(k)} - 1\right] \tag{8.4.13}$$

where $\varepsilon_e(k)$ is the dielectric constant of the degenerate Fermi gas of interacting conduction electrons, and is related to the static response function $\chi_{ee}(k)$ by equation (8.2.34). In equation (8.4.13), $\hat{v}_{ii}(k)$ is the direct ion–ion interaction. This reduces to the Coulomb repulsion $4\pi Z^2 e^2/k^2$, which is sufficient to prevent the ion cores from touching; $\hat{w}(k)$ is the electron-induced indirect ion–ion interaction, which will be treated as a perturbation. The ion–electron pseudopotential $v_{ei}(r)$ can be approximated with reasonably accuracy by the Ashcroft empty-core model[48]:

$$v_{ei}(r) = 0, \qquad r < r_c$$

$$= -\frac{Ze^2}{r}, \qquad r > r_c$$

$$\hat{v}_{ei}(k) = -\frac{4\pi Ze^2}{k^2}\cos(kr_c) \tag{8.4.14}$$

where the core radius is fitted to solid-state data.

The ionic-structure factor is finally calculated by a generalized RPA.[45] If $S_0(k)$, $\hat{c}_0(k)$ and $S(k)$, $\hat{c}(k)$ are the structure factor and the direct correlation function for a reference system (in the present case, the OCP) and of the system of interest, these quantities are related by the exact expression, which follows trivially from equation (8.2.16b):

$$S(k) = \frac{S_0(k)}{1 - [\hat{c}(k) - \hat{c}_0(k)]S_0(k)} \tag{8.4.15}$$

The RPA (8.2.55a) is now applied to the *difference* $\hat{c}(k) - \hat{c}_0(k)$ and yields

$$S(k) = \frac{S_0(k)}{1 + \beta\hat{w}(k)} \tag{8.4.16}$$

The small-$k$ limit of the resulting structure factor is obtained from equations (8.4.13) and (8.4.14), and the limiting forms

$$\varepsilon_e(k) = 1 + k_e^2/k^2 \tag{8.4.17a}$$

$$S_0(k) = \frac{k^2}{k_D^2}\left(1 + \frac{k^2}{k_s^2}\right)^{-1} \tag{8.4.17b}$$

where $k_e$ is the inverse screening length of the electron gas, $k_s$ is the inverse screening length of the OCP and given by equation (8.2.51), and $k_D = (4\pi nZ^2 e^2/k_B T)^{1/2}$ is the Debye wave number. The resulting expression for the isothermal compressibility is

$$n_i k_B T\chi_T = \lim_{k\to 0} S(k) = \frac{k_D^2}{k_e^2} + \frac{k_D^2}{k_s^2} + k_D^2 r_c^2 \tag{8.4.18}$$

which leads to values for liquid alkali in satisfactory agreement with experiment (typically, $n_i k_B T \chi_T$ is about 0.02).[46]

Historically, the point of view of considering a simple liquid metal as a perturbed OCP goes back to Bohm and Staver,[49] who identified sound waves in liquid metals with screened plasmons. The screened ionic-plasma frequency yields an acoustic phonon-like dispersion:

$$\omega(k) = \left[\frac{\omega_p^2}{\varepsilon_e(k)}\right]^{1/2} = \left[\frac{4\pi n_i Z^2 e^2}{m_i \varepsilon_e(k)}\right]^{1/2} \underset{k \to 0}{=} ck$$

where, in view of the limit (8.4.17a), the speed of sound is given by

$$c = \left(\frac{4\pi n_i Z^2 e^2}{k_e^2}\right)^{1/2} = \left(\frac{1}{3} Z \frac{m_e}{m_i}\right)^{1/2} v_F$$

$v_F$ being the Fermi velocity and the inverse screening length $k_e$ is taken to be that of an ideal Fermi gas.

### 8.4.3. Charged Colloidal Dispersions

Macroionic solutions contain large ions, carrying up to several hundred elementary charges, and small counterions, which form essentially a screening cloud around the much larger macroions. Typical examples are polyelectrolytes, micelles, and colloids. In this section we briefly examine the latter, which have been the object of intense experimental research in recent years.

Colloids cover a wide range of colloidal particles, containing a large number of ionizable sites, dispersed in a solvent (generally water). Typical examples are organic colloids, like certain globular viruses or polymer microspheres (e.g., polystyrene balls) or inorganic crystallites, like the AgI sol or certain suspended silica. Colloidal particles have sizes ranging between $10^2$ and $10^4$ Å and acquire a large electric charge in solution. Because of the large size of the colloidal particles, interfacial phenomena play a dominant role. The main characteristic of this interface is its electric polarization: the charged colloidal particles are surrounded by a cloud of counterions, thus giving rise to an electric double layer having a spatial extension determined by the Debye screening length ($\lambda_D$ is typically of the order of $10^3$–$10^4$ Å in colloids, depending on the counterion concentration). The colloidal particles interact, hence, via screened electrostatic repulsion. It is this repulsion that stabilizes the suspension and prevents agglomeration. The addition of small amounts of electrolyte to the suspension results in a decrease of the screening length, and hence of the Coulomb repulsion, and can provoke coagulation (or flocculation) of the suspension. In this section we restrict ourselves to monodisperse colloids, comprising spherical particles of practically identical diameters; suspensions of polystyrene balls come very close to such an ideal situation. Light or small-angle neutron-scattering experiments on the static structure factor of such colloids have revealed a considerable amount of liquid-like short-range order[50] and in many situations such suspensions have been observed to solidify into colloidal *crystals.*[51] This crystallization is a Coulomb correlation effect, because the particles are observed to occupy regular lattice sites long before they are closely packed.

In the immediate vicinity of a large colloidal particle its curvature can be neglected in a first approximation and the double layer can be considered as planar. The electric potential in the double layer and the charge profile of the counterions can be determined from the one-dimensional Poisson–Boltzmann equation (Gouy–Chapman theory[52]). If $x$ denotes the coordinate perpendicular to the surface, $\Phi(x)$ the electrostatic potential, and $\rho(x)$ the charge profile, the Poisson equation (8.2.56b) here assumes the form

$$\frac{d^2\Phi(x)}{dx^2} = -\frac{4\pi e^2}{\varepsilon}\rho(x) \tag{8.4.19}$$

where $\varepsilon$ is the dielectric constant of the medium. Assuming that the positive and negative counterions have the same absolute valence $Z$, the density profile is related to the potential by

$$\rho(x) = Z[\rho_+(x) - \rho_-(x)]$$

$$= Zn \sinh[Ze\Phi(x)/k_B T] \tag{8.4.20}$$

Contrary to their three-dimensional counterpart, the set of equations (8.4.19) and (8.4.20) can be solved analytically without linearization.[52] For large distances $x$ the potential decays exponentially, and the screening length is the Debye length given by

$$\lambda_D = [4\pi n Z^2 e^2/(\varepsilon k_B T)]^{-1/2} \tag{8.4.21}$$

The mutual interaction of two interpenetrating double layers (and hence of two colloidal particles surrounded by their clouds of counterions) is then calculated to be[52]

$$v(r) = \pi\varepsilon\sigma^2\psi_0^2\exp[-(r-\sigma)/\lambda_D]/r \tag{8.4.22}$$

where $\psi_0$ is the potential at the surface of the colloidal particles and $\sigma$ their diameter. The potential (8.4.22) is of the screened Coulomb form, as expected. The simplest model of a colloid is consequently a collection of hard spheres interacting via the potential (8.4.22). The counterions appear only through the screening length $\lambda_D$ and the solvent through the dielectric constant. Such semimacroscopic assumptions are justified in view of the large size of the colloidal particles compared to that of the counterions and the solvent molecules. The MSA, introduced in Section 8.2.5, has been solved analytically for that model.[53] However, since the MSA yields poor results for low packing fractions ($\eta = \pi n_c \sigma^3/6 \ll 1$, where $n_c$ is the number of colloidal particles per unit volume), the physical diameter $\sigma$ is increased to an effective diameter $\sigma' > \sigma$, which does not affect the structure (because of the strong Coulomb repulsion that prevents the particles from coming into contact) but strongly improves the accuracy of the MSA; $\sigma'$ is chosen such that $v(\sigma')$ remains large compared to $k_B T$.[54] This procedure yields colloid structure factors in excellent agreement with experimental data.[50] The dynamics of interacting "Brownian" colloidal particles have been successfully studied by a generalized Fokker–Planck equation.[55]

## 8.5. Selected Applications: Part 2

The systems considered in the previous section were *effectively* one-component fluids, the second (neutralizing) component being essentially reduced to a continuum. The validity of such an approximation is basically a consequence of the large mass ratio of the two components. This justifies a Born–Oppenheimer approximation: the lighter species readjusts quasi-instantaneously to the slower motion of the heavier particles, and a partial averaging over the degrees of freedom of the lighter particles for any given configuration of the heavier species results in an effective *screened* interaction between the heavier particles.

In this section we address ourselves to genuine two-component ionic liquids in which both positive and negative ions have similar masses and must be treated on an equal footing. We shall be essentially concerned with the case of molten salts and aqueous solutions of strong electrolytes. We shall end by very briefly considering two-dimensional Coulomb systems, in order to illustrate the influence of space dimensionality on fluids of charged particles.

### 8.5.1. Molten Salts

The simplest molten salts are the alkali halides in which the ions are spherical and have opposite valences $Z = \pm 1$. If the ions are assumed not to be polarized by the strong local electric fields (*rigid* ion model), they interact essentially via pairwise additive forces deriving from the pair potentials:

$$v_{\alpha\beta}(r) = v^s_{\alpha\beta}(r) + \frac{Z_\alpha Z_\beta e^2}{r} \tag{8.5.1}$$

The Coulomb term leads to a very large cohesive energy of ionic crystals, exemplified by the high melting temperatures of all alkali halides ($T_m \gtrsim$ 1000 K!). The excess internal energy of the ionic melt is typically 90% Coulomb in origin. The short-range term $v^s(r)$ in equation (8.5.1) accounts essentially for the short-range repulsion between oppositely charged ions due to their impenetrability (which is crucial to prevent the collapse of + − pairs!). It includes also the attractive van der Waals interactions, which are relatively insignificant in ionic melts, compared to the strong Coulomb interactions. A widely used short-range potential is of the Huggins–Mayer form:

$$v^s_{\alpha\beta}(r) = A_{\alpha\beta}\exp(-a_{\alpha\beta}r) - \frac{C_{\alpha\beta}}{r^6} - \frac{D_{\alpha\beta}}{r^8} \tag{8.5.2}$$

where the exponential term describes the overlap repulsion while the two attractive terms correspond to the dipole–dipole and dipole–quadrupole dispersion forces, respectively. The coefficients $A_{\alpha\beta}$, $C_{\alpha\beta}$, $D_{\alpha\beta}$, and $a_{\alpha\beta}$ ($\alpha, \beta = +$ or $-$) are generally determined from solid-state data.[56] If the van der Waals terms in equation (8.5.2) are neglected, ion pairs of the same charge (i.e., + + and − − pairs) interact essentially via their Coulomb repulsion, which is sufficiently strong to mask the overlap repulsion in equation (8.5.2). Under these conditions the overlap repulsion need only be specified for + − pairs, and complete symmetry between anions and cations has been achieved (charge-conjugation symmetry). Due to this symmetry, there will be only two independent pair distribution functions, which we shall denote by "like" ($l$) and "unlike" (u):

$$g_l(r) = g_{++}(r) \equiv g_{--}(r) = \tfrac{1}{2}[g_{++}(r) + g_{--}(r)] \tag{8.5.3a}$$

$$g_u(r) = g_{+-}(r) \tag{8.5.3b}$$

In fact, both computer simulations[57] and neutron-scattering experiments[58] do show that $g_{++}(r)$ and $g_{--}(r)$ are rather close for the alkali halides; the small, but significant differences are due to the van der Waals terms and, more importantly, to the polarizability of the ions, particularly of the anions. This similarity justifies the use of the "restricted primitive model" (RPM) or of a simple "soft-core" model to study the essential structural and dynamic features of alkali halides. A "soft-core" model that has been extensively studied ("symmetric molten salt")[9] is based on the very simple potential

$$v_{\alpha\beta}(r) = \frac{e^2}{\lambda}\left[\frac{1}{n}\left(\frac{\lambda}{r}\right)^n + Z_\alpha Z_\beta\left(\frac{\lambda}{r}\right)\right] \tag{8.5.4}$$

With the choice $n = 9$ for the repulsive exponent and $\lambda = 2.34$ Å for the length scale, this potential model yields a semiquantitative description of NaCl. The essential results of the simulation work[9] are the following (the reader is referred to the original paper for the figures):

1. The pair structure exhibits marked charge ordering. The partial structure factors are very well reproduced by HNC theory, except for small wave numbers where the compressibility is overestimated by a factor of three.
2. The dynamics were investigated for the fully symmetric equal-mass case ($m_+ = m_-$); the two velocity ACFs are identical in that case. Correlations between velocities of different ions at different times lead to an electric current ACF $J_Z(t)$, which decays slightly faster than the velocity ACF, and exhibits a slightly more pronounced negative region. This leads to a significant deviation from the Nernst–Einstein relation (8.3.37), which must be corrected in the form ($Z_1 = -Z_2 = 1$)

$$\sigma = \tfrac{1}{2}n\,\frac{e^2}{k_B T}(D_+ + D_-)(1 - \Delta) \tag{8.5.5}$$

where the correction factor $\Delta$ is about 0.2.
3. Due to charge-conjugation symmetry, the mass-charge cross-correlation function is identically zero for the symmetric molten salt, i.e.,

$$F_{MZ}(k, t) = 0$$

so that charge- and mass-density fluctuations are completely decoupled in this model. This simplifying feature is obviously not shared by more realistic models (where $v_{++}(r) \neq v_{--}(r)$ and $m_+ \neq m_-$), but it can be expected to be approximately valid as long as the anion and cation masses are close and $g_{++}(r) \simeq g_{--}(r)$. The decoupling always occurs in the $k \to 0$ limit.[38]

The simulation results for the charge dynamic structure factor $S_{ZZ}(k, \omega)$ exhibit a pronounced peak close to the plasma frequency, characteristic of a well-defined plasmon mode, reminiscent of the "optic" mode observed in alkali-halide crystals. Due to the finite conductivity, the mode is considerably more damped as in the OCP under comparable coupling conditions, and the long-wavelength-mode frequency is shifted by nearly 30% above $\omega_p$.[12] Note that $S(k = 0, \omega)$ can be directly deduced from the simulation results for $J_z(t)$ via equation (8.3.70), and is also experimentally accessible by optical reflectivity measurements.[12]

The plasmon mode is shifted to lower frequencies with increasing $k$ (negative dispersion as in the OCP) and vanishes for $k \simeq 1$ Å$^{-1}$. A striking feature is that the damping (i.e., the imaginary part of the plasmon frequency) appears to go through a *minimum* as $k$ increases from zero. This unusual behavior has recently been investigated by detailed mode-coupling calculations,[59] but its physical origin is not very clear.

The qualitative features of the "symmetric molten salt" have been essentially confirmed by simulations of more "realistic" models of RbCl[60] and NaCl.[6] The influence of the ion polarizability on the dynamic properties of molten salts has been examined in the case of NaCl and KI.[61] These simulations show that the extra freedom in the rearrangement of the local charge density due to ion polarization leads to a sizeable increase of the self-diffusion coefficients $D_+$ and $D_-$, while the electric conductivity $\sigma$ remains practically unchanged.

Neutron inelastic-scattering experiments have been performed to detect the optic mode in molten alkali halides. The differential cross-section for inelastic scattering from a molten alkali halide is given by

$$
\begin{aligned}
\frac{d^2\sigma}{d\Omega d\omega} &\sim \sum_\alpha \sum_\beta f_\alpha f_\beta S_{\alpha\beta}(k, \omega) \\
&\sim [(f_+ + f_-)^2 S_{NN}(k, \omega) + 2(f_+^2 - f_-^2) S_{NZ}(k, \omega) \\
&\quad + (f_+ - f_-)^2 S_{ZZ}(k, \omega)]
\end{aligned}
\tag{8.5.6}
$$

where $f_+$ and $f_-$ are the scattering lengths of the cations and anions, and the dynamic structure factors $S_{NN}$, $S_{NZ}$, and $S_{ZZ}$ are linear combinations of the partial dynamic structure factors $S_{\alpha\beta}(k, \omega)$ given by equation (8.2.20). In an ideal situation, where $f_+ = -f_-$, the scattering experiment would measure $S_{ZZ}(k, \omega)$ directly. Careful experiments have been performed on KBr for which $f_+ = 0.37 \times 10^{-12}$ cm and $f_- = 0.68 \times 10^{-12}$ cm.[62] Such an experiment yields a linear combination of $S_{NN}$, $S_{NZ}$, and $S_{ZZ}$ from which it is not very easy to extract $S_{ZZ}(k, \omega)$, particularly at small $k$, where the total intensity $S_{ZZ}(k) \sim k^2$ of the charge-fluctuation spectrum vanishes. The published data do show a shoulder on the high-frequency side of the observed spectrum at roughly the correct frequency ($\simeq \omega_p$), but the experimental evidence for the existence of the optic mode at finite $k$ is not very conclusive.[71]

It is noteworthy that the simulations of molten NaCl[6] have also given convincing evidence of propagating *transverse* mass (i.e., shear) and charge (i.e., optic) modes, which appear in the spectra of the corresponding transverse-current correlation functions.

## 8.5.2. Electrolyte Solutions

The presence of polar solvent molecules surrounding the anions and cations in ionic solutions makes these electrolytes much more complex systems than molten salts. This complexity is illustrated strikingly by the extensive neutron-diffraction experiments with isotopic substitution carried out by Enderby and his co-workers to determine the static structure in aqueous solutions of strong electrolytes. The "primitive model" of electrolytes, in which the solvent is replaced by a dielectric continuum, is totally incapable of rendering the details of the pair structure of such solutions, although it does yield reasonable values of the osmotic pressure. The failure of the "primitive model" is due to strong coupling between the ionic charges and the dipole moments of the solvent molecules that reflects itself in relatively well-defined coordination shells of solvent molecules around the ions (ion solvation). For this reason there is at present considerable interest in developing models for electrolyte solutions that go beyond the primitive model by incorporating the discrete molecular nature of the solvent.[72] A very successful step in this direction was the solution of the MSA for a model fluid in which the ions and solvent particles are taken to be charged and dipolar hard spheres, respectively.[63] In the general case we have a three-component fluid; each species carries an electric charge $Z_\alpha e$, a point dipole moment $\mu_\alpha$, and has diameter $d_\alpha$ ($1 \leq \alpha \leq 3$). Conventionally, $\alpha = 1$ and 2 refer to the cations and anions ($\mu_\alpha = 0$), while $\alpha = 3$ refers to the dipolar hard spheres ($Z_3 = 0$). Under these conditions we have three types of interaction between particles $i$ and $j$ at distance $r_{ij}$:

*ion–ion*:

$$v_{\alpha\beta}(r_{ij}) = \infty, \qquad r_{ij} < d_{\alpha\beta}$$

$$= \frac{Z_\alpha Z_\beta e^2}{r_{ij}}, \quad r_{ij} > d_{\alpha\beta} \quad (\alpha, \beta = 1, 2) \qquad (8.5.7a)$$

*ion–dipole*:

$$v_{\alpha\beta}(r_{ij}, \mathbf{\Omega}_j) = \infty, \qquad\qquad r_{ij} < d_{\alpha\beta}$$

$$= \frac{-Z_\alpha e \mu_\beta}{r_{ij}^2} F_1(\mathbf{\Omega}_j), \quad r_{ij} > d_{\alpha\beta} \quad (\alpha = 1, 2; \beta = 3) \qquad (8.5.7b)$$

*dipole–dipole*:

$$v_{\alpha\beta}(r_{ij}, \mathbf{\Omega}_i, \mathbf{\Omega}_j) = \infty, \qquad\qquad r_{ij} < d_{\alpha\beta}$$

$$= \frac{-\mu_\alpha \mu_\beta}{r_{ij}^3} F_3(\mathbf{\Omega}_i, \mathbf{\Omega}_j), \quad r_{ij} > d_{\alpha\beta} \quad (\alpha = \beta = 3) \qquad (8.5.7c)$$

where $d_{\alpha\beta} = \frac{1}{2}(d_\alpha + d_\beta)$, $\Omega_i = (\theta_i, \varphi_i)$ denotes the polar angles of the unit vector $\hat{\mu}_i = \boldsymbol{\mu}_i / \mu_i$ along the dipole moment of particle $i$, and the $F_\nu$ denote the angular factors:

$$F_1(\Omega_j) = \hat{\mu}_j \cdot \hat{r}_{ij} \qquad\qquad (8.5.8a)$$

$$F_2(\Omega_i, \Omega_j) = \hat{\mu}_i \cdot \hat{\mu}_j \qquad\qquad (8.5.8b)$$

$$F_3(\Omega_i, \Omega_j) = 3(\hat{\mu}_i \cdot \hat{r}_{ij})(\hat{\mu}_j \cdot \hat{r}_{ij}) - \hat{\mu}_i \cdot \hat{\mu}_j \qquad\qquad (8.5.8c)$$

The MSA amounts to solving the coupled OZ relations (8.2.12) (where the convolution integral now also includes an integration over the orientation $\Omega$ of the dipole) together with the closure relations

$$g_{\alpha\beta}(r) = 0, \qquad r < d_{\alpha\beta} \qquad\qquad (8.5.9a)$$

$$c_{\alpha\beta}(r, \Omega, \Omega') = -v_{\alpha\beta}(r, \Omega, \Omega')/k_B T, \qquad r > d_{\alpha\beta} \qquad\qquad (8.5.9b)$$

The two-component, purely ionic case (i.e., $n_3 = 0$) corresponding to the "primitive model" has been discussed in Section 8.2.5. The purely dipolar case ($n_1 = n_2 = 0$) has been solved by Wertheim,[64] who showed that the angular dependence of the resulting correlation function could be entirely described in terms of the angular functions $F_0 \equiv 1$, $F_2$, and $F_3$. In the general mixed case the ion–dipole and dipole–dipole correlation functions are combinations of the four angular functions $F_\nu$ $(0 \leq \nu \leq 3)$ with $r$-dependent coefficients. The ion–ion correlation functions remain, of course, spherically symmetric but they turn out to differ considerably from their "primitive-model" counterparts. In particular, the solvent interaction leads to pronounced oscillations in the $g_{\alpha\beta}(r)$ $(\alpha, \beta = 1, 2)$, which are completely absent in the "primitive model".[65] All six correlation functions exhibit exponential screening, with the classical Debye screening length (8.1.3) at low ionic concentration.

The HNC equations introduced in Section 8.2.5 have also been extended to anisotropic situations involving dipole interactions, and solved numerically for the model of oppositely charged hard spheres in a dipolar hard-sphere solvent.[66] In order to handle the angular dependence of the correlation functions, the HNC closure (8.2.58) is expanded to first (LHNC) or second (QHNC) order in the angular-dependent terms, and only those terms involving the angular functions $F_\nu$ $(0 \leq \nu \leq 3)$ are retained. The results for the ion–ion pair distribution functions $g_{\alpha\beta}(r)$ confirm the total failure of the "primitive model" even at relatively low ion concentration. The $g_{\alpha\beta}(r)$ always exhibit a pronounced oscillatory structure, with the period of the oscillations determined by the solvent diameter; the maxima correspond to ion pairs separated by integral numbers of solvent molecules. Under certain conditions $g_{+-}(r)$ is very large at contact, indicating a low degree of dissociation, while in other cases solvent-separated ion pairs are found to be the more common species. HNC theory also predicts that for relatively low ion concentration, the dielectric constant of the solution decreases with increasing ion concentration, in agreement with experimental findings.[66]

### 8.5.3. Two-Dimensional Coulomb Fluids

To conclude this chapter we briefly survey some recent results concerning two-dimensional (2D) Coulomb fluids. A careful distinction must be made between two-dimensional systems of particles interacting via the three-dimensional (3D) Coulomb potential $e^2/r$, and genuine 2D Coulomb systems made up of particles interacting via the 2D Coulomb potential $-e^2 \ln(r/a)$, which is the solution of the 2D Poisson equation

$$\nabla^2 v(r) = -2\pi e^2 \delta(\mathbf{r}) \qquad (8.5.10)$$

While the latter systems behave *qualitatively* very much like their 3D counterparts (exhibiting exponential screening and collective plasma oscillations), the former have a very different behavior, which can be traced back to a different small-$k$ singularity. For a $1/r$ potential this singularity goes like $1/k$ in 2D, while the logarithmic potential following from equation (8.5.10) in 2D has the same $1/k^2$ singularity as the 3D Coulomb potential considered so far. Although the genuine 2D Coulomb fluids are of great theoretical importance, we shall restrict ourselves to the $1/r$ case in 2D, in order to illustrate the influence of the exact nature of the small-$k$ singularity on static and dynamic properties. The system we have in mind is the 2D electron gas, which is obtained experimentally in MOS devices or under the form of monolayers of electrons trapped on the surface of liquid helium. In this latter case the electrons are pressed against the liquid helium surface by the electrostatic field of a condenser with plates parallel to the helium surface. Areal densities of typically $10^8$ cm$^{-2}$ can be achieved, and it is easily verified that even at 1 K the de Broglie thermal wavelength of the surface electrons is short compared to their mean spacing (which is of the order of 1 micron!), so that the electron gas can be regarded as essentially classical. The Coulomb coupling constant $\Gamma = e^2/ak_BT$, with $a = (1/n)^{1/2}$ is however large, and the system has in fact been observed to crystallize into a triangular lattice for $\Gamma \approx 130$ (2D Wigner transition).[67] If the coupling to the helium surface waves ("ripplons") is neglected, the electron monolayer can be considered as a perfect realization of a 2D OCP, where the role of the uniform background is played by the charges on the condenser plates that render the system electrically neutral. The 2D Fourier transform of the 3D Coulomb potential is

$$\hat{v}(k) = 2\pi e^2/k \tag{8.5.11}$$

By analogy with the 3D case, we define the Debye wave number

$$k_D = 2\pi ne^2/k_B T$$

and the plasma frequency

$$\omega_p = v_0 k_D = 2\pi ne^2/(mk_B T)^{1/2}$$

The RPA for the structure factor is derived directly from equations (8.2.55a) and (8.5.11), with the result

$$S_{RPA}(k) = \frac{k}{k + k_D} \tag{8.5.12}$$

indicating a linear behavior of $S(k)$ at small $k$, in contrast to the quadratic behavior in 3D. The corresponding Debye–Hückel pair distribution function decays *algebraically* ($\sim 1/r^3$) at large $r$, rather than exponentially as in the 3D case. The screening is hence much less efficient in 2D.

From the small-$k$ limit of the characteristic longitudinal frequency (8.4.2),

$$\lim_{k\to 0} \omega_{0l}(k) = \lim_{k\to 0} \left(\frac{v_0^2 k^2}{S(k)}\right)^{1/2}$$

$$= \lim_{k\to 0} \left(\frac{v_0^2 k^2}{k/k_D}\right)^{1/2}$$

$$= \omega_p \left(\frac{k}{k_D}\right)^{1/2} \tag{8.5.13}$$

It is immediately apparent that the plasmon mode is a *low*-frequency mode in 2D in the long-wavelength limit.

Computer-simulation results for $S(k, \omega)$ indeed show the existence of a well-defined collective mode up to $k \times a \simeq 2$ at various values of $\Gamma$.[68] For strong couplings, the resulting dispersion curve lies below its small-$k$ limit (8.5.13) and flattens out before the plasmon peak vanishes. This rather flat dispersion curve, which allows the determination of a relatively well-defined frequency over an extensive range of wave numbers, yields a natural explanation of the pronounced oscillatory behavior observed in the velocity ACF of the 2D OCP for strong couplings, in analogy with the case of the three-dimensional OCP.[69]

# References

1. P. Debye and E. Hückel, *Z. Phys.* **24**, 185, 305 (1923).
2. J. P. Hansen, *J. Phys. C* **14**, L-151 (1981).
3. J. P. Hansen and I. R. McDonald, *Theory of Simple Liquids*, Academic Press, London (1976).
4. J. P. Hansen, G. M. Torrie and P. Vieillefosse, *Phys. Rev. A* **16**, 2153 (1977).
5. F. G. Edwards, J. E. Enderby, R. A. Rowe, and D. I. Page, *J. Phys. C* **8**, 3483 (1975).
6. E. M. Adams, I. R. McDonald, and K. Singer, *Proc. R. Soc. London, Ser. A* **357**, 37 (1977).
7. J. E. Enderby, D. M. North, and P. A. Egelstaff, *Philos. Mag.* **14**, 961 (1966).
8. L. V. Woodcock and K. Singer, *Trans. Faraday Soc.* **67**, 12 (1971).
9. J. P. Hansen and I. R. McDonald, *Phys. Rev. A* **11**, 2111 (1975).
10. A. B. Bhatia and D. E. Thornton, *Phys. Rev. A* **2**, 3004 (1970).
11. R. Kubo, *Rep. Prog. Phys.* **29**, 255 (1966).
12. M. Parrinello and M. P. Tosi, *Riv. Nuovo Cimento* **2**, No. 6 (1979).
13. J. G. Kirkwood and F. Buff, *J. Chem. Phys.* **19**, 774 (1951).
14. P. Vieillefosse and J. P. Hansen, *Phys. Rev. A* **12**, 1106 (1975).
15. See, e.g., D. Pines and Ph. Nozières, *Theory of Quantum Fluids*, Benjamin, New York (1966).
16. G. Stell, in: *Statistical Mechanics, Part A* (B. J. Berne, ed.), Plenum Press, New York (1977).
17. E. Salpeter, *Aust. J. Phys.* **7**, 353 (1954).
18. P. Vieillefosse, *J. Phys. (Paris)* **42**, 723 (1981).
19. M. Baus and J. P. Hansen, *Physics Reports* **59**, 1 (1980).
20. J. C. Rasaiah, D. N. Card, and J. P. Valleau, *J. Chem. Phys.* **56**, 248 (1972).
21. G. M. Abernethy, M. Dixon, and M. J. Gillan, *Phil. Mag. B* **43**, 1113 (1981).
22. B. Larsen, *J. Chem. Phys.* **68**, 4511 (1978).
23. Y. Rosenfeld and N. W. Ashcroft, *Phys. Rev. A* **20**, 1208 (1979).
24. H. Iyetomi and S. Ichimaru, *Phys. Rev. A* **25**, 2434 (1982).
25. E. Waisman and J. Lebowitz, *J. Chem. Phys.* **56**, 3086, 3093 (1972).
26. L. Blum, *Mol. Phys.* **30**, 1529 (1975); L. Blum and J. S. Høye, *J. Phys. Chem.* **81**, 1311 (1977).
27. M. C. Abramo, C. Caccamo, G. Pizzimenti, M. Parrinello, and M. P. Tosi, *J. Chem. Phys.* **68**, 2889 (1978).
28. J. S. Høye and G. Stell, *J. Chem. Phys.* **67**, 524 (1977).
29. M. C. Abramo, C. Caccamo, and G. Pizzimenti, *Lettere al N. Cim.* **30**, 297 (1981).
30. P. Martin, in: *Many-Body Physics* (C. De Witt and R. Balian, eds.), Gordon and Breach, New York (1967).
31. D. Forster, *Hydrodynamic Fluctuations, Broken Symmetry and Correlation Functions*, Benjamin, Reading, MA (1975).
32. J. P. Hansen, in: *Microscopic Structure and Dynamics of Liquids* (J. Dupuy and A. J. Dianoux, eds.), Plenum Press, New York (1978).
33. H. Mori, *Prog. Theor. Phys.* **33**, 423 (1965); **34**, 399 (1965).
34. See, e.g., T. E. Faber, *An Introduction to the Theory of Liquid Metals*, Cambridge University Press (1972).
35. H. Minoo, C. Deutsch, and J. P. Hansen, *J. Phys. (Paris), Lett.* **38**, L 191 (1977).
36. G. Jacucci and I. R. McDonald, *Physica A* **80**, 607 (1975).
37. P. C. Martin, *Phys. Rev.* **161**, 143 (1967).
38. P. V. Giaquinta, M. Parrinello, and M. P. Tosi, *Phys. Chem. Liq.* **5**, 305 (1976).
39. W. L. Slattery, G. D. Doolen, and H. E. De Witt, *Phys. Rev. A* **21**, 2087 (1980).
40. J. P. Hansen, I. R. McDonald, and E. L. Pollock, *Phys. Rev. A* **11**, 1025 (1975).

41. See M. Baus and J. P. Hansen, *Phys. Rep.* **59**, 1 (1980) for references; a very recent approach is that of T. Gaskell, *J. Phys. C* **15**, 1601 (1982).
42. J. Wallenborn and M. Baus, *Phys. Rev. A* **18**, 1737 (1978).
43. S. W. Lovesey, *J. Phys. C* **4**, 3057 (1971).
44. H. M. Van Horn, *Phys. Today* (January 1979).
45. S. Galam and J. P. Hansen, *Phys. Rev. A* **14**, 816 (1976).
46. M. P. Tosi, in *Electron Correlations in Solids, Molecules and Atoms* (J. Devreese and F. Brosens, eds.) Plenum, 1983.
47. M. J. Huijben and W. Van der Lugt, in: *Liquid Metals, 1976*, p. 141, Conférence séries N° 30, The Institute of Physics, Bristol (1977).
48. N. W. Ashcroft, *J. Phys. C* **1**, 232 (1968).
49. D. Bohm and T. Staver, *Phys. Rev.* **84**, 836 (1951).
50. J. C. Brown, P. N. Pusey, J. W. Goodwin, and R. H. Ottenvill, *J. Phys. A* **8**, 664 (1975).
51. P. Pieranski, in: *Physics of Defects, Les Houches Session XXXV* (R. Balian *et al.*, eds.), North-Holland, Amsterdam (1981).
52. E. J. W. Verwey and J. T. G. Overbeek, *Theory of the Stability of Lyophobic Colloids*, Elsevier, Amsterdam (1948).
53. See, e.g., J. B. Hayter and J. Penfold, *Mol. Phys.* **42**, 109 (1981).
54. J. P. Hansen and J. B. Hayter, *Mol. Phys.* **46**, 651 (1982).
55. See, e.g., W. Hess and R. Klein, *Adv. in Phys.* **32**, 173 (1983).
56. M. P. Tosi and F. G. Fumi, *J. Phys. Chem. Solids* **25** 45 (1964).
57. M. J. L. Sangster and M. Dixon, *Adv. Phys.* **25**, 247 (1976).
58. J. E. Enderby and G. W. Neilson, *Adv. Phys.* **29**, 323 (1980).
59. J. Bosse and T. Munakata, *Phys. Rev. A* **24**, 2261 (1981).
60. J. R. D. Copley and A. Rahman, *Phys. Rev. A* **13**, 2276 (1976).
61. G. Jacucci, I. R. McDonald, and A. Rahman, *Phys. Rev. A* **13**, 1581 (1976).
62. J. R. D. Copley and G. Dolling, *J. Phys. C* **11**, 1259 (1978).
63. L. Blum, *Chem. Phys. Lett.* **26**, 200 (1974). S. A. Adelman and J. M. Deutch, *J. Chem. Phys.* **60**, 3935 (1974).
64. M. S. Wertheim, *Mol. Phys.* **25**, 211 (1973).
65. D. Y. C. Chan, D. J. Mitchell, and B. W. Ninham, *J. Chem. Phys.* **70**, 2946 (1979).
66. D. Levesque, J. J. Weis, and G. N. Patey, *J. Chem. Phys.* **72**, 1887 (1980).
67. C. C. Grimes and G. Adams, *Phys. Rev. Lett.* **42**, 795 (1979).
68. H. Totsuji and H. Kaleya, *Phys. Rev. A* **22**, 1220 (1980).
69. J. P. Hansen, D. Levesque, and J. J. Weis, *Phys. Rev. Lett.* **43**, 979 (1979).
70. P. Carini and G. Kalman, *Phys. Lett.* **105A**, 229 (1984).
71. See, however, R. L. McGreevy, E. W. J. Mitchell and F. M. A. Margaca, *J. Phys. C* **17**, 775 (1984).
72. For a recent review, see J. P. Hansen, *J. Phys. (Paris)* **45**, C7–97 (1984).

# Electric Transport in Liquid Metals

## H. Beck

### 9.1. Basic Notions

A metal is a two-component system of ions of charge $Ze$ and valence electrons, described by the following Hamiltonian:

$$H = H_{el} + H_{ion} + H_{el-ion} \tag{9.1.1}$$

$$H_{el} = \sum_i \frac{p_i^2}{2m} + \frac{1}{2} \sum_{i,j}' \frac{e^2}{|\mathbf{r}_i - \mathbf{r}_j|} \tag{9.1.2}$$

$$H_{ion} = \sum_n \frac{P_n^2}{2M_n} + \frac{1}{2} \sum_{n,m}' \frac{(Ze)^2}{|\mathbf{R}_n - \mathbf{R}_m|} \tag{9.1.3}$$

$$H_{el-ion} = \sum_{i,n} V(\mathbf{r}_i - \mathbf{R}_n) + \sum_{i,n} J(\mathbf{r}_i - \mathbf{R}_n)\boldsymbol{\sigma}_i \cdot \mathbf{S}_n \tag{9.1.4}$$

Here and in the following $\mathbf{P}_n$, $\mathbf{R}_n$, and $\mathbf{S}_n$ denote ionic momenta, positions, and spins, respectively, while position, momentum, and spin of the $i$th electron are denoted by $\mathbf{r}_i$, $\mathbf{p}_i$, and $\boldsymbol{\sigma}_i$. Before specifying the electron–ion interaction in more detail, we shall outline the aims of a comprehensive theory of liquid metals.

1. *Equilibrium properties.* Hamiltonian (9.1.1) should provide the

**H. Beck** · Institut de Physique, Université de Neuchâtel, Rue A.-L. Breguet 1, CH-2000 Neuchâtel, Switzerland.

electronic density of states, the equilibrium charge densities, and information about the one-electron stationary states. Moreover, the electronic screening will lead to effective interactions between the ions, which determine the ionic structure and their dynamics.

2. *Nonequilibrium phenomena.* This field comprises electric conductivity and other transport coefficients like thermopower and the Hall coefficient, and ionic transport.

When discussing electric transport in a disordered metal, we may assume the ionic structure to be given. For many purposes it is sufficient to know the static ionic-structure factor defined by

$$S(q) \equiv \frac{1}{N} \sum_{n,m} \langle e^{i q \cdot (R_n - R_m)} \rangle - N \delta_{q,0} \qquad (9.1.5)$$

which is the Fourier transform of the pair correlation function $g$ ($\Omega_0$ is the atomic volume):

$$S(q) - 1 = \frac{1}{\Omega_0} \int d^3 r\, e^{i q \cdot r} [g(r) - 1] \qquad (9.1.6)$$

The symbol $\langle\ \rangle$ denotes an average over an ensemble of samples with the same macroscopic properties (such as density or concentration of constituents) and, at finite temperature, an average over thermal fluctuations.[1]

For a crystal at zero temperature $S(\mathbf{q})$ consists of Bragg peaks at all the reciprocal lattice vectors $\mathbf{Q}_i$, which is characteristic of long-range translation order. For a liquid metal of macroscopic volume, $S(q)$ is a smooth function whose structure describes the short-range order of such a system (see Fig. 9.1). The temperature dependence of $S(q)$, which will be of importance later on, is shown in Figure 9.2.

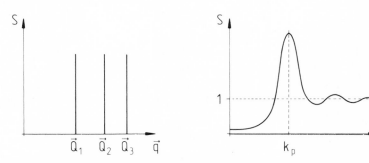

Figure 9.1. Static structure factors. Left: Bragg peaks for an ideal crystal in some direction of reciprocal space. Right: typical structure factor of a liquid metal ($k_p$ denotes the position of the first peak).

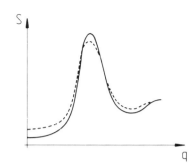

Figure 9.2. Schematic $T$-dependence of $S(q)$:
——— low $T$ ($T \gtrsim T_{\text{melt}}$), --- high $T$.

For an alloy containing $N_\alpha$ ions of species $\alpha$, partial structure factors can be defined by[1]

$$S_{\alpha\beta}(q) \equiv (N_\alpha N_\beta)^{-1/2} \sum_{n_\alpha}^{N_\alpha} \sum_{n_\beta}^{N_\beta} \langle e^{i\mathbf{q}\cdot(\mathbf{R}_{n_\alpha} - \mathbf{R}_{m_\beta})} \rangle - (N_\alpha N_\beta)^{1/2} \delta_{\mathbf{q},0} \qquad (9.1.7)$$

Later on, dynamic structure factors

$$S(q,\omega) \equiv \frac{1}{N} \sum_{n,m} \int dt\, e^{-i\omega t} \langle e^{i\mathbf{q}\cdot[\mathbf{R}_n(t) - \mathbf{R}_m(0)]} \rangle \qquad (9.1.8)$$

(and, correspondingly, the partials) will also be important.

Structure factors can be measured by neutron or X-ray diffraction. Where such experimental results are unavailable (especially for alloys, where only few partials have been measured) model forms can be used, such as the Percus–Yevick result for hard spheres.

When the ionic positions are regarded as given parameters (at some time $t$), the electrons move in a fixed random potential given by equation (9.1.4). For the electron–ion interaction we distinguish various cases:

1. *Simple metals* (weak scatterers), namely alkali metals (Li, Na, K, Rb, Cs), alkaline earths (Be, Mg, Ca, etc.), some of the heavier ones actually being strong scatterers!, and polyvalent metals (Al, Ga, In, Si, Ge, Sn, Pb, Sb, Bi, etc.). In this case $V(r)$ is given by a (weak) pseudopotential. In its simplest (local and energy-independent) form it is given by a Coulomb tail and some constant value inside the ionic core.

2. *Nonsimple metals* (strong electron scatterers), namely noble metals (Cu, Ag, Au), early transition metals (TM) (Sc, Y, Ti, Zr, V, Nb, etc.), late transition metals (Fe, Co, Ni, Pd, etc.), rare earths (RE), and actinides. Here, $V(r)$ will be an atomic potential (like that used

for band-structure calculations) containing the nuclear attraction and repulsion, exchange and correlation with the core electrons.

3. *Magnetic ions.* If the ions carry a magnetic moment, due to localized $f$- (or $d$-) electrons, its effect on the conduction electrons can be represented by an exchange function $J(r)$. More details will be given in the applications to liquid Gd in Section 9.4.

Finally, in the spirit of a local density functional approach, the explicit electron–electron interaction in (9.1.2) is replaced by introducing screened electron–ion interactions. For simple metals linear screening amounts to replacing the Fourier transform $\hat{V}(q)$ of the pseudopotential by $\hat{V}(q)/\varepsilon(q)$, where $\varepsilon$ is a suitable dielectric function of the homogeneous interacting electron gas.[2] In the atomic potentials, screening by the core electrons is built in from the beginning through electrostatic, exchange, and correlation interactions. Screening by the conduction electrons can be taken into account using atomic wave functions or self-consistently screened potentials obtained from band-structure calculations for the corresponding crystalline solids. In practice, one usually adopts the "muffin-tin" picture: each ion is surrounded by a sphere and the potential within such a sphere is calculated on the basis of the total charge density contained in it. The potential outside the spheres (which should not overlap) is assumed to be constant. For the practical evaluation of such potentials, atomic wave functions tabulated by Herman and Skillman[3] can be used, following the Mattheiss procedure.[4] Alternatively, self-consistent potentials from band-structure calculations are now also available.[5]

We have thus reduced our problem to one of noninteracting electrons in a given external potential $U$:

$$H = \sum_i \left[ \frac{P_i^2}{2m} + U(\mathbf{r}_i, \boldsymbol{\sigma}_i) \right] \tag{9.1.9}$$

with

$$U(\mathbf{r}, \boldsymbol{\sigma}) = \sum_n [V(\mathbf{r} - \mathbf{R}_n) + J(\mathbf{r} - \mathbf{R}_n)\boldsymbol{\sigma} \cdot \mathbf{S}_n] \tag{9.1.10}$$

The price to be paid for the simplicity of (9.1.10) is that $U$ should be determined self-consistently through the equilibrium electronic-charge densities.

Before dealing with transport problems we should now determine the equilibrium electronic structure on the basis of equation (9.1.9). This is already a difficult problem owing to the disorder in the ionic positions (and possibly their spins). In the absence of systematic theoretical results the following standpoint is adopted:

1. For simple metals we shall use the free-electron approximation for the conduction electrons. Photoemission experiments have indeed shown that in many cases the density of states is well represented by its free-electron form. Deviations could in principle be calculated by a perturbation expansion with respect to $V$ or by introducing phenomenologically some parameters, like an effective mass.

2. In the case of metals with unfilled atomic $d$-states it will turn out to be crucial that at least some general features of the electronic structure are accounted for in order to develop a consistent picture of transport phenomena. Typically, the band structure of, say, a crystalline transition or noble metal exhibits the following overall features (see Figure 9.3): (a) At low energies (near the bottom of what is called the conduction band) and high energies, the dispersion is quite free-electron-like. The corresponding density of states forms some sort of background extending throughout the whole band. (b) At intermediate energies we find many rather flat bands comprising predominantly $d$-states. They contribute to a high $d$-like density of states.

Such behavior is easly understood, at least qualitatively, in a tight-binding scheme, where we start from atomic $s$- (and possibly $p$-) and $d$-states and allow for hybridization.

In a disordered metal we expect to find an electron structure that is not totally different since, again in the tight-binding approach, many features of the electron states are determined by short-range rather than long-range ionic order.[6] This is corroborated by the fact that the electronic density of states $D(E)$ does not change drastically when going from a crystalline to a

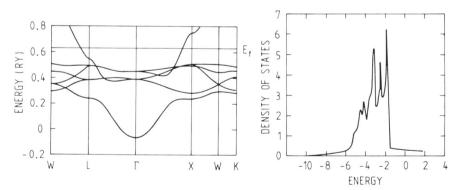

Figure 9.3. Band structure and electronic density of states of crystalline Cu (after Moruzzi et al.[5]).

disordered form of the same metal (except that fine structures caused by van Hove singularities are usually washed out), as photoemission experiments often show. Indeed, experimental data for $D(E)$ are often reproduced qualitatively by band-structure calculations for the crystalline phase.[7] Moreover, various multiple-scattering calculations (see Section 9.3) for $D(E)$ have produced some kind of isotropic "average band structure" for liquid transition or noble metals, which shows the main properties of a hybridized $s$-$d$ band.[8]

## 9.2. General Transport Theory

Transport coefficients are measured by applying external fields (electric field **E**, temperature gradient $\nabla T$, etc.) to the system and measuring the current **j** flowing in response to this external perturbation. For our system, described by equation (9.1.9), the electric-current operator can be defined in second quantization by

$$\mathbf{j}(\mathbf{r}) = -e \sum_{\mathbf{k},\mathbf{q}} \sum_{s} e^{i\mathbf{q}\cdot\mathbf{r}} \frac{\mathbf{k}}{m} a^{+}_{\mathbf{k}+\frac{1}{2}\mathbf{q},s} a_{\mathbf{k}-\frac{1}{2}\mathbf{q},s} \qquad (9.2.1)$$

where $a^{+}_{\mathbf{p}s}$ ($a_{\mathbf{q}s}$) is a creation (annihilation) operator of an electron with momentum **p** and spin $s$ (i.e., the $Z$-component of the spin with respect to some axis). Therefore, the basic quantity to be calculated is the electronic distribution function

$$f_{s}(\mathbf{k}, \mathbf{r}, t) \equiv \sum_{\mathbf{q}} e^{i\mathbf{q}\cdot\mathbf{r}} \langle a^{+}_{\mathbf{k}+\frac{1}{2}\mathbf{q},s} a_{\mathbf{k}-\frac{1}{2}\mathbf{q},s} \rangle_{t} \qquad (9.2.2)$$

where $\langle \cdot\cdot \rangle_{t}$ is a quantum-mechanical average over a density matrix whose time evolution includes the effect of the external field. We briefly sketch two different approaches to evaluating expression (9.2.2): a (nonequilibrium) scattering approach and a linear-response approach involving equilibrium correlation functions.

### 9.2.1. Scattering Approach

Suppose (see Figure 9.4) the metallic sample M, described by equation (9.1.9), is embedded in a "free-electron sea." The electron states outside M are plane waves, which are scattered by M. The probability per unit time of a transition from $(\mathbf{k}s)$ to $(\mathbf{k}'s')$ is given by

$$W(\mathbf{k}s, \mathbf{k}'s') = \frac{v(\mathbf{k}s)}{\Omega} \frac{d\sigma}{d\Omega} \qquad (9.2.3)$$

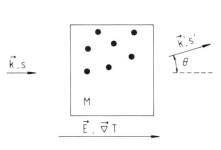

Figure 9.4. Plane wave scattered from a disordered metal.

where $\Omega$ is the volume, $d\sigma/d\Omega$ is the scattering cross-section of M, and $v(\mathbf{k}s) = \hbar k/m$ is the velocity of the incoming electron stream. Now our distribution function $f_s$ will be given by the Fermi distribution $N_0$ plus a deviation $\varphi_s$. For small external fields the latter will obey a linear Boltzmann equation (see, e.g., Ziman's book[9])

$$\frac{\partial \varphi_s}{\partial t} + \mathbf{v} \cdot \frac{\partial \varphi_s}{\partial \mathbf{r}} - e\mathbf{E} \cdot \frac{\partial N_0}{\partial \mathbf{k}} + \mathbf{v} \cdot \nabla T \frac{\partial N_0}{\partial T}$$
$$= \int d^3k' \sum_{s'} W(\mathbf{k}s, \mathbf{k}'s')[\varphi_s(\mathbf{k}) - \varphi_{s'}(\mathbf{k}')] \qquad (9.2.4)$$

This equation is simplified if we consider a homogeneous steady-state situation and introduce the spin sum, which must be done in equation (9.2.1):

$$\varphi \equiv \sum_s \varphi_s, \qquad \bar{W} \equiv \frac{1}{2} \sum_{s,s'} W(s, s') \qquad (9.2.5)$$

Then, the transport equation

$$-e\mathbf{E} \cdot \frac{\partial N_0}{\partial \mathbf{k}} + \mathbf{v} \cdot \nabla T \frac{\partial N_0}{\partial T} = \int d^3k' \bar{W}(\mathbf{k}, \mathbf{k}')[\varphi(\mathbf{k}) - \varphi(\mathbf{k}')] \qquad (9.2.6)$$

describes the balance between the external forces and the collisions inside M. If the scatterer M is macroscopically isotropic (i.e., $\bar{W}$ depends only on $k$ and $q = |\mathbf{k} - \mathbf{k}'|$), equation (9.2.6) can be solved exactly[9] by introducing the relaxation time

$$\tau^{-1}(k) \equiv \frac{2\pi}{\Omega} \int_{-1}^{+1} d(\cos\theta)(1 - \cos\theta) \frac{\overline{d\sigma}}{d\Omega}(k, q) \qquad (9.2.7)$$

where $\theta$ is the scattering angle between $\mathbf{k}$ and $\mathbf{k}'$. The current is then linearly related to the driving forces:

$$\mathbf{j} = L_1\mathbf{E} + L_2\nabla T \tag{9.2.8}$$

with

$$L_1 = -|e| \sum_{\mathbf{k}} \mathbf{v} \cdot \frac{dN_0}{\partial \mathbf{k}} \tau(k) \tag{9.2.9}$$

and

$$L_2 = |e| \sum_{\mathbf{k}} v^2 \frac{\partial N_0}{\partial T} \tau(k) \tag{9.2.10}$$

For calculating the electric (isothermal) conductivity $\sigma$, we put $\nabla T = 0$ and identify $\sigma$ with $L_1$, while the thermopower $Q$ is given by $L_2/L_1$. The Hall coefficient will be discussed below.

This result is (in some sense) exact.[10] Approximations for $d\sigma/d\Omega$ must be found for practical calculations.

### 9.2.1.1. Weak-Scattering Limit

If the electron–ion interaction is given by a weak (screened) pseudopotential $\hat{V}$, as assumed for simple metals, the cross-section can be evaluated in a Born approximation. If equation (9.1.10) is used with $J = 0$, then

$$\frac{\overline{d\sigma}}{d\Omega}(|\mathbf{k}|, q) \propto |\langle \mathbf{k}|U|\mathbf{k}'\rangle|^2$$

$$= \left| \int d^3r\, e^{i\mathbf{q}\cdot\mathbf{r}} U(\mathbf{r}) \right|^2$$

$$= \int d^3r\, d^3r'\, e^{i\mathbf{q}\cdot(\mathbf{r}-\mathbf{r}')} \sum_{n,m} \hat{V}^*(\mathbf{r}-\mathbf{R}_n)\hat{V}(\mathbf{r}'-\mathbf{R}_m)$$

$$= \sum_{n,m} e^{i\mathbf{q}\cdot(\mathbf{R}_n - \mathbf{R}_m)} |\hat{V}(\mathbf{q})|^2$$

$$= NS(q)|\hat{V}(q)|^2 \tag{9.2.11}$$

When this equation is substituted into equations (9.2.9) and (9.2.10) and the free-electron relation $k_F^3 = 3\pi^2 Z/\Omega_0$ (where $\Omega_0$ is the atomic volume and $Z$ the valence) is used, we obtain for low $T$ the following expressions for the resistivity and thermopower, respectively:

$$\rho \equiv \sigma^{-1} = \frac{3\pi\Omega_0 m^2}{4e^2\hbar^3 k_F^6} \int_0^{2k_F} dq\, q^3 S(q)|\hat{V}(q)|^2 \tag{9.2.12}$$

$$Q = \frac{\pi^2 k_B T}{3|e|} \frac{\partial \ln\rho(E)}{\partial E}\bigg|_{E=E_F} \tag{9.2.13}$$

where

$$\rho(E) = E^{-3} \int_0^{2\sqrt{(\hbar E/m)}} dq\, q^3 S(q) |\hat{V}(q)|^2 \qquad (9.2.14)$$

These are the results of what is called "Ziman theory," or the "diffraction model," or the "weak-scattering limit" of electronic transport in liquid metals. There are two immediate generalizations.

(a) For a liquid alloy, partial structure factors arise in the "Ziman integral." For example, in the resistivity

$$\rho = \frac{3\pi^2 \Omega_0 m^2}{4e^2 \hbar^3 k_F^6} \int_0^{2k_F} dq\, q^3 \sum_{\alpha\beta} (x_\alpha x_\beta)^{1/2} \hat{V}_\alpha(q) \hat{V}_\beta^*(q) S_{\alpha\beta}(q) \qquad (9.2.15)$$

where $\hat{V}_\alpha$ and $x_\alpha$ are the pseudopotential and concentration of species $\alpha$.

(b) A more complete theory takes account of inelastic election–ion collisions.[11] The result is the same as equations (9.2.12)–(9.2.14), but the static structure factor is replaced by an integral over the dynamic one:

$$S(q) \rightarrow \int d\omega S(q, \omega) \frac{\hbar\beta\omega}{e^{\hbar\beta\omega} - 1} \qquad \left(\beta \equiv \frac{1}{k_B T}\right) \qquad (9.2.16)$$

This generalization, due to Baym,[11] is used to calculate transport coefficients for amorphous metals, especially at low $T$. In liquid metals $T$ is above the Debye temperature, so the factor $\hbar\beta\omega/(e^{\beta\hbar\omega} - 1)$ can be replaced by unity and we are back to $S(q)$. It is interesting to note that in this temperature domain, where the restrictions on the inelastic processes due to the Bose factor in equation (9.2.16) are no longer important, integrating over all inelastic collisions (all $\omega$) yields the same result as our static theory, which led to equation (9.2.11).

### 9.2.1.2. Multiple-Scattering Expansion

In scattering theory, the cross-section $d\sigma/d\Omega$ is determined formally by the $T$-matrix of the system, where

$$\frac{d\sigma}{d\Omega}(\mathbf{k} - \mathbf{k}') = \frac{16\pi^4 m^2}{\hbar^4} |T(\mathbf{k}, \mathbf{k}')|^2 \qquad (9.2.17)$$

The many-ion $T$-matrix is expanded in terms of single-ion $t$-matrices and the free-electron Green function $G_0$:

$$T = \sum_n t_n + \sum_{n,m}' t_n G_0 t_m + \cdots \qquad (9.2.18)$$

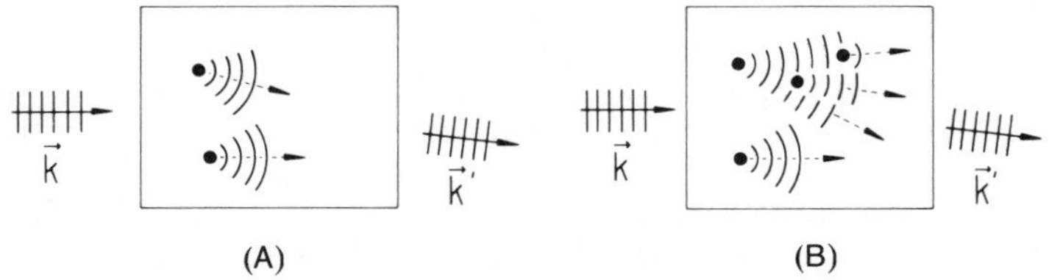

(A)                                                    (B)

Figure 9.5. (A) Single-site scattering; only wavelets that have been scattered from one ion are superposed. (B) Multiple scattering; wavelets are allowed to scatter again from other ions.

If we neglect multiple scattering, represented by case (A) in Figure 9.5, then we obtain

$$\frac{d\sigma}{d\Omega} = \sum_{n,m} e^{i\mathbf{q}\cdot(\mathbf{R}_n - \mathbf{R}_m)} |t(q, E)|^2 \qquad (9.2.19)$$

a result that has the same structure as expression (9.2.11). The single-ion $t$-matrix, which replaces the pseudopotential in the previous formulas, depends on $E = \hbar^2 k^2/2m$ and on the momentum transfer $q = 2k \sin \theta/2$. For spherical potentials, the result can be expressed in terms of the phase shifts $\delta_l$:

$$t(q, E) = -\frac{2\pi\hbar^3}{\Omega_0 m \sqrt{2mE}} \sum_l (2l + 1) P_l (\cos \theta) \sin \delta_l (E) e^{i\delta_l(E)} \qquad (9.2.20)$$

where $P_l$ is the Legendre polynomial.

The final results for $\rho$ and $Q$ have the same form as equations (9.2.12)–(9.2.14), but with $\hat{V}(q)$ replaced by $t(q, E)$ in equation (9.2.14) and by $t(q, E_F)$ in equation (9.2.12), $E_F$ being the Fermi energy.

For simplicity we have omitted spin indexes in these equations. However, it is straightforward to include exchange scattering in this approach. The term $\sum_n J(\mathbf{r} - \mathbf{R}_n)\boldsymbol{\sigma} \cdot \mathbf{S}_n$ acts like a random magnetic field on the electron spin. The $t$-matrix is then really a $2 \times 2$ matrix in spin space and, in the single-site approximation, equation (9.2.17) has the form

$$\overline{\frac{d\sigma}{d\Omega}} = \frac{16\pi^4 m^2}{\hbar^4} \frac{1}{2} \sum_{s,s'} \sum_{n,m} t_{ss'}^{(n)} t_{ss'}^{(m)*} \qquad (9.2.21)$$

The spin $\mathbf{S}_n$ of an individual ion sets up an orientation in space, such that the most general form of $t$ is[12]

$$t_{ss'}^{(n)} = a\delta_{ss'} + b\boldsymbol{\sigma}_{ss'} \cdot \mathbf{S}_n \qquad (9.2.22)$$

Coefficients $a$ and $b$ are best evaluated in the representation where the total spin $\mathbf{J} = \boldsymbol{\sigma} + \mathbf{S}_n$ is diagonal. There are then two scattering amplitudes $t_J$ for $J = S \pm \frac{1}{2}$, each given by a series like (9.2.20) where the phase shifts depend on $J$. The relation to equation (9.2.22) is found by

$$b = \frac{2}{2S+1}(t_+ - t_-), \qquad a = \frac{1}{2S+1}[(S+1)t_+ - St_-] \qquad (9.2.23)$$

where $S$ is the quantum number of the ionic spin and $t_\pm = t_J$ for $J = S \pm \frac{1}{2}$. Inserting equation (9.2.22) into (9.2.21) yields, instead of (9.2.19),

$$\frac{d\sigma}{d\Omega} = \frac{1}{2}\sum_{s,s'}\sum_{n,m}\langle e^{i\mathbf{q}\cdot(\mathbf{R}_n - \mathbf{R}_m)}(a^*\delta_{ss'} + b^*\boldsymbol{\sigma}_{ss'}\cdot\mathbf{S}_n)(a\delta_{ss'} + b\boldsymbol{\sigma}_{ss'}\cdot\mathbf{S}_m)\rangle \qquad (9.2.24)$$

Besides the usual (positional) structure factor for the ions, we must introduce spin correlation functions

$$\langle \mathbf{S}_n \cdot \mathbf{S}_m \rangle = \begin{cases} S(S+1), & n = m \\ S(S+1)\frac{1}{3}\hat{M}(|\mathbf{R}_n - \mathbf{R}_m|), & n \neq m \end{cases} \qquad (9.2.25)$$

where we have assumed that the spin–spin correlation function depends only on the interion distance.

The final result for the resistivity is [compare with equation (9.2.12)]

$$\rho = \frac{3\pi\Omega_0 m^2}{4e^2\hbar^3 k_F^6}\int_0^{2k_F} dq\, q^3\{S(q)R_1(q) + [1 + M(q)][R_2(q) - R_1(q)]\} \qquad (9.2.26)$$

where $M(q)$ is the Fourier transform of the product $\hat{M}(r)g(r)$, while

$$R_1(q) = \left|\sum_J \frac{2J+1}{2(2S+1)}t_J\right|^2 \qquad (9.2.27a)$$

and

$$R_2(q) = \sum_J \frac{2J+1}{2(2S+1)}|t_J|^2 \qquad (9.2.27b)$$

are "coherent" and "incoherent" averages of the $t_J$s.

In principle, it would be straightforward to include some multiple-scattering terms from expansion (9.2.18) into this formalism. However, they would involve higher-order ionic correlation functions, which are generally unknown and have to be approximated in some way. A favorite choice is the quasi-crystalline approximation, which amounts to factorizing successively the higher-order ionic correlation function such that the result is again of the form (9.2.12), but the quantity replacing $\hat{V}(q)$ is now a structure-dependent $t$-matrix.[10]

The Hall coefficient is also easily evaluated in the framework of a relaxation-time approximation to the collision operator (which was exact in our previous analysis!). When the Lorentz force is included, the transport equation becomes

$$-e(\mathbf{E} + \mathbf{v} \times \mathbf{B})\frac{\partial f}{\partial \mathbf{k}} = -\frac{\varphi}{\tau} \qquad (9.2.28)$$

In a typical Hall geometry $\mathbf{B} = (0, 0, B)$, $\mathbf{E} = (E_x, E_y, 0)$, and $\mathbf{j} = (j_x, 0, 0)$. The Hall coefficient for small $B$ is given by

$$R_H \equiv \frac{E_y}{Bj_x} = \frac{3\pi^2}{|e|} \frac{\int_0^\infty dk k^2 \tau^2(k) v^2(k) \partial N_0/\partial k}{\left[\int_0^\infty dk k^2 \tau(k) v(k) \partial N_0/\partial k\right]^2} \qquad (9.2.29)$$

At low $T$, $\partial N_0/\partial k \approx -\delta(k - k_F)$. Due to the isotropy of the system, the relaxation time then drops out of equation (9.2.29) and we are left with the free-electron value, namely

$$R_H = -\frac{3\pi^2}{|e|k_F^3} = -\frac{1}{n_c|e|} \qquad (9.2.30)$$

Thus the Hall coefficient appears to be a direct indicator of the density $n_c$ of conduction electrons (see Section 9.4). Unfortunately, the measured values of $R_H$ often do not seem to give the "correct" values of $n_c$ and are even positive in many instances. For magnetic materials this may be due to the extraordinary contribution to the Hall field $E_y$, measured by $R_2$,

$$E_y = (R_1 B + R_2 M)j_x \qquad (9.2.31)$$

and due to the (permanent or field-induced) magnetic moment $M$. If $M$ is induced by $B$, then $M = \chi B$ and the total Hall coefficient

$$R_H \equiv \frac{E_y}{Bj_x} = R_1 + R_2\chi \qquad (9.2.32)$$

contains the magnetic susceptibility $\chi$. Other special features of disordered metals, such as electron lifetimes, nonmonotonic dispersion, or multiband electron structure (see the end of Section 9.1), might also yield deviations from equation (9.2.30) by introducing, for instance, holelike conduction; but these possibilities have not yet been checked in detail.

## 9.2.2. Linear-Response Approach

The expectation value of expression (9.2.1), $\langle \mathbf{j}(\mathbf{r}) \rangle_t$, must be calculated as an average over a density operator $\rho$ whose time evolution includes the external field:

$$i \frac{\partial \rho}{\partial t} = \left[ H_{el} - e \sum_i \mathbf{E}(t) \cdot \mathbf{r}_i, \rho \right] \tag{9.2.33}$$

By splitting $\rho$ into a quantity $\rho_{eq}$ proportional to $e^{-\beta H_{el}}$ and a deviation linear in $\mathbf{E}$, we find for an isotropic system

$$\langle \mathbf{j} \rangle_\omega = \sigma(\omega) \mathbf{E}(\omega) \tag{9.2.34}$$

The frequency-dependent conductivity

$$\sigma(\omega) = \int_0^\infty dt\, e^{-i\omega t} \int_0^\beta d\lambda \, \langle \mathbf{j}(0) \cdot \mathbf{j}(t + i\lambda) \rangle_{eq} \tag{9.2.35}$$

is given by an equilibrium current correlation function. Equation (9.2.35), called the Kubo equation for $\sigma$, is again exact for weak fields. The difficulty lies in evaluating this correlation function for a disordered system. Since equation (9.1.9) treats the electrons as independent, the two-electron correlation function

$$\langle \mathbf{j}(0) \cdot \mathbf{j}(t) \rangle = \sum_{\mathbf{k}, \mathbf{k}'} \mathbf{v}(\mathbf{k}) \cdot \mathbf{v}(\mathbf{k}') \langle a_\mathbf{k}^+ a_\mathbf{k} a_{\mathbf{k}'}^+(t) a_{\mathbf{k}'}(t) \rangle \tag{9.2.36}$$

immediately factorizes into a product of two one-electron correlation functions. Taking the configurational average over an ionic ensemble yields translationally invariant correlation functions, but the average of equation (9.2.36) is not equal to the product of the averages of the corresponding one-electron functions. We note two approaches to the evaluation of equation (9.2.36):

1. Derive a rigorous Bethe–Salpeter equation for (9.2.36) involving the average one-electron functions and a kernel ("vertex part").[13] In the limit where the lifetime of an average plane-wave state near the Fermi surface is long, one can derive the same results for $\rho$ as our equations (9.2.12), (9.2.19), and (9.2.26) by solving the

Bethe–Salpeter equation (which is actually equivalent to a linearized transport equation). This is reassuring, since it justifies the scattering approach of Section 9.3.1 from a different point of view.

2. The coherent-potential approximation (CPA), which has proved very useful for evaluating equilibrium properties of disordered systems,[14] is used to calculate equation (9.2.36). However, this approach has been limited mainly to disordered crystalline alloys, since it is not easy to incorporate the structure factor of a topologically disordered system like a liquid metal. Moreover, even for weak scattering potentials the so-called "vertex corrections" yielding the $\cos \theta$ term in equation (9.2.7) are usually missing.[15]

Finally, we mention that there exist more sophisticated approaches, which try to extend beyond the single-site $t$-matrix result[16] using heavy "many-body machinery." However, the formal results seem to be too complicated in most cases to be amenable to numerical calculations. In Section 9.4 we will address ourselves specifically to the problem of strong scattering in disordered metals and present some new, recently developed ideas.

## 9.3. Application to Liquid Metals

Our results for the resistivity $\rho$ and thermopower $Q$ for disordered metals, determined in the scattering approach of Section 9.2 (diffraction model), can be summarized as follows:

$$\rho = \frac{3\pi\Omega_0 m^2}{4e^2\hbar^3 k_F^6} I, \qquad I = \int_0^{2k_F} dq \, q^3 S(q) A(q, E_F) \tag{9.3.1}$$

$$Q = -\frac{\pi^2 k_B T}{3|e|E_F} \left\{ 3 - \frac{1}{I} \left[ 8k_F^4 S(2k_F) A(2k_F, E_F) + \int_0^{2k_F} dq \, q^3 S(q) \frac{\partial A(q, E_F)}{\partial \ln E_F} \right] \right\}$$

$$\tag{9.3.2}$$

where

$$A(q, E_F) = \begin{cases} |\hat{V}(q)|^2 & \text{for simple metals} \\ |t(q, E_F)|^2 & \text{for TM, RE, etc.} \end{cases} \tag{9.3.3}$$

and spin-disorder scattering has been omitted for the moment.

Before reporting actual computations, let us focus on the $T$-dependence of these quantities. Besides the explicit $T$-factor in $Q$, the main $T$-dependence originates from the structure factor, as shown in

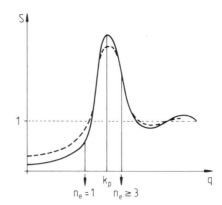

Figure 9.6. Function $S(q)$ at two temperatures (see Figure 9.2) and two possible positions of $2k_F$ (marked by arrows) corresponding to two different values of $n_e$.

Figure 9.2. The temperature coefficient (TC) of the resistivity

$$\alpha \equiv \frac{1}{\rho}\frac{d\rho}{dT} \tag{9.3.4}$$

depends sensitively on the value of $2k_F$; see Figure 9.6.

Since the factor $q^3$ in $I$ gives most weight to the $q$-region just below $2k_F$, we expect

$$\begin{aligned} \alpha > 0 \qquad &\text{if } 2k_F < k_p \text{ or } 2k_F > k_p \\ \alpha < 0 \qquad &\text{if } 2k_F \approx k_p \end{aligned} \tag{9.3.5}$$

Negative $\alpha$s, which are exceptional for metallic conductivity from a crystalline point of view, can therefore be expected for liquid metals when the number of conducting electrons per ion ($n_e$) is about 1.6 to 2, using the relation $k_F^3 = 3\pi^2 n_e/\Omega_0$. Then the most important momentum transfer ($2k_F$) for electronic backscattering coincides with the most important wave vector in the ionic structure. At higher $T$ the ionic structure becomes more blurred, which decreases the scattering probability. Equation (9.3.5) is an important test for the validity of the diffraction model.

We now discuss some applications of equations (9.3.1)–(9.3.3) to various classes of liquid metals.

### 9.3.1. Simple Metals

Now $n_e$ is taken to be the valence of the ions, and $k_F$ and $E_F$ are found in the quasi-free electron model. Many calculations of $\rho$ and $Q$ have been published using various pseudopotentials and structure factors (experimental and hard-sphere model). On the whole agreement with experiment is reasonable, but one has to admit that the results depend rather sensitively on the choice of $\hat{V}$ and $S(q)$. For pure metals $\alpha$ is usually positive except, e.g., for Zn, where $2k_F \approx k_p$.

Rather than draw conclusions from the comparison of *one* theoretical result with *one* experimental point, it is usually more informative to check general trends of $\rho(x, T)$ for alloys (binary, say) of concentration $x$. To this end the alloy generalization of equations (9.3.1)–(9.3.3), discussed in Section 9.2, must be used. The most sophisticated calculations are conducted by constructing pseudopotentials from first principles for each concentration (thus allowing for charge transfer). Screening is done by the best available $\varepsilon(q)$. The partial structure factors are taken from experiment wherever available or modeled by hard spheres, whereby packing fractions, etc., are determined from thermodynamics using first-principles ionic pair potentials. (For details see the book cited elsewhere,[2] or original papers, such as Wang and Lai.[17]) The $T$-dependence (9.3.5) can be checked very nicely on such alloys, in which case we will denote by $k_p$ the position of the main peak of the *total* structure factor, for simplicity.

### 9.3.1.1. Components of Equal Valence

Examples are alloys of alkalis. Here, $2k_F$ does not change appreciably with concentration $x$. Thus $\alpha > 0$ for all $x$, if $\alpha > 0$ for both pure components. The behavior of $\rho$ as a function of $x$ is usually parabolic for monovalent, and rather linear for polyvalent, alloys (cf. Figure 9.7); this is easily understood (see Section 707 of the book cited elsewhere[2]).

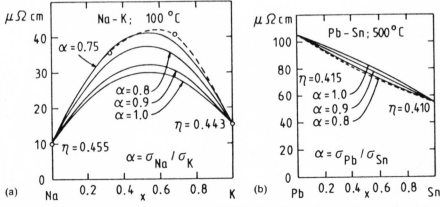

Figure 9.7. Resistivity of (a) Na–K alloys and (b) Pb–Sn alloys, where $\alpha$ is the ratio of the hard-sphere diameters used in the calculations and dashed curves represent experimental results (after Ashcroft and Langreth[18]).

## 9.3.1.2. Components of Different Valence

Examples are alloys of noble metals with simple metals which are polyvalent. In this case $2k_F$ changes with $x$. For an example like Cu–Sn (see Figure 9.8) $n_e$ varies from 1 (Cu) to 4 (Sn). There is an intermediate $x$-range where $2k_F$ crosses $k_p$. Here $\alpha < 0$ as expected, and moreover $\rho$ itself has a rather high value, which is understandable from (9.3.1) and Figure 9.6: for $2k_F < k_p$ the integral $I$ increases strongly with $2k_F$ (and thus with $x$); for $2k_F > k_p$, $I$ still grows slowly but the premultiplier $k_F^{-6}$ tends to suppress $\rho$ again. This correlation between $\alpha < 0$ and high $\rho$-values will be discussed in Section 9.4 below.

The thermopower is somewhat more delicate to predict since, according to equation (9.3.2), it depends sensitively on the precise value of $S$ and $\hat{V}$ at $2k_F$. Nevertheless, in many cases the $x$-dependence is well reproduced (see Figure 9.8), especially if the values of $S_{\alpha\beta}(2k_F)$ can be taken from experiment.

### 9.3.2. Strong Scatterers

In these systems the equilibrium electronic structure is already quite different from the free-electron version. A thorough transport theory should therefore be based on a good approach to what we called the average-band structure in Section 9.2, e.g., by summing sufficient terms of (9.2.18). In the absence of a systematic attempt in this direction the following simplified concept has been used.

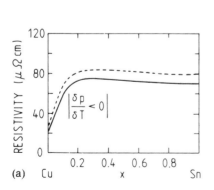 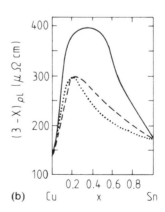

Figure 9.8. (a) Resistivity of liquid CuSn, full curve: theory, dashed curve: experiment (after Dreirach *et al.*[19]). (b) Thermopower of liquid CuSn, where the dotted line represents experiments, —— calculation with a hard-sphere structure factor, and --- calculation with experimental $S(q)$ (after Shimoji[2]).

Once the atomic muffin-tin potential is calculated, energy-dependent phase shifts $\delta_l(E)$ for $l = 0, 1, 2, 3$ are evaluated. What we need in equations (9.3.1)–(9.3.3) are the various $\delta_l$ near $E_F$, giving the scattering probability of a conduction electron near the Fermi surface, and $2k_F$, the maximum momentum transfer. The Bristol group (see, e.g., Dreirach[19]) proceeded as follows:

1. The bottom energy $E_B$ of the lowest conduction band (see Figure 9.3), which is usually free-electron-like, is evaluated by a Wigner–Seitz condition.
2. $E_F$ is taken from solid-band structure data.
3. $k_F$ is determined by $n_e$, using the free-electron relation

$$k_F^3 = 3\pi^2 n_e / \Omega_0 \qquad (9.3.6)$$

4. An effective mass determined by $E_F = E_B + \hbar^2 k_F^2 / 2m^*$ is then introduced into equations (9.3.1)–(9.3.3).

In some sense, this amounts to choosing from the complicated band structure of the type in Figure 9.3 one free-electron-like band (call it the "$s$–$p$ band"), which starts at $E_B$ and intersects the Fermi energy at a wave vector $k_F$. Its shape is determined such that the usual free-electron relations remain valid for a given $n_e$. Other bands (mostly $d$-like), which also intersect $E_F$, are neglected. This may be justified by invoking the lower mobility of $d$-electrons and has found some corroboration in first-principles calculations of $\rho$ for crystalline metals.[20] The crucial quantity is then $n_e$. For the alkaline earths an obvious and simple choice is $n_e = 2$ (valence). Calculations of $\rho$ and $Q$ for Ca, Sr, and Ba conducted in this way have yielded reasonable agreement with experiment.[21] The high $\rho$-value of 306 $\mu\Omega$ cm for Ba is explained by large phase shifts, especially for $l = 2$. Cs, under pressure, also acquires important $d$-phase shifts, leading to high resistivity.[22]

For late TM, $n_e$ is on the order of 1 or somewhat smaller. This choice is supported by two main pieces of evidence:

1. The Hall constant. For Ni, $R_H < 0$ and consistent with such a value of $n_e$. Co has $R_H > 0$. However, the extraordinary part can be subtracted by plotting $R_H$ versus $\chi$, according to equation (9.2.32). The remaining, normal part corresponds to $n_e \approx 0.5$.[23]
2. Systematic studies of $\rho$ for alloys of TM with monovalent and polyvalent metals have confirmed $n_e \lesssim 1$; $\alpha > 0$ for all $x$ in the first case, while there is a region of $x$ where $\alpha < 0$ in the second (such as FeGe), namely when $2k_F \approx k_p$. This shows that late TM behave like monovalent as far as electric transport is concerned.

The strong variation of $\rho$ through the late TM is a consequence of $\rho_2(E_F)$ passing through a resonance. At resonance $\delta_2 = \pi/2$, leading to a large value of $t$. The $T$-dependence of $\rho$ is mainly governed by $S(q)$, but the pressure dependence is also strongly influenced by the phase shifts, changing due to density variations. Thus the resistivity of liquid Fe is predicted to decrease substantially at pressures that may be found in the interior of the earth.[22]

More recently it has been suggested[24] that a consistent theory should also provide information about $n_e$ and $E_F$, using the same type of approximation as for the calculation of $\rho$. Moreover, the only relation between these quantities consistent with the free-electron boundary conditions in evaluating phase shifts is

$$E_F = \frac{\hbar^2 k_F^2}{2m}, \qquad k_F^3 = 3\pi^2 n_e/\Omega_0 \qquad (9.3.7)$$

Imposing a reasonable value for $E_F$ in equation (9.3.7) would, however, lead to $n_e \approx 1.2$ for Fe, say, and $\rho$ would be much too high owing to the relatively large value of $2k_F$ in the Ziman integral (9.3.1).[24] Actually, requiring equation (9.3.7) may be too stringent, since even in a KKR band-structure calculation free-electron boundary conditions are used for calculating $\delta_l(E)$.

In the rare-earth series the elements Eu and Yb behave as divalent in the solid state. Their resistivities have been calculated by choosing $n_e = 2$. Their very different values of $\rho$ (see Figure 9.9) can be understood from the difference in the muffin-tin potentials. Since $2k_F \approx k_p$ the negative $\alpha$ is easily reproduced.[25]

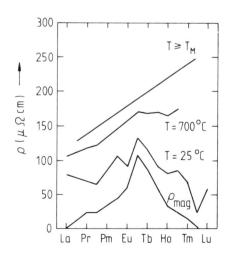

Figure 9.9. Resistivities of the rare earths at various temperatures (after Güntherodt et al.[23]).

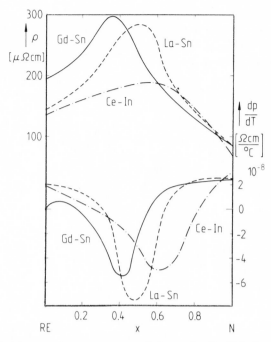

Figure 9.10. Resistivity $\rho$ and temperature coefficient $\alpha$ for liquid LaSn, GdSn, and CeIn (after Güntherodt et al.[23]).

For the remaining lanthanides, which have three $s$–$d$ electrons in the outer shells, $n_e = 3$ has yielded reasonable values of $\rho$,[26] but it is inconsistent with the temperature coefficients of alloys: $\alpha > 0$ for alloys of RE with monovalent metals, but $\alpha < 0$ in some domain for alloys with polyvalent metals (see Figure 9.10).

The criticism of Esposito et al.[24] has been taken into account in a different approach to the trivalent RE.[27]

(a) From the phase shifts, the total integrated density of states containing the first scattering correction to the free-electron value $N_{FE}$, where

$$N(E) = N_{FE}(E) + \frac{2}{\pi} \sum_l (2l + 1)\delta_l(E) \qquad (9.3.8)$$

is used to find $E_F$ by putting $N(E_F) = 3$ ( = valence).

(b) The partial $s$–$p$ integrated density of states

$$N_{sp}(E) \equiv \frac{2}{\pi} \sum_{l=0}^{1} (2l + 1)\delta_l(E) + N_{FE}(E) \qquad (9.3.9)$$

yields the number of effective current-carrying electrons:

$$n_e = N_{sp}(E_F), \qquad k_F^3 \equiv 3\pi^2 n_e / \Omega_0 \qquad (9.3.10)$$

In general, this choice of $k_F$ is not consistent with requirement (9.3.7) and yields overestimated values for $\rho$.

(c) $E_B \equiv E_F - \hbar^2 k_F^2 / 2m$ is the bottom of the conduction band. The situation $E_B \neq 0$ is a sign that the ionic potentials change sufficiently the electron structure such that incoming electrons possess an approximate dispersion relation

$$E(k) = E_B + \frac{\hbar^2 k^2}{2m} \qquad (9.3.11)$$

rather than simply $\hbar^2 k^2 / 2m$. In order to take into account this effect of the medium outside a muffin tin, the single-ion potential was then given the value $E_B$ (rather than zero) outside the ionic spheres. Hence the modified consistency relation

$$E_F - E_B = \frac{\hbar^2 k_F^2}{2m} \qquad (9.3.12)$$

was fulfilled for the scattered electrons at the Fermi surface.[27] The calculated values of $\rho$ were then in good agreement with experiment. Moreover, the trend of $\rho$ and $\alpha$ through the series was traced back to a systematic variation of $n_e$, from about 0.6 for La to about 1.3 for Lu. This is in good agreement with the number of $(s-p)$-like Bloch states in crystalline band-structure calculations.[28] For Lu it leads to $2k_F \leqslant k_p$, which explains the slightly negative value of $\alpha$ in that case. Thus a consistent picture of $\rho$ and $\alpha$ for RE and their alloys has been achieved on the basis of equation (9.3.1). It is of interest that incorporating multiple-scattering effects into the calculation of $\rho$ for transition metals,[10] as discussed in Section 9.2, has also produced a reduction of $\rho$ in many cases, like the above-mentioned incorporation of an effective medium into the ionic potential.

The effect of $s-f$ exchange scattering has also been investigated for Gd.[29] The exchange function $J(r)$ in equation (9.1.10) was taken from solid-state band-structure calculations. The main result is that the total resistivity of Gd is only about 5 to 10% higher than without $s-f$ scattering, which is consistent with the rather monotonic behavior of $\rho$ through the RE series, without exhibiting any higher values for the magnetic elements in the liquid state. At room temperature $\rho$ attains its maximum at Gd, which was previously explained by spin-disorder scattering. It still remains unclear why the situation changes so drastically between the solid at ambient $T$ and the liquid phase.

## 9.4. Beyond Weak (Single-Site) Scattering

In this final section we briefly sketch some new ideas on how to calculate transport coefficients when the electron–ion interaction is strong. Most of these approaches can equally well be applied to both liquid and solid disordered metals, such as random alloys or metallic glasses.

In order to illustrate that even our single-site $t$-matrix approach is questionable in many cases, we estimate the mean free path $\bar{l} \equiv v_F \bar{\tau}$ of a conduction electron with Fermi velocity, $\bar{\tau}$ being a mean relaxation time entering the kinetic equation $\rho = m/ne^2\bar{\tau}$ with $n = n_e/\Omega_0$. We find typically $\bar{l} \approx 250/\rho$, given in Å, when $\rho$ is inserted in $\mu\Omega$ cm. Thus the mean free path becomes as short as an average interionic distance $d$ when $\rho$ exceeds about 100 to 200 $\mu\Omega$ cm. It is clear that under these circumstances neglect of multiple scattering is difficult to justify. (On the other hand, one should be a bit cautious about the statement "$\bar{l} < d$ makes no sense," since the scattering potential is, of course, nonzero also between ions.)

There are therefore two main problems with high-resistivity disordered metals:

1. A sound theoretical approach should extend beyond single-site scattering. Such methods are not yet readily available for actual calculations.
2. The fact that many different materials, liquid or solid, exhibit negative $\alpha$ calls for a universal explanation. More precisely, there seems to be some correlation between $\alpha < 0$ and high values of $\rho$, often referred to as Mooij's rule, although Mooij[30] had originally observed it for disordered transition metal alloys only. Figure 9.11 presents the data of many such alloys and shows that $\alpha < 0$ typically when $\rho > \rho_c \approx 150$ $\mu\Omega$ cm.

For the present we shall disregard the above formal criticism (1) and review the success and failure of the diffraction model of Section 9.2 as regards (2).

(a) The values of $\rho$ calculated this way are reasonable and often allow one to understand the basic trends. The price to pay is a possibly inconsistent $k_F \leftrightarrow E_F$ relation, or the use of an *ad hoc* renormalization of the electron–ion potential (see RE calculations) in order to save the $k_F \leftrightarrow E_F$ relation. Moreover, the inclusion of some multiple scattering[10] seems to improve the results more or less systematically.

(b) Many negative values of $\alpha$ — for pure metals like Ba, Eu, Yb, and Lu, or for heterovalent alloy series — can be understood consistently on the basis of the structure factor and the number of conduction electrons, which are two very general features of a disordered metal. Correlations also exist between $\rho$ and $Q$: the second term in the bracket of equation

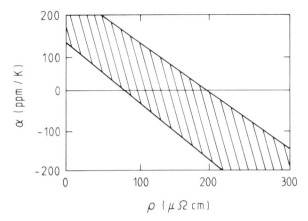

Figure 9.11. Mooij's rule: correlation between $\rho$ and $\alpha \equiv (1/\rho)d\rho/dT$ for various transition metal alloys (after Mooij[30]).

(9.3.2) is expected to be important when $2k_F \approx k_p$ (if the energy variation of the third term is small). Therefore condition (9.3.5) for $\alpha < 0$ should yield $Q > 0$. This is indeed observed, for example, in glassy BeTiZr.[31]

(c) In some sense Mooij's rule is also predicted: if in an alloy, like Ce–Sn in Figure 9.10 and many others, the resistivity has a high value in the intermediate concentration range, this is also typically the domain where $\alpha < 0$. On the other hand, exceptions to this rule, like liquid Eu and Yb that both have $\alpha < 0$ but very different values of $\rho$ (Figure 9.9), can also be understood (see Section 9.3).

(d) There are however counterexamples, such as the alloy $Gd_{67}Co_{33}$,[7] which has a pronounced negative $\alpha$ both in the liquid and glassy phase, but according to our rules (Section 9.3) for the number of conduction electrons, $2k_F$ is expected to lie below $k_p$. Thus either $\alpha$ is negative against the rules (9.3.5), or there is sufficient rearrangement of the electron structure in the alloy such that $n_e$ is not the weighted sum of the values for pure Gd and Co.

The remainder of this section is devoted to a very brief presentation of some recent developments toward a more rigorous transport theory. (Other models, like Mott's $s–d$ scattering[32] or tunneling models for low $T$,[33] will not be considered here.)

### 9.4.1. Precursor Effects of Localization

It is well known that strong disorder in a system described by (9.1.9) can lead to the existence of localized eigenstates (Anderson localization).[34] In a one-band model, density of states $D(E)$ and mobility $\sigma(E)$ are thought to possess the behavior illustrated in Figure 9.12.

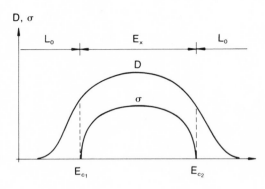

Figure 9.12. Electronic density of states $D$ and mobility $\sigma$ in a schematic one-band model of a disordered solid, where $L_0$ and $E_x$ denote energy domains of localized and extended electronic states, respectively.

If $E_F$ lies in the region of localized states ($E_F < E_{c_1}$ or $> E_{c_2}$), the $T = 0$ conductivity of the system vanishes. If $E_F$ lies between the mobility edges $E_{c_i}$ the conductivity is metallic, as in our liquid and amorphous metals or in disordered alloys. At $T > 0$ the conductivity will be finite even in the first case, due to two mechanisms:

1. Electrons are thermally activated into extended states (respectively, holes are created in such states). This leads to thermally activated transport, as in crystalline semiconductors.
2. Interaction with phonons allows the electron to hop from one localized state to another (thermally activated hopping).

In a disordered metal we may see precursors of these effects even though the states of $E_F$ are extended, if $E_F$ is close enough to $E_{c_1}$, say. Both effects can be visible.

1. The conductivity $\sigma(T)$ can be calculated as

$$\sigma(T) = \int_{E_{c_1}}^{E_{c_2}} dE \sigma(E) \left[ 4T \cosh^2 \left( \frac{E - E_F}{2k_B T} \right) \right]^{-1} \qquad (9.4.1)$$

If $\sigma(E)$ varies strongly near $E_F$ this may yield appreciable $T$-dependence of $\sigma$, especially $\alpha > 0$.

2. Although the eigenvalues near $E_F$ are in principle extended, their shape is very different from a plane wave with constant amplitude. The probability density $|\psi|^2$ will be large only in some regions, which are linked by bottlenecks of small amplitude (see Figure 9.13). If the electron is in region I, electron–phonon collisions tend to attenuate the electric current and contribute to a finite resistivity, while in regions of type II, phonons may help the electron to pass through the bottleneck, thereby effectively decreasing $\rho$.

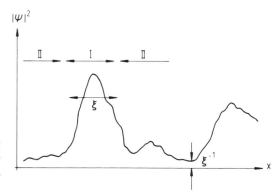

Figure 9.13. Probability density of a one-electron state in a strongly disordered system. Symbols $\xi$, I, and II are explained in the text.

### 9.4.2. Scaling Theory

Various scaling formalisms have been developed to describe localization.[35] Imry[36] has incorporated the latter mechanism (2) into this scaling approach. The main quantity is some correlation length $\xi$, which diverges when the Anderson transition is approached by, for instance, shifting $E_F$ through $E_{c_1}$. In the localized phase ($E_F < E_{c_1}$) $\xi$ is the localization length, while in the extended phase $\xi$ has the geometric meaning shown in Figure 9.13, characterizing the typical extent of region I and the width of the bottlenecks.[36] At $T = 0$ scaling yields

$$\sigma \propto \frac{e^2}{\hbar\xi} \tag{9.4.2}$$

in the extended phase. When $T > 0$ electron–phonon collisions introduce another length $l_{ph}$, the inelastic mean free path. If $l_{ph} > \xi$ there is a weak $T$-dependence, as in usual metals, while if $l_{ph} < \xi$, relation (9.4.2) has to be replaced by

$$\sigma \propto \frac{e^2}{\hbar l_{ph}} \tag{9.4.3}$$

which yields $\alpha < 0$.

### 9.4.3. Nonadiabatic Treatment of Electron–Phonon Interaction

Electron–ion interactions are usually treated in the adiabatic approximation.[9] However, Girvin and Jonson[37] have recently noted that, owing to strong disorder, the electronic self-energy may vary rapidly on an energy scale of the order of a phonon energy, so invalidating the adiabatic approximation. They have evaluated the Kubo expression (9.2.35) for a tight-binding model on a Cayley tree and found that Mooij's rule could be reproduced when the frequency dependence of the electron self-energy accounted for the energy transfer between electrons and phonons.

### 9.4.4. Self-Consistent Transport Equations

This approach, elaborated by Götze and co-workers,[38] is perhaps the most promising, since it covers the weak-scattering limit (Ziman theory), it produces localization for strong disorder, and it seems to be amenable to numerical calculations and various generalizations. The main idea is to derive a transport equation similar to equation (9.2.4) for linear response functions such as equation (9.2.36), but with a collision kernel $W$ [see equation (9.2.34)], which depends self-consistently on the two-electron correlation function (9.2.36). Götze achieved this goal by using Mori's projector technique for deriving equations of motion for dynamic correlation functions, but other many-body techniques, such as Green functions, would also be adequate. The key quantity is the electronic density response function

$$\phi_{\mathbf{k}\mathbf{k}'}(\mathbf{q}z) \equiv (-i) \int_0^\infty dt \, e^{izt} \langle [f_{\mathbf{k}}(\mathbf{q}, t), f_{\mathbf{k}'}^+(\mathbf{q}, 0)] \rangle \qquad (9.4.4)$$

with

$$f_{\mathbf{k}}(\mathbf{q}) = a_{\mathbf{k}+\frac{1}{2}\mathbf{q}}^+ a_{\mathbf{k}-\frac{1}{2}\mathbf{q}} \qquad (9.4.5)$$

[see equation (9.2.2)]. Function $\phi$ satisfies a linear integral equation (since the electrons are noninteracting), which involves the random potential $U$ of equation (9.1.9). When equation (9.4.4) is averaged over ionic configurations, the average of products like $\langle U\phi \rangle$ can be expressed in terms of a self-energy $C$:

$$\left( z - \frac{\mathbf{q} \cdot \mathbf{k}}{m} \right) \hat{\phi}_{\mathbf{k}\mathbf{k}'}(\mathbf{q}, z) + \sum_{\mathbf{k}''} C_{\mathbf{k}\mathbf{k}''}(\mathbf{q}z) \hat{\phi}_{\mathbf{k}''\mathbf{k}'}(\mathbf{q}, z)$$

$$= \delta_{\mathbf{k}\mathbf{k}'} \left[ N_0 \left( \mathbf{k} + \frac{\mathbf{q}}{2} \right) - N_0 \left( \mathbf{k} - \frac{\mathbf{q}}{2} \right) \right] \qquad (9.4.6)$$

where $\hat{\phi}$ is the averaged $\phi$ and $N_0(\mathbf{k})$ the equilibrium electronic distribution function. To lowest order in $U$, $C$ is given by

$$C_{\mathbf{k}\mathbf{k}''}(\mathbf{q}z) = \sum_{\mathbf{p}} \langle |U(\mathbf{k}-\mathbf{p})|^2 \rangle [H_{\mathbf{k}+(\mathbf{q}-\mathbf{p})/2, \mathbf{k}''+(\mathbf{q}-\mathbf{p})/2}(\mathbf{p}z) + \cdots] \qquad (9.4.7)$$

The missing terms $(\cdots)$ involve the same function $H$ — the solution of equation (9.4.6) with the right-hand side equal to $\delta_{kk'}$ — but with different arguments. The aim now is to solve the generalized collision equation (9.4.6) together with equation (9.4.7). This is carried out conveniently by projecting $\hat{\phi}$ onto the two important hydrodynamic variables, density $\hat{\rho}$ and current $\mathbf{j}$, namely

$$\hat{\rho}(\mathbf{q}z) = \sum_{kk'} \hat{\phi}_{kk'}(\mathbf{q}z), \qquad j(\mathbf{q}z) \equiv \sum_{kk'} \frac{\mathbf{k} \cdot \mathbf{k}'}{m^2} \hat{\phi}_{kk'}(\mathbf{q}z) \qquad (9.4.8)$$

and by evaluating the current-relaxation time $\tau \equiv M^{-1}$ with

$$M = B \sum_k \langle |U(\mathbf{k})|^2 \rangle k^2 \hat{\rho}(\mathbf{k}, 0) \qquad (9.4.9)$$

where $B$ involves various constants. Quantity $M$ is directly proportional to the resistivity of the system. However, when the linear transport equation (9.2.4) or (9.4.6) is solved in a relaxation-time approximation like expression (9.2.28) by taking into acount the conserved quantity, density, one finds that $\hat{\rho}(q, z)$ has a rather different behavior, depending on whether $\Omega\tau \ll 1$ or $\Omega\tau \gg 1$, where $z = \Omega + i\varepsilon$. We shall examine these two cases.

(a) When $\Omega\tau \gg 1$ (few collisions),

$$\hat{\rho}(q, \Omega) \propto \left[ \Omega^2 \left( 1 - \frac{i}{\Omega\tau} \right) - q^2 v_F^2 \right]^{-1} \qquad (9.4.10)$$

This corresponds to wave-like propagation of density disturbances ("zero sound").

(b) When $\Omega\tau \ll 1$ (many collisions),

$$\hat{\rho}(q, \Omega) \propto (\Omega - iq^2 v_F^2 \tau)^{-1} \qquad (9.4.11)$$

This is a hydrodynamic mode of diffusive character.

Substitution of a result like (9.4.10) into (9.4.9) in the weak-scattering limit reproduces Ziman's results (9.2.7) and (9.2.10) for the resistivity. For very strong scattering, however, substitution of equation (9.4.11) into (9.4.9) yields a nonlinear equation for $M$ given by

$$M = \alpha M + \beta M^{-1} + o(M^{-3}) \qquad (9.4.12)$$

$\alpha$ and $\beta$ being related to $\langle |U|^2 \rangle$. For large $M$ the result is

$$M = \left( \frac{\beta}{1 - \alpha} \right)^{1/2} \qquad (9.4.13)$$

i.e., $M$ diverges when $\alpha \to 1$, corresponding to localization ($\sigma \to 0$, $\rho \to \infty$) as a result of strong disorder. Hence this theory allows one in an approximate way to cover the whole range between weak and strong disorder scattering.

The following are some results, which can be obtained within this framework.

1. For a given potential $U$ the resistivity $\rho$ can be evaluated in a self-consistent way, extending beyond second-order perturbation theory. The results for $\rho$ turn out to be higher than in second order (because the system tends toward localization), by up to 30–40% for systems like polyvalent liquid metals or alkaline earths.

2. The function $\sigma(E)$ needed in equation (9.4.1) can be evaluated by varying the Fermi energy. This has been used when calculating $\sigma(T)$ for alloys like $La_{1-x}Sr_x VO_3$ or Sb-doped Ge, systems that exhibit a metal-insulator transition, which is probably due to disorder.[39]

3. By including in equation (9.2.9) an additional term describing randomness in the kinetic energy of electrons, it has been shown[40] that an increase in the disorder can lead to a decrease in $\rho$ if the latter is relatively high. Such behavior would be compatible with Mooij's rule, the increase in disorder then being due to higher temperatures.

It seems evident that these new ideas promise new approaches to strongly disordered electron systems, but they definitely need more refining. Further numerical calculations must be conducted and the influence of electron–phonon as well as electron–electron interactions has to be investigated.

# References

1. N. W. Ashcroft, in: *Liquid Metals, 1976*, p. 39, Conf. Ser. No. 30, The Institute of Physics, Bristol and London (1976). The appendix of this article contains a useful summary of relations between structure factors and pair correlation functions.
2. Note that there are two forms of $\varepsilon(q)$ that differ in the way exchange and correlation terms enter. One is used for screening bare pseudopotentials in electric transport, while the calculation of ionic pair potentials requires the other one. For current approximations for exchange and correlation, see M. Shimoji, *Liquid Metals*, p. 107, Academic Press, New York (1977).
3. F. C. Herman and S. Skillman, *Atomic Structure Calculations*, Prentice-Hall, New Jersey (1963).
4. L. F. Mattheiss, *Phys. Rev. A*, **133**, 3199 (1964).
5. V. L. Moruzzi, J. F. Janak, and A. R. Williams, *Calculated Electronic Properties of Metals*, Pergamon Press, New York (1978).

6. There may, however, be strong changes when this short-range structure changes, e.g., solid tetrahedrally coordinated Si or Ge is a semiconductor, while liquid Si or Ge is a metal with a much larger number of nearest neighbors.

7. H.-J. Güntherodt, P. Oelhafen, R. Lapka, H. U. Künzi, G. Indlekofer, J. Krieg, T. Laubscher, H. Rudin, U. Gubler, F. Rösel, K. P. Ackermann, B. Delley, M. Fischer, F. Greuter, E. Hauser, M. Liard, M. Müller, J. Kübler, K. H. Bennemann, and C. F. Hague, *J. Phys. (Paris) C8* **41**, 381 (1980).

8. See, for example, A. Bansil, H. K. Peterson, and L. Schwartz, in: *Liquid Metals, 1976*, p. 313, Conf. Ser. No. 30, The Institute of Physics, Bristol and London (1976).

9. J. M. Ziman, *Electrons and Phonons*, Clarendon Press, Oxford (1960).

10. H. N. Dunleavy and W. Jones, *J. Phys. F* **8**, 1477 (1978).

11. G. Baym, *Phys. Rev. A* **135**, 1691 (1964).

12. M. Parrinello, N. H. March, and M. P. Tosi, *Nuovo Cimento B* **39**, 233 (1977).

13. J. Rubio, *J. Phys. C* **2**, 288 (1969); N. W. Ashcroft and W. Schaich, *Phys. Rev. B* **1**, 1370 (1970).

14. R. J. Elliott, J. A. Krumhansl, and P. L. Leath, *Rev. Mod. Phys.* **46**, 465 (1974).

15. For a recent review of electric transport in glassy metals, see P. J. Cote and L. V. Meisel, in: *Glassy Metals I*, Vol. 46 of *Topics in Applied Physics* p. 141, Springer-Verlag, Berlin (1981).

16. See, e.g., F. Yonezawa, *J. Phys. (Paris) C8* **41**, 447 (1980).

17. S. Wang and S. K. Lai, *J. Phys. F* **10**, 2717 (1980).

18. N. W. Ashcroft and D. C. Langreth, *Phys. Rev.* **159**, 500 (1967).

19. O. Dreirach, R. Evans, H.-J. Güntherodt, and H.-U. Künzi, *J. Phys. F* **2**, 709 (1972).

20. F. J. Pinski, P. B. Allen, and W. H. Butler, *Phys. Rev. Lett.* **41**, 431 (1978).

21. V. K. Ratti and R. Evans, *J. Phys. F* **3**, L238 (1973).

22. R. Evans and A. Jain, *Phys. Earth Planet. Inter.* **6**, 141 (1972).

23. H.-J. Güntherodt, H. U. Künzi, M. Liard, R. Müller, R. Oberle, and H. Rudin, in: *Liquid Metals, 1976*, p. 342, Conf. Ser. No. 30, The Institute of Physics, Bristol and London (1976).

24. E. Esposito, H. Ehrenreich, and C. D. Gelatt Jr., *Phys. Rev. B* **18**, 3913 (1978).

25. B. Delley *et al. J. Phys. F* **9**, 505 (1979).

26. Y. Waseda, A. Jain, and S. Tamaki, *J. Phys. F* **8**, 125 (1978).

27. B. Delley and H. Beck, *J. Phys. F* **9**, 517 (1979).

28. J. C. Duthie and D. G. Pettifor, *Phys. Rev. Lett.* **38**, 564 (1977).

29. B. Delley and H. Beck, *J. Phys. F* **9**, 2231 (1979).

30. J. H. Mooij, *Phys. Status Solidi A* **17**, 521 (1973).

31. S. R. Nagel, *Phys. Rev. Lett.* **41**, 990 (1978).

32. N. F. Mott, *Philos. Mag.* **26**, 1249 (1972).

33. J. L. Black, in: *Glassy Metals I*, Vol. 46 of *Topics in Applied Physics*, p. 167, Springer-Verlag, Berlin (1981).

34. J. M. Ziman, *Models of Disorder*, Cambridge University Press (1979).

35. E. Abrahams, P. W. Anderson, D. C. Licciardello, and T. V. Ramakrishnan, *Phys. Rev. Lett.* **42**, 673 (1979).

36. Y. Imry, *Phys. Rev. Lett.* **44**, 469 (1980).

37. S. M. Girvin and M. Jonson, *Phys. Rev. B* **22**, 3583 (1980).

38. The main features of this theory can be found in W. Götze, *Philos. Mag., Ser. B* **43**, 219 (1981), which contains references to previous publications.

39. D. Belitz and W. Götze, *Philos. Mag. Ser. B* **43**, 517 (1981).

40. D. Belitz and W. Götze, preprint (1981).

# 10

# Transport Theory for Extended and Localized States

## P. N. Butcher

## 10.1. Introduction

The study of electron transport in severely disordered systems is only just beginning. A general understanding of the behavior of simple liquid metals may be obtained from Boltzmann transport theory, which we review in Section 10.2. This is a good place to start because, apart from having application to liquid metals, we may also put the results in a form having wider validity than their derivation might suggest. In particular we find that the static conductivity and thermopower may be expressed as Kubo–Greenwood equations.[1] These equations have been used a great deal to analyze transport data for amorphous semiconductors, as we discuss in Section 10.3.

It is characteristic of disordered systems that some of the electron states are localized. Transport through these states takes place through a hopping mechanism, which has received much attention in recent years. We review the general formalism in Section 10.4 and make simple applications to ac conductivity and dc conductivity in Sections 10.5 and 10.6. Finally, in Section 10.7 we outline a unified theory of both ac and dc hopping conductivity, which has been developed in the last two years.

Hopping transport is one characteristic feature of amorphous semiconductors. Another is anomalous carrier pulse propagation. In

---

**P. N. Butcher** · University of Warwick, Coventry CV4 7AL, England.

crystalline semiconductors, subjected to a steady electric field, injected carrier pulses are usually found to be drifted Gaussians. In amorphous semiconductors such simple behavior is seldom observed. We discuss the marked anomalies that arise and the reasons for them in Section 10.8. At one time a hopping theory was used to discuss the origin of the anomalies.[2] More recent treatments point to trapping as the most likely mechanism.[3-7]

Boltzmann transport theory is treated in most books on crystalline solid-state physics. The author has written a review[8] and applications to liquid metals have recently been reviewed by March.[9] The book by Mott and Davis[1] is outstanding in its treatment of many topics in noncrystalline materials. Hopping conductivity has been reviewed by several authors[10-16] and anomalous carrier pulse propagation is discussed at length by Pfister and Scher[2] on the basis of a hopping model.

## 10.2. Boltzmann Transport Theory

### 10.2.1. Introduction

Suppose the electron states may be labeled by a wave vector **k** and that scattering from **k** to **k'** is weak. Then we may describe the electron system by a semiclassical electron distribution function $f(\mathbf{k}, \mathbf{r}, t)$, which depends on position **r** and time $t$ as well as **k**, and satisfies the Boltzmann equation

$$\frac{\partial f}{\partial t} + \mathbf{v} \cdot \nabla f + \frac{1}{\hbar} \mathbf{F} \cdot \nabla_k f = \left(\frac{\partial f}{\partial t}\right)_c \qquad (10.2.1)$$

Here $f = f(\mathbf{k}, \mathbf{r}, t)$, **v** is the velocity of an electron in state **k**, and **F** is the classical force. The right-hand side of (10.2.1) describes the effect of collisions and has the form

$$\left(\frac{\partial f}{\partial t}\right)_c = \int [f(1-f)P(\mathbf{k}', \mathbf{k}) - f(1-f')P(\mathbf{k}, \mathbf{k}')]d\mathbf{k} \qquad (10.2.2)$$

where $f' = f(\mathbf{k}', \mathbf{r}, t)$ and $P(\mathbf{k}', \mathbf{k})$ describes the scattering rate from **k'** to **k**. We suppose that $f$ is normalized so as to be equal to the probability that state **k** (with spin up) is occupied and that spin flips are forbidden. Then $P(\mathbf{k}', \mathbf{k}) = \Omega/8\pi^3$ times the transition rate from **k'** to **k** in a volume $\Omega$ and is independent of $\Omega$. The physical interpretation of the solution of equation (10.2.1) is provided by the following expressions for the electron density $n$, the electric-current density $\mathbf{J}$, and the heat-flux vector $\mathbf{Q}$:

$$n = \frac{1}{4\pi^3} \int f d\mathbf{k} \tag{10.2.3}$$

$$\mathbf{J} = -\frac{e}{4\pi^3} \int f \mathbf{v} d\mathbf{k} \tag{10.2.4}$$

$$\mathbf{Q} = \frac{1}{4\pi^3} \int f(\varepsilon - \varepsilon_F)\mathbf{v} d\mathbf{k} \tag{10.2.5}$$

In these equations we ignore spin splitting and give equal weights to the contributions from both spin orientations; $\varepsilon$ is the energy of state $\mathbf{k}$ and $\varepsilon_F$ is the chemical potential.

### 10.2.2. Solution of Boltzmann's Equation

The formalism comprised in equations (10.2.1)–(10.2.5) is often applied to crystalline metals and semiconductors and details are given in most textbooks on solid-state physics. A review has been given previously by the author.[8] To determine the transport properties of the electrons in a particular energy band we use the functional dependence of $\varepsilon$ on $\mathbf{k}$ for that band and interpret $\mathbf{v}$ as the corresponding group velocity $\hbar^{-1}\nabla_k\varepsilon$. The formalism becomes particularly simple for an isotropic system in which details of the band structure are averaged out so that $\varepsilon$ depends only on $k = |\mathbf{k}|$. We confine our attention to this case. Our results are then particularly appropriate to liquid metals, which were the subject of Chapter 9. They also provide a useful, but idealized, description of crystalline metals and semiconductors. In the interest of simplicity we consider only static situations, so that $\partial f/\partial t = 0$. We also suppose that a weak, uniform electric field $\mathbf{E}$ is applied and that the magnetic field vanishes. Finally, we allow weak gradients of $\varepsilon_F$ and temperature. Then we may solve equation (10.2.1) by expanding in powers of the small quantities $\mathbf{E}$, $\nabla\varepsilon_F$, and $\nabla T$.

In the zero order we write $f = f_0$ and determine $f_0$ from

$$\left(\frac{\partial f_0}{\partial t}\right)_c = 0 \tag{10.2.6}$$

Now, $P(\mathbf{k}', \mathbf{k})$ satisfies the detailed balance condition

$$\frac{P(\mathbf{k}', \mathbf{k})}{P(\mathbf{k}, \mathbf{k}')} = \exp[\beta(\varepsilon' - \varepsilon)] \tag{10.2.7}$$

where $\beta = (k_B T)^{-1}$ with $k_B$ denoting Boltzmann's constant and $T$ the temperature at point $\mathbf{r}$; $\varepsilon = \varepsilon(\mathbf{k})$ and $\varepsilon' = \varepsilon(\mathbf{k}')$. It follows immediately that equation (10.2.6) may be satisfied by making the integrand in equation (10.2.2) vanish. Thus we find that $f_0$ is just the Fermi–Dirac function of the energy:

$$f_0(\varepsilon) = \{\exp[\beta(\varepsilon - \varepsilon_F)] + 1\}^{-1} \tag{10.2.8}$$

where $\beta$ and $\varepsilon_F$ are arbitrary functions of $\mathbf{r}$.

In first order we write

$$f = f_0 + f_1 \tag{10.2.9}$$

and find from equation (10.2.1) that $f_1$ is determined by the linearized Boltzmann equation

$$\mathbf{v} \cdot \mathbf{\nabla} f_0 - \frac{e\mathbf{E}}{\hbar} \cdot \mathbf{\nabla}_k f_0 = \left(\frac{\partial f_1}{\partial t}\right)_c \tag{10.2.10}$$

In this equation $(\partial f_1 / \partial t)_c$ denotes the linearized collision term:

$$\left(\frac{\partial f_1}{\partial t}\right)_c = \int \Gamma(\mathbf{k}', \mathbf{k}) \left(\frac{f_1'}{f_0'(1 - f_0')} - \frac{f_1}{f_0(1 - f_0)}\right) d\mathbf{k}' \tag{10.2.11}$$

where $f_1 = f_1(\mathbf{k}, \mathbf{r}, t)$, $f_1' = f_1(\mathbf{k}', \mathbf{r}, t)$, $f_0 = f_0(\varepsilon)$, $f_0' = f_0(\varepsilon')$, and the function $\Gamma(\mathbf{k}', \mathbf{k})$ is the thermal equilibrium electron flux from $\mathbf{k}'$ to $\mathbf{k}$ per unit volume of $\mathbf{k}$-space:

$$\Gamma(\mathbf{k}', \mathbf{k}) = f_0'(1 - f_0)P(\mathbf{k}', \mathbf{k}) \tag{10.2.12}$$

We see immediately from expressions (10.2.7) and (10.2.8) that $\Gamma(\mathbf{k}', \mathbf{k})$ is symmetrical, i.e., $\Gamma(\mathbf{k}', \mathbf{k}) = \Gamma(\mathbf{k}, \mathbf{k}')$.

In solving equation (10.2.10) it is convenient to use the relations

$$\mathbf{\nabla} f_0 = -\frac{df_0}{d\varepsilon} [\mathbf{\nabla}\varepsilon_F + T^{-1}\mathbf{\nabla} T(\varepsilon - \varepsilon_F)] \tag{10.2.13a}$$

$$\frac{1}{\hbar} \mathbf{\nabla}_k f_0 = \frac{df_0}{d\varepsilon} \mathbf{v} \tag{10.2.13b}$$

on the left-hand side. To simplify the subseqent analysis we introduce on the right-hand side the relaxation time $\tau = \tau(\varepsilon)$ by writing

$$\left(\frac{\partial f_1}{\partial t}\right)_c = -\frac{f_1}{\tau} \tag{10.2.13c}$$

Then we have

$$f_1 = \tau \frac{df_0}{d\varepsilon} \mathbf{v} \cdot [e\mathbf{E}' + T^{-1}\nabla T(\varepsilon - \varepsilon_F)] \tag{10.2.14}$$

where $\mathbf{E}' \equiv \mathbf{E} + e^{-1}\nabla\varepsilon_F$ is the electromotive force.

### 10.2.3. Equations for the Transport Coefficients

We obtain equations for the transport coefficients by substituting expression (10.2.9) into equations (10.2.4) and (10.2.5). Thus we find with the aid of equation (10.2.14) that

$$\left.\begin{array}{l} \mathbf{J} = \sigma \cdot \mathbf{E}' + L \cdot \nabla T \\[2mm] \mathbf{Q} = M \cdot \mathbf{R}' + N \cdot \nabla T \end{array}\right\} \tag{10.2.15a}$$

where

$$\left.\begin{array}{l} \sigma = -\dfrac{e^2}{4\pi^3} \displaystyle\int \mathbf{vv}\tau \dfrac{df_0}{d\varepsilon}\, d\mathbf{k} \\[4mm] L = -\dfrac{e}{4\pi^3 T} \displaystyle\int \mathbf{vv}\tau \dfrac{df_0}{d\varepsilon}(\varepsilon - \varepsilon_F)\, d\mathbf{k} \\[4mm] M = \dfrac{e}{4\pi^3} \displaystyle\int \mathbf{vv}\tau \dfrac{df_0}{d\varepsilon}(\varepsilon - \varepsilon_F)\, d\mathbf{k} \\[4mm] N = \dfrac{1}{4\pi^3 T} \displaystyle\int \mathbf{vv}\tau \dfrac{df_0}{d\varepsilon}(\varepsilon - \varepsilon_F)^2\, d\mathbf{k} \end{array}\right\} \tag{10.2.16}$$

In crystals $\sigma$, $L$, $M$, and $N$ are tensors. For liquid metals they reduce to scalars, which are obtained from expressions (10.2.16) by replacing the dyadic $\mathbf{vv}$ by the scalar $v^2/3$. Equations (10.2.15a) are the "theorists' form" of the transport equations in which $\mathbf{E}'$ and $\nabla T$ are taken as the independent variables. In the "experimentalists' form" $\mathbf{J}$ and $\nabla T$ are the independent variables. Thus we have

$$\left.\begin{array}{l} \mathbf{E}' = \rho\mathbf{J} + S\nabla T \\[2mm] \mathbf{Q} = \Pi\mathbf{J} - \kappa\nabla T \end{array}\right\} \tag{10.2.15b}$$

where the (scalar) resistivity $\rho$, thermopower $S$, Peltier coefficient $\Pi$, and thermal conductivity $\kappa$ are given by

$$\begin{array}{ll} \rho = \sigma^{-1}, & S = -\sigma^{-1}L \\[2mm] \Pi = \sigma^{-1}M, & \kappa = \sigma^{-1}ML - N \end{array} \tag{10.2.17}$$

The Onsager relation

$$M = -TL \tag{10.2.18a}$$

i.e.,

$$\Pi = TS \tag{10.2.18b}$$

is trivially obvious from equations (10.2.16) and (10.2.17).

### 10.2.4. Conductivity and Thermopower

We are primarily concerned with the most commonly measured transport coefficients $\sigma$ and $S$. By writing $d\mathbf{k} = 4\pi k^2 (dk/d\varepsilon)d\varepsilon$ and $v = \hbar^{-1} d\varepsilon/dk$ in the scalar form of the above equations we readily find that

$$\sigma = -\int \frac{df_0}{d\varepsilon} \sigma(\varepsilon)d\varepsilon \tag{10.2.19}$$

$$S = \frac{k_B}{e} \int \frac{df_0}{d\varepsilon} \frac{\sigma(\varepsilon)}{\sigma} \frac{(\varepsilon - \varepsilon_F)}{k_B T} d\varepsilon \tag{10.2.20}$$

where

$$\sigma(\varepsilon) = \frac{e^2}{12\pi^3 \hbar} 4\pi k^2 v\tau \tag{10.2.21}$$

is the "conductivity at energy $\varepsilon$." For liquid metals the electron statistics are degenerate and we find from equation (10.2.8) that $df_0/d\varepsilon \simeq -\delta(\varepsilon - \varepsilon_F)$. Hence, we easily obtain the familiar results

$$\sigma \simeq \sigma(\varepsilon_F) \tag{10.2.22}$$

$$S \simeq -\frac{\pi^2}{3e} k_B^2 T \frac{d\sigma(\varepsilon_F)/d\varepsilon}{\sigma(\varepsilon_F)} \tag{10.2.23}$$

For n-type semiconductors with a conduction band edge at $\varepsilon_c$, which is several $k_B T$ away from $\varepsilon_F$, the electron statistics are nondegenerate. Hence equation (10.2.8) reduces to $f_0 \simeq \exp[\beta(\varepsilon_F - \varepsilon)]$ and we have

$$\sigma = \beta \int \exp[\beta(\varepsilon_F - \varepsilon)]\sigma(\varepsilon)d\varepsilon$$

$$S = -\frac{k_B}{e} \left( \frac{\varepsilon_c - \varepsilon_F}{k_B T} \right) + A \tag{10.2.24}$$

where

$$A = -\int \frac{df_0}{d\varepsilon} \frac{\sigma(\varepsilon)}{\sigma} \frac{\varepsilon - \varepsilon_c}{k_B T} d\varepsilon \tag{10.2.25}$$

is the "heat of transport" in units of $k_B T$.

We see that equation (10.2.20) specifies $S$ as $-k_B/e$ times the average carrier energy measured relative to $\varepsilon_F$ calculated with the weighting factor $-df_0/d\varepsilon\sigma(\varepsilon)$. This result is thought to have validity outside the domain of Boltzmann transport theory in which we have derived it.[1]

### 10.2.5. The Einstein Relation

In equations (10.2.15) $\mathbf{E}'$ denotes the emf $\mathbf{E} + e^{-1}\nabla\varepsilon_F$. Hence, under isothermal conditions the current density takes the form

$$\mathbf{J} = \sigma\mathbf{E} + e^{-1}\sigma\nabla\varepsilon_F \tag{10.2.26}$$

where $\varepsilon_F$ is determined from the equation for the electron density

$$n = \frac{1}{4\pi^3}\int f_0 d\mathbf{k} \tag{10.2.27}$$

to which $f_1$ makes no contribution. The first term in equation (10.2.26) is the current density due to the electric field. We identify the second term as the current density $eD\nabla n$ due to diffusion and, by writing $\nabla n = \nabla\varepsilon_F(dn/d\varepsilon_F)$, we obtain the Einstein relation

$$\sigma = e^2 D(dn/d\varepsilon_F) \tag{10.2.28}$$

Equation (10.2.28) is the most general form of the Einstein relation and it is valid for both degenerate and nondegenerate statistics. For nondegenerate statistics, however, the relation is usually expressed in an alternative form by noting that $f \simeq \exp[\beta(\varepsilon_F - \varepsilon)]$. Hence $n$ is proportional to $\exp(\beta\varepsilon_F)$ and $dn/d\varepsilon_F = \beta n$. Thus we have

$$\mu = eD/k_B T \tag{10.2.29}$$

where $\mu = \sigma/en$ is the electron drift mobility. It should be emphasized, as a preamble to the next subsection, that equation (10.2.29) is valid *only* for nondegenerate statistics.

### 10.2.6. Energy-Dependent Diffusivity and Mobility[17]

We may rewrite equation (10.2.27) for $n$ in the form

$$n = \int f_0 N(\varepsilon)d\varepsilon \tag{10.2.30a}$$

where

$$N(\varepsilon) = \frac{4\pi k^2}{4\pi^3 d\varepsilon/dk} \tag{10.2.30b}$$

is the density of states. When equation (10.2.30a) is differentiated with respect to $\varepsilon_F$ we see from equation (10.2.8) that

$$\frac{dn}{d\varepsilon_F} = -\int \frac{df_0}{d\varepsilon} N(\varepsilon) d\varepsilon \tag{10.2.31}$$

By substituting this result into equation (10.2.28) and using equation (10.2.19) we find that the diffusion coefficient $D$ may be expressed in the form

$$D = \langle D(\varepsilon) \rangle \tag{10.2.32}$$

where

$$D(\varepsilon) = \frac{\sigma(\varepsilon)}{e^2 N(\varepsilon)} \tag{10.2.33}$$

is the "energy-dependent diffusivity" and the angle brackets signify an average taken with the weighting factor $-N(\varepsilon)\, df_0/d\varepsilon$. We see immediately from equations (10.2.21) and (10.2.30b) and (10.2.33) that

$$D(\varepsilon) = \tfrac{1}{3} v^2 \tau$$
$$= \tfrac{1}{3} \lambda^2 \tau^{-1} \tag{10.2.34}$$

where $v$, $\tau$, and the mean free path $\lambda = v\tau$ are all evaluated at energy $\varepsilon$.

We may rewrite equation (10.2.33) in a form that is often used in discussions of amorphous semiconductors:

$$\sigma(\varepsilon) = eN(\varepsilon)k_B T\mu(\varepsilon) \tag{10.2.35}$$

where

$$\mu(\varepsilon) = \frac{eD(\varepsilon)}{k_B T} \tag{10.2.36}$$

is usually referred to as the "mobility at energy $\varepsilon$." This terminology is a little unfortunate but has now become firmly established. The factors $k_B T$ actually cancel out of equations (10.2.35) and (10.2.36) to yield the simpler equation

$$\sigma(\varepsilon) = e^2 N(\varepsilon) D(\varepsilon) \tag{10.2.37}$$

which has the same physical content. Thus $\mu(\varepsilon)$ as given by equation (10.2.36) is simply $D(\varepsilon)$ scaled by a factor $e/k_B T$ to give it the *dimensions* of mobility. The scaling factor does not change the physical interpretation of $D(\varepsilon)$, which is that of a diffusion constant. Indeed, we may readily verify from equations (10.2.22), (10.2.28), (10.2.31), and (10.2.37) that, for degenerate statistics, the macroscopic diffusion constant is just $D(\varepsilon_F)$ while $\mu(\varepsilon_F)$ as given by equation (10.2.36) has no simple macroscopic interpretation. By way of illustration we consider electrons with a constant effective mass $m^*$. Then

$$\mu(\varepsilon_F) = \frac{e\tau(\varepsilon_F)}{m^*} \frac{2\varepsilon_F}{3k_B T} \qquad (10.2.38)$$

in which we recognize the first factor as the drift mobility.

For nondegenerate statistics $\mu(\varepsilon)$ has a more reasonable interpretation. We readily verify that the macroscopic mobility in this case is $\sigma/en = \langle \mu(\varepsilon) \rangle$, where the angle brackets have the same significance as in equation (10.2.32). Unfortunately, this simple result becomes invalid as soon as the statistics become degenerate to any degree.

### 10.2.7. Hall Mobility and ac Conductivity

To determine the low-field Hall mobility $\mu_H = \sigma_{xy}/B\sigma_{xx}$ we must include a Lorentz force term $-e\mathbf{v} \times \mathbf{B}$ in the classical force $\mathbf{F}$ entering Boltzmann's equation (10.2.1). The calculation of $\mu_H$ is elementary but tedious for isotropic systems.[8] We find that $\mu_H = \mu$ for degenerate statistics and

$$\mu_H = \mu \langle\langle \tau^2 \rangle\rangle / \langle\langle \tau \rangle\rangle^2 \qquad (10.2.39)$$

for general statistics and a constant effective mass. In equation (10.2.39) the double angle brackets indicate an average calculated with a weighting factor $-df_0/d\varepsilon$ times the *integrated* density of states.

The current density response to an ac electric field $\text{Re}[\mathbf{E} \exp(-i\omega t)]$ may be calculated by keeping the time derivative in Boltzmann's equation (10.2.1). For degenerate statistics the result is $\text{Re}[\sigma(\omega)\mathbf{E}\exp(-i\omega t)]$, where $\sigma(\omega)$ is given by the Drude equation

$$\sigma(\omega) = \sigma[1 - i\omega\tau(\varepsilon_F)]^{-1} \qquad (10.2.40)$$

For general statistics the general behavior is the same as that predicted by the Drude formula, but $\tau(\varepsilon_F)$ is replaced by an appropriate, frequency-dependent average of $\tau(\varepsilon)$.

### 10.2.8. Conclusion

To evaluate the equations for the transport coefficients we require $\tau$. For elastic scattering between plane wave states $\mathbf{k}$ and $\mathbf{k}'$ the behavior of each individual scatterer is usually described by the differential scattering cross-section $\sigma(\theta)$, where $\theta$ is the angle between $\mathbf{k}$ and $\mathbf{k}'$. The correlations in the positions of an assembly of scatterers with density $n_s$ may be allowed for by multiplying $\sigma(\theta)$ by the static structure factor $a(K)$, where $K = 2k \sin \theta/2$ is the magnitude of the scattering wave vector $\mathbf{k}' - \mathbf{k}$. With this notation, $\tau^{-1}$ is given by the well-known equation

$$\frac{1}{\tau} = n_s v 2\pi \int \sigma(\theta)(1 - \cos \theta)a(K)\sin \theta d\theta \qquad (10.2.41)$$

In semiconductors we usually assume completely random scatterers, so that $a(K) = 1$, and calculate $\sigma(\theta)$ in the Born approximation. In liquid metals $a(K)$ is usually taken from $X$-ray or neutron-scattering data and $\sigma(\theta)$ is calculated from the ion-core phase shifts. This topic has been discussed in Chapter 9.

When we turn to inelastic scattering, we find that $\tau$ is not very well-defined.[8] Nevertheless, very often an approximate relaxation time may be introduced that provides good physical insight. When this is not possible it becomes necessary to solve the linearized Boltzmann equation by numerical or variational methods. Some examples are given elsewhere.[8]

## 10.3. Applications of Kubo–Greenwood Formulas

### 10.3.1. Introduction

Equations (10.2.19) and (10.2.20) are usually known as Kubo–Greenwood formulas.[1] In the one-electron approximation they have validity outside the domain of Boltzmann transport theory. When the transport mechanism is different from that assumed in Section 10.2, what changes is $\sigma(\varepsilon)$; equations (10.2.19) and (10.2.20) remain intact. We quote them again here for convenience:

$$\sigma = - \int \frac{df_0}{d\varepsilon} \sigma(\varepsilon) d\varepsilon \qquad (10.3.1)$$

$$S = \frac{k_B}{e} \int \frac{df_0}{d\varepsilon} \frac{\sigma(\varepsilon)}{\sigma} \frac{(\varepsilon - \varepsilon_F)}{k_B T} d\varepsilon \qquad (10.3.2)$$

The first relation is easily justified for electrons moving in a disordered static potential. We give a derivation in Section 10.3.2. Accepting equation (10.3.1) for the moment we may immediately give a heuristic derivation of equation (10.3.2). Equation (10.2.18), expressing Onsager symmetry, and equation (10.2.15), expressing the linear transport relations, have general validity. Hence $S = \Pi/T$, where the Peltier coefficient is the ratio of $\mathbf{Q}$ to $\mathbf{J}$ when $\nabla T = 0$. For the electric field $\mathbf{E} = (E_x, 0, 0)$, $\mathbf{J} = (\sigma E_x, 0, 0)$ and we see from equation (10.3.1) that the contribution to $\sigma E_x$ from energies between $\varepsilon$ and $\varepsilon + d\varepsilon$ is $- E_x (df_0/d\varepsilon)\sigma(\varepsilon)d\varepsilon$. The corresponding contribution to $Q_x$ in the heat flux vector $\mathbf{Q} = (Q_x, 0, 0)$ is obtained by dividing by $- e$ and multiplying by $\varepsilon - \varepsilon_F$. Thus we obtain equation (10.3.2). The argument used here is obviously not rigorous. Nevertheless equation (10.3.2) is usually assumed to be generally valid.[1]

### 10.3.2. Derivation of the Conductivity Equation[1,8]

As the nomenclature implies, equation (10.3.1) may be derived from the Kubo equation[18] for the conductivity $\sigma(\omega)$ at frequency $\omega$. A very simple approach is possible when we recognize that the real part $\sigma_1(\omega)$ of $\sigma(\omega)$ is equal to $2/|E|^2$ times the time-averaged power absorption density $P$ in an electric field $(E_x, 0, 0)$ with $E_x = \mathrm{Re}[E \exp(- i\omega t)]$. The perturbation of the Hamiltonian produced by $E_x$ for one electron is $eE_x x$ and this produces a transition rate $W_{mn}$ between any pair of one-electron states $m$ and $n$ that we may calculate using Fermi's Golden Rule:

$$W_{mn} = \frac{2\pi e^2}{\hbar} \frac{|E|^2}{4} |\langle n | x | m \rangle|^2 [\delta(\varepsilon_n - \varepsilon_m - \hbar\omega) + \delta(\varepsilon_n - \varepsilon_m + \hbar\omega)]$$

(10.3.3)

The corresponding power dissipation if $m$ is known to be occupied and $n$ is known to be empty is $W_{mn}(\varepsilon_n - \varepsilon_m)$. In thermal equilibrium the probability of the initial condition necessary for the absorption to take place is $f_m^0(1 - f_n^0)$, where $f_m^0 = f_0(\varepsilon_m)$ is the Fermi–Dirac function (10.2.8). Hence we have, considering a volume $\Omega$,

$$\sigma_1(\omega) = \frac{\pi e^2}{\hbar\Omega} \sum_{m,n} (\langle n | x | m \rangle)^2 (\varepsilon_n - \varepsilon_m) f_m^0 (1 - f_n^0)$$

$$\times [\delta(\varepsilon_n - \varepsilon_m - \hbar\omega) + \delta(\varepsilon_n - \varepsilon_m + \hbar\omega)]$$

$$= \frac{\pi e^2 \omega}{\Omega} \sum_{m,n} |\langle n | x | m \rangle|^2 \delta(\varepsilon_n - \varepsilon_m - \hbar\omega)[f_m^0(1 - f_n^0) - f_n^0(1 - f_m^0)]$$

$$= \frac{\pi e^2 \omega}{\Omega} \sum_{m,n} |\langle n | x | m \rangle|^2 \delta(\varepsilon_n - \varepsilon_m - \hbar\omega)(f_m^0 - f_n^0)$$

(10.3.4)

In the second line we have interchanged the dummy variables $m$ and $n$ and have used the properties of the $\delta$-function and the hermiticity of $\langle m \mid x \mid n \rangle$.

In the one-electron approximation

$$[x, H] = \frac{1}{2m_e}[x, p_x^2] = \frac{i\hbar}{m_e} p_x \qquad (10.3.5)$$

where $H$ is the one-electron Hamiltonian and $m_e$ is the free electron mass. It follows that

$$\langle n \mid x \mid m \rangle (\varepsilon_n - \varepsilon_m) = \frac{i\hbar}{m_e} \langle n \mid p_x \mid m \rangle \qquad (10.3.6)$$

and equation (10.3.4) may be written in the equivalent form

$$\sigma_1(\omega) = \frac{\pi e^2 \hbar}{\Omega m_e^2} \sum_{m,n} |\langle n \mid p_x \mid m \rangle|^2 \delta(\varepsilon_n - \varepsilon_m - \hbar\omega) \times \frac{[f_0(\varepsilon_m) - f_0(\varepsilon_m + \hbar\omega)]}{\hbar\omega}$$

In the limit $\omega \to 0$ we therefore have

$$\sigma = \sigma_1(0) = -\frac{\pi\hbar}{\Omega m_e^2} \sum_{m,n} |\langle n \mid p_x \mid m \rangle|^2 \delta(\varepsilon_n - \varepsilon_m) \frac{df_0(\varepsilon_m)}{d\varepsilon_m}$$

$$= -\int \frac{df_0}{d\varepsilon} \sigma(\varepsilon) d\varepsilon \qquad (10.3.7)$$

with

$$\sigma(\varepsilon) = \frac{\pi e^2 \hbar}{\Omega m_e^2} \sum_{m,n} |\langle n \mid p_x \mid m \rangle|^2 \delta(\varepsilon - \varepsilon_n) \delta(\varepsilon - \varepsilon_m)$$

$$= \frac{\pi e^2 \hbar \Omega}{m_e^2} N^2(\varepsilon) p_x^2(\varepsilon)_{av} \qquad (10.3.8)$$

where

$$N(\varepsilon) = \frac{1}{\Omega} \sum_n \delta(\varepsilon - \varepsilon_n) \qquad (10.3.9)$$

is the density of states (counting both spin orientations) and $p_x^2(\varepsilon)_{av}$ is the average of $|\langle n \mid p_x \mid m \rangle|^2$ calculated with the weighting factor $\delta(\varepsilon - \varepsilon_m)\delta(\varepsilon - \varepsilon_n)$.

### 10.3.3. Localized and Extended States

We see from the above discussion that $\sigma(\varepsilon)$ is determined by the matrix elements of $p_x$ between states with energy $\varepsilon$. For this reason the distinction between extended and localized states is of great importance in the theory of electron transport in disordered systems. As the name implies, extended states, like Bloch functions, run right over the macroscopic volume $\Omega$ of the system while localized states, like impurity wave functions, are confined to a microscopic volume. The two types of state cannot coexist at the same energy because any coupling of a localized state to an extended one will lead to delocalization.[1]

Let us consider a single energy band for which $N(\varepsilon)$ has the form shown in Figure 10.1. For weak disorder most of the states are extended. Localized states arise only in the tails of $N(\varepsilon)$, as indicated by the cross-hatching, and are due to extreme fluctuations of the potential. The localized states are separated from the extended states by "mobility edges" $\varepsilon_c$ and $\varepsilon_c'$. The reason for this nomenclature is made clear below. As the disorder increases $\varepsilon_c$ and $\varepsilon_c'$ move toward the center of the band and coalesce at a critical value of the disorder. Thereafter *all* the states in the band are localized. This is called an Anderson transition.[1]

### 10.3.4. Behavior of Conductivity as Dependent on Energy

When $T \to 0$ the conductivity $\sigma$ in equation (10.3.1) approaches $\sigma(\varepsilon_F)$. Now, we see from the discussion of Section 10.3.2 that $\sigma(\varepsilon_F)$ is determined by the matrix elements of $p_x$ between states having energy $\varepsilon_F$. Moreover, it follows from equation (10.3.6) that all these matrix elements vanish if $\langle n \mid x \mid m \rangle$ is well defined — which it is for localized states. Consequently $\sigma(\varepsilon_F) = 0$ when $\varepsilon_F$ lies in the region of localized states. This argument breaks down when $\varepsilon_F$ lies in the region of extended states because $\langle n \mid x \mid m \rangle$ diverges. Thus $\sigma(\varepsilon_F) \neq 0$ in this case.

We are therefore led to picture $\sigma(\varepsilon)$ as vanishing when $\varepsilon$ lies outside the region between $\varepsilon_c$ and $\varepsilon_c'$ and the reason for calling these energies "mobility edges" is apparent. In the extended-state region $\sigma(\varepsilon) > 0$. The behavior of $\sigma(\varepsilon)$ as $\varepsilon$ approaches a mobility edge remains controversial.

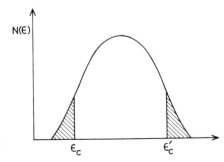

Figure 10.1. Density of states for a single band showing localized state regions (cross-hatched) bounded by mobility edges at $\varepsilon_c$ and $\varepsilon_c'$.

Mott supposes that it attains a minimum value

$$\sigma_{\min} = 0.026e^2/\hbar a = 610/a \ \Omega^{-1} \ cm^{-1} \qquad (10.3.10)$$

where $a$ is a typical interatomic dimension measured in Angstroms. We may derive this result very easily from equation (10.2.21), by arguing that the minimum value of the mean free path $v\tau \sim a$ and $k \sim 1/a$, which gives $\sigma_{\min} = e^2/3\pi^2\hbar a$. There is considerable experimental evidence to support this contention.[1] Nevertheless, it may be that $\sigma(\varepsilon)$ reaches a value in the order of $\sigma_{\min}$ near the mobility edge and then drops rapidly to zero. It would not be easy to tell the difference from experimental data.

Since equation (10.3.8) must agree with equation (10.2.21) for weakly scattered free electrons with $N(\varepsilon) = 4\pi k^2/4\pi^3\hbar v$, $v = \hbar k/m_e$, and $\lambda = v\tau$, we have

$$p_x^2(\varepsilon)_{av} = \frac{\pi\hbar^2}{3}\frac{\lambda}{\Omega} \qquad (10.3.11)$$

Mott supposes that a result of this form has general validity for extended states with $\lambda$ being the phase coherence length of the wave function.[1] If this is the case, and if $\lambda$ reaches a minimum value of $a$, $p_x^2(\varepsilon)_{av}$ is fixed and we see from equation (10.3.8) that $\sigma(\varepsilon)$ becomes proportional to the square of the density of states.

For temperatures above absolute zero the coupling of electrons to the phonons gives $\sigma(\varepsilon)$ a small but nonvanishing value in the localized-state regions. Transport through these states proceeds by a hopping mechanism as we discuss in detail later on. For the moment we concentrate on general applications of the Kubo–Greenwood formulas without detailed discussion of the transport mechanism.

### 10.3.5. n-Type Semiconductor Transport

In amorphous semiconductors the transport properties are determined by carriers in states near the "pseudogap" in the density of states shown schematically in Figure 10.2. At room temperature the transport properties

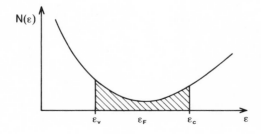

Figure 10.2. Density of states near the pseudogap in an amorphous semiconductor showing the localized-state region (cross-hatched) extending from a mobility edge $\varepsilon_v$ in the valence band to a mobility edge $\varepsilon_c$ in the conduction band. The Fermi level $\varepsilon_F$ is near the middle of the pseudogap.

are usually thought to be dominated by either holes in extended states in the valence band or electrons in extended states in the conduction band. In either case the statistics are nondegenerate because $\varepsilon_F$ is located near the middle of the gap.[1] To be definite we consider n-type material. Then we may write $f_0(\varepsilon) \simeq \exp[\beta(\varepsilon_F - \varepsilon)]$ in the conduction band and ignore the valence-band states altogether. The Kubo–Greenwood formulas (10.3.1) and (10.3.2) therefore become

$$\sigma = \int \sigma_D(\varepsilon) \exp[\beta(\varepsilon_F - \varepsilon)] d\varepsilon \qquad (10.3.12)$$

$$S = -\frac{k_B}{e} \int \frac{\sigma_D(\varepsilon)}{\sigma k_B T} \exp[\beta(\varepsilon_F - \varepsilon)](\varepsilon - \varepsilon_F) d\varepsilon \qquad (10.3.13)$$

where, by expressing $\sigma(\varepsilon)$ in the form (10.2.35), we have introduced a "differential conductivity"

$$\sigma_D(\varepsilon) = \sigma(\varepsilon)/k_B T = eN(\varepsilon)\mu(\varepsilon) \qquad (10.3.14)$$

There have been a number of different treatments of these equations and we mention some of them briefly here.

In the most elementary treatment we set $\sigma(\varepsilon) = \sigma(\varepsilon_c)$ for $\varepsilon > \varepsilon_c$ and $\sigma(\varepsilon) = 0$ for $\varepsilon < \varepsilon_c$. Then we have immediately

$$\sigma = \sigma(\varepsilon_c) \exp[-\beta(\varepsilon_\sigma - \varepsilon_F)] \qquad (10.3.15)$$

$$S = -\frac{k_B}{e}\left(\frac{\varepsilon_S - \varepsilon_F}{k_B T} + A\right) \qquad (10.3.16)$$

with $\varepsilon_\sigma = \varepsilon_S = \varepsilon_c$ and $A = 1$. If $\varepsilon_c - \varepsilon_F = W - \gamma T$, where $W$ and $\gamma$ are constants, we see that a plot of $\log \sigma$ against $\beta$ has an activation energy (slope) $W$ and $\beta = 0$ intercept $\sigma(\varepsilon_c)\exp(\gamma/k_B)$, while a plot of $Se/k_B$ against $\beta$ has an activation energy $W$ and $\beta = 0$ intercept $1 - \gamma/k_B$. A great deal of experimental data for many amorphous semiconductors exhibits this sort of general behavior but the slopes and intercepts are very sensitive to the method of preparation.[1]

A material that has been studied very extensively in recent years is a-Si prepared by the glow-discharge decomposition of silane and the experimental data on $\sigma$ and $S$ have been investigated by several authors for both doped and undoped samples. Le Comber and Spear made the first detailed investigations of this material at Dundee and the work of the Dundee group is reviewed elsewhere.[19] They have often analyzed their data on the assumption that there are three transport channels: extended states, tail states, and donor states.[20] Friedman[21] has made a careful analysis of Dundee data on the basis of equations (10.3.12)–(10.3.14) using the values of $N(\varepsilon)$ determined by Le Comber and Spear[20] and making reasonable assumptions about the variation of $\mu(\varepsilon)$ across the mobility edge. Spear and co-workers have also emphasized the importance of the temperature dependence of both $\varepsilon_F$ and possibly $\varepsilon_c$ in determining the behavior of $\sigma$ and $S$ and the variation of this behavior from sample to sample.[22,23]

The transport problem, which has received most attention in recent years, is the observation that while equations (10.3.15) and (10.3.16) are valid for a-Si, $\varepsilon_\sigma$ is frequently greater than $\varepsilon_S$ and the difference $\varepsilon_\sigma - \varepsilon_S$ varies from one sample to another. The simplest explanation is that the mobility involved in $\sigma(\varepsilon_c) = eN(\varepsilon_c)\mu(\varepsilon_c)k_BT$ is activated with an activation energy $\varepsilon_\mu = \varepsilon_\sigma - \varepsilon_S$, but then the dependence on preparation method is difficult to understand. Döhler[24] has emphasized that one may actually determine the differential conductivity (10.3.14) from experimental data on $S$ and $\sigma$ provided it is supposed that $\sigma_D(\varepsilon)$ is independent of $\beta$. To do this we write equations (10.3.12) and (10.3.13) in the form

$$\sigma(\beta)e^{-\beta\varepsilon_F} = \int_{-\infty}^{\infty} \sigma_D(\varepsilon)e^{-\beta\varepsilon}d\varepsilon \tag{10.3.17}$$

$$\frac{eS}{k_B} = -\frac{e^{\beta\varepsilon_F}}{k_BT}\int_{-\infty}^{\infty} \frac{\sigma_D(\varepsilon)e^{-\beta\varepsilon}}{\sigma}(\varepsilon - \varepsilon_F)d\varepsilon \tag{10.3.18}$$

We see from equation (10.3.17) that $\sigma(\beta)\exp(-\beta\varepsilon_F)$ is the Laplace transform of $\sigma_D(\varepsilon)$. Hence we may determine $\sigma_D(\varepsilon)$ by inverting the transform once $\varepsilon_F$ is known as a function of $\beta$. To determine that we differentiate equation (10.3.17) with respect to $\beta$ and use equation (10.3.18) to show that

$$\frac{d\varepsilon_F}{d\beta} = -\frac{1}{\beta}\left[\frac{eS}{k_B}k_BT - \frac{d\log\sigma(\beta)}{d\beta}\right] \tag{10.3.19}$$

The right-hand side of this equation is known from experiment. Hence $\varepsilon_F$ may be determined to within an additive constant, which depends on the origin of energy. Consequently $\sigma_D(\varepsilon)$ may be determined apart from an arbitrary shift on the energy scale. Döhler finds that $\sigma(\varepsilon)$ increases rapidly by several orders of magnitudes over a few tenths of an electron volt and depends on doping. This is consistent with the sort of behavior to be expected near a mobility edge but the details are difficult to understand because no transport model is assumed.

Grünwald and Thomas[25] have shown that it is also possible to explain some of the data using a variable-range hopping model, but to do so requires what are probably unrealistic assumptions about the distribution of tail states and its doping dependence.

The most recent and most promising treatment of the problem is due to Overhof and Beyer,[26] who suggest that $\varepsilon_\mu$ is due to long-range fluctuations of $\varepsilon_c$ caused by charged defects in the material. The value of $\varepsilon_\sigma$ is determined by extreme fluctuations while $\varepsilon_S$ involves an average and is consequently less. These authors emphasize the importance of the quantity

$$Q = \log \sigma - eS/k_B$$
$$= \log \sigma(\varepsilon_c) - \varepsilon_\mu \beta + A \qquad (10.3.20)$$

where we have used equations (10.3.15) and (10.3.16). We see that $Q$ does not involve $\varepsilon_F$ explicitly and is much more indicative of the transport mechanism (as opposed to the electron distribution) than either $\sigma$ or $S$ individually, which do involve $\varepsilon_F$. A plot of $Q$ against $\beta$ is usually a straight line over a much wider temperature range than is the case for either $\log \sigma$ or $S$. Moreover the slope $\varepsilon_\mu$ depends on preparation, which one would expect if it is determined by fluctuations of $\varepsilon_c$.

### 10.3.6. Conductivity near the Mobility Edge

Götze and co-workers have developed an unusual and interesting approach to the problem of conductivity in random systems.[27-30] The theory provides equations for both ac and dc conductivity. We outline the dc calculation here. Free electrons with an effective mass $m^*$ are considered and it is supposed that they are scattered by a random potential. A relaxation time $\tau$ depending on frequency and the location of the Fermi level is introduced and this determines the conductivity when $T = 0$. In the Boltzmann-transport regime $\tau^{-1}$ is given by (10.2.41) with $\varepsilon = \varepsilon_F = \hbar^2 k_F^2/2m^*$ and it is supposed, for simplicity, that there is a cutoff in the Fourier transform of the scattering potential such that $a(K)\sigma(\theta)$ is constant for $K < K_0$ and zero for $K > K_0$. Then

$$\tau \propto \begin{cases} \varepsilon_F^{-1/2}, & 2k_F < K_0 \\ \varepsilon_F^{3/2}, & 2k_F > K_0 \end{cases} \qquad (10.3.21)$$

Götze goes beyond the Boltzmann transport regime by relating the frequency-dependent $\tau$ to the density fluctuations in the system, as perturbed by the scattering, that also involve $\tau$ in the simplest approximation. Thus a self-consistency condition arises and can be solved numerically. For weak coupling the solution falls slightly below the Boltzmann

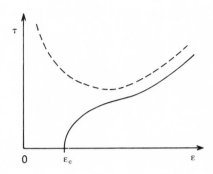

Figure 10.3. Schematic plot of the relaxation time at zero frequency calculated by Belitz and Götze.[30] The dashed line refers to the Boltzmann transport regime and the solid line is the solution of the self-consistent equation.

result (10.3.21) at high energies but, and this is the important point, $\tau$ drops to zero at a mobility edge $\varepsilon_c$, varying as $(\varepsilon - \varepsilon_c)^{1/2}$, as indicated in Figure 10.3. For $\varepsilon < \varepsilon_c$, $\tau$ becomes imaginary and the dc conductivity vanishes.

A mobility edge is thus predicted by the theory, but $\sigma(\varepsilon)$ falls continuously to zero as $\varepsilon$ approaches $\varepsilon_c$ from above and a minimum metallic conductivity does not appear in the formalism. Belitz and Götze have therefore looked again at experimental data on the dc conductivity of Sb-doped Ge, $La_{1-x}Sr_xVO_3$, and $Gd_{1-x}Sr_xVO_3$ in which variation of $\varepsilon_F$ shows up a transition from semiconducting (activated) to metallic (temperature-independent) behavior. This has been interpreted as an Anderson transition in the sense that $\varepsilon_F$ moves through the mobility edge $\varepsilon_c$. There seems little doubt that this is the case.[1] However, the data does not seem to provide good evidence for a minimum metallic conductivity (as has sometimes been suggested[1]) because Belitz and Götze can fit it with reasonable parameter values by using the $\sigma(\varepsilon)$ derived from their model in the Kubo–Greenwood formula (10.3.1).

## 10.4. Hopping Transport Theory: General Formalism

### 10.4.1. Introduction

The states in the tails of the conduction band and valence band of an amorphous semiconductor are localized. Electrons move among these states by "hopping" from full states to empty states. The treatment of electron transport involving this mechanism is usually described by a rate equation, which is the analogue of Boltzmann's equation for localized states. We use this approach here. As in all transport problems the same results may be obtained by considering the expectation values of appropriate correlation functions in thermal equilibrium. For hopping conductivity the appropriate equilibrium variable is the mean-square distance moved by a hopping electron in time $t$. Thus we may calculate the diffusivity from a study of a random-walk problem and then use the Einstein relation to obtain the conductivity. This approach has attracted the interest of several authors.[12,31-38] We do not use it here because for degenerate statistics the random walk is necessarily an artificial one since, in the real system, the electrons get in each other's way. Instead we proceed as in Section 10.2 to introduce a weak field into the rate equation and calculate the perturbations of the occupation probabilities of the localized states that it produces. We shall see how the random-walk formalism arises in this context. However, for calculating transport coefficients the Boltzmann-equation approach is more direct and is easier to use because, for weak fields, it reduces to Kirchhoff's equations for an equivalent RC network.[39]

### 10.4.2. Miller–Abrahams Equivalent Circuit

We follow previous treatments given by the author and regard the localized states as defining sites that may be occupied by one, and only one, electron. Let $f_m$ and $R_{mn}$ be respectively the occupation probability of site $m$ and the transition rate from $m$ to $n$. Then the $f_m$s are determined by the rate equations

$$\frac{df_m}{dt} = \sum_n [f_n(1 - f_m)R_{nm} - f_m(1 - f_n)R_{mn}] \tag{10.4.1}$$

Let us write $\varepsilon_m$ for the unperturbed energy of site $m$ and $U_m$ for the perturbation of $\varepsilon_m$ produced by the applied field. We suppose that $R_{mn}$ satisfies the detailed balance relation

$$\frac{R_{mn}}{R_{nm}} = \exp[\beta(\varepsilon_m + U_m - \varepsilon_n - U_n)]$$

$$\simeq \exp[\beta(\varepsilon_m - \varepsilon_n)][1 + \beta(U_m - U_n)] \tag{10.4.2}$$

for small applied fields. When $U_m = 0$ for all $m$, $f_m$ reduces to the thermal-equilibrium form

$$f_m^0 = \{\tfrac{1}{2}\exp[\beta(\varepsilon_m - \varepsilon_F)] + 1\}^{-1} \tag{10.4.5}$$

This is just the Fermi function (10.2.8) modified by a factor of 0.5 in front of the exponential to take account of the fact that only one electron, but with either spin orientation, may occupy site $m$. When $U_m \neq 0$ it is convenient to set

$$f_m = f_m^0 - \frac{df_m^0}{d\varepsilon_m}\phi_m \tag{10.4.4}$$

where $\phi_m$ remains to be determined. When equations (10.4.2) and (10.4.4) are substituted into equation (10.4.1) we find that, to first order in $\phi_m$ and $U_m$, the linearized rate equations are

$$C_m \frac{d}{dt}(V_m + Ex_m) = \sum_n g_{mn}(V_n - V_m) \qquad (10.4.5)$$

where

$$C_m = -e^2 df_m^0/d\varepsilon_m \qquad (10.4.6a)$$

$$g_{mn} = e^2 \beta f_m^0 (1 - f_n^0) R_{mn}^0 \qquad (10.4.6b)$$

$$V_m = -[Ex_m + e^{-1}\phi_m] \qquad (10.4.6c)$$

In equations (10.4.5) and (10.4.6) we have written

$$U_m = eEx_m \qquad (10.4.7)$$

to specialize to the case when a uniform electric field $E$ is applied in the positive $x$-direction. In equation (10.4.6b) $R_{mn}^0$ is the equilibrium value of $R_{mn}$.

Equations (10.4.4), (10.4.5) and (10.4.6) have a simple interpretation. We see from equations (10.4.3) and (10.4.4) that $\phi_m$ may be regarded as the local change of chemical potential at site $m$. Hence $V_m$ in equation (10.4.6c) is the local change of the electrochemical potential $\phi_m + eEx_m$ multiplied by $-e^{-1}$ for convenience of notation. We refer to $V_m$ simply as the "voltage" at site $m$. The particle current, which flows from $n$ to $m$, is proportional to the electrochemical difference between the two sites. In equation (10.4.5) we have introduced a factor $-e$ so that the *electric* current from $n$ to $m$ appears as $g_{mn}(V_n - V_m)$. The quantity $g_{mn}$ is given by equation (10.4.6b). It follows from the detailed balance relation (10.4.2) in zero field that $g_{mn} = g_{nm}$. Thus $g_{mn}$ is a positive, symmetrical conductance. Finally, we note that the left-hand side of equation (10.4.5) is the time-rate of change of the electric charge on site $m$ and that $C_m$ in equation (10.4.6a) is positive and may be interpreted as a capacitance. Hence equations (10.4.5) are Kirchhoff's equations for the RC network shown in Figure 10.4,

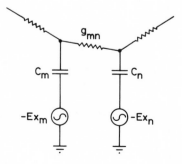

Figure 10.4. The Miller–Abrahams equivalent circuit.

where the ground potential is zero and the voltage generators sustain the applied potential at each site. This point of view was originally introduced by Miller and Abrahams[39] and it has proven to be very fruitful.

### 10.4.3. Conductivity Equations

There are several alternative ways to write a general equation for the conductivity $\sigma(\omega)$ at frequency $\omega$. Suppose that $E$ has a time factor $\exp(-i\omega t)$. Consider a large cube of side $L$ inside an infinite specimen. The current density across the coordinate plane at $x$ is

$$J(x) = L^{-2} \sum_{m,n} g_{mn} (V_m - V_n) \theta(x - x_m) \theta(x_n - x) \qquad (10.4.8)$$

where the unit step functions exclude contributions from pairs of sites for which the inequalities $x_m < x < x_n$ are not satisfied. The current density is independent of $x$ when $L \to \infty$. We may therefore average over $x$ from $0$ to $L$ and divide by $E$ to obtain the conductivity equation

$$\sigma(\omega) = \frac{1}{\Omega E} \sum_{m,n} (x_n - x_m) \theta(x_n - x_m) g_{mn} (V_m - V_n)$$

$$= \frac{1}{2\Omega E} \sum_{m,n} g_{mn} V_{mn} x_{nm} \qquad (10.4.9)$$

where $\Omega = L^3$, $V_{mn} = V_m - V_n$, and $x_{nm} = x_n - x_m$. In the first line of equation (10.4.9) terms with $x_n < x_m$ are excluded. In the second line we have used the symmetry of $g_{mn}$ to include them.

We may also obtain a simple equation for the real part $\sigma_1(\omega)$ of $\sigma(\omega)$ by noting that the Joule heat generated in $\Omega$ is the sum of the Joule heats developed in the individual conductances. Thus we find that

$$\sigma_1(\omega) = \frac{1}{2\Omega |E|^2} \sum_{m,n} g_{mn} |V_{mn}|^2 \qquad (10.4.10)$$

when $\omega \to 0$, $E$ and $V_{mn}$ are real, and the modulus signs may be removed from equation (10.4.10). It will be noted that there is no difficulty in taking the limit $\omega \to 0$ in either equation (10.4.9) or (10.4.10). However, to use either equation we must first obtain an approximation to $V_{mn}$. We discuss ways of doing this in Sections 10.5, 10.6, and 10.7.

Our final task in this subsection is to obtain an explicit expression for $\sigma(\omega)$ by introducing a Green's matrix to solve equations (10.4.5). Let us write

$$q_m = C_m (V_m + E x_m) \qquad (10.4.11)$$

for the charge on site $m$. We consider a finite system and introduce row matrices $\mathbf{q}$ and $\mathbf{x}$ whose $n$th columns are $q_n$ and $x_n$, respectively, and the square matrices $W$ and $C$ whose $(mn)$th elements are respectively

$$W_{mn} = - g_{mn}/C_m + \delta_{mn} \sum_n g_{mn}/C_m \qquad (10.4.12)$$

and

$$C_{mn} = C_m \delta_{mn} \qquad (10.4.13)$$

Then, assuming a time factor $\exp(-i\omega t)$, the linearized rate equations (10.4.5) take the matrix form

$$-i\omega \mathbf{q} = -\mathbf{q} W + E\mathbf{x} C W \qquad (10.4.14)$$

Writing

$$G = (W - i\omega)^{-1} \qquad (10.4.15)$$

for the Green's matrix we have

$$\mathbf{q} = E\mathbf{x} C W G \qquad (10.4.16)$$

The conductivity is equal to $-i\omega/E$ times the $x$-component of the dipole-moment density. Hence

$$\sigma(\omega) = -\frac{i\omega}{\Omega} \mathbf{x} C W G \bar{\mathbf{x}} \qquad (10.4.17)$$

where $\bar{\mathbf{x}}$ is the transpose of $\mathbf{x}$. It only remains to write this formal result in a more recognizable form.

### 10.4.4. The Einstein Relation

To rewrite equation (10.4.17) in the form of an Einstein relation we note some elementary properties of $G$. We see from equation (10.4.15) that $W$ commutes with $G$ and

$$1 + i\omega G = W G = G W \qquad (10.4.18)$$

Taking the $mn$th matrix elements of this equation and summing over $n$ we obtain

$$1 + i\omega \sum_n G_{mn} = 0 \qquad (10.4.19)$$

since, from equation (10.4.12),

$$\sum_n W_{mn} = 0 \tag{10.4.20}$$

Finally, we note that the detailed balance relation $g_{mn} = g_{nm}$ means that the conductance matrix $g = CW$ is symmetric. It follows that $S = C^{-1/2}gC^{-1/2} = C^{1/2}WC^{-1/2}$ is symmetric. Hence

$$CG = -\frac{C}{i\omega}\left(1 + \frac{W}{i\omega} - \frac{W^2}{\omega^2} + \cdots\right)$$

$$= -\frac{C^{1/2}}{i\omega}\left(1 + \frac{S}{i\omega} - \frac{S^2}{\omega^2} + \cdots\right)C^{1/2} \tag{10.4.21}$$

is also symmetric, i.e.,

$$C_m G_{mn} = C_n G_{nm} \tag{10.4.22}$$

To carry out the algebra we substitute for $WG$ in equation (10.4.17) from equation (10.4.18) and eliminate the diagonal elements of $G$ with the aid of equation (10.4.19). Thus we obtain

$$\sigma(\omega) = -\frac{i\omega}{\Omega}\left[\sum_m x_m C_m \sum_{n \neq m}(1 + i\omega G)_{mn}x_n - \sum_m x_m C_m \sum_{n \neq m}(1 + i\omega G)_{mn}x_m\right]$$

$$= \frac{\omega^2}{\Omega}\sum_{m,n} x_m C_m G_{mn}(x_n - x_m)$$

$$= -\frac{\omega^2}{2\Omega}\sum_{m,n} C_m G_{mn}(x_m - x_n)^2 \tag{10.4.23}$$

In the last line of equation (10.4.23) we have symmetrized the summand and made use of equation (10.4.22).

Equation (10.4.23) provides a formal expression for $\sigma(\omega)$. To put it in its most easily interpretable form we note that the electron density is

$$n = \Omega^{-1}\sum_m f_m^0 \tag{10.4.24}$$

where $f_m^0$ is given by equation (10.4.3). Hence

$$\frac{dn}{d\varepsilon_F} = -\Omega^{-1}\sum_m \frac{df_m^0}{d\varepsilon_m} \tag{10.4.25}$$

which, with equation (10.4.6a), allows us to express equation (10.4.23) in the form of an Einstein relation:

$$\sigma(\omega) = e^2 \frac{dn}{d\varepsilon_F} D(\omega) \tag{10.4.26}$$

with

$$D(\omega) = -\frac{1}{2}\,\omega^2\,\frac{\displaystyle\sum_m \left|\frac{df_m^0}{d\varepsilon_m}\right| \sum_n G_{mn}(x_m - x_n)^2}{\displaystyle\sum_m \left|\frac{df_m^0}{d\varepsilon_m}\right|} \tag{10.4.27}$$

being a frequency-dependent diffusivity.

### 10.4.5. The Associated Random Walk

It follows from equation (10.4.15) that $G_{mn}$ is the causal Fourier transform of $P_{mn}(t)$, where

$$\frac{d}{dt}P_{mn} = -\sum_l [W_{ln}P_{ml} - W_{nl}P_{mn}], \quad t > 0, \quad P_{mn} = \delta_{mn}, \quad t = 0 \tag{10.4.28}$$

We see that the diagonal elements of $W$ are not involved in this equation. Moreover $-W_{ln} = g_{ln}/C_l$ is positive and may be regarded as a hop rate. Hence $P_{mn}(t)$ may be interpreted as the probability that a single particle will be found on $n$ at time $t$, given that it was on $m$ at time $0$ and that it hops between sites with the rate $-W_{ln}$. Consequently, the sum over $n$ in equation (10.4.27) is the Fourier transform of the mean-square distance moved by the particle in time $t$. The sum over $m$ averages this quantity over sites with a weighting factor $df_m^0/d\varepsilon_m$.

The connection of the hopping-conductivity problem with random walks is contained in (10.4.26) and (10.4.27). For nondegenerate statistics $df_m^0/d\varepsilon_m = -\beta f_m^0$ and we readily find that $-W_{ln} = R_{ln}^0$. Hence the particle is hopping with the thermal-equilibrium hop rate and the site averaging in equation (10.4.27) is done with a Boltzmann weighting factor. In that case the relevant random-walk problem is the actual random walk of any one electron in the hopping system because nondegenerate statistics imply that the electron density is so low that the electrons never get in each other's way. For degenerate statistics $df_m^0/d\varepsilon_m \neq -\beta f_m^0$ and $W_{mn} \neq R_{mn}^0$. The relevant random walk is therefore an artificial one and the site averaging is also done in an artificial way, which takes proper account of the degenerate statistics. A random-walk picture, albeit an artificial one, makes an appearance even in this case because of the particular structure of the linearized rate equations.[12]

## 10.5. Hopping Conductivity: ac Conductivity

### 10.5.1. The Pair Approximation

Assuming a time factor $\exp(-i\omega t)$, we see from equation (10.4.5) that $V_m \to Ex_m$ when $\omega \to \infty$ so that both equations (10.4.9) and (10.4.10) yield the same asymptotic limit

$$\sigma(\infty) = \frac{1}{2\Omega} \sum_{m,n} g_{mn} x_{mn}^2 \qquad (10.5.1)$$

The existence of a nonzero high-frequency limit to $\sigma(\omega)$ is due to our tacit assumption that the hops take place instantaneously, which introduces $\delta$-functions into the current response. It is easy to expand $\sigma(\omega)$ in powers of $\omega^{-1}$ but the convergence is too slow for the series to be of any value. A much more successful expansion when $\omega \neq 0$ is in powers of the site density $n_s$. Since double sums are involved in the general equations for $\sigma(\omega)$, the leading term in the expansion is proportional to $n_s^2$ and it is obtained by summing the contributions from all pairs of sites calculated by treating each pair as being isolated from all other sites. This is the "pair approximation" originally introduced by Pollak and Geballe.[40]

Let us consider an arbitrary pair of sites $m$ and $n$. Then Kirchhoff's equations (10.4.5) reduce to

$$-i\omega C_m (V_m + Ex_m) = g_{mn} V_{nm} \qquad (10.5.2a)$$

$$-i\omega C_n (V_n + Ex_n) = -g_{mn} V_{nm} \qquad (10.5.2b)$$

so that

$$V_{mn} = -\frac{Ex_{mn}}{g_{mn} Z_{mn}^P} \qquad (10.5.3)$$

where

$$Z_{mn}^P = \frac{1}{g_{mn}} - \frac{1}{i\omega C_m} - \frac{1}{i\omega C_n} \qquad (10.5.4)$$

When this result is substituted into equation (10.4.9) we obtain the pair approximation to $\sigma(\omega)$:

$$\sigma(\omega) = \frac{1}{2\Omega} \sum_{m,n} x_{mn}^2 / Z_{mn}^P \qquad (10.5.5)$$

which remains to be system-averaged. Before doing that we note that equation (10.5.5) yields the exact result (10.5.1) when $\omega \to \infty$ and that it gives $\sigma(0) = 0$ because $Z_{mn}^{p} \to \infty$ when $\omega \to 0$. Thus the pair approximation becomes progressively worse as $\omega$ is reduced. Nevertheless, it is frequently used to analyze data for audio and video frequencies provided $\sigma_1(\omega) \gg \sigma(0)$.

### 10.5.2. The r-Hopping Model

A special case that is often considered is what we shall call the r-hopping model. It is defined by

$$g_{mn} = g_a \exp(-2\alpha r_{mn})$$
$$C_m = C \tag{10.5.6}$$

where $\alpha$, $g_a$, and $C$ are constants. To arrive at a model of this type we consider nondegenerate electrons in a band of states whose width is small compared to $k_B T$. Then $f_m^0 \to f$, a constant much less than unity so that $C = e^2 \beta f$. Moreover, if we suppose that the hop rate between two sites distance $r_{mn}$ apart is $R_0 \exp(-2\alpha r_{mn})$, then $g_a = R_0 C$, i.e., $R_0 = g_a C^{-1}$ is the natural unit of frequency in the problem. For sites randomly distributed with density $n_s$ the system average of equation (10.5.5) yields

$$\sigma(\omega) = \frac{2\pi}{3} n_s^2 g_a(-i\omega\tau_0) \int_0^\infty \frac{r^4 dr}{1 - i\omega\tau} \tag{10.5.7}$$

where

$$\tau = \tau_0 e^{2\alpha r} \tag{10.5.8a}$$

with

$$\tau_0 = (2R_0)^{-1} \tag{10.5.8b}$$

When $\omega \to \infty$ we obtain

$$\sigma(\infty) = (g_a \alpha) \frac{\pi}{2} (n_s \alpha^{-3})^2 \tag{10.5.9}$$

At lower frequencies we use this result to normalize $\sigma(\omega)$ so that

$$\frac{\sigma(\omega)}{\sigma(\infty)} = \frac{4}{3} \alpha^5 (-i\omega\tau_0) \int_0^\infty r^4 dr \left( \frac{1}{1 + \omega^2 \tau^2} + \frac{i\omega\tau}{1 + \omega^2 \tau^2} \right) \tag{10.5.10}$$

The real and imaginary parts of the integral in equation (10.5.10) are easily approximated at low frequencies.[40–42] When $\alpha n_s^{-1/3} \gg 1$ and $\omega\tau_0 \ll 1$

we find that $\omega\tau/(1 + \omega^2\tau^2)$ has a sharp peak at

$$r_\omega = \frac{1}{2\alpha} \log(\omega\tau_0)^{-1} \qquad (10.5.11)$$

which is the value of $r$ at which $\omega\tau = 1$. We may readily integrate $\omega\tau/(1 + \omega^2\tau^2)$ over all $r$ to obtain the value $\pi/4\alpha$. Thus $\omega\tau/(1 + \omega^2\tau^2)$ may be approximated by $\pi/4\alpha\delta(r - r_\omega)$. Similarly $(1 + \omega^2\tau^2)^{-1} \simeq \theta(r_\omega - r)$. When these results are substituted into equation (10.5.10) we obtain immediately

$$\frac{\sigma_1(\omega)}{\sigma(\infty)} = \frac{\pi}{3}(\omega\tau_0)(\alpha r_\omega)^4 \qquad (10.5.12a)$$

$$\frac{\sigma_2(\omega)}{\sigma(\infty)} = -\frac{4}{15}(\omega\tau_0)(\alpha r_\omega)^5 \qquad (10.5.12b)$$

$$\frac{\sigma_2(\omega)}{\sigma_1(\omega)} = -\frac{2}{5\pi}\log(\omega\tau_0) = -0.293\log_{10}(\omega\tau_0) \qquad (10.5.12c)$$

It is usually supposed that $\tau_0 \sim 10^{-13}$ s.[1] Then $r_\omega$ decreases slowly with increasing $\omega$ for frequencies in the typical experimental range of 1 kHz to 1 MHz. Consequently both $\sigma_1(\omega)$ and $\sigma_2(\omega)$ show a slightly sublinear frequency dependence, which is often approximated by a $\omega^s$ with

$$s = \frac{d \log \sigma_1(\omega)}{d \log \omega} = 1 - \frac{4}{\log \omega\tau_0} \qquad (10.5.13)$$

being the slope of the real part of $\sigma_1(\omega)$ on a log–log plot. The frequency dependence of $s$ and the ratio $\sigma_2(\omega)/\sigma_1(\omega)$ is very weak and is usually not discernible in experimental data.

### 10.5.3. The AHL Model

In the $r$-hopping model we consider a very narrow band. It is of interest to go to the opposite extreme of a very wide band. The simplest model for this case is one introduced several years ago by Ambegaokar et al.[43] (AHL model) to describe dc conductivity. It is defined by

$$g_{mn} = g_a \exp(-s_{mn}) \qquad (10.5.14)$$

where the conductivity exponent $s_{mn}$ is given by

$$s_{mn} = 2\alpha r_{mn} + \tfrac{1}{2}\beta(|\varepsilon_m| + |\varepsilon_n| + |\varepsilon_m - \varepsilon_n|) \qquad (10.5.15)$$

the site energies being measured from the Fermi level plus $k_B T \log 2$. The energy-dependent exponential factor in equation (10.5.14) arises from asymptotic approximations to Fermi and Bose factors in equation (10.4.6b) for $g_{mn}$. To carry out the configuration average we suppose that the sites are uniformly distributed in space with density $n_s$ and uniformly distributed in energy over a wide bandwidth $W$ with density of states $\rho_F = n_s/W$. Then the system average of equation (10.5.5) becomes

$$\sigma(\omega) = \frac{2\pi}{3} \rho_F^2 g_a \int d\varepsilon_1 \int d\varepsilon_2 \int_0^\infty r^4 \frac{(-i\omega\tau)Qe^{-2\alpha r}}{1 - i\omega\tau} \, dr \quad (10.5.16)$$

where

$$Q = \exp[-\tfrac{1}{2}\beta(|\varepsilon_1| + |\varepsilon_2| + |\varepsilon_1 - \varepsilon_2|)] \quad (10.5.17)$$

and

$$\tau^{-1} = \left(\frac{1}{C_1} + \frac{1}{C_2}\right) g_a Q e^{-2\alpha r} \quad (10.5.18)$$

We suppose that $W \gg k_B T$. Then the major contribution to the energy integrals in equation (10.5.16) comes from the neighborhood of $\varepsilon_1 = \varepsilon_2 = 0$ and the limits of integration may be extended to infinity.

When $\omega \to \infty$ we have

$$\sigma(\infty) = (g_a \alpha) \frac{\pi}{2} (n_s \alpha^{-3})^2 \bar{Q} \quad (10.5.19)$$

where

$$\bar{Q} = W^{-2} \int_{-\infty}^\infty d\varepsilon_1 \int_{-\infty}^\infty d\varepsilon_2 Q = 6(k_B T/W)^2 \quad (10.5.20)$$

Equation (10.5.19) differs from our result (10.5.9) for $r$-hopping only by the factor $\bar{Q}$. Moreover, the frequency dependence of $\sigma(\omega)$ obviously still has the same general character as before and we may roughly evaluate expression (10.5.14) by giving $\tau$ its value when $\varepsilon_1 = \varepsilon_2 = 0$. Thus we write $\tau$ as in equation (10.5.8) with $R_0$ equal to the transition rate in thermal equilibrium for $r = 0$ and $\varepsilon_1 = \varepsilon_2 = 0$. Then our previous evaluation of $\sigma(\omega)/\sigma(\infty)$ goes through unaltered except for the reinterpretation of $\sigma(\infty)$ and $\tau_0$. At low frequencies the final result is

$$\sigma_1(\omega) = (g_a \alpha)\pi^2(\rho_F k_B T \alpha^{-3})^2 \omega \tau_0 (\alpha r_\omega)^4 \quad (10.5.21)$$

with $r_\omega$ given by equation (10.5.11) and $\sigma_2(\omega)/\sigma_1(\omega)$ by equation (10.5.12c). This result differs from equation (10.5.12a) only in that $n_s^2$ is replaced by

$6(\rho_F k_B T)^2$. It was originally derived by Austin and Mott with a slightly different numerical factor.[44] The simple approximation used to evaluate the integral in equation (5.16) has been criticized recently by Pramanik and Islam.[45] The integral has therefore been evaluated numerically. The approximate results quoted here are found to be substantially correct.[46]

### 10.5.4. General Debye Models

The basic structure of the pair approximation is of the Debye type[47]:

$$\sigma(\omega) = -i\omega \left\langle \frac{\chi_0}{1 - i\omega\tau} \right\rangle \qquad (10.5.22)$$

where the angle brackets signify an average and the choice of $\chi_0$, $\tau$, and the average defines the model. In the energy-dependent case we took $\tau = \tau_0 \exp(2\alpha r)$, $\chi_0 \sim r^2$, and the probability distribution in $r$ was proportional to $4\pi r^2$. In the AHL model we again, in the end, put $\tau = \tau_0 \exp(2\alpha r)$ but took $\chi_0 \sim Qr^2$ and the probability distribution of $r$, $\varepsilon_1$, and $\varepsilon_2$ was proportional to $4\pi r^2 W^{-2}$. Another variant, which has attracted considerable interest in recent years, is the "correlated barrier hopping model" originally introduced by Pike[48] but discussed extensively by Elliott[47,49,50] for chalcogenide glasses. In these materials the absence of spin resonance signals suggests that one is concerned with the hopping of *pairs* of electrons between $D^+$ and $D^-$ centers. (An account of the defect centers found in amorphous semiconductors is given in Chapter 15 and in the book by Mott and Davis.[1]) Elliott supposes that the hops are thermally activated over a Coulomb barrier of the form

$$W = W_m - 2e^2/\varepsilon\pi r \qquad (10.5.23)$$

where $\varepsilon$ is the electric permittivity of the material and $r$ is the spatial separation of the two sites. Then we have

$$\tau = \tau_1 \exp(-\beta/r) \qquad (10.5.24)$$

where $\tau_1 = \tau_0 \exp(W_m/k_B T)$ and $\beta = 2e^2/\varepsilon\pi k_B T$. The real part of the conductivity has the form $\omega(r_\omega)^6$ in this case, the additional two powers of $r_\omega$ arising in the evaluation of the Debye integral because the exponent in $\tau$ involves $r^{-1}$. The optimum distance for ac absorption $r_\omega$ is determined as before by the equation $\omega\tau = 1$, which gives for Elliott's model

$$r_\omega = \frac{\beta}{\log(\omega\tau_1)} = \frac{\beta}{W_m/k_B T + \log \omega\tau_0} \qquad (10.5.25)$$

Consequently $\sigma_1(\omega) \sim \omega^s$ with

$$s = 1 - \frac{6}{W_m/k_B T + \log \omega \tau_0} \tag{10.5.26}$$

where, for most chalcogenides, below about 200 K the frequency-dependent term may be neglected in comparison to $W_m/k_B T$. Consequently $s = 1 - 6k_B T/W_m$, which increases with decreasing $T$ as is often observed.[47,49,50] Recently Shimakawa[51] has pointed out that at high temperatures it is possible to identify additional components in the ac conductivity that he associates with single electron hops between $D^0$ and $D^+$ or $D^-$ centers. The $D^0$ centers are thermally activated from $D^+$ and $D^-$ centers as the temperature rises.[1,17]

### 10.5.5. Conclusion

The strong frequency dependence associated with ac hopping conductivity is due to the highly disordered state of the materials under consideration. For a random walk on a regular Bravais lattice the mean-square distance moved in time $t$ is easily seen to be strictly proportional to $t$ for all $t$. The Fourier transform of this ramp function is proportional to $\omega^{-2}$ and consequently, from equation (10.4.27), we see that $D(\omega)$ is independent of $\omega$. The fact that $\sigma_1(\omega)$, *increases* with $\omega$ is also indicative of hopping in a disordered system. This behavior is to be contrasted with the decreases of $\sigma_1(\omega)$ with increasing $\omega$, which we see from (10.2.40) is characteristic of Boltzmann transport involving extended states.

## 10.6. Hopping Transport: dc Conductivity and Thermopower

### 10.6.1. Introduction: The $r$-Hopping Model

In the early 1970s several groups realized that the calculation of dc hopping conductivity involves ideas taken from percolation theory.[43,52,53] To illustrate the sort of considerations involved we discuss first the $r$-hopping model for which the ac conductivity has already been calculated in Section 10.5.2 by using the pair approximation. We have to go outside this approximation to calculate the dc limit. Taking *finite* clusters containing two or more sites can never yield a nonzero value of $\sigma(0)$ because the mean-squared distance moved by a particle hopping in such a cluster must remain finite for large times. To calculate $\sigma(0)$ it is therefore necessary to consider an infinite system.

We easily obtain a formal expression for $\sigma(0)$ by taking the dc limit of equation (10.4.10):

$$\sigma(0) = \frac{1}{2\Omega E^2} \sum_{m,n} g_{mn} V^2_{mn} \tag{10.6.1}$$

This equation has been used to evaluate $\sigma(0)$ for large systems modeled on a computer.[14] To proceed with an analytic derivation of $\sigma(0)$ we take the system average of (10.6.1) to obtain

$$\sigma(0) = \frac{2\pi}{E^2} n_s^2 \int_0^\infty r^2 g(r) \Delta V^2(r) dr \tag{10.6.2}$$

where $g(r) = g_a \exp(-2\alpha r)$ and $\Delta V^2(r)$ is the mean-square voltage drop between two sites separated by distance $r$. The quantity $P(r) = g(r)\Delta V^2(r)$ is the mean power dissipated in the conductance $g(r)$ joining the sites. When $r$ is large, $\Delta V^2(r) = E^2 r^2/3$ because there is an electric field of magnitude $E$ oriented in the $x$-direction. Hence $P(r)$ grows exponentially as $r$ decreases. Eventually, however, an optimum separation $r_p$ for power absorption will be reached and $P(r)$ falls off again when $r$ is reduced still further. This is because $g(r)$ behaves essentially like a short circuit when $r \to 0$. Thus $\Delta V^2(r) \to 0$ but a nonzero current flows through $g(r)$ that is determined by the currents in the neighboring conductances. For $r < r_p$ it is therefore convenient to rewrite $P(r)$ in the form $I^2(r)/g(r)$, where $I^2(r)$ is the mean-square current through $g(r)$. To make $P(r)$ continuous at $r_p$ we must have $I^2(r_p) = g^2(r_p)E^2 r_p^2/3$. For simplicity we neglect the variation of $I^2(r)$ when $r < r_p$. Thus we finally arrive at the ansatz

$$g(r)\Delta V^2(r) = g(r)E^2 r^2/3, \qquad r > r_p$$

$$= \frac{g^2(r_p)}{g(r)} E^2 r_p^2/3, \qquad r < r_p \tag{10.6.3}$$

When this is substituted into equation (10.6.2) the integral may be evaluated immediately. When $\alpha r_p$ is large we obtain the simple result

$$\sigma(0) = \frac{2\pi}{3} n_s^2 r_p^4 (g_a/\alpha) e^{-2\alpha r_p} \tag{10.6.4}$$

The behavior of $\sigma(0)$ is dominated by the final exponential factor in equation (10.6.4). For dimensional reasons $r_p$ is proportional to $n_s^{-1/3}$. Percolation theory is introduced to determine the constant of proportionality. The argument is an asymptotic one concerning the behavior of the system as $\alpha \to \infty$ with $n_s$ fixed. There is then an enormous spread in conductance values. As $\alpha$ increases, all the conductances with $r$ greater than a chosen value, $r_0$ say, can be omitted from the network without affecting $\sigma(0)$ because they become exponentially small compared with those that are retained. This argument is valid provided that removing all the conductances with $r > r_0$ does not cause the infinite network to break into isolated islands so that dc conductivity becomes impossible. The critical value of $r_0$ for which the network just falls apart is called the critical percolation radius. It is identical to the value of $r$ for which $P(r)$ has its maximum value when $\alpha r_p$ is large.

Henceforth we interpret $r_p$ as the critical percolation radius. Its value may be determined on a computer by shrinking identical spheres centered on random sites until no continuous paths of overlapping spheres remain.[53,54] Thus we find that the percolation criterion determining $r_p$ takes the following form: the mean number of neighbors with $r < r_p$ is equal to $N_p$, where $N_p = 2.7$ in 3D, 4.5 in 2D, and 2.1 in 4D (which we shall need later in energy-dependent problems). Our current problem is three-dimensional and we therefore have

$$\frac{4\pi}{3} r_p^3 n_s = N_p = 2.7 \tag{10.6.5}$$

which determines $r_p$ completely. In Figure 10.5 the solid curve was calculated from equation (10.6.2) by using ansatz (10.6.3) with $r_p$ given by equation (10.6.5); the points were obtained by solving Kirchhoff's equations for a system containing 2500 sites and substituting the voltage drops calculated in this way into the sum (10.6.1).[56] Agreement between the

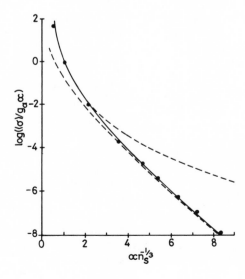

Figure 10.5. dc-Conductivity as a function of density for the $r$-hopping model. The solid curve refers to the full analytic equations (10.6.2), (10.6.3), and (10.6.5), the upper dashed curve is calculated from equation (10.6.2) with $r_p = 0$, the lower dashed curve is calculated from equation (10.6.4) and (10.6.5), and the points are computed directly for a system of 2500 sites.[55]

analytic and computed results is extremely good and should serve to remove any lingering doubts about the validity of the above argument. The lower dashed curve in Figure 10.5 is the low-density approximation (10.6.4); the upper dashed curve is obtained by setting $r_p = 0$ in equation (10.6.3).

## 10.6.2. The AHL Model

We may extend the discussion in Section 10.6.1 to the case when the conductance between two sites involves the site energies as well as the intersite distance. To be definite we consider the AHL model discussed in Section 10.5.3. Then the conductance exponent $s_{mn}$ given in equation (10.5.15) replaces $2\alpha r$ and the configuration average of equation (10.6.1) reduces to

$$\sigma(0) = \frac{2\pi\rho_F^2}{E^2} \int d\varepsilon_1 \int d\varepsilon_2 \int r_{12}^2 g_{12} \Delta V_{12}^2 dr_{12} \qquad (10.6.6)$$

where $\Delta V_{12}^2$ is the mean-square voltage drop across the conductance $g_{12}$. To evaluate expression (10.6.6) we generalize ansatz (10.6.3):

$$g_{12}\Delta V_{12}^2 = g_a \exp(-s_{12})E^2 r^2/3, \qquad s_{12} > s_p$$

$$= g_a \exp(s_{12} - 2s_p)E^2 r_p^2/3, \qquad s_{12} < s_p \qquad (10.6.7)$$

where the critical percolation exponent $s_p$ is determined by the criterion that the mean number of conductances emanating from site 1 with $s_{12} < s_p$ is $N_p = 4.5$ (the value for 4D hyperspheres). Once $s_p$ is fixed, $r_p$ is determined as a function of $\varepsilon_1$ and $\varepsilon_2$ by setting $s_{12} = s_p$ in equation (10.5.15).

A problem arises in identifying the mean number of neighbors emanating from site 1 with $s_{12} < s_p$ because, as we see from equation (10.5.15), there can be no neighbors at all that satisfy the percolation criterion when $\varepsilon_1 > k_B T s_p$. These sites must therefore play a negligible role in the dc conductivity and we solve the problem by excluding them altogether in calculating $s_p$. Then the mean number of sites with $\varepsilon_1 < s_p k_B T$ per unit volume is $2\rho_F s_p k_B T$ while the mean number of conductances per unit volume with $s_{12} < s_p$ is

$$B = \frac{1}{2}\rho_F^2 \int d\varepsilon_1 \int d\varepsilon_2 \int 4\pi r^2 dr \qquad (10.6.8)$$

where the factor of $\frac{1}{2}$ prevents double counting. Since each conductance is connected to two sites, the percolation criterion is

$$\frac{2B}{2\rho_F s_p k_B T} = N_p = 4.5 \tag{10.6.9}$$

Thus we find that[8]

$$s_p = \left(\frac{40 N_p \alpha^3 \beta}{\pi \rho_F}\right)^{1/4} \tag{10.6.10}$$

which leads to Mott's famous $T^{1/4}$ law for hopping conductivity by degenerate electrons.[1] It is usual to speak of "variable range" hopping in this case because of the interplay of distance and energy variations that determines the critical percolation exponent $s_p$. When the ansatz (10.6.7) is used to evaluate $\sigma(0)$ in equation (10.6.6) the final result is[8]

$$\sigma(0) = 2\alpha g_a \frac{4\pi}{3} (\rho_F k_B T)^2 \frac{1}{s} \left(\frac{s_p}{2\alpha}\right)^6 e^{-s_p} \tag{10.6.11}$$

when $s_p$ is large. In Figure 10.6 the dashed curve is a plot of $\sigma(0)$ calculated from equation (10.6.11). The dots are calculated by solving Kirchhoff's equations for 2197 sites.[56] They are believed to be overestimates because

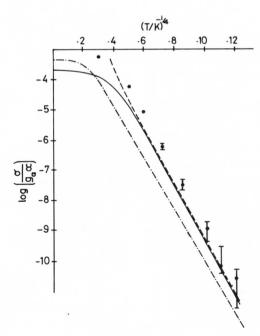

Figure 10.6. dc-Conductivity as a function of $T^{-0.25}$ for an AHL model with $\alpha n_s^{-1/3} = 3.4$ and $W = 10$ meV. The dashed curve is calculated from equation (10.6.11), the dots are computed using a full numerical solution of Kirchhoff's equations for 2197 sites,[56] the solid curve is calculated from the extended pair-approximation equations,[57] and the dash-dot curve is calculated from the barrier hopping model.[37]

only twelve neighbors to each site were taken into account in the calculations.[55,56] The significance of the solid and dash-dot curves is explained in Section 10.7.

### 10.6.3. More General Models

The general idea behind the application of percolation theory to the calculation of $\sigma(0)$ is easily extended to more general cases. We may vary the density of states so as to deal with tail states and two-dimensional impurity bands may be handled as well.[12,14] The idea that percolation theory determines the exponent in $\sigma(0)$ is now over a decade old. The notion that the simple ansatz that we have used for the mean-square voltage drop between two sites could also be successful in predicting the prefactor is taking longer to gain acceptance. Suffice it to say that it has worked well in every case for which numerical results are available.

### 10.6.4. Thermopower

An entirely satisfactory theory has yet to be developed because of the intimate involvement of the phonons in the hopping process that should be taken into account.[58] There has therefore been some controversy in the literature.[59,60] However, it is usually assumed that the mean energy $\frac{1}{2}(\varepsilon_m + \varepsilon_n)$ is transported when an electron hops from $m$ to $n$.[61,62] Then, on the basis of the argument given in Section 10.3.1, we find that the thermopower for a hopping system is given by

$$S = -\frac{1}{eT} \frac{\sum_{m,n} g_{mn} V_{mn} x_{nm} \left[\frac{1}{2}(\varepsilon_m + \varepsilon_n) - \varepsilon_F\right]}{\sum_{m,n} g_{mn} V_{mn} x_{nm}} \qquad (10.6.12)$$

where $V_{mn}$ is the voltage drop between sites $m$ and $n$ produced by an applied electric field. For nondegenerate statistics a result of the standard type (10.2.24) is obtained. At low temperatures the statistics become degenerate. For extended-state conduction in metals the result is that $S$ is proportional to $T$ as in equation (10.2.23) because the mean energy transported relative to the Fermi level is proportional to $(k_B T)^2$. In the hopping case the significant range is $s_p k_B T$ leading to a $T^{1/2}$ dependence of $S$[61,62] because $s_p$ is proportional to $T^{-1/4}$. Moreover, we see from equation (10.4.3) that the peak of $-df_m^0/d\varepsilon_m$ occurs at $\varepsilon_F + k_B T \log 2$ when the hopping sites can only be singly occupied. This leads to an additional constant term $-k_B \log 2/e$ when $T \to 0$.[1,63] When the Hubbard $U$ is finite (rather than infinite as we have assumed), $S$ again tends to a constant value when $T \rightarrow 0$ but the value of the constant depends on the details of the model.[58,63]

## 10.7. Hopping Conductivity: Unified Theory

In Sections 10.6 and 10.7 we have approached the calculation of ac and dc hopping conductivity along entirely different routes. There is a clear need for a unified treatment. This problem has been treated recently by Movaghar and co-workers[35–37] through the random-walk formalism and by Butcher and Summerfield[64,57] via the Miller and Abrahams equivalent circuit. We outline the latter approach because it is very simple. The pair model equations (10.5.2) are modified by the addition of extra terms $- Y(\varepsilon_m)(V_m + Ex_m)$ and $- Y(\varepsilon_n)(V_n + Ex_n)$ on the right-hand sides of equations (10.5.2a) and (10.5.2b), respectively, where $Y(\varepsilon_m)$ and $Y(\varepsilon_n)$ are average admittances, which take into account the effect of the rest of the network on the pair $(m, n)$. They are determined from a modified mean-field equation

$$Y(\varepsilon_m) = B^{-1} \int d\varepsilon_n \rho(\varepsilon_n) \int 4\pi r_{mn}^2 \left[ \frac{1}{g_{mn}} + \frac{1}{Y(\varepsilon_n) - i\omega C_n} \right]^{-1} dr_{mn}$$

(10.7.1)

Once this equation is solved, the pair equations (10.5.2), modified as indicated above, may be solved immediately to yield $V_{mn}$ for insertion into equation (10.4.9) prior to system averaging. The correction factor $B$ is chosen so that the theory yields the correct percolation threshold in percolation problems involving cutoffs in the conductances. We find that $B = N_p$ or two- and three-dimensional $R$-hopping, while $B = 4.4$ for the AHL model.[57]

Details of the calculations are given elsewhere.[57] By way of illustration we show in Figures 10.7 and 10.8 the real and imaginary parts of $\sigma(\omega)$ for a 3D $r$-hopping model. The model is studied numerically by McInnes et al.[65] It should be noted that it differs from the $r$-hopping model defined by (10.5.6) in that $g_{mn}$ is multiplied by an additional factor $(\alpha r_{mn})^{3/2}$. The solid curves in Figures 10.6 and 10.7 are derived as indicated here. The points are obtained by solving Kirchhoff's equations for 1600 sites[65] and the dashed curve is taken from the unified random-walk theory.[37] In Figure 10.6 we make a corresponding comparison for the dc conductivity calculated for the AHL model. The solid curve is the unified equivalent circuit-theory result,[57] the dots are computer points,[56] the dashed curve is obtained using percolation theory,[8] and the dot-dash curve is obtained from the barrier hopping model of Movaghar et al.[37]

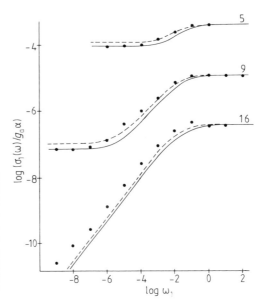

Figure 10.7. Real part $\sigma_1(\omega)$ of the ac conductivity for a three-dimensional $R$-hopping system as a function of $\omega_1 = \omega/R_0$. The solid curve is calculated using the extended-pair approximation,[57,64] the dashed curve is obtained from the unified random-walk theory,[37] and the points are obtained using a full numerical solution of Kirchhoff's equations for 1600 sites.[65]

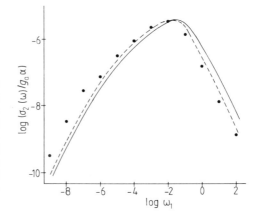

Figure 10.8. Imaginary part $\sigma_2(\omega)$ of the ac conductivity for a three-dimensional $R$-hopping system as a function of $\omega_1 = \omega/R_0$. The solid curve is calculated using the extended-pair approximation,[57,67] the dashed curve is obtained from the unified random-walk theory,[37] and the points are obtained using a full numerical solution of Kirchhoff's equations for 1600 sites.[65]

The availability of a unified theory is very useful for the interpretation of experimental data. It is all too easy to fit either ac data *or* dc data, but not both. A preliminary study[66] indicates that the data of Pollak and Geballe[40] for impurity bands in c-Si is in fair agreement with the theory. The data of Kahlert[67] for impurity bands in c-GaAs is in excellent agreement with the theory as far as dc conductivity is concerned, but the theory fails completely to account for the observed rapid rise of ac conductivity. In these cases the only unknown is the density-of-states curve. Excellent experimental data are also available[68,69] for a-Ge which exhibits $T^{1/4}$ behavior when $\omega = 0$. Very few of the relevant parameters are known in this case. The dc behavior is easily fitted using the AHL model, but the characteristic hopping frequency has an unacceptably high value in the order of $10^{21}$ Hz.[66] Again the observed rapid rise of the ac conductivity is not predicted by the theory. In this connection it should be noted that the excellent agreement between theory and experiment for a-Ge shown in Figure 10.8 of the paper by Movaghar et al.[37] is incorrect; the frequency axis has been scaled inappropriately to achieve it.

## 10.8. Anomalous Carrier-Pulse Propagation and Trap-Controlled Transport

### 10.8.1. Introduction

Let us consider a plate-like specimen with the spatial coordinate $x$ measured across the plate, as shown in Figure 10.9. Suppose a pulse of electrons with number density $n(x, t)$ is created at one end by uniform illumination over the cross section of the plate. This may be achieved by shining light through a transparent electrode (see Figure 10.9). In this section we address the question of how the carrier pulse will move and distort when an electric field $E(x, t)$ is applied in the negative $x$-direction.

To study carrier-pulse propagation experimentally it is usual to monitor the total current density in the positive $x$-direction:

$$I(t) = J(x, t) - \varepsilon \frac{\partial E(x, t)}{\partial t} \qquad (10.8.1)$$

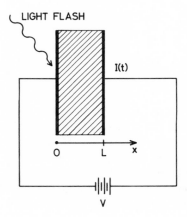

LIGHT FLASH

I(t)

O     L     x

V

Figure 10.9. Schematic diagram of the experimental arrangement for observing carrier-pulse propagation. The potential difference $V$ across the specimen (cross-hatched) is held constant and the total current $I(t)$ is monitored.

where $J(x, t)$ is the electric current density due to the electrons and $\varepsilon$ is the electric permittivity. The minus sign appears in front of the displacement current because $E(x, t)$ has been taken in the negative $x$-direction to make the electron pulse move in the positive $x$-direction. Maxwell's equations show that the total current density is equal to the curl of the magnetic field and is therefore nondivergent. For the one-dimensional situation under discussion here this means that $I(t)$ is independent of $x$ (as our notation implies) and may be monitored outside the specimen. The behavior of $I(t)$ depends on the boundary conditions produced by the circuit external to the specimen. We suppose, for simplicity, that the potential difference between the ends of the specimen is held constant. Then we may average equation (10.8.1) over the specimen length $L$ to express $I(t)$ as the spatial average of $J(x, t)$:

$$I(t) = \frac{1}{L} \int_0^L J(x, t)dx \tag{10.8.2}$$

where $I(t)$ is the quantity that we wish to calculate. To do so we use the charge conservation equation

$$\frac{\partial J(x, t)}{\partial x} - e\frac{\partial n(x, t)}{\partial t} = 0 \tag{10.8.3}$$

where $n(x, t)$ is the *total* electron density. We stress the word *total* here because we will subsequently be concerned with situations in which some electrons cease to contribute to $J(x, t)$ because they drop into traps. Nevertheless, the trapped electrons continue to contribute to $n(x, t)$. Theories of carrier-pulse propagation differ from one another in the relation that is supposed to exist between $J(x, t)$ and $n(x, t)$ in equation (10.8.3). We review the conventional theory in the next subsection. The pulse is a drifted Gaussian; $I(t)$ is constant while the pulse remains in the specimen and falls away after a transit time proportional to $L/E$ at which the pulse passes out of the far end. This sort of behavior is observed in most crystalline semiconductors. It is the basis of the method for measuring carrier-drift mobility originally pioneered by Shockley and Haynes[70] for crystalline semiconductors and developed by Spear and co-workers for amorphous semiconductors.[1]

In amorphous semiconductors $I(t)$ often behaves completely differently. Indeed on a linear–linear plot it frequently drops to zero in an apparently featureless way. Replotting on log–log paper reveals small-time behavior proportional to $t^{-1+\alpha}$, and large-time behavior proportional to $t^{-1-\alpha}$, where $\alpha$ is a constant between 0 and 1. One may identify a "transit time" $t$ by interpolating between these two regimes and it is found that

$t_T \sim (L/T)^{1/\alpha}$. Finally, $\log I(t)$ against $\log t$ plots for different values of $t_T$ (i.e., for different values of $E$ or $L$) can be superimposed by scaling.

What we have described above may be called archetypal anomalous carrier-pulse propagation. We show a good example in Figure 10.10. More complicated data can arise and there may be a switch to conventional carrier-pulse propagation as the temperature varies.[1,2] Nevertheless, archetypal behavior is observed often enough for us to seek a general explanation of it in the properties of amorphous semiconductors. A possible explanation was advanced by Scher and Montroll,[32] who suggested that the anomalies were to be expected if one was concerned with hopping carriers. We review the theory in Section 10.8.3. Qualitatively it makes the right predictions, but quantitatively it seems unlikely to be correct. More recently it has been emphasized by several authors[3-7] that trapping is a more likely explanation. It is a remarkable fact that the central equation in the theory is formally the same in both cases, which makes the two possibilities difficult to distinguish on a qualitative basis. Nevertheless, the bulk of the quantitative evidence would appear to favor trapping effects as being dominant. We review the trap-controlled case in Section 10.8.4 and the theory of archetypal anomalous carrier-pulse propagation in Section 10.8.5.

### 10.8.2. Conventional Carrier-Pulse Propagation

It is assumed in the conventional theory of carrier-pulse propagation that

$$J(x, t) = - en(x, t)\mu E + eD \frac{\partial n(x, t)}{\partial x} \qquad (10.8.4)$$

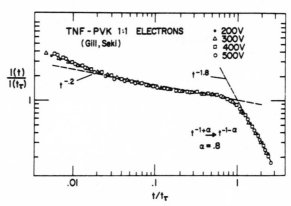

Figure 10.10. Superposed $\log I$–$\log t$ plots for 1:1 TNF:PVK with the applied voltages indicated.[32]

where $\mu$ and $D$ are respectively the electron mobility and diffusivity. These quantities are taken to be constant and are related by the Einstein relation (assuming nondegenerate statistics)

$$\mu = \frac{eD}{k_B T} \qquad (10.8.5)$$

We also neglect the effect of carrier space charge on $E$, which is identified with the constant applied potential difference divided by $L$.

When equation (10.8.4) is substituted into equation (10.8.3) we therefore obtain the familiar equation

$$\frac{\partial n}{\partial t} + v_D \frac{\partial n}{\partial x} = D \frac{\partial^2 n}{\partial x^2} \qquad (10.8.6)$$

where $v_D = \mu E$ is the electron-drift velocity. We impose the boundary condition

$$n(x,0) = n_0 \delta(x), \qquad t = 0 \qquad (10.8.7)$$

Then the appropriate solution of equation (10.8.6) may be obtained by elementary methods:

$$n(x,t) = \frac{n_0}{(4\pi Dt)^{1/2}} \exp\left[ -\frac{(x - v_D t)^2}{4Dt} \right] \qquad (10.8.8)$$

To determine $I(t)$ we substitute this result into equation (10.8.4) and then use expression (10.8.2). If the peak of the pulse is well away from the ends of the specimen, the diffusion term in equation (10.8.4) integrates to zero while the conduction-current term yields $-en_0 v_D$. The pulse approaches the far end of the specimen at a transit time $t_T = L/v_D = L/\mu E$. The spatial width of the pulse is then proportional to $t_T^{1/2}$ and this will be reflected in a trailing edge on $I(t)$ with time-width proportional to $(L/E)^{1/2}$.

### 10.8.3. Carrier-Pulse Propagation Involving Hopping

Let us suppose that the electrons in the pulse reach a local equilibrium at temperature $T$ and move by hopping. We have seen in Section 10.4 that the diffusivity $D(\omega)$ becomes strongly frequency-dependent in this case. Moreover, assuming nondegenerate statistics, the Einstein relation is preserved at frequency $\omega$ so that the frequency-dependent mobility $\mu(\omega)$ is given by (see Section 10.4.4)

$$\mu(\omega) = eD(\omega)/k_B T \qquad (10.8.9)$$

These observations suggest that we modify equation (10.8.4) by taking its Fourier transform and replacing constants $\mu$ and $D$ by the frequency-dependent quantities $\mu(\omega)$ and $D(\omega)$. Thus we obtain

$$J(\omega) = -eE\mu(\omega)n(\omega) + eD(\omega)\frac{\partial}{\partial x}n(\omega) \qquad (10.8.10)$$

where $J(\omega)$ and $n(\omega)$ are the Fourier transforms of $J(x,t)$ and $n(x,t)$, respectively. Equation (10.8.4) is regained by neglecting the frequency dependence of $\mu(\omega)$ and $D(\omega)$. Inverting the transform we have

$$J(x,t) = -\frac{e^2 E}{k_B T}\int_{-\infty}^{t} d(t-\tau)n(x,\tau)d\tau$$

$$+ e\frac{\partial}{\partial x}\int_{-\infty}^{t} d(t-\tau)n(x,\tau)d\tau \qquad (10.8.11)$$

where

$$d(t-\tau) = \frac{1}{2\pi}\int D(\omega)e^{-i\omega(t-\tau)}d\omega \qquad (10.8.12)$$

Thus we must make equation (10.8.4) nonlocal in time because of the strong frequency-dependence of $D(\omega)$. It is not difficult to verify the validity of equation (10.8.11) by studying in detail the behavior of the hopping Green's function in the long-wavelength limit.[71-77]

### 10.8.4. Carrier-Pulse Propagation Involving Trapping

We suppose that the current density $J(x,t)$ is carried by electrons in a conduction band according to the equation

$$J(x,t) = -en_c(x,t)\mu_c E + eD_c\frac{\partial n_c(x,t)}{\partial x} \qquad (10.8.13)$$

where $n_c(x,t)$ is the carrier density in the conduction band, $\mu_c$ is their mobility, and $D_c$ is their diffusivity, which we take to be constants related through the Einstein relation. Without traps, equation (10.8.13) reduces to the conventional relation (10.8.4). The traps make themselves felt because $n_c(x,t)$ is different from the total electron density $n(x,t)$, which appears in the particle conservation equation (10.8.3). We are here discussing trap-controlled band conduction for simplicity. The argument may be generalized to deal with trap-controlled hopping.[2,6]

To describe the traps we use the linear rate equations[4,6]

$$\frac{\partial p_j}{\partial t} = n_c C_j - p_j R_j \tag{10.8.14}$$

where $C_j$ is the capture rate and $R_j$ is the release rate of the $j$th type of trap. To solve these equations we suppose that the carrier pulse is injected into the conduction band at $x = 0$ when $t = 0$ at which time all the traps are empty. Then the Fourier transform of equation (10.8.14) yields

$$-i\omega p_j(\omega) = n_c(\omega)C_j - p_j(\omega)R_j \tag{10.8.15}$$

in an obvious notation. Hence

$$p_j(\omega) = \frac{n_c(\omega)C_j}{R_j - i\omega} \tag{10.8.16}$$

The total electron density is

$$n(x, t) = n_c(x, t) + \sum_j p_j \tag{10.8.17}$$

By taking the Fourier transform of equation (10.8.17) and using relation (10.8.16) we obtain immediately

$$n_c(\omega) = n(\omega)\left(1 + \sum_j \frac{C_j}{R_j - i\omega}\right)^{-1} \tag{10.8.18}$$

When this result is inserted into the Fourier transform of expression (10.8.13) we obtain an equation of identical structure to equation (10.8.10), but where now

$$\mu(\omega) = \mu_c\left(1 + \sum_j \frac{C_j}{R_j - i\omega}\right)^{-1} \tag{10.8.19}$$

and $D(\omega)$ is determined by the Einstein relation (10.8.9).

There is seen to be a formal identity between the equation determining the carrier-pulse shape for hopping and trap-controlled carriers.

### 10.8.5. Archetypal Anomalous Carrier-Pulse Propagation

For very long times the carrier-pulse propagation is controlled by the dc value $\mu(0)$ of $\mu(\omega)$. Thus the pulse eventually converts to the drifted Gaussian predicted by the conventional transport equations. Anomalous carrier-pulse propagation arises at times prior to that for which the approximation $\mu(\omega) \simeq \mu(0)$ is valid. The nature of the propagation is then determined by the frequency dependence of $\mu(\omega)$. In the hopping case we have seen in Section 10.4 that there is a large frequency range in which the ac conductivity has a power-law dependence on $\omega$. The ac mobility therefore exhibits similar behavior. The work of Nolandi[4] may be interpreted as showing that quite simple trap distributions can lead to power-law behavior for the effective mobility defined in equation (10.8.19). We show below that such a power-law dependence of $\mu(\omega)$ on $\omega$ leads immediately to the archetypal anomalous carrier-pulse propagation described in Section 10.6.1.

Let us therefore suppose that

$$\mu(\omega) = \mu_1(-i\omega\tau_0)^{1-\alpha} \tag{10.8.20}$$

where $\mu_1$, $\tau_0$, and $\alpha$ are constants with $0 < \alpha < 1$. We shall see below that the frequency dependence of the mobility will, of itself, cause an injected carrier pulse to spread out. The diffusion term in equation (10.8.10) is not necessary for this purpose and may be neglected at sufficiently long times. This situation is in complete contrast to the conventional one when the diffusion term can never be neglected. We note that our neglect of any dc contribution to $\mu(\omega)$ means that the conversion to conventional carrier-pulse propagation, discussed above, will never take place for the model under discussion. When diffusion is neglected the Fourier transform of equation (10.8.3) yields

$$\frac{d}{dx}[\mu(\omega)n(\omega)E] - i\omega n(\omega) = n_0\delta(x) \tag{10.8.21}$$

where we have imposed the boundary condition (10.8.7). Equation (10.8.21) may be integrated immediately to give

$$n(\omega) = \frac{n_0}{\mu(\omega)E} \exp[i\omega x/E\mu(\omega)]\theta(x) \tag{10.8.22}$$

The unit step function $\theta(x)$ appears in $n(\omega)$ because, with the field pointing in the negative $x$-direction (as we suppose), the electrons must move in the positive $x$-direction.

When the Laplace transform (10.8.22) is inverted we have the pulse-shape equation

$$n(x, t) = \theta(x)\frac{n_0}{2\pi}\int_{-\infty}^{\infty} \frac{\exp\{-i\omega[t - x/E\mu(\omega)]\}}{E\mu(\omega)} \, d\omega \tag{10.8.23}$$

For arbitrary $\mu(\omega)$ this integral must be evaluated numerically. However, there are two especially simple cases. First, when $\alpha \to 1$, equation (10.8.20) reduces to $\mu(\omega) = \mu_1$, a constant. Then we immediately obtain the expected result

$$n(x, t) = n_0 \delta(x - \mu_1 E t) \qquad (10.8.24)$$

Second, when $\alpha = \frac{1}{2}$, we may write equation (10.8.22) in the form

$$n(\omega) = -2 \frac{n_0}{2ikD_e} \exp(ikx)\theta(x) \qquad (10.8.25a)$$

where

$$D_e = (\mu_1 E)^2 \tau_0 \qquad (10.8.25b)$$

is an effective diffusion constant and

$$k = (i\omega/D_e)^{1/2} \qquad (10.8.25c)$$

We may immediately verify from equation (10.8.8) with $v_D = 0$ (or from the conventional diffusion equation, which is rather easier) that

$$n(x, t) = \frac{2n_0}{(4\pi D_e t)^{1/2}} \exp(-x^2/4D_e t)\theta(x) \qquad (10.8.26)$$

We begin to see the seeds of archetype anomalous carrier-pulse propagation in the simple result (10.8.26). It will be recalled that we have neglected diffusion altogether. Nevertheless the pulse diffuses rather than drifts into the region $x > 0$ with an effective diffusion constant $D_e$ determined by the electric field. The peak of the pulse remains at $x = 0$ for all $t$. The factor of 2 in equation (10.8.26) keeps $n(x, t)$ normalized to $n_0$ for all $t$ since none of the electrons moves into the region $x < 0$. It is easy to verify that $n(x, t)$ is proportional to $t^{-\alpha}$ times a function of $x$ and $t$ in the combination $x/t^\alpha$. In Figure 10.11[78] we show pulse shapes calculated for various values of $\alpha$ as functions of $(x/t^\alpha)/E\mu_1\tau_0^{1-\alpha}$.

To study the behavior of $I(t)$, which is what is measured experimentally, we substitute expression (10.8.22) into the first term of equation (10.8.10) and invert the transform to obtain the conduction-current density

$$J(x, t) = -\frac{n_0 e}{2\pi} \int_{-\infty}^{\infty} d\omega e^{-i\omega t} \theta(x) \exp[i\omega x/E\mu(\omega)] \qquad (10.8.27)$$

Then $I(t)$ follows from equation (10.8.2) when we invert the order of

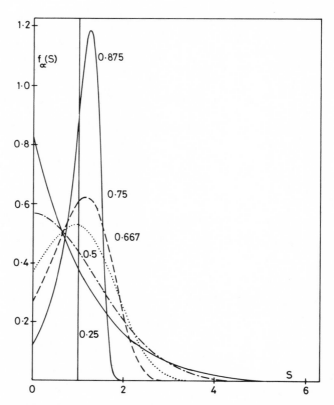

Figure 10.11. Function $f_\alpha(s) = n(x, t)/(n_0 E \mu_1 t^\alpha \tau_0^{1-\alpha})$ plotted against $s = (x/t^\alpha)/E \mu_1 \tau_0^{1-\alpha}$ for the values of $\alpha$ shown.[78]

integration. Hence

$$I(t) = -\frac{n_0 e}{2\pi} \int_{-\infty}^{\infty} d\omega e^{-i\omega t} \frac{e^{iqL} - 1}{iqL} \qquad (10.8.28)$$

where $q = \omega/E\mu(\omega)$, which is proportional to $\omega^\alpha$ when $\mu(\omega)$ has the power-law form (10.8.20). We consider times that are long enough for diffusion to be neglected but still short enough to ensure that no significant part of the carrier pulse has reached the far end of the specimen. Then we may let $L \to \infty$ in equation (10.8.27), in which case the exponential term is removed. The change of variable $u = \omega t$ yields

$$I(t) \sim \frac{E}{L} t^{\alpha - 1}, \qquad t \to 0 \qquad (10.8.29)$$

On the other hand, when $t \to \infty$ we are concerned with small $\omega$, i.e., with small $q$. The leading terms in the factor of the integrand in equation (10.8.28) involving $q$ are

$$\frac{e^{iqL} - 1}{iqL} \sim 1 + \tfrac{1}{2}iqL \tag{10.8.30}$$

The first term yields a $\delta$-function, which may be ignored. The second term gives

$$I(t) \sim \frac{L}{E}\, t^{-1-\alpha} \tag{10.8.31}$$

Finally, we may define a transit time $t_T$ by equating relation (10.8.29) and (10.8.30), in which case

$$t_T \sim \left(\frac{L}{E}\right)^{1/\alpha} \tag{10.8.32}$$

Equations (10.8.29), (10.8.31), and (10.8.32) are the salient qualitative features observed in archetypal anomalous carrier-pulse propagation. The ability to superpose curves of $I(t)$ for different transit times follows immediately from the power-law behavior, which is exhibited in these equations and stems from the power-law behavior assumed for $\mu(\omega)$.

## 10.8.6. Comparison with Experiment

The addition of a dc term to the mobility function modifies the behavior found above.[76,77,79] As an example we show in Figure 10.12 $n(x, t)$ calculated as a function of $x/\mu_0 E\tau_0$ for various values of $T = t/\tau_0$ when

$$\mu(\omega) = \mu_0 + \mu_1(-i\omega\tau_0)^{1-\alpha} \tag{10.8.33}$$

with $\alpha = 0.5$ and $\mu_1/\mu_0 = 100$.[79] We note in particular that the peak of the pulse drifts away from the origin in contrast to the behavior found in subsection 10.8.5, when $\alpha = 0.5$ and $\mu_0 = 0$. This is always the case when $\mu_0 \neq 0$.[79] Consequently, the total current settles down to a constant value after a time $t_2$ determined by the relative strengths of the ac and dc contributions to $\mu(\omega)$. For anomalous carrier-pulse propagation to occur we must have $t_T \ll t_2$. For a hopping mechanism it is possible to estimate the ac and dc contributions to $\mu(\omega)$ from ac and dc conductivity data. The values of $t_2$ calculated on this basis for a-Se are always very large compared to estimates of $t_T$, and yet this material shows a transition from conven-

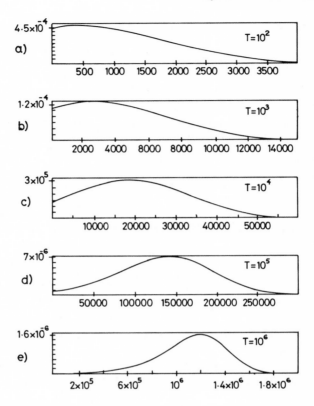

Figure 10.12. Carrier density as a function of $x/\mu_0 E\tau_0$ for various values of $T = t/\tau_0$ when $\mu(\omega) = \mu_0[1 + 100(-i\omega\tau_0)^{1/2}]$.[79]

tional to anomalous carrier-pulse propagation at 143 K.[78] Moreover, the values of $\alpha$ should correlate with the powers of $\omega$ arising in ac conductivity. The degree of correlation is very poor[78] and we conclude that trapping is more likely to be the mechanism involved in most cases in which anomalous carrier-pulse propagation has been observed.

## 10.9. Conclusion

In this review of the theory of electron transport in disordered solids we have given most weight to the hopping mechanisms, because its theory is now well developed within the framework of the rate-equation formalism. The rate equations themselves have been taken as intuitive. They may of course be derived from first principles[16] or from a master equation.[80]

There has been a great deal of work on calculations of the Hall effect for hopping electrons.[82-87] The Hall mobility is small but not zero. However, the calculations remain of academic interest because the Hall effect has so far defeated all attempts to measure it in a hopping system.

Our discussion of transport by electrons in extended states has necessarily been more tenuous, because the simple picture afforded by a rate equation is not generally valid. We have concentrated on concepts commonly used in the analysis of experimental data. The treatment of conductivity for electrons near a mobility edge, which has been developed recently by Götze and co-workers,[27-30] has not yet been used very much. Nevertheless, it would appear to be the most promising new approach to a more general theory. None of the properties of amorphous materials is understood very well from a quantitative point of view. There are always a great many unknown parameters in any transport theory that can be adjusted so as to give apparent agreement between theory and experiment. The most obvious case in which this obstacle to progress can be easily overcome is impurity bands in crystalline semiconductors. Further experimental work on these systems would be extremely valuable. This is particularly so at the present time, because a theoretical understanding of the effect of Coulomb interaction on their transport properties is beginning to emerge.[1,15,16,88,89]

# References

1. N. F. Mott and E. A. Davis, *Electronic Processes in Non-Crystalline Materials*, The University Press, Oxford (1979).
2. G. Pfister and H. Scher, *Adv. Phys.* **27**, 747 (1978).
3. M. Pollak, *Philos. Mag.* **36**, 1157 (1977).
4. J. Nolandi, *Phys. Rev. B* **16**, 4466, 4474 (1977).
5. J. M. Marshall, *Philos. Mag. B* **38**, 335 (1978).
6. F. W. Schmidlin, *Philos. Mag. B* **41**, 535 (1980).
7. P. N. Butcher and J. D. Clark, *Philos. Mag. B* **42**, 191 (1980).
8. P. N. Butcher, *Electrons in Crystalline Solids*, p. 103, IAEA, Vienna (1973).
9. N. H. March, in: *The Metal–Non-Metal Transition in Disordered Systems* (L. R. Friedman and D. P. Turnstall, eds.), p. 1, SUSSP, Edinburgh (1978).
10. B. I. Shklovskii and A. L. Efros, *Sov. Phys. Usp.* **18**, 845 (1976).
11. H. Böttger and V. V. Bryksin, *Phys. Status Solidi B* **78**, 11, 415 (1976).
12. P. N. Butcher, in: *Linear and Nonlinear Electronic Transport in Solids* (J. T. DeVreese and V. E. van Doren, eds.), p. 348, Plenum Press, New York (1976).
13. M. Pollak, in: *The Metal–Non-Metal Transition in Disordered Systems* (L. Friedman and D. P. Turnstall, eds.), p. 95, SUSSP, Edinburgh (1978).
14. P. N. Butcher, *Philos. Mag. B* **42**, 789 (1980).
15. B. I. Shklovskii and A. L. Efros, 1980, *Sov. Phys. Semicond.* **14**, 487 (1980).
16. I. P. Zvyagin, 1980, *Phys. Status Solidi B* **101**, 9 (1980).
17. P. N. Butcher, *J. Phys. C* **5**, 3164 (1972).

18. R. J. Kubo, *J. Phys. Soc. Jpn.* **12**, 570 (1957).
19. W. E. Spear, paper presented at Spring College, International Centre for Theoretical Physics, Miramare, Trieste (1982).
20. P. G. Le Comber and W. E. Spear, in: *Amorphous Semiconductors* (M. H. Brodsky, ed.), p. 251, Springer-Verlag, Berlin (1979).
21. L. Friedman, *Philos. Mag.* **36**, 553 (1977).
22. W. E. Spear and P. G. Le Comber, *Philos. Mag.* **33**, 935 (1976).
23. W. E. Spear, D. Allen, P. G. Le Comber, and A. Ghaith, *Philos. Mag. B* **41**, 419 (1980).
24. G. Döhler, *Phys. Rev. B* **19**, 2083 (1979).
25. M. Grünwald and P. Thomas, *Phys. Status Solidi B* **94**, 125 (1979).
26. H. Overhof and W. Beyer, *Philos. Mag. B* **43**, 433 (1981).
27. W. Götze, *Solid State Commun.* **27**, 1393 (1978).
28. W. Götze, *J. Phys. C* **12**, 1279 (1979).
29. W. Götze, *Philos. Mag. B* **43**, 219 (1981).
30. D. Belitz and W. Götze, *Philos. Mag. B* **43**, 517 (1981).
31. H. Scher and M. Lax, *Phys. Rev. B* **7**, 4491, 4502 (1973).
32. H. Scher and E. W. Montroll, *Phys. Rev. B* **12**, 2455 (1975).
33. C. R. Gochanour, H. C. Andersen, and M. D. Fayer, *J. Chem. Phys.* **70**, 4254 (1979).
34. J. Klafter and R. Silbey, *Phys. Rev. Lett.* **44**, 55 (1980).
35. B. Movaghar, B. Pohlman, and W. Schirmacher, *Philos. Mag. B* **41**, 49 (1980).
36. B. Movaghar, B. Pohlman, and W. Schirmacher, *Solid State Commun.* **34**, 451 (1980).
37. B. Movaghar, B. Pohlman, and G. W. Sauer, *Phys. Status Solidi B* **97**, 533 (1980).
38. B. Movaghar and W. Schirmacher, *J. Phys. C* **14**, 859 (1981).
39. A. Miller and E. Abrahams, *Phys. Rev.* **120**, 745 (1960).
40. M. Pollak and T. H. Geballe, *Phys. Rev.* **122**, 1742 (1961).
41. P. N. Butcher and P. L. Morys, *J. Phys. C* **6**, 2147 (1973).
42. P. N. Butcher and P. L. Morys, in: *Proceedings of the 5th International Conference on Amorphous and Liquid Semiconductors* (J. Stuke and W. Brenig, eds.), p. 153, Taylor and Francis, London (1974).
43. V. Ambegaokar, B. I. Holperin, and J. S. Langer, *Phys. Rev. B* **4**, 2612 (1971).
44. I. G. Austin and N. F. Mott, *Adv. Phys.* **18**, 41 (1969).
45. M. H. A. Pramanik and D. Islam, *Philos. Mag. B* **42**, 311 (1980).
46. P. N. Butcher and B. Reis, *Philos. Mag. B* **44**, 179 (1981).
47. S. R. Elliott, *Philos. Mag. B* **40**, 507 (1979).
48. G. E. Pike, *Phys. Rev. B* **6**, 1572 (1972).
49. S. R. Elliott, *Philos. Mag.* **36**, 1291 (1977).
50. S. R. Elliott, *Philos. Mag.* **37**, 553 (1978).
51. K. Shimakawa, *J. Phys. (Paris)* **42**, C4-167 (1981).
52. B. I. Shklovskii and A. L. Efros, *Sov. Phys. JETP* **33**, 468 (1971).
53. M. Pollak, *J. Non-Cryst. Solids* **11**, 1 (1972).
54. G. E. Pike and C. H. Seager, *Phys. Rev. B* **10**, 1421, 1435 (1974).
55. P. N. Butcher and J. A. McInnes, *Philos. Mag. B*, **37**, 249 (1978).
56. J. A. McInnes, private communication (1982).
57. S. Summerfield and P. N. Butcher, *J. Phys. C* **15**, 7003 (1982).
58. L. Banyai, A. Aldea, and P. Gartner, *Rev. Roumaine Phys.* **26**, 923 (1981).
59. D. Emin, *Solid State Commun.* **22**, 409 (1977).
60. P. N. Butcher and L. Friedman, *J. Phys. C* **10**, 3803 (1977).
61. I. P. Zvyagin, *Phys. Status Solidi B* **58**, 443 (1973).
62. H. Overhof, *Phys. Status Solidi B* **67**, 709 (1975).
63. A. Aldea, L. Benyai, and V. Capek, *Czech J. Phys. B* **26**, 717 (1976).
64. P. N. Butcher and S. Summerfield, *J. Phys. C* **14**, L1099 (1981).

65. J. A. McInnes, P. N. Butcher, and J. D. Clark, *Philos. Mag. B* **41**, 1 (1980).
66. S. Summerfield and P. N. Butcher, to be published (1982).
67. H. Kahlert, *J. Phys. C* **9**, 491 (1976).
68. A. R. Long and N. Balkan, *Philos. Mag. B* **41**, 287 (1980).
69. A. R. Long, W. R. Hogg, N. Balkan, and R. P. Ferrier, *J. Phys. (Paris)* **42**, C4-107 (1981).
70. W. Shockley, *Electrons and Holes in Semiconductors*, Van Nostrand, Princeton (1950).
71. G. F. Leal Ferreira, *Phys. Rev. B* **16**, 4719 (1977).
72. P. N. Butcher, *Philos. Mag. B* **37**, 653 (1978).
73. P. N. Butcher, in: *Modern Trends in the Theory of Condensed Matter*, Proc. 16th Karpacz Winter School of Theoretical Physics, p. 195, Springer-Verlag, Berlin (1980).
74. J. Klafter and R. Silbey, *Phys. Rev. Lett.* **44**, 55 (1980).
75. P. N. Butcher and J. D. Clark, *Philos. Mag. B* **43**, 1029 (1981).
76. W. Schirmacher, *Solid State Commun.* **39**, 893 (1981).
77. K. Godzik and W. Schirmacher, *J. Phys. (Paris)* **42**, C4-127 (1981).
78. P. N. Bucher and J. D. Clark, *Philos. Mag. B* **42**, 191 (1980).
79. J. D. Clark and P. N. Butcher, *Philos. Mag. B* **43**, 1017 (1981).
80. L. Benyai and A. Aldea, *Fortschr. Phys.* **27**, 435 (1979).
81. H. Bottger and V. V. Bryksin, *Phys. Status Solidi B* **81**, 433 (1977).
82. L. Friedman and M. Pollak, *Philos. Mag. B* **38**, 173 (1978).
83. P. N. Bucher and A. A. Kumar, *Philos. Mag. B* **42**, 201 (1980).
84. P. N. Butcher and J. A. McInnes, *Philos. Mag. B* **44**, 595 (1981).
85. B. Movaghar, B. Pohlmann, and D. Wuertz, *J. Phys. C* **14**, 5127 (1981).
86. H. Grunewald, P. Müller, and D. Würtz, *Solid State Commun.* **43**, 419 (1982).
87. L. Friedman and M. Pollak, *J. Phys. (Paris)* **42**, C4–87 (1981).
88. M. Pollak, and M. L. Knotek, *J. Non-Cryst. Solids* **32**, 141 (1979).
89. C. J. Hearn, J. A. McInnes, and P. N. Butcher, *J. Phys. C* **15**, 5013 (1982).

# III

# Some Miscellaneous Topics

# Exafs Studies of
# Amorphous Solids and Liquids

## P. Lagarde

## 11.1. General Presentation of Exafs

The name Exafs (Extended X-Ray Absorption Fine Structure) was given around 1970 to a phenomenon discovered in the 1930s. What was called for forty years "Kronig's oscillations" became the Exafs technique when Sayers et al.[1] realized the usefulness of this experiment for structural analysis.

If the X-ray absorption coefficient of an element is plotted (Figure 11.1) around the characteristic energy of this element — a K or L absorption edge — this coefficient, after a monotonic decrease and strong jump at the edge, exhibits oscillations after the edge that extend up to several hundreds of electron volts (eV) before dying out. The electronic processes corresponding to three characteristic situations are also shown in Figure 11.1. When the incoming X-ray photon has too low an energy (before the edge) there is no unoccupied state into which an electron can be excited by the photon. When the energy of the photon is just sufficient to promote the $1s$ electron to an empty state, the absorption coefficient increases strongly, the width of this edge being typically a few eV. Above this energy, the photoelectron is left in the medium with some kinetic energy $E_c$, which is the difference between the photon energy and that of the edge.

**P. Lagarde** · LURE, Université de Paris-Sud, Bâtiment 209 C, 91405 Orsay Cedex, France.

The oscillations of the absorption coefficient after the edge have been related by Stern, Sayers, and Lytle to the presence around the X-ray absorbing atom of other atoms at a definite distance. The experimental result described in Figure 11.2 definitely demonstrates this relationship: the K-edge absorption spectrum of neon has been taken in two different physical situations, gas and solid phases. Both edges look similar at the same energy, but while in the gas phase the post-edge absorption decreases monotonically, in the solid phase it shows Exafs oscillations due to neighbors at a fixed distance in this case. Another example is given by a careful comparison of the X-ray absorptions of bromine and krypton in the gas phase. Krypton is a monatomic rare gas and the absorption coefficient does not show any wiggles, while bromine is a diatomic molecule, and each atom has another at a definite distance and the absorption coefficient possesses a quasi-sinusoidal oscillation.

These two examples demonstrate the validity of the basic hypotheses for the physical model of Exafs formulated around 1970 by Lytle et al.[2,3] and at the same time by Ashley and Doniach[4] and Lee and Pendry.[5]

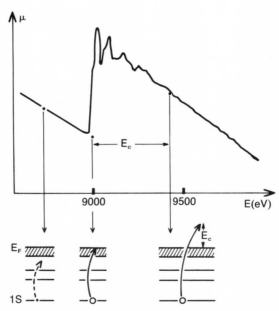

Figure 11.1. X-ray absorption spectrum of a solid solution Al–Cu 2% at around the copper K-edge at 8980 eV. The three electronic processes corresponding to three photon energies are shown in the lower part of the figure.

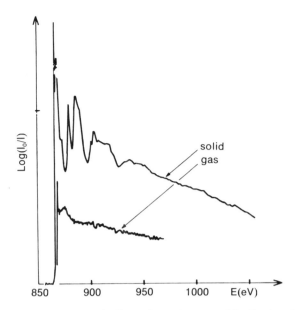

Figure 11.2. Comparison between the X-ray absorption around the K-edge of neon in the gas phase and in the solid phase.

## 11.2. Physical Interpretation of Exafs — The Equation

The basic idea behind the model is the following. The oscillations of the absorption coefficient above the edge are due to interference effects between the wave function of the final state of the photoelectron and that part being scattered by the neighbors of the excited (also called central) atom. These oscillations are superimposed on an "atomic" absorption coefficient, which would be the result of an experiment on the same element in an atomic situation (monatomic gas phase).

We concern ourselves with the quantity

$$X(E) = \frac{\mu - \mu_0}{\mu_0}$$

where $\mu$ is the actual absorption coefficient and $\mu_0$ is the ideal atomic part.

The absorption coefficient $\mu$ is given by Fermi's golden rule

$$\mu = \frac{2\pi}{\hbar} \sum_f |\langle i | H | f \rangle|^2 \delta(E_i - E_f - \hbar\omega)$$

$$\propto |\langle i | H | f \rangle|^2 n(E) \qquad (11.2.1)$$

where $n(E)$ is the density of states of the photoelectron of energy $E$. This photoelectron has several hundreds of eV, and can then be considered as a free electron in the medium with wave vector $k = \sqrt{2mE/\hbar^2}$ and a smooth density of states without oscillations. Modulations of $\mu$ are due to the final wave function $|f\rangle$, which contains a scattered part, while $\langle i|$ is a deep atomic state ($1s$ state, for instance), very concentrated around the central nucleus. Quantity $H$ is the dipolar Hamiltonian for optical absorption.

If $|f\rangle$ is assumed to be composed of an atomic part $|f_0\rangle$ and a scattered one $|\delta f\rangle$, we write $|f\rangle = |f_0 + \delta f\rangle$ and expand to second order. This yields

$$\mu \propto n(E)(|\langle i|H|f_0\rangle|^2 + 2R_e\langle i|H|f_0\rangle\langle i|H|\delta f\rangle^*) \qquad (11.2.2)$$

$$\mu_0 \propto n(E)|\langle i|H|f_0\rangle|^2$$

therefore

$$x = \frac{\mu - \mu_0}{\mu_0} = 2R_e \frac{\langle i|H|\delta f\rangle^*}{\langle i|H|f_0\rangle^*} \qquad (11.2.3)$$

We must then express $|\delta f\rangle$ vs. $|f_0\rangle$ for a photoelectron ejected from the central atom, backscattered by a neighbor at some distance $R$, and calculated at the site of the central atom, since $\langle i|$ is a deep core level (Figure 11.3).

For a K-edge issuing from a $1s$ level, the wave function is a Hankel function $h_1(kr) \times Y_{10}(\mathbf{r}/r)$. This photoelectron first experiences a phase shift $\delta_1$ due to the atomic potential of the central atom. In order to calculate scattering by a neighbor, we use the asymptotic limit for $h_1(kr)$ and then assume that the wave function of the photoelectron outside the central atom can be described by a plane wave (free-electron hypothesis) with amplitude $h_1(kr) \sim ie^{ikr}/2kr$. At the site of the neighboring atom, the amplitude of the wave function is $iY_{10}(e^{ikR}/2kR)e^{i\delta_1}$, where $R$ is the interatomic distance.

The scattering of this photoelectron is described in terms of the scattering function $f(\theta)$. When the photoelectron returns to the central

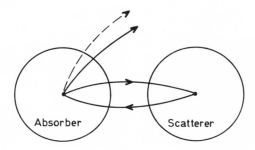

Figure 11.3. Two photoabsorption processes giving rise to $\mu$ and $\mu_0$. The solid line corresponds to the photoelectron backscattered by a neighbor with corresponding absorption coefficient $\mu$, and the dashed line represents the absorbing atom being free and the photoelectron ejected directly, leading to $\mu_0$.

Absorber                    Scatterer

atom, the amplitude compared to the "atomic" case becomes

$$| \delta f \rangle = i \, | f_0 \rangle \frac{e^{2ikR}}{2kR^2} f(\pi) e^{2i\delta_1} \tag{11.2.4}$$

where $f(\theta)$ is the scattering function given by

$$f(\theta) = \frac{1}{2ik} \sum_l (2l + 1)(e^{i\delta_l} \sin \delta_l) P_l(\cos \theta) \tag{11.2.5}$$

We have to evaluate it for $\theta = \pi$. In expression (11.2.4), we have taken into account that the photoelectron wave function must be calculated at the center of the central atom since $\langle i |$ is a core state, so $r = R$, the interatomic distance.

Quantity $f(\pi)$ is a complex number with amplitude $|f(\pi)|$ and argument $\varphi$, hence

$$| \delta f \rangle = | f_0 \rangle \frac{i}{2kR^2} | f(\pi) | e^{i(2kR + 2\delta_1 + \varphi)}$$

and expression (11.2.3) gives

$$X(E) = \frac{-1}{kR^2} | f(\pi) | \sin(2kR + 2\delta_1 + \varphi) \tag{11.2.6}$$

This case has dealt with just one scattering atom at distance $R$. For different atoms, with various scattering powers $f_j(\pi)$ at different distances $R_j$, expression (11.2.6) must be summed over subscript $j$. At this stage, however, we have to take into account that the mean free path of a photoelectron in the medium is limited: its value $\lambda(k)$ is $k$-dependent and of the order of a few Å. We describe its effect by a multiplicative term $\exp[-2R/\lambda(k)]$, which expresses the fact that at large distances from the central atom, inelastic processes make the photoelectron useless for the interference phenomenon of Exafs.

The interatomic distances $R$ fluctuate around their average value because of static or thermal disorder. We will see later that this effect can be expressed by a Debye–Waller-like term, $\exp(-2\sigma^2 k^2)$, analogous to the one used in X-ray diffraction, where $\sigma$ is related to the mean fluctuation of $R$.

Therefore, the general equation for Exafs oscillations of the absorption coefficient around the monotonic "atomic" value is

$$X(k) = - \sum_j \frac{N_j}{kR_j^2} \exp(-2\sigma_j^2 k^2)$$

$$\times \exp\left[ -2 \frac{R_j}{\lambda(k)} \right] | f_j(\pi) | \sin[2kR_j + 2\delta_1 + \varphi_j(k)] \tag{11.2.7}$$

where, around the central atom, $N_j$ atoms of type $j$ with backscattering function $f_j(\pi)$ are located at distance $R_j$. We recall here that the wave vector $k$ is related to the photon energy $E$ by

$$E - E_0 = \frac{\hbar^2 k^2}{2m}$$

where $E_0$ is the energy of the edge.

## 11.3. Experimental Technique

Exafs is an X-ray spectroscopic technique, therefore it needs a white X-ray source, a monochromator, and a detection system in order to collect the intensities $I_0$ and $I$, before and after the sample.

### 11.3.1. The Source

The renewed interest in Exafs is based on two main reasons. The first reason was explained in the preceding section and is related to the physical interpretation of the phenomenon in terms of scattering interferences from the neighbors. The second one, which also originated in the 1970s, is its use as an X-ray source of synchrotron radiation. Figure 11.4 gives the power spectrum, in photons per second, of some machines used now for low-energy physics; the gain in intensity over conventional X-ray tubes is seen to be several orders of magnitude. Therefore, an experiment that required days or even weeks twenty years ago is now conducted in fifteen minutes with better statistics. This has altogether changed the approach of physicists to the technique.

### 11.3.2. Monochromators and Detection

The synchrotron radiation is very collimated vertically (with a divergence of a few milliradians), so the best design for a monochromator is obtained by two Bragg reflections in the $(1, -1)$ mode.

The basic system is called "channel cut" and Figure 11.5 illustrates its principle while Figure 11.6 gives the actual shape for a Si (400) device. The two reflecting planes are cut in a single crystal of Si or Ge parallel to low-Miller-index planes [typically (111), (220), (400)]. While the first reflection gives the monochromaticity, the second is used just to keep the outgoing beam parallel to the white one. Typical resolutions of these devices are of the order of $10^{-4}$. The energy is swept by rotating the channel cut around an axis perpendicular to the figure in Figure 11.5, and different cuts are used in

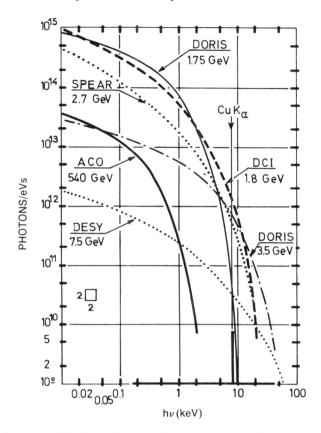

Figure 11.4. Intensity of the X-ray beam vs. the photon energy delivered by some storage rings in use for synchrotron radiation studies.

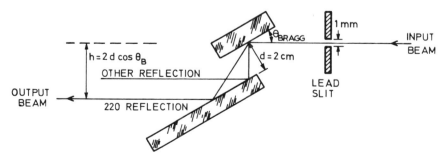

Figure 11.5. Principle of an X-ray monochromator based on a channel-cut device.

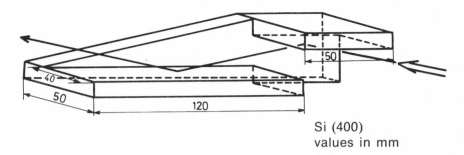

Figure 11.6. Design of a high-energy channel-cut Si (400) monochromator.

order to obtain the whole X-ray energy domain with good resolution and efficiency.

The flux of photons is of the order of $10^{10}$ photons s$^{-1}$ eV$^{-1}$, so the detectors used are ion chambers, the current being digitalized by voltage-to-frequency converters. The layout of an Exafs setup is given in Figure 11.7.

In the conventional technique, the absorption coefficient is measured as a function of energy. Another solution, with better efficiency in cases of a very dilute sample, is to detect the intensity of the X-ray fluorescence, which is directly proportional to the absorption. Finally, Exafs studies restricted to surface atoms use the detection of the photoelectron yield, or the Auger electrons, which are also proportional to the absorption coefficient.

### 11.3.3. Samples

The typical cross section of the beam is 40 mm wide and 5 mm high. The absorption coefficient of matter for X-rays is high, so the thickness is of the order of a few microns of the element of interest; for instance, the correct thickness of copper foil is about 10 $\mu$.

In the fluorescence mode, used to discard background absorption due to the medium in a very dilute sample, the ultimate concentration for the element of interest is of the order of $10^{-5}$ M.

Surface Exafs experiments, where adsorbed atoms are examined, require only $10^{13}$ atoms in the beam with the synchrotron sources now in use. Such a result was obtained, e.g., for iodine deposited on copper.[6] These values should be compared with the equivalent quantities for another structural technique, neutron scattering, which needs a much larger quantity of matter.

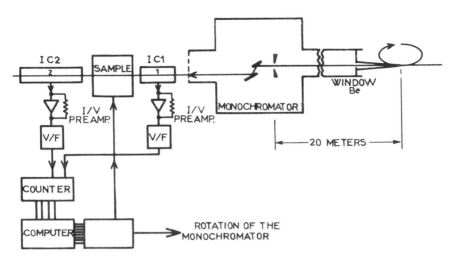

Figure 11.7. Schematic diagram of an Exafs experiment on a storage ring.

## 11.3.4. New Developments

Compared to the conventional technique described above, two major improvements are now in use or under test:

1. The channel-cut system has the disadvantage to give, together with the correct monochromatic beam, all the harmonics, which can be quite intense with synchrotron radiation. Moreover, the exit beam translates vertically when the energy is swept. A way to cure that is to use the so-called "two-crystals monochromator,"[7] where the two reflecting planes are physically separated. By slightly detuning one crystal with respect to the other one can decrease drastically the intensity of the harmonics, since the rocking curves narrow when the order of the reflection increases. If one crystal is translated relative to the other by a quantity depending on the Bragg angle, the outgoing beam can be kept fixed. Finally, one of the crystals can be curved in order to focus the beam and then increase the density of photons at the sample.

2. A new system, Exafs in dispersive mode, is now under test and is sketched in Figure 11.8. This system makes use of the small size of the point source from a storage ring. The main goal, with the use of new solid-state detectors, is to be able to collect an Exafs spectrum in a few hundredths of a second, thus allowing kinetic experiments.

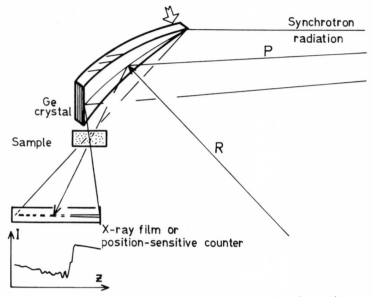

Figure 11.8. Principle of an Exafs experiment in dispersive mode.

## 11.4. Data Analysis — Strength and Limitation of the Technique

We recall here the Exafs equation

$$X(k) = -\sum_j \frac{N_j}{kR_j^2} \exp(-2R_j/\lambda)\exp(-2\sigma_j^2 k^2)|f_j(\pi)|\sin(2kR_j + \phi_j)$$

$$(11.4.1)$$

$$\phi_j = 2\delta_1 + \varphi_j$$

which contains two types of terms:

1. "Electronic terms," $\lambda$, $f_j(\pi)$, and $\delta_1$, which are dependent on the type of atoms, the atomic potentials, but not on the structure.
2. "Structural terms," $N_j$, $R_j$, and $\sigma_j$, which are the parameters defining the mutual arrangement of the neighbors around the central atom.

Therefore, if the atomic terms (1) are somehow known, then the Exafs signal is directly related to the spatial structure of the atoms described by the terms (2). In particular, if we assume that the term $\phi_j(k)$ in equation (11.4.1) is a linear function of $k$, where $\phi_j(k) = ak + b$, then a simple Fourier transform of $X(k)$ in $k$-space will give, in the corresponding

$R$-space, peaks at distances $R_j - a$.[1] A simple hypothesis on the phase shifts and a Fourier transform then gives, from the Exafs data, a quantity closely related to the radial distribution function of the material.

An analysis of Exafs data can therefore be schematized in the

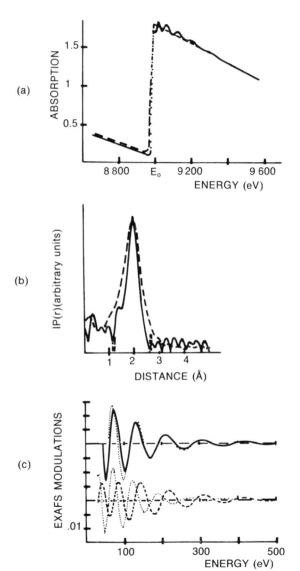

Figure 11.9. The three major steps of an Exafs analysis: (a) extraction of the data from the absorption coefficient, (b) Fourier transform of $X(E)$, (c) fit of the Exafs data.

following manner (Figure 11.9):

1. Extract the Exafs data by assuming an atomic absorption $\mu_0$.
2. Transform into $R$-space by a Fourier transform and filter the contribution made by the shell of neighbors of interest.
3. Transform back to $k$-space and try to model the result using equation (11.4.1) with experimental or calculated phase shifts.

Before proceeding, we shall review the intrinsic qualities of the method, comment critically on the basic hypothesis of the model, and indicate the limitations of the technique, especially in the case of disordered systems.

### 11.4.1. Selectivity — Short-Range Order

The value of the mean free path $\lambda$ is low (about 6 Å), so the number of shells of neighbors sampled by the photoelectron is limited to three or four. Therefore, Exafs is a structural probe but restricted to the short range only. This limitation has some compensation, since there is no need for a long-range periodicity in the medium. Exafs is therefore a local structural probe for systems without long-range order.

This order is obtained around the central atom that absorbs the photon and is defined in the absorption experiment by its absorption edge. Since all the absorption edges are well separated by a few hundred eV, Exafs is strictly related to the radial distribution function around one type of element of the medium. In the case of multicomponent systems, this advantage must be compared with results from other structural techniques, such as X-ray or neutron scattering, where the whole radial distribution function is obtained. This is typically the case in metallic alloys,[9–12] amorphous systems,[13,14] molecular solutions,[15,16] catalysts,[17,18] and metalloproteins,[19,20] where only the element of interest is examined and the contribution of the rest of the medium is discarded.

The results of Section 11.2 indicate that the mathematics associated with the analysis of the data are quite simple.

### 11.4.2. Phase Shifts — Transferability

The analysis of Exafs data requires knowledge of the so-called "atomic parameters," namely $\lambda$, $f(\pi)$, and $\delta_1$. One postulates that these phase shifts are transferable from one pair of absorber–scatterer atoms to another, provided the mutual distance and the chemistry do not change much from one situation to the other. Therefore a set of values, which gives a good result in a well-known model compound, can be transferred to an unknown situation, thus allowing a good structural determination.

This postulate has been carefully checked in numerous situations.[21] It works quite well for the phase $\phi(k)$, allowing a typical precision in distance measurements of the order of 0.02 Å. It is less accurate for $|f(\pi)|$, giving about a 10% accuracy in coordination numbers. In both cases the situation improves when the atomic number of the scatterer increases.

Lee and Beni[22] and then Teo and Lee[23] calculated every phase shift for all the elements in an atomic approximation. In order to make suitable allowance in the case of a condensed material, one uses a possible shift of the origin of energies $E_0$ over a $\pm 20$ eV range. This shift has to be checked on a model compound. Anyway, Exafs is mainly a comparative method between a well-known situation and an unknown compound of interest close to it.

Figure 11.10 shows the calculated phase shift and scattering amplitude for two elements, zinc and aluminum. Note the difference of almost $\pi$ between the two phase shifts, which can be used in some analysis.

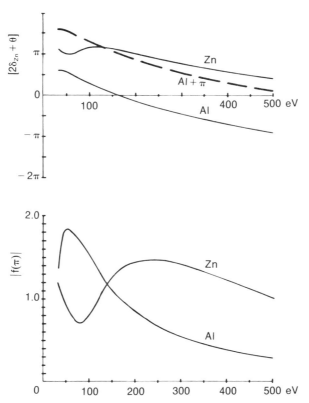

Figure 11.10. Calculated modulus and argument of the scattering function $f(\pi)$ for aluminum and zinc. Note the phase shift of almost $\pi$ between the two arguments.

### 11.4.3. Mean Free Path

Various experiments have given an overall shape of the mean free path of a free electron vs. the energy. Between 50 and 1000 eV this variation can be expressed as $\lambda \sim k/\Gamma$, where $\Gamma$ is a constant. Therefore, the mean-free-path parameter is often written as $e(-2\Gamma R/k)$.

At low energy the mean free path increases and the hypothesis that the photoelectron is free becomes invalid, because valence-band effects assume importance. Therefore, at low energy the model presented here is not valid and the corresponding data (between 0 and around 50 eV) must be discarded. Because of the Fourier-transform relationship, this means that the information is lost at large values of $R$, which is coherent with a short mean free path of the photoelectron. This problem has been carefully considered by Stern et al.[24]

The meaning of the above is that Exafs only studies the local order around the central atom, and that the long-range order is irrelevant in the phenomenon.

### 11.4.4. Other Hypotheses

The Exafs equation was obtained in a plane-wave approximation for the photoelectron. It has been proved by Pettifer[25] that this approximation is less satisfactory for small values of $k$ and for light elements. Care must therefore be taken in the analysis of such situations, since light elements have rapidly-dying backscattering amplitudes and the Exafs signal is very energy-limited.

Another hypothesis is that of a single scattering process only. Multiple scattering, such as low-energy electron diffraction, would involve large paths for the photoelectron, with scattering angles differing from $\pi$. Therefore these processes are usually of less importance.

Finally, the problem of the screening of the core hole by the surrounding electrons has been analyzed by Noguera et al.[26] Its effect has been found to be limited to the low-energy domain.

In conclusion, all the assumptions made in the basic formulation of Exafs have some effect on the low-energy photoelectron domain, and beyond that have only a slight influence, at least when we are interested in sufficiently heavy elements.

### 11.4.5. Effect of Disorder on the Exafs Spectrum

In the case of disordered systems or liquids, the influence of fluctuations of the interatomic distance around their mean value must be treated with special care. We concern ourselves first with the simplest case of harmonic motion, leading to the Debye–Waller-like term, following a work by Beni and Platzman.[27] In Section 11.6 we introduce the case of nonharmonic radial distribution functions from either static or thermal origin.

The time scale of the Exafs process is that of the photon absorption, namely $10^{-16}$ s, while the interatomic vibrations are of the order of $10^{-12}$ s. An Exafs experiment then averages over all possible distances. The following calculations are a simplified approach to the problem.

Let $\mathbf{r}_0 = \mathbf{r}_0^0 + \mathbf{u}_0$ be the instantaneous position of the central (absorbing) atom, where $\mathbf{r}_0^0$ is the mean position and $\mathbf{r}_j = \mathbf{r}_j^0 + \mathbf{u}_j$ is the same quantity for a neighbor. Then

$$r = |\mathbf{r}_{0j}|$$
$$= |\mathbf{r}_j^0 + \mathbf{u}_j - \mathbf{r}_0^0 - \mathbf{u}_0|$$
$$\simeq r_0^0 + |(\mathbf{u}_j - \mathbf{u}_0)_\parallel| + |(\mathbf{u}_j - \mathbf{u}_0)_\perp|^2/2r_{0j}^0$$
$$= r_{0j}^0 + u_\parallel + u_\perp^2/2r_{0j}^0$$

where $\parallel$ and $\perp$ denote parallel and perpendicular to the interatomic bond.

We take the average:

$$\langle \sin(2kr + \phi) \rangle = \sin(2kr_{0j}^0 + \phi) \left\langle \cos\left(2ku_\parallel + k\frac{u_\perp^2}{r_{0j}}\right) \right\rangle$$
$$+ \cos(2kr_{0j}^0 + \phi) \left\langle \sin\left(2ku_\parallel + k\frac{u_\perp^2}{r_{0j}^0}\right) \right\rangle \qquad (11.4.2)$$

We now consider the first term:

$$\left\langle \cos\left(2ku_\parallel + k\frac{u_\perp^2}{r_{0j}^0}\right) \right\rangle = \langle \cos 2ku_\parallel \rangle \left\langle \cos k\frac{u_\perp^2}{r_{0j}^0} \right\rangle - \langle \sin 2ku_\parallel \rangle \left\langle \sin k\frac{u_\perp^2}{r_{0j}^0} \right\rangle$$

since $u_\parallel$ and $u_\perp$ are independent. Then

$$\langle \cos 2ku_\parallel \rangle \simeq 1 - 2k^2\langle u_\parallel^2 \rangle = 1 - 2k^2\sigma_\parallel^2$$

$$\left\langle \cos k\frac{u_\perp^2}{r_{0j}^0} \right\rangle \sim 1 + \text{fourth order terms}$$

$$\langle \sin 2ku_\parallel \rangle \sim 2k\langle u_\parallel \rangle + \text{third order terms}$$

$$= 0 \qquad \text{since the motion is harmonic}$$

$$\left\langle \sin k\frac{u_\perp^2}{r_{0j}^0} \right\rangle \sim k\frac{\langle u_\perp^2 \rangle}{r_{0j}^0} = k\frac{\sigma_\perp^2}{r_{0j}^0}$$

Therefore, to second order

$$\left\langle \cos\left(2ku_\parallel + k\frac{u_\perp^2}{r_{0j}^0}\right)\right\rangle \sim 1 - 2\sigma_\parallel^2 k^2 \sim \exp(-2\sigma_\parallel^2 k^2)$$

Similarly, it can be easily shown that the second term of equation (11.4.2) averages to zero. Therefore, in the harmonic approximation, the effect of the disorder can be described by a Debye–Waller term. Note, however, that the term entering this Debye–Waller form involves only the relative displacements $u_\parallel$; only correlated motion contributes to the damping of Exafs. It can be shown for a three-dimensional system that the approximate relation

$$\sigma^2(\text{Exafs}) \sim \sum_q 2\langle u^2\rangle (1 - \cos qR)$$

holds, where $q$ runs over all the phonon modes. Because of the $\cos qR$ term the acoustic phonons do not contribute to $\sigma$ since, in this case, the relative displacement between two atoms is zero. The Debye–Waller factor in Exafs and that for X-ray diffraction are related approximately as follows:

$$\sigma^2(\text{Exafs}) \sim 2\sigma^2(\text{X-rays})$$

which shows that the damping of Exafs by the disorder is much more important than in the case of X-ray diffraction.

Experiments on simple systems at various temperatures have shown that a Debye or an Einstein model for the phonon density of states is already a good approximation. Figure 11.11 gives some results on metallic copper obtained by Greegor and Lytle,[28] while Crozier[29] studied the case of amorphous and crystalline germanium. The variation of $\sigma^2$ vs. $T$ is well fitted by an Einstein model with $\Theta_E = 360$ K for the amorphous Ge and $\Theta_E = 351$ K for the crystal, the difference indicating a change in the static disorder.

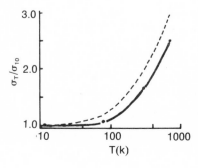

Figure 11.11. Variation of the Debye–Waller factor for Exafs in the case of metallic copper. The dotted line corresponds to a Debye theory while the solid line is from the correlated Debye model by Beni and Platzman.[27] Points are experimental values (after Greegor and Lytle[28]).

## 11.5. Domains of Application of Exafs and Examples

Since Exafs is a selective, local, and quite precise method for structural analysis, it has been used in systems where the local order is of importance, with a lack of long-range order, making some of the other methods (such as X-ray diffraction and microscopy) inefficient.

This is the case for amorphous systems, especially in multicomponent alloys, since the partial radial distribution function around each type of atom can be obtained separately.* The selectivity is also widely used in the study of the metal site in metalloproteins, or the metal clusters of metal-supported catalysts. In these cases we make full use of the qualities of Exafs, since the site of interest is limited to the close environment of a generally heavy metal (such as copper, platinum, or molybdenum) embedded in a light medium (such as protein, alumina, or silica) without any long-range order. At the K-edge of the metal, the absorption of the surrounding support is quite low and does not contribute much to the signal.

We shall give here two examples of direct application of Exafs to determine local structure.

### 11.5.1. Electrolytic Solutions

A very concentrated aqueous solution of $CuBr_2$ or $ZnBr_2$ shows in Raman scattering, for instance, some indication that there is clustering of the ions in the solution. Exafs has provided a direct answer to this problem, as illustrated by Figure 11.12 in the case of $ZnBr_2$[15,16] for which Exafs spectra were taken around the same Zn edge in three cases: anhydrous solid $ZnBr_2$, a concentrated solution (8.08 $M$), and a dilute solution (1.29 $M$). At the high-energy side, above 130 eV, the spectra of both the solid and the 0.08 $M$ solution look very similar, indicating that the local order around the Zn atoms is the same in both cases and involves the bromine atoms. In the low-energy range, both solutions have a similar spectrum different from that of the solid. Over this range, the contribution to the Exafs spectrum comes mainly from the oxygen atoms of the water: this part is then a signature of the hydration shell of the metal in the solution. These results are also corroborated by studying the Exafs spectra on the bromine edges.

---

* The determination of partial structure factors for liquids and amorphous systems is discussed at length in Chapters 1 and 2.

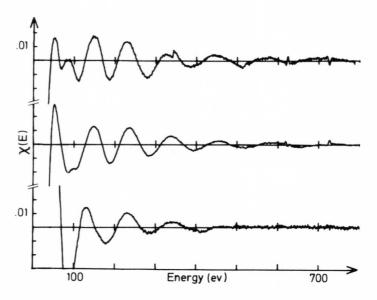

Figure 11.12. Exafs spectra above the zinc edge for three different Zn–Br situations: top, anhydrous solid $ZnBr_2$; middle, aqueous concentrated $ZnBr_2$ solution (8.08 $M$); bottom, dilute solution (1.29 $M$).

## 11.5.2. Core Effect in Al–Cu Alloys

A problem that can be solved by Exafs is the so-called "core effect," which is the local, structural effect of one impurity whose size differs from that of the atom which it substitutes. This is the case of the Al–Cu solid solution with a 2% concentration of copper. The problem is to measure the variation of the distance between the Cu impurity and its 12 neighbors in the Al structure, compared to the normal value for pure aluminum metal. This has been done[9] by direct comparison between two situations: a quite well-known one, the $\theta'$ phase, which is $Al_2Cu$ ordered with eight Al at 2.49 Å around one Cu; and the unknown situation of Cu as a solid solution in Al. Figure 11.13 gives the result of the two Fourier transforms, and a direct measure of the Al–Cu distance in the solid solution can be obtained from the shift of the main first peak (the Cu-Al contribution) with respect to the same peak in the $Al_2Cu$ phase. Therefore, the Cu–Al distance has been found to be 2.725 Å, while the Al–Al in pure metal is 2.851 Å at 15 K.

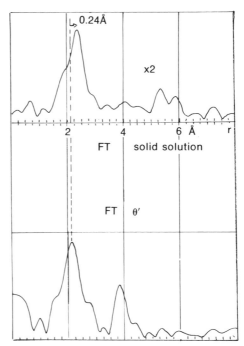

Figure 11.13. Comparison of the Fourier transforms of Exafs data above the copper edge in two cases: the Al–Cu solid solution and the $Al_2Cu$ ordered $\theta'$ phase. The shift of the first peak is a direct measure of the core effect in this alloy.

## 11.6. The Case of Nonharmonic Systems

Radial distribution functions in amorphous disordered systems or in liquids are hardly symmetric, the limit case being the distribution function $g(R)$ in the "dense random packing of hard spheres" model for amorphous alloys, where $g(R)$ possesses a sharp leading edge at a distance equal to the diameter of the spheres, and a smooth tail at large $R$.

Since the Exafs "free-electron" model is invalid below $k \sim 2 \text{ Å}^{-1}$, by the Fourier-transform relationship, the information at large $R$ is permanently lost. Let us consider, for example, a function $g(R)$ composed of two Gaussian distributions of the same type of atom (this could be the case in amorphous metallic alloys):

$$g(R) = g_1(R) + g_2(R)$$

where

$$g_1(R) \simeq \exp\left[ -\frac{(R - R_1)^2}{\sigma_1^2} \right] \quad \text{and} \quad g_2(R) \simeq \exp\left[ -\frac{(R - R_2)^2}{\sigma_2^2} \right]$$

with $R_2 > R_1$ and $\sigma_2$ equal to about 2 or 3 times $\sigma_1$. This function $g(R)$ will look like a Gaussian curve with width $\sigma_1$ and a tail at large $R$ due to $\sigma_2$.

The corresponding Debye–Waller terms in Exafs will be $\exp(-2\sigma_1^2 k^2)$ and $\exp(-2\sigma_2^2 k^2) \ll \exp(-2\sigma_1^2 k^2)$. Therefore, the contribution of $g_2(R)$ to the Exafs data can completely disappear and Exafs will measure a distance $R_1$, which is shorter than the average value of the overall distribution. For instance, in zinc, Eisenberger and Brown[30] have found along the $C$-direction a contraction of 5% of the first interatomic distance between 20 and 300 K, while X-ray diffraction shows a 2% expansion. The same behavior has also been found by Crozier and Seary[31] above room temperature in zinc.

The reason for this phenomenon is the following. At low energy, and therefore at low values of $k$, the basic hypothesis of the Exafs models — namely the photoelectron is free in the medium — is not valid, since band effects begin to be of importance. Therefore the low-$k$ information must be discarded and the large-$R$ information is lost; the long tail of the asymmetric distribution function is not seen by Exafs, which is mainly sensitive to sharp features at small distances.

Following Eisenberger and Brown,[30] the correct form of the Exafs spectrum for neighboring atoms distributed (statically or thermally) with radial distribution function $g(R)$ is given by

$$X(k) = -\frac{f(\pi)}{k} \int_0^\infty F(r)\sin(2kr + \phi)dr \qquad (11.6.1)$$

with

$$F(r) = F(\bar{r} + x) = g(r)\frac{e^{-2r/\lambda}}{r^2}$$

where $\bar{r}$ is the average value.

If the sine function in equation (11.6.1) is expanded, we obtain

$$X(k) = -\frac{f(\pi)}{k}\sqrt{A_k^2 + S_k^2}\sin(2k\bar{r} + \Sigma_k)$$

where

$$A_k = \int_{-\infty}^{+\infty} F(\bar{r} + x)\sin 2kx\, dx$$

$$S_k = \int_{-\infty}^{+\infty} F(\bar{r} + x)\cos 2kx\, dx$$

$$\Sigma_k = \arctan(A_k/S_k)$$

If $F(r)$ is a pair function around $\bar{r}$, which is almost the case if $g(r)$ is symmetric, the terms $A_k$ and $\Sigma_k$ equal zero and Exafs gives the correct answer. However, if $g(r)$ is not symmetric around the average value $\bar{r}$, then $\Sigma_k$ will give a shift in the measured distance and $\sqrt{A_k^2 + S_k^2}$ will change the measured coordination number. The criterion that these authors recommend in order to safely apply the simplest Exafs equation is

$$\frac{\langle x^2 \rangle}{R} < 0.01 \text{ Å}, \qquad \frac{\langle x^2 \rangle}{\lambda} < 0.01 \text{ Å} \quad \text{and} \quad k^3 \langle x^3 \rangle \ll 1$$

What are the solutions to this problem in the case of disordered materials, and where is it likely that the radial distribution functions are not sharp and symmetric?

The first thing to bear in mind is that Exafs results must be coherent with other structural results when they are available. With these constraints, Exafs has shown in numerous examples (to be described below) that it can provide a very precise picture of the real system, while an Exafs analysis alone could be ambiguous.

## 11.7. Examples of Exafs Application in Disordered Systems

When no other structural result is available, the first approach is to expand the function $g(R)$ in moments. Dubois[32] did that in the case of amorphous Co–B, using also a careful comparison with corresponding crystalline materials.

Another, now more widely used approach is to model analytically the radial distribution function $g(R)$ bearing in mind that the same set of values must fit other structural results. In the case of liquid zinc, Crozier and Seary[31] used the following form of $g(R)$:

$$g(R) = 0 \qquad\qquad\qquad\qquad \text{if } R < R_0$$
$$g(R) = A(R - R_0)^2 \exp[-B(R - R_0)] \qquad \text{if } R > R_0$$

Haensel et al.[33] suggested the following expression for modeling the first peak of a dense packing of hard spheres:

$$g(R) = 0 \qquad\qquad\qquad\qquad \text{if } R < R_0$$
$$g(R) = \frac{1}{\sigma_D} \exp[-(R - R_0)/\sigma_D] \qquad \text{if } R \geqslant R_0$$

An analytic expression for the Exafs oscillations can be obtained[34] by using this form, and assuming that the mean-free-path term can be extracted from the integral (11.6.1). The mean value of $g(R)$ turns out to be $\bar{R} = R_0 + \sigma_D$.

Finally, another way to overcome the problem of a non-Gaussian $g(R)$ is to decompose it into two Gaussian subshells. Exafs is an additive phenomenon, so expression (11.2.7) can be used for each subshell.

### 11.7.1. Amorphous Co–P

This alloy, with the classical composition $Co_{80}P_{20}$, has already been studied by X-ray and neutron scattering.[35] All the metal–metalloid alloys seem to follow the same rule: there is no metalloid as first nearest neighbor. Moreover, optical measurements indicate that the local ordering could be very similar to that of the crystalline counterpart.

Exafs has the ability to examine separately the radial distribution functions around each type of element. This feature was applied in the case of Pd–Ge,[36] most important being the local order around the metalloid.

Introducing the X-ray results into an Exafs analysis definitely gives bad agreement, as shown in Figure 11.14. In order to obtain a correct fit, we have used two different analytic approaches.[37]

1. A function $g(R)$ possessing the form introduced by Haensel *et al.*,[33] with $R_0 = 2.2$ Å, $N = 9$, and $\sigma_D = 0.1$ Å (Figure 11.15), in which case the mean value of the distribution is $\bar{R} = 2.3$ Å.
2. Two Gaussian subshells with respective parameters $R_1 = 2.23$ Å, $N_1 = 4.5$ and $R_2 = 2.33$ Å, $N_2 = 4.5$ (Figure 11.16).

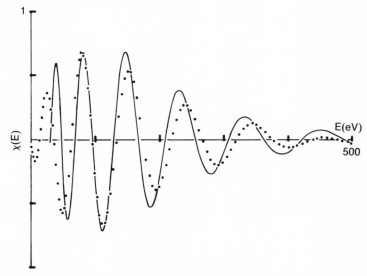

Figure 11.14. Filtered first shell of amorphous Co–P on the P-edge (dots) and an attempt to fit it using X-ray scattering results (solid line): 9 Co at 2.3 Å.

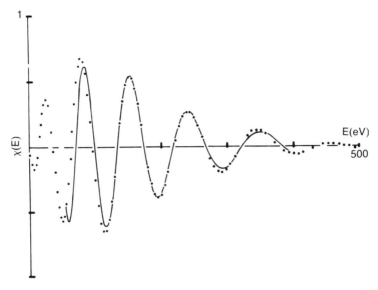

Figure 11.15. Filtered first shell of a CoP on the P-edge (dots) and the fit (solid line) using an asymmetric distribution (see text).

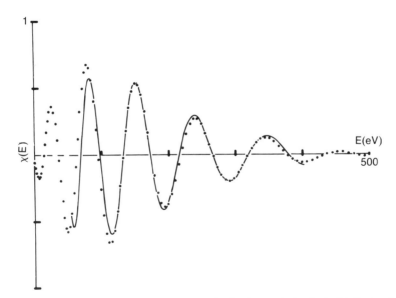

Figure 11.16. Filtered first shell of a CoP on the P-edge and the fit (solid line) using a two-subshell model (see text).

Since the two fits are equally good, it means that the overall shape of $g(R)$, which is the only physical meaningful quantity, should be the same in both cases. This has been verified and illustrated in Figure 11.17, where the expression of the form 1) has been convoluted with thermal Debye–Waller broadening used in the fit. Comparing the results obtained in the amorphous case with the local structure of $Co_2P$, Exafs definitely shows similarity between the two local orders, a result that X-ray scattering could not achieve. Nevertheless, the mean values of the Exafs results are identical to those of scattering experiments.

### 11.7.2. Metallic Alloys

Another example of the ability of Exafs to extract detailed information, provided other structural results are used as guidelines, is the case of the amorphous alloy $Ni_{66}Y_{33}$.[38]

As in the preceding case, full use has been made of the coherence needed between the results on the two edges of Ni and Y. Figure 11.18(a) shows that the X-ray results are unable to fit the Exafs data. In order to obtain agreement, shown in Figure 11.18(b), each shell of each type of atom must be split into two subshells, indicating that the local structure is slightly more complicated than a simple sphere of nearest neighbors.

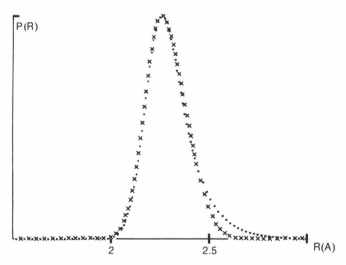

Figure 11.17. Comparison of the two radial distribution functions around a phosphorus atom in a CoP using the two models of Figures 11.15 and 11.16 dots refer to the analytic asymmetric function and crosses to the two-subshell model.

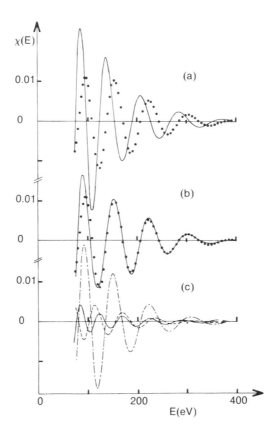

Figure 11.18. Exafs data of amorphous $Ni_{66}Y_{33}$ on the yttrium edge: filtered first shell result (dots). (a) Theoretical comparison with a model (solid line) based on X-ray scattering results. (b) Fit obtained with the two-subshell model. (c) Partial contributions to the result (b). The solid line refers to 4Y at 3.40 Å, the dashed line to 3 Ni at 3.05 Å, and the dot-dashed line to 9 Ni at 2.71 Å.

Figure 11.19 shows the final result obtained for the partial distribution functions, and the asymmetry of these functions is clearly evident. Table 11.1 summarizes these results for $Ni_{66}Y_{33}$. Note here too that X-ray results appear to be the mean values of the Exafs results. By comparing the Exafs numbers we obtained with, on the one hand, the statistical values where the two atoms are mixed at random and, on the other hand, the coordination numbers of the crystal, it can be seen that the structure of this amorphous alloy is not random but has a chemical ordering which bears some resemblance to that of the crystal.

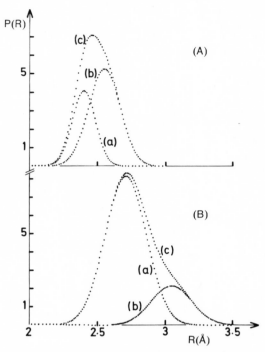

Figure 11.19. Radial distribution functions in $Ni_2Y$. (A) The nickel atoms around one nickel: (a) 2 Ni at 2.40 Å, (b) 4 Ni at 2.55 Å, (c) overall distribution. (B) The nickel atoms around one yttrium: (a) 9 Ni at 2.71 Å, (b) 3 Ni at 3.05 Å, (c) overall distribution.

### 11.7.3. Covalent Glasses

Exafs has so far been mainly applied to germanate or silicate glasses. Beside a study of the local environment of the main component (Ge or Si), minority elements acting as modifiers (Si, Na, Al, Fe, and Mn) have been investigated by using the selectivity of this technique.

In both pure germanium oxide[39,40] and silica,[41] the close environment of the metalloid due to its oxygen neighbors is found to be identical to that of the crystalline material. The influence of two different glass-formers, such as $SiO_2$ or $B_2O_3$ in vitreous $GeO_2$, is found to have a different influence on the mutual ordering of the elementary tetrahedral units $GeO_4$; this is shown by the behavior of the second Ge–Ge shell.

Greaves et al.[41] have also studied the local order around sodium in sodium disilicate or soda–lime–silica glasses. The presence of a well-defined Exafs structure around the minority component confirms that the

Table 11.1. Structural Results on Two Bimetallic Amorphous Alloys

| | Distances | | | Coordination numbers | | | |
| | Amorphous | | | Amorphous | | | |
| | Exafs | X-ray scattering | Crystal | Exafs | X-ray scattering | Crystal | Statistical model |
|---|---|---|---|---|---|---|---|
| $Ni_{66}Y_{33}$ | | | | | | | |
| Ni–Ni | 2.40 / 2.55 | 2.45 | 2.54 | 2 / 4 | 5.5 | 6 | 8 |
| Ni–Y | 2.71 / 3.05 | 2.80 | 2.98 | 4.5 / 1.5 | 5 | 6 | 4 |
| Y–Y | 3.40 | 3.40 | 3.11 | 4 | 4 | 4 | 5.3 |
| $Cu_3Zr_2$ | | | | | | | |
| Cu–Cu | 2.52 / 3.00 | 2.50–2.55 | | 3.75 / 2 | 4–8 | 4.2 | 7.35 |
| Cu–Zr | 2.71 / 3.05 | 2.80 | | 4.5 / 2 | 4–6 | 6.4 | 4.9 |
| Zr–Zr | 3.15 | 3.05–3.25 | | 4 | 4 | 5.7 | 5.6 |

local structure of alkali ions is well defined. Therefore a model for such glasses could be made up of two interpenetrating sublattices, the covalent network of $SiO_2$ being intercalated at random by an ionic fraction of the modifier component $Na_2O$. In that sense, the structure of the glass somewhat resembles that of the crystalline counterpart.

## 11.8. Conclusion

Though Exafs is a relatively new technique, it has proved to be very valuable for solving numerous problems of local structure in domains as varied as catalysts, metalloproteins, solutions, or amorphous materials.

In the case of disordered systems, the lack of information at low values of $k$ can cause some errors if only a crude analysis is performed. However, by careful modeling of the assumed radial distribution function together with the aid of other structural techniques, when available, the final result of an Exafs analysis can be very detailed due to its high sensitivity. Therefore it is, and should remain, a pretty useful tool in the study of these systems, where the need for a precise description of the structure is now more and more evident.

# References

1. D. Sayers, E. Stern, and F. Lytle, *Adv. X-Ray Anal.* **13**, 248 (1970); *Phys. Rev. Lett.* **27**, 1204 (1971); **30**, 174 (1973).
2. F. W. Lytle, D. Sayers, and E. Stern, *Phys. Rev. B* **11**, 4825 (1975).
3. F. W. Lytle, D. Sayers, and E. Stern, *Phys. Rev. B* **11**, 4836 (1975).
4. C. A. Ashley and S. Doniach, *Phys. Rev. B* **11**, 1279 (1975).
5. P. A. Lee and J. B. Pendry, *Phys. Rev. B* **11**, 2795 (1975).
6. P. H. Citrin, P. Eisenberger, and R. C. Hewitt, *Phys. Rev. Lett.* **45**, 1948 (1980).
7. A. Fontaine, P. Lagarde, D. Raoux, and J. M. Esteva, *J. Phys. F* **9**, 2143 (1979).
8. A. M. Flank, A. Fontaine, A. Jucha, M. Lemonnier, and C. Williams, *J. Phys. (Paris), Lett.* **43**, L315 (1982).
9. A. Fontaine, P. Lagarde, A. Naudon, D. Raoux, and D. Spanjaard, *Philos. Mag. B* **40**, 17 (1979).
10. J. Mimault, A. Fontaine, P. Lagarde, D. Raoux, A. Sadoc, and D. Spanjaard, *J. Phys. F* **11**, 1311 (1980).
11. D. Raoux, A. Fontaine, P. Lagarde, and A. Sadoc, *Phys. Rev. B* **24**, 5547 (1980).
12. B. Lengeler and P. Eisenberger, *Phys. Rev. B* **21**, 4507 (1980); *Phys. Rev. B* **22**, 3551 (1980).
13. T. M. Hayes, *J. Non-Cryst. Solids* **31**, 57 (1978).
14. S. J. Gurman, *J. Mater. Sci.* **17**, 1541 (1982).
15. D. R. Sandstrom, H. W. Dodgen, and F. W. Lytle, *J. Chem. Phys.* **67**, 473 (1977).
16. P. Lagarde, A. Fontaine, D. Raoux, A. Sadoc, and P. Migliardo, *J. Chem. Phys.* **72**, 3061 (1980).
17. J. H. Sinfelt, G. H. Via, and F. W. Lytle, *J. Chem. Phys.* **76**, 2799 (1982).
18. R. B. Greegor and F. W. Lytle, *J. Catal.* **63**, 476 (1980).
19. S. P. Cramer and K. O. Hodgson, *Prog. Inorg. Chem.* **25**, 1 (1979).
20. P. Eisenberger, R. G. Shulman, B. M. Kincaid, G. S. Brown, and S. Ogawa, *Nature* **274**, 30 (1978).
21. P. Citrin, P. Eisenberger, and B. Kincaid, *Phys. Rev. Lett.* **36**, 1346 (1976).
22. P. A. Lee and G. Beni, *Phys. Rev. B* **15**, 2862 (1979).
23. B. K. Teo and P. A. Lee, *J. Am. Chem. Soc.* **101**, 2815 (1979).
24. E. A. Stern, B. A. Bunker, and S. M. Heald, *Phys. Rev. B* **21**, 5521 (1980).
25. R. F. Pettifer, Thesis, University of Warwick (1978).
26. C. Noguera, D. Spanjaard, and J. Friedel, *J. Phys. F* **9**, 1189 (1979).
27. G. Beni and P. M. Platzman, *Phys. Rev. B* **14**, 9514 (1976).
28. R. B. Greegor and F. W. Lytle, *Phys. Rev. B* **20**, 4902 (1979).
29. E. D. Crozier, in: *Exafs Spectroscopy: Techniques and Applications* (B. K. Teo and D. C. Joy, eds.), p. 89, Plenum Press, New York (1980).
30. P. Eisenberger and G. Brown, *Solid State Commun.* **29**, 481 (1979).
31. E. D. Crozier and A. J. Seary, *Can. J. Phys.* **58**, 1388 (1980).
32. J. M. Dubois, Thesis, Université de Nancy (1981).
33. R. Haensel, P. Rabe, G. Tolkiehn, and A. Werner, in: *Proc. NATO Adv. Study Inst: Liquid and Amorphous Metals* (E. Luscher, ed.), p. 467, D. Reidel, Publ. Co., Dodrecht (1980).
34. M. De Crescenzi, A. Balzarotti, F. Comin, L. Incoccia, S. Mobilio, and N. Motta, *Solid State Commun.* **37**, 921 (1981).
35. J. F. Sadoc and J. Dixmier, The Structure of Non-Crystalline Materials (P. H. Gaskell, ed.), p. 85, Taylor and Francis, London (1977).
36. T. M. Hayes, J. W. Allen, J. Tauc, B. C. Giessen, and J. J. Hauser, *Phys. Rev. Lett.* **40**, 1282 (1978).

37. P. Lagarde, J. Rivory, and G. Vlaic, *J. Non-Cryst. Solids* (to appear).
38. A. Sadoc, D. Raoux, A. Fontaine, and P. Lagarde, *J. Non-Cryst. Solids* **50**, 331 (1982).
39. C. Lapeyre, Thesis, Université de Paris (1982).
40. C. Lapeyre, J. Petiau, and G. Calas, *Proc. The Structure of Non-Crystalline Materials II*, Cambridge (1982).
41. G. N. Greaves, A. Fontaine, P. Lagarde, D. Raoux, and S. J. Gurman, *Nature* **293**, 611 (1981).

## Books and Review Papers

B. K. Teo and D. C. Joy, eds., *Exafs Spectroscopy*, Plenum Press, New York (1981).

J. Wong, in: *Metallic Glasses* (H. J. Guntherod, ed.), Springer-Verlag, Berlin (1980).

P. A. Lee, P. H. Citrin, P. Eisenberger, and B. M. Kincaid, *Rev. Mod. Phys.* **53**, 769 (1981).

D. Raoux, J. Petiau, P. Bondot, G. Calas, A. Fontaine, P. Lagarde, P. Levitz, G. Loupias, and A. Sadoc, *Rev. Phys. Appl.* **15**, 1079 (1980).

T. M. Hayes, *J. Non-Cryst. Solids* **31**, 57 (1978).

# The Glass Transition

## A. E. Owen

### 12.1. Definitions and Experimental Characteristics

#### 12.1.1. Definitions

In this subsection we present various definitions of glass.

1. Definition given by the American Society for Testing Materials (ASTM): "An inorganic product of fusion which has cooled to a rigid condition without crystallizing."

2. Definition given by Morey:[1] "A glass is an inorganic substance in a condition which is continuous with, and analogous to, the liquid state but which, as the result of having been cooled from a fused condition, has attained so high a degree of viscosity as to be for all practical purposes rigid."

3. Definition given by Jones:[2] "A glass, or a substance in the glassy or vitreous state, is a material which has been formed by cooling from the normal liquid state and which has shown no discontinuous change in first order thermodynamic properties such as volume ($V$), heat content ($H$) or entropy ($S$) but has become rigid (i.e. solid) through a progressive increase in its viscosity. Discontinuities are observed, however, in derivative or second order thermodynamic properties such as specific heat capacity and thermal expansivity."

4. Definition given by the U.S. National Research Council:[3] "Glass is an X-ray amorphous material which exhibits the glass transition, this being

A. E. Owen · Department of Electrical Engineering, School of Engineering, University of Edinburgh, Edinburgh EH9 3JL, Scotland, UK.

defined as that phenomenon in which a solid amorphous phase exhibits with changing temperature a more or less sudden change in the derivative thermodynamic properties, such as heat capacity and expansion coefficient, from crystal-like to liquid-like values."

### 12.1.2. Experimental Characteristics

Suppose a liquid has been cooled rapidly enough to prevent crystallization (the kinetics of nucleation and crystal growth are considered in Section 12.2). The subsequent behavior on further cooling is usually represented either in terms of the temperature variation of a first-order extensive thermodynamic variable (i.e., volume $V$, enthalpy $H$, or entropy $S$), or in terms of the temperature dependence of a viscosity. The former emphasizes the thermodynamic (equilibrium) aspects of the glass transition, the latter the kinetic (nonequilibrium) aspects.

The variation of volume with temperature is illustrated in Figure 12.1

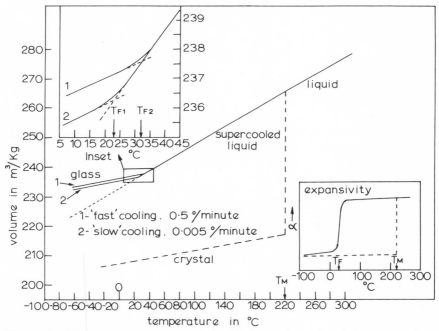

Figure 12.1. The volume–temperature relationship for the formation of vitreous selenium. The inset at the top left shows, on an enlarged scale, the transition region from supercooled liquid to glass and illustrates the effect of cooling rate. In the bottom right inset the expansion coefficient (i.e., temperature derivative of volume) is plotted to show the almost discontinuous change at the glass transition. These diagrams are typical of all glass-forming materials (after Dzhalilov and Rzaev[4]).

for vitreous selenium.[4] This is typical in its general form for all glasses. The experimental characteristics explicit or implicit in Figure 12.1 are as follows:[5]

1. First-order extensive thermodynamic variables, namely volume $V$, enthalpy $H$, and entropy $S$, are continuous with temperature (and pressure) but show a change of slope on the transition liquid → glass. The *range* of temperature over which this transition takes place is called the glass transition (or transformation) temperature $T_g$.

2. Although first-order variables are continuous, there are discontinuities in the second-order thermodynamic quantities, namely expansion coefficient $\alpha$, specific heat $C_p$, and compressibility $\kappa$ (see bottom right insert to Figure 12.1). For this reason the glass transformation has sometimes been regarded as a second-order thermodynamic transition. This is probably incorrect, or at best is a gross oversimplification of the situation (see Sections 12.3 and 12.4).

3. It is important to realize that, $T_g$ is not an accurately defined temperature. In the first place, the glass transition is not sharp but occurs over a range of temperature. It is sometimes convenient to define as a specific temperature the point at which the extrapolated liquid and glass lines cross (see top left insert to Figure 12.1). In glass technology this temperature is known as the fictive temperature $T_F$.[2] It is the temperature at which the glass would be in (metastable) equilibrium if it could be brought to that temperature instantaneously.

4. Furthermore, $T_g$ (and hence $T_F$) depends on the rate of cooling and subsequent thermal history. In typical silicate glasses, which normally have transformation temperatures in the range 400–700 °C, $T_g$ can be changed by 100–200 °C, i.e., by 20–30% or more.[6] The effect of cooling rate on the glass transition can sometimes be expressed empirically in the form

$$R = R_0 \exp\left[ -\frac{1}{C}\left(\frac{1}{T_g} - \frac{1}{T_m}\right) \right] \qquad (12.1.1)$$

where $R$ is the cooling rate, and $R_0$ and $C$ are constants. Expressed in this way, $R_0$ is the cooling rate at which $T_g = T_m$.

5. A glass is thermodynamically unstable and if held at a temperature below $T_g$ it will tend to approach the (meta-) stable equilibrium of the supercooled liquid at that temperature. Consider in Figure 12.2(a), for example, two glasses prepared by different rates of cooling and therefore with different values of $T_g$, namely $T_{g_1}$ (slow cooling) and $T_{g_3}$ (fast cooling). If they are both held at a temperature $T'$, the volume of each will tend toward that of the supercooled liquid at that temperature, i.e., $V'$. The value of $T_g$ will also change, of course. This process is known as "stabilization" and the rate at which it occurs will depend very sensitively

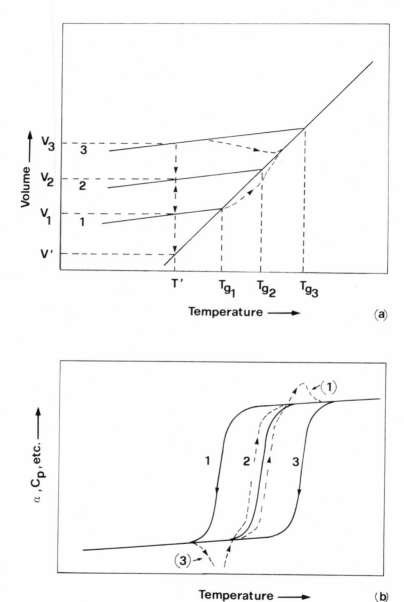

Figure 12.2. (a) Schematic volume–temperature diagram illustrating the effect of slow cooling (1), a medium cooling rate (2), and fast cooling (3). The dotted curves show the effect of reheating along curve 3 at a slower rate than the original rate of cooling, and along curve 1 at a faster rate. (b) The derivative curves corresponding to Figure 2(a). Again the dotted curves are for a reheating experiment and they illustrate the anomalies caused by differences between heating and cooling rates (after Jones[2]).

on temperature. Within a few degrees of $T_g$ it will normally take place on a time scale of minutes or hours. Far removed from $T_g$ it will take many years.

6. If a glass is reheated in order to determine $T_g$ by differential thermal-analysis measurements, for example, anomalies will occur unless the *rate of heating is the same as the original rate of cooling*. This is illustrated by Figure 12.2.[2] Suppose glass 1 (prepared by *slow* cooling) with $T_g = T_{g_1}$ is reheated at the (faster) rate corresponding to glass 2 (intermediate rate of cooling). Its volume will tend to undershoot the supercooled liquid line shown by the dotted curve in Figure 12.2, because it is being reheated too rapidly to attain equilibrium. On the other hand, if glass 3 (fast cooling) is reheated at the intermediate rate its volume will tend to decrease before $T_{g_3}$ is reached (dotted curve in Figure 12.2) because its rate of *heating* is slow enough to allow volume changes (corresponding to stabilization) to take place at a temperature lower than its own $T_g$. The derivative curves ($\alpha, C_p$) will be as shown in Figure 12.2(b), and if we were using these experiments to determine $T_g$, say, the measured value would be closer to $T_{g_2}$ rather than the values corresponding to the original rates of cooling.

7. It is possible to define and observe a glass transformation pressure $P_g$.[6] The derivative $dV/dP$ is negative, of course, and shows a change of slope but no discontinuity at $P_g$. The experimental $V-P-T$ relationship for vitreous selenium is shown in Figure 12.3, in which it is noteworthy that $T_g$ *increases* with pressure.

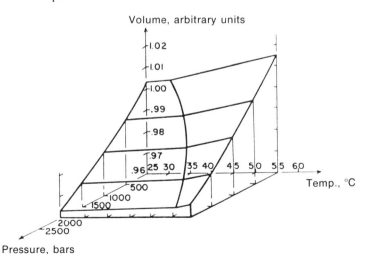

Figure 12.3. The pressure–volume–temperature surface for the formation of vitreous selenium. Note that the glass transformation temperature increases as pressure increases (after Eisenberg[7]).

8. In polymeric materials, which would include Se and possibly other chalcogenide glasses, a glass transformation molecular weight $M_g$ can also be defined.[7] The three quantities $T_g$, $P_g$, and $M_g$ are empirically related:

$$T_g = T_g^\infty(0) - \left[\frac{\partial T_g}{\partial (1/M)}\right] \frac{1}{M_g} + \left(\frac{\partial T_g}{\partial P}\right) P_g$$

where $T_g^\infty(0) = T_g$ at $P = 0$ and $M = \infty$. In polymeric *liquids* like Se the average molecular weight and distribution is a function of temperature. Thus in addition to the parameters mentioned earlier one might also expect the value of $T_g$ for a glass cooled from a polymeric melt to depend on the thermal history of *the liquid above* $T_m$ as well as on rate of cooling, etc., below $T_m$.

The temperature dependence of the inverse of viscosity (i.e., of fluidity $\phi$) is illustrated in Figure 12.4, in which the fluidity (in poise$^{-1}$) is plotted

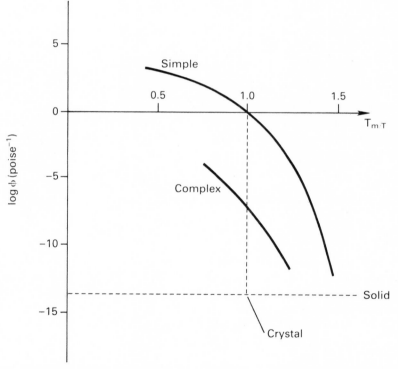

Figure 12.4. Typical fluidity–temperature relationships for simple and complex liquids. The temperature scale is plotted inversely, normalized to the thermodynamic melting point $T_m$ (after Turnbull and Cohen[8]).

against a reduced temperature $T_m/T$, where $T_m$ is the thermodynamic melting point.[8] In Figure 12.4 the curve labeled "simple" corresponds to liquids such as the intermediate alcohols and glycols, which are well-known organic glass-forming materials. The curve labeled "complex" corresponds to materials such as fused silica, silicates, and borates. The points to note about Figure 12.4 are as follows:

1. Glass-forming liquids show a smooth decrease in fluidity (increase in viscosity). It is also illustrated in Figure 12.4 that a liquid which crystallizes has a marked discontinuity in fluidity at its melting point.
2. The fluidity $\phi \simeq 10^{-13.5}$ poise$^{-1}$ is traditionally taken as the boundary between the supercooled liquid and the (solid) glass. At that value of fluidity configurational relaxation times are in the range of a few minutes to a few hours (i.e., in the range of typical experimental times).
3. For many glass-forming liquids, and over a wide range of temperature, the viscosity satisfies the equation

$$\eta = C \exp[E_\eta/k(T - T_0)] \qquad (12.1.2)$$

where $C$ is a constant, $E_\eta$ is the activation energy for viscous flow, and $T_0$ is some characteristic temperature always less than $T_g$. Equation (12.1.2) is known as the Fulcher or Doolittle equation.

It should be recognized that linking glass formation to a nonequilibrium transport property such as viscosity (fluidity) implies that equilibrium thermodynamics cannot be applicable in a straightforward way.

## 12.2. Kinetics of Nucleation and Crystal Growth

### 12.2.1. Introduction

It follows from Section 12.1 that the first requirement for glass formation is to avoid the nucleation and growth of crystals as a liquid is supercooled. In this sense at least, the glass transition is clearly a kinetic (nonequilibrium) phenomenon.

### 12.2.2. Formation of Critical Nuclei — Descriptive Account

A liquid has a disordered structure best represented by a radial distribution function (RDF). Since a glass is structurally continuous with the liquid from which it is formed (Section 12.1) it has a similar structure.

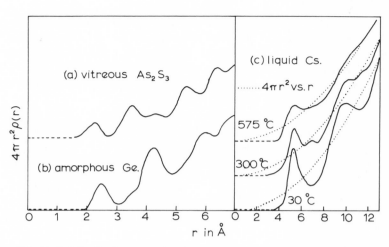

Figure 12.5. Radial distribution curves for vitreous $As_2S_3$ and amorphous germanium (on the left) and, for comparison, for liquid cesium (on the right). Note that as the temperature is reduced the features on the radial distribution function of liquid cesium, particularly the first peak, become sharper.

Typical RDFs are illustrated in Figure 12.5 for vitreous (glassy) $As_2Se_3$ [Figure 12.5(a)], amorphous Ge [Figure 12.5(b)], and liquid Cs at three different temperatures [Figure 12.5(c)]. Note that in Figure 12.5(c) the RDF becomes sharper as the temperature decreases, i.e., the local (near-neighbor) configuration becomes more ordered. However, the RDF represents *a time and space average*; the distribution (structure) about any particular atom changes rapidly with time. At some times the local order is more pronounced than at others, i.e., the configuration about any given atom is sometimes more like that in a crystalline solid, sometimes more like that in a gas. Near the melting point larger ordered groupings are potential nuclei for the start of crystallization.

The free-energy change for the isothermal formation of a spherical nucleus of radius $r$ is the sum of two terms, one proportional to its volume and the other proportional to surface area, i.e.,

$$\Delta G = f(T)r^3 + g(T)r^2 \qquad (12.2.1)$$

Alternately, at a particular temperature,

$$\Delta G = ar^3 + br^2 \qquad (12.2.2)$$

where $a$ and $b$ are constants. The form of the two terms and the net $\Delta G$ is illustrated schematically in Figure 12.6 for four temperatures, $T > T_m$

[Figure 12.6(a)], $T = T_m$ [Figure 12.6(b)], $T \leqslant T_m$ [Figure 12.6(c)], and $T < T_m$ [Figure 12.6(d)]. The surface term $br^2$ is always positive; the volume term is positive for $T > T_m$, zero for $T = T_m$, and negative for $T < T_m$. Thus the net $\Delta G$ passes through a maximum for any $T < T_m$ at a critical radius $r = r_c$. A nucleus with $r < r_c$ will tend to disappear ($\Delta G$ decreases spontaneously) while a nucleus with $r > r_c$ will grow ($\Delta G$ also decreases). As the temperature decreases below $T_m$, $r_c$ also decreases; at $T = T_m$, $r_c = \infty$. In the range of temperature where potential nuclei actually grow, $r_c$ is usually about 10 atomic diameters. At a temperature just below $T_m$, $r_c$ is still sufficiently large that suitable nuclei are rarely formed by thermal motion; the rate of nucleation, and hence of crystallization, is therefore small. As the temperature is lowered $r_c$ decreases and the rate of

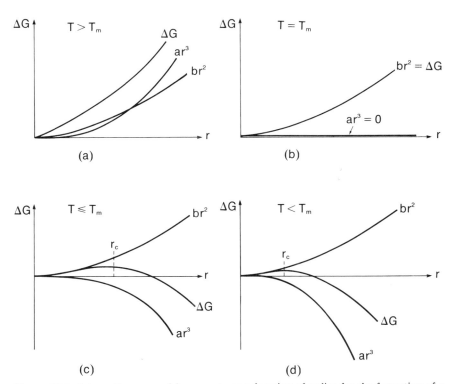

Figure 12.6. Schematic curves of free energy as a function of radius for the formation of a critical nucleus, and at different temperatures relative to the thermodynamic melting point $T_m$. The term $ar^3$ expresses the contribution from the volume and the term $br^2$ the contribution from the surface of the nucleus. The free energy $\Delta G$ involves the sum of the two terms. At temperatures less than $T_m$, the volume term is negative while the surface term remains positive and hence there is a maximum in $\Delta G$ when $T < T_m$.

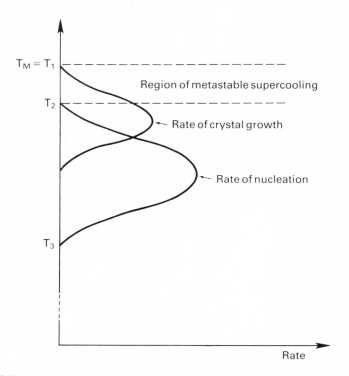

Figure 12.7. A schematic diagram illustrating the variation of the rate of crystal growth and the rate of nucleation as a function of temperature below the thermodynamic melting point $T_m$. Note that below some critical temperature, both rates decrease as temperature decreases.

nucleation (and crystallization) increases rapidly. At still lower temperatures $r_c$ does not become very much smaller and thermal motion also decreases, *hence the rate of nucleation (and crystallization) passes through a maximum* (cf. Figure 12.7).

### 12.2.3. Elementary Theory of Nucleation and Crystal Growth[8–10]

Suppose a spherical crystal nucleus containing $i$ atoms is formed in a supercooled liquid. The free energy involved is

$$\Delta G_i = Ai^{2/3} - Bi \tag{12.2.3}$$

where the constant $A$ is determined by the interfacial (surface) free energy and $B$ by the excess of the bulk free energy of the supercooled liquid over the crystalline solid. The quantity $\Delta G_i$ attains a maximum when

$$i = i^* = \left(\frac{2A}{3B}\right)^3 \tag{12.2.4}$$

i.e.,

$$G_{i^*} = \frac{4A^3}{27B^2} \tag{12.2.5}$$

Thus, if a nucleus reaches the critical size $i^*$ it can grow spontaneously. The rate of homogeneous nucleation is the rate at which critically sized nuclei are formed and the probability of their formation is proportional to $\exp(-\Delta G_{i^*}/kT)$. However, for a nucleus to grow, atoms or molecules must be added to it and this will also be a process involving an activation energy, such as diffusion and reorientation. If $\Delta G_D$ is the activation energy associated with the growth process, then the rate of formation and growth of critically sized nuclei is given by

$$\chi = C \exp(-\Delta G_D/kT)\exp(-\Delta G_{i^*}/kT) \tag{12.2.6}$$

Evaluation of the constants $A$ and $B$ in equation (12.2) and $C$ gives

$$\chi = \frac{nkT}{h} \exp\left(-\frac{\Delta G_D}{kT}\right) \exp\left(-\frac{16\pi\sigma^3 T_m^2}{3\lambda^2 \Delta T^2 kT}\right) \tag{12.2.7}$$

Here $n$ is the total number of atoms (or molecules), $h$ is Planck's constant, $\sigma$ is the interfacial free energy of a nucleus (per unit area), $\Delta T = T_m - T$, i.e., the supercooling, and $\lambda = \Delta H_f/V_m$, where $\Delta H_f$ is the latent heat of fusion and $V_m$ is the molar volume.

From equation (12.2.7), $\chi$ rises very rapidly as $T$ decreases, passes through a maximum, and then decreases (NB at $T = 0$ K, $\chi = 0$). Differentiation yields the following information:

$$\text{if } \Delta G_D = 0, \qquad \chi = \chi_{max} \text{ when } T = T_m/3$$

As $\Delta G_D$ increases, the temperature of $\chi_{max}$ is displaced toward $T_m$.

Thus, if a liquid has a sufficiently high $\Delta G_D$ to enable it to be cooled below the temperature at which $\chi = \chi_{max}$ without crystallizing, then crystallization is *less* likely to occur on further cooling. These are the circumstances that favor glass formation.

For a wide range of glasses, organic and inorganic, prepared in quantitites of at least several grams, and by normal quenching from the melt, it is found experimentally that[11]

$$T_g \simeq \tfrac{2}{3} T_m$$

For $\chi_{max}$ to occur at $T = \frac{2}{3}T_m$

$$\left(\frac{\Delta G_D}{kT}\right) \text{ must be equal to approximately 40}$$

and if $T = 300$ K

$$\Delta G_D \text{ must be equal to approximately 24 kcal/mol}$$

We compare this with $E_\eta$, the activation energy for viscous flow, in the following table.

|                  | $E_\eta$ (kcal/mol) |
|------------------|---------------------|
| Water            | 5.1 at $T_m$        |
| Propyl alcohol   | 20 at $T_g$         |
| Glucose          | 125 at $T_g$        |
| Boric oxide      | 75 at $T_g$         |
| Silicate glasses | 150 at $T_g$        |

## 12.2.4. The Critical Cooling Rate

A necessary and sufficient condition for glass formation is that the liquid should be cooled rapidly enough so that detectable nucleation and crystal growth cannot occur.

It is possible to make rough estimates of the cooling rate that must be exceeded, namely the critical cooling rate $R_c$, in order to encourage glass formation in particular liquids. It will obviously be a function of diffusion and flow processes in the liquid. Sarjeant and Roy[12] made an empirical estimate of $R_c$ by first measuring the velocity of crystallization in a small melt of NaCl subjected to very rapid cooling ($10^4$–$10^5$ deg/s). They then assumed that for NaCl the spontaneous nucleation and growth rates become extremely low at $10^0$ of supercooling and that the minimum detectable crystalline ordering occurs when the crystallite size is approximately $10^{-7}$ cm. Thus $R_c$ is the rate which will not allow the growth of a crystallite of that size in a $10^0$ interval. For NaCl Sarjeant and Roy find this to be $1.2 \times 10^9$ deg/s. They further suggest that $R_c$ can be written as

$$R_c = Z \frac{T_m}{\tau} \tag{12.2.8}$$

where $Z$ is a constant, $T_m$ is the thermodynamic melting point, and $\tau$ is the appropriate jump rate of "structural units," which for NaCl is equated with the Restrahlen frequency. Hence, from known parameters, $Z$ equals approximately $2.0 \times 10^{-6}$. To relate $R_c$ to viscosity $\eta$, the simplest general relationship is assumed, i.e.,

$$\tau = \frac{V\eta}{NkT_m} \tag{12.2.9}$$

where $V$ is the molar volume and $N$ is Avogadro's number. Hence

$$R_c = 2.0 \times 10^{-6} \frac{T_m^2 R}{V\eta} \tag{12.2.10}$$

where $R$ is the gas constant. This relationship is shown schematically in Figure 12.8 with $T_m$ as a parameter and is roughly in accord with experience.

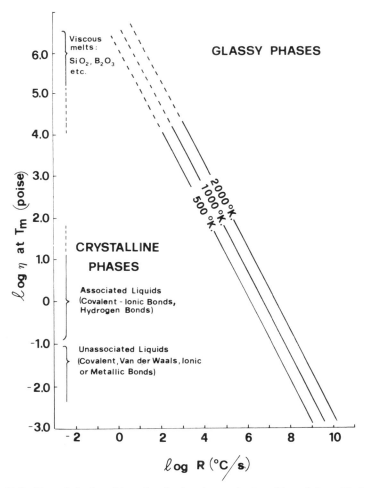

Figure 12.8. Plots of the logarithm of melt viscosity vs. the logarithm of the critical cooling rate at various melting temperatures. The diagram also indicates the ranges of viscosity for different types of liquid. A glassy phase is obtained only at cooling rates to the right of the lines; at lower cooling rates the melt crystallizes (after Sarjeant and Roy[12]).

A more quantitative evaluation of $R_c$ has been derived by Davies.[13] From an extension of the theory of nucleation and crystal-growth theory described in Section 12.2.3, he showed that the time $t$ (in seconds) required to form a small volume fraction $X$ of crystal is given by

$$t \simeq \frac{9.3\eta}{kT} \left\{ \frac{a_0^9 X}{f^3 \bar{N}_v} [\exp(1.07/T_r^3 \Delta T_r^2)][1 - \exp(-\Delta H_m \Delta T_r / RT)]^3 \right\}^{1/4} \quad (12.2.11)$$

where $a_0$ is the average atomic diameter, $f$ is the fraction of sites at the crystal/liquid interface where growth can occur, $\bar{N}_v$ is the average volume concentration of atoms, $T_r$ is the reduced temperature $T/T_m$, and $\Delta T_r$ is the reduced supercooling $(T_m - T)/T_m$.

If the temperature dependence of $\eta$ below $T_m$ is known, then equation (12.2.11) can be plotted in the form of a time–temperature–transformation (T–T–T) curve expressing the time required for a just detectable volume fraction $X$ of crystal to grow, as a function of temperature. An example, for $X = 10^{-6}$, is shown in Figure 12.9 for the elements Ge and Te.[13] The critical cooling rate $R_c$ is then given approximately by the linear cooling

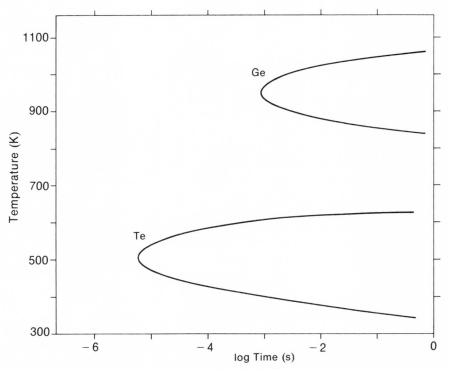

Figure 12.9. Time–temperature–transformation (T–T–T) curves for germanium (Ge) and tellurium (Te) (after Davies[13]).

## Table 12.1. Cooling Rates in Various Techniques

| | | |
|---|---|---|
| Annealing $\begin{cases} \text{Large telescope mirrors} \\ \text{Optical glasses} \\ \text{Ordinary glasses} \end{cases}$ | | $10^{-5}$ deg/s<br>about $3 \times 10^{-4}$ deg/s<br>$10^{-3}$–$10^{-2}$ deg/s |
| Air quenching — typical experimental<br>batches (10s of grams) | | 1–10 deg/s |
| Small drops (1–100 mg) into liquid Hg | | $10^2$–$10^3$ deg/s |
| Strip furnace (mgm batches) | | $10^3$–$10^4$ deg/s |
| Splat cooling | | $10^5$–$10^7$ deg/s |
| Evaporation, sputtering | | ? |

curve required to just miss the nose of the T–T–T curve, in the form

$$R_c \simeq \frac{T_m - T_n}{t_n} \tag{12.2.12}$$

where $T_n$ and $t_n$ are the temperature and time, respectively, corresponding to the nose. The value obtained for Te is $10^8$ K/s and for Ge is $10^6$ K/s. The latter is about two orders of magnitude less than the value estimated experimentally.

### 12..2.5. Universality of Glass Formation

To exceed the critical cooling rate $R_c$ is a necessary and sufficient condition for glass formation. It is also a *minimum* condition, i.e., *any* cooling rate in excess of $R_c$ will result in a liquid solidifying as a glass. In principle, therefore, it should be possible to quench any liquid into a glassy (vitreous) state and glasses are known to form from liquids in which the bonding is basically covalent, ionic, metallic, hydrogen bonded, or van der Waals bonded. Some materials, notably pure metals and simple ionic compounds such as NaCl, have not been prepared in a glassy form but so far as is known this must be a practical limitation of the cooling technique. The range of cooling rates available in a variety of techniques is listed in Table 12.1.

## 12.3. Kinetic (Nonequilibrium) and Thermodynamic (Equilibrium) Aspects of the Glass Transition

### 12.3.1. Liquid Supercooling

As explained in Section 12.2, a prerequisite for glass formation is the prevention of nucleation and crystal growth as a liquid is cooled below its melting (freezing) point. Once the temperature has been reduced below that at which the rate of crystal growth is a maximum, then further supercooling makes it more and more likely that the liquid will solidify to form a glass. The ease, or otherwise, of glass formation is therefore very sensitive to the dynamics of flow and diffusion processes in liquids and to the rate of cooling. In this sense glass formation is clearly a kinetic process but this says nothing about what happens when, having avoided crystallization, the supercooled liquid solidifies to a glass, i.e., the phenomenon associated with the glass transition temperature $T_g$, described in Section 12.1. Is this an equilibrium thermodynamic transition or a nonequilibrium kinetic (or relaxational, or dynamic) process? The two possibilities are fundamentally distinct. A thermodynamic mechanism arises from a structural or configurational change in the system (or from quantum-mechanical discreteness of energy levels) at some definable equilibrium temperature. A relaxational mechanism is a consequence of a deficiency in the experimental procedure; it results from changing the external forces acting on the system and making measurements before the system has had time to reestablish thermodynamic equilibrium.

Thermodynamic and relaxational mechanisms can in *principle* be distinguished experimentally. If some degree of freedom seems not to be contributing to a property of a system when a certain time has been allowed for equilibrium to be reached, the equilibration time is simply prolonged more and more to determine the limiting value of the property in question.[14,2] If the limiting value is *less* than that expected for full participation of all degrees of freedom of the system, then a thermodynamic change must have taken place. Such tests have been applied to some organic glasses that can be studied at relatively low temperatures, such as glycol and glycerol, and the results always seem to indicate that the glass transition is a relaxational phenomenon. For example, as illustrated schematically in Figure 12.10, in stabilization experiments on glasses with different values of $T_g$, the limiting equilibrium $V$ (or $H$, or $S$) of the metastable supercooled liquid can be approached from both sides.[14,2]

### 12.3.2. Kinetic or Relaxational (Nonequilibrium) View of the Glass Transition

Probably all experimental observations on the glass transformation are at least qualitatively consistent with the view that the phenomenon occurs because, as the temperature is reduced, certain "configurational" changes associated with variations in the volume of the supercooled liquid become so slow that they can no longer take place *on the time scale of the experiment.* Below the glass transition any further changes of $V$ (or $H$ or $S$) with temperature are associated with the temperature dependence of the

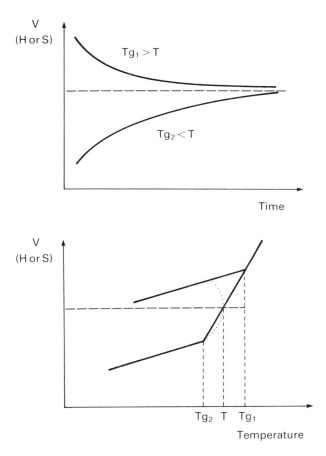

Figure 12.10. A glass may be brought into a metastable equilibrium state (i.e., stabilized) by a temperature treatment either above or below the glass transformation temperature. The dotted lines in the volume (or enthalpy or entropy) vs. temperature diagram (lower) illustrate this for $T_{g1} > T$ and $T_{g2} < T$. The time dependence of the approach to equilibrium is shown in the upper diagram.

(anharmonic) vibrational and rotational modes of the atoms or molecules. The expansion coefficient of the glass is therefore practically identical with that of the cyrstalline form of the material (as shown in the insert to Figure 12.1). If $X$ represents the extensive variables of a *liquid* ($V$, $H$, or $S$) then

$$\left(\frac{dX}{dT}\right)_{\text{liquid}} = \left(\frac{\delta X}{\delta T}\right)_{\substack{\text{vib,} \\ \text{rot,} \\ \text{etc.}}} + \left(\frac{\delta X}{\delta T}\right)_{\text{config}} \tag{12.3.1}$$

and it is the second term that is ineffective ("frozen-in") in the glass. Equivalent statements could be made about the effects of pressure and its derivative properties. It is very difficult, however, to quantify the *kinetics* of the process. Indeed the very nature of the configurational changes, which are supposed to be "frozen-in" at the glass transformation, are ill-understood but some indication of the processes contributing to the configurational relaxation (or prevented from contributing in the case of glass) is given by four experimental observations, three of which are generally true for a wide variety of glasses:

1. The transformation temperature $T_g$ determined from measurements of $\alpha$ and $C_p$ are very nearly the same.[14,2] There is no *a priori* reason for supposing that the relaxation rates for changes in heat content and volume should be related. The fact that they are apparently the same in the liquid $\leftrightarrow$ glass transition suggests that the molecular or atomic motions involved in the readjustment of energy and volume, following a change in temperature, are very similar.

2. The temperature coefficient for the relaxation rate in the glass-transformation region is very large (e.g., for volume changes in experiments such as those illustrated in Figure 12.10). This is a crucial factor but it is not easy to get quantitative information. Douglas[15] pointed out, however, that for *stabilized* glasses, at least, the stress relaxation is a simple linear phenomenon obeying the superposition principle. This allows a viscosity to be calculated from the stress-relaxation modulus and for stabilized oxide glasses at any rate, the activation energy for this viscosity is very high, e.g., approximately 7 eV. Jones[2] has also emphasized that viscosity activation energies *in the transformation range* are large and can be double the relevant bond energy. Thus the relaxation cannot represent an atomic or simple molecular diffusion process.

3. At $T_g$ most glasses have viscosities of about $10^{13.5}$ poise. From the simple rate theory of viscous flow this implies a "jump" rate of perhaps one every few minutes. This is about the same as the expected relaxation rate at $T_g$ since, according to the kinetic view, the transformation occurs when the relaxation rate is the inverse of the experimental observation time, which will normally be a few minutes or more.[14,16] Hence the relaxation processes in glass formation probably have a close similarity with the mechanisms of viscous flow. It must be recalled, however (cf. previous paragraph), that the activation energies for viscous flow in the transformation range of typical oxide glasses are very large.

4. There is sometimes a close relationship between dielectric relaxa-

tion and the glass transformation, i.e., the dielectric relaxation time is about one minute to one hour at $T_g$.[17] This observation has only been made, however, in certain organic glasses.

The inference from the above observations is that at the glass ↔ liquid transition, atomic or molecular motions associated with phenomena as diverse as the redistribution of energy and volume, viscous flow and diffusion processes, and dipole reorientation, all occur at about the same frequency. One can account for this by assuming that for *any* type of molecular motion to occur a temporary disruption of rearrangement of the liquid configuration is required in the neighborhood of the atom or molecule. Once this rearrangement has taken place any type of molecular motion is equally likely. In a sense, therefore, the process is cooperative, probably involving a large number of atoms or molecules. The very large activation energies for relaxation in the transformation region are particularly relevant to this point. When the temperature is reduced below $T_g$ the configurational degrees of freedom are ineffective and it is obvious that $T_g$ will depend on the relaxation rate and the duration of the "experiment" (i.e., the rate of cooling).

The situation can be aptly summarized by the use of a dimensionless number familiar in rheology, namely the Deborah number, DN,[18] where

$$DN = \frac{\text{Time of relaxation, } \tau}{\text{Time of observation}}$$

The difference between "solids" and fluids is then defined by differences in DN. If the time of observation is large, *or* $\tau$ is small, DN is small and the consequences of the relaxation process (whatever they may be, e.g., plastic flow) can be observed. On the other hand, if DN is large the material behaves as a solid. Let us consider this in terms of the $V-T$ diagram for a glass (such as Figure 12.1) and the relaxation time for the structure to adapt to a change in temperature. In the liquid, the relaxation time will be perhaps $10^{-18}$–$10^{-13}$ s (depending on the type of liquid) and hence DN is very small for normal experimental times. As the temperature is reduced to $T_g$ the relaxation time approaches a few minutes, *i.e.*, $T_g$ *is defined as the temperature where* DN = 1. In the glassy state ($T < T_g$) $\tau$ increases rapidly, the structure cannot change to its appropriate equilibrium configuration, and DN becomes very large. In a normal glass-formation experiment the time of observation is kept fixed, and DN is changed by varying $\tau$ through its temperature dependence. An analogous experiment, however, would be to change DN by keeping the temperature, and hence $\tau$, fixed and varying the time of observation. This would be most conveniently done by conducting measurements as a function of frequency,

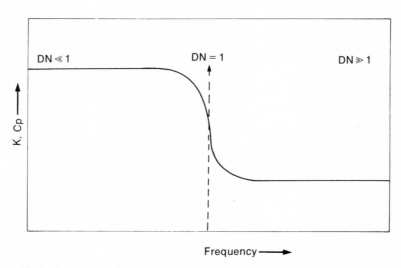

Figure 12.11. A schematic diagram for the compressibility $\kappa$ or specific heat $C_p$ for a glass (or a liquid) measured as function of frequency, where DN denotes the Deborah number (after Stevels[18]).

e.g., measuring the specific heat $C_p$ or compressibility $\kappa$ of a liquid by ultrasonic absorption.[19] The expected results of such an experiment are shown schematically in Figure 12.11. At low frequencies, the time of observation ($= 1/2\pi f$) is long compared with $\tau$ and DN is very small. When $\tau = 1/2\pi f$, DN = 1, and if the frequency is increased further a typical sigmoid-type relaxation curve is obtained. At high frequencies DN becomes very large. The particular temperature at which the experiment is performed could be regarded as a glass-transformation temperature corresponding to the observation time $1/2\pi f_{mid}$, where $f_{mid}$ is the frequency of the midpoint in the relaxation curve. This is completely consistent with the kinetic view of the liquid ↔ glass transition. As in all relaxation processes, the sharp decrease in $\kappa$ or $C_p$ as the frequency increases through $f_{mid}$ is caused by the inability of the system to respond to the external stimulus. In this situation, however, the liquid is at all times in internal thermal equilibrium, i.e., the "glassy" state corresponding to $f > f_{mid}$ is an *equilibrium* glass.[19] It should also be realized that several different mechanical and electrical relaxation processes are possible in liquids and glasses[15,19,20]; only when the relaxation involves configurational or structural changes is it realistic to equate it with a glass transformation. In organic polymers, there are often structural relaxation processes connected with, for instance, side groups. These processes usually freeze-in at temperatures below the main glass transformation and are referred to as secondary glass transitions.

### 12.3.3. Limitation of the Relaxational Approach

Although the available experimental evidence favors the relaxational model, the enormous temperature coefficients of relaxation make tests such as that illustrated in Figure 12.10 almost impossible to apply unequivocally.[14,2] It is difficult to eliminate the possibility that there is a real thermodynamic transition to a glassy state at temperatures below normally attainable values of $T_g$.

Kauzmann[17] posed the following question: what behavior would liquids show at low temperatures if enough time could always be allowed to avoid vitrification (glass formation)? Would it be found that a "nonvitreous" liquid could exist in some kind of metastable equilibrium close to 0 K?

Figure 12.12 compares the trends in temperature dependence of some crystals and supercooled melts above $T_g$ and extrapolated to lower

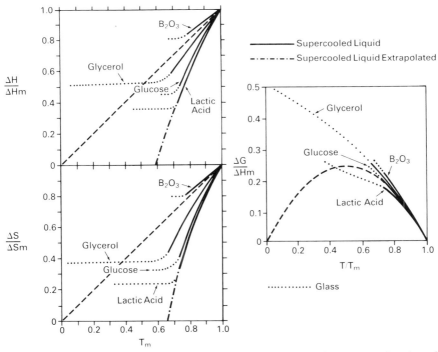

Figure 12.12. Enthalpy ($\Delta H$), entropy ($\Delta S$), and free-energy ($\Delta G$) plots as a function of temperature for some typical glass-forming liquids. In each case the parameters are normalized to the appropriate quantity at the thermodynamic melting point (after Kauzmann[17]).

temperature.[17] Note that for glucose, $S$ for the liquid phase rapidly approaches that of the crystal (similarly for $H$ and $V$). This is even more marked for lactic acid but less so for glycerol, and probably not at all for boric oxide ($B_2O_3$), although the latter becomes glassy at relatively high temperatures. It seems unlikely that $S_{liq}$ could ever become *less* than $S_{cryst}$, but it could by argued that these results do suggest that the supercooled liquid may pass continuously into the crystalline state (analogous to the liquefaction of gases at the critical pressure). There seems little justification for this, however, as the $S$-curves do not approach the abscissa at the same temperature as the $H$-curves, and the free energies of the two phases show no tendency to approach each other down to $T_g$.[17]

How, therefore, can the curves in Figure 12.12 be extrapolated below $T_g$? Probably the "nonvitreous" liquid curves of $\Delta S$ vs. $T$ [$\Delta \equiv$ (liquid − crystal)] must become horizontal not far below $T_g$, and since

$$\frac{\delta \Delta H}{\delta T} = T \frac{\delta \Delta S}{\delta T} \tag{12.3.2}$$

this would imply a similar change in the slope of $\Delta H$ vs. $T$. In turn, this means that

$$\Delta C_p = 0$$

i.e.,

$$(C_p)_{liquid} = (C_p)_{crystal}$$

But this is exactly what happens at the glass transition, which has been interpreted as a relaxation process unconnected with equilibrium thermodynamics! Is this a coincidence, or might the glass transition be a thermodynamic phenomenon, at least in some circumstances?

## 12.4. Thermodynamics of the Glass Transition

### 12.4.1. Introduction

There are two problems involved in the thermodynamics of the glass transition. First, can the nonequilibrium (thermodynamically unstable) glassy state, which solidifies at normally attainable values of $T_g$, be treated by a thermodynamic formalism? Second, does a truly equilibrium transition occur (presumably at lower values of $T_g$) if cooling rates are made slower and slower (i.e., the problem posed in Section 12.3), and if so, what is its thermodynamic status? The first question is relatively easy to answer.

### 12.4.2. Irreversible Thermodynamics of the Nonequilibrium Glass[21]

The problem is to relate the thermodynamic properties of the non-equilibrium glass to those of the *equilibrium* liquid from which it was formed. Two relevant properties are, for example, the coefficient of expansion

$$\alpha_{liq} = \frac{1}{V}\left(\frac{\partial V}{\partial T}\right)_P$$

and the compressibility

$$\kappa_{liq} = -\frac{1}{V}\left(\frac{\partial V}{\partial P}\right)_T$$

where the subscript "liq" is used to denote "liquid." (Later the subscript "g" will be used to denote "glass.") For liquids these definitions are sufficient since $V = V(P, T)$. To describe the glass, at least one extra parameter must be introduced and this is denoted by $z$. A necessary property of $z$ is that it shall be constant for a *given* glass [e.g., it will be different along the lines 1, 2, and 3 in Figure 12.2(a)] and vary with $T$ (and $P$) in the liquid.

Thus

$$V = V(P, T, z)$$

for a glass while

$$V = V(P, T, z) \qquad \text{but } z = z(P, T)$$

for the liquid. The physical significance of $z$ is unimportant at the moment but it could be regarded as a measure of, say, the configurational order, or viscosity, or "free volume"; or we could equate it to the fictive temperature $T_F$ (see Section 12.1.2). It is usually referred to simply as an "ordering parameter." The statements made above imply that a single ordering parameter is adequate to define the state of a glass. This further implies that two glasses brought to the same volume and temperature by different routes [e.g., the glass at $V_2$, $T_{g_2}$ in Figure 12.2(a)] are in the same thermodynamic state with the same $H$ and $S$ as well as $V$, i.e., a state completely defined by $P$, $T$, and $z$.

There are several ways to proceed from here. In the formalism of irreversible thermodynamics developed by de Donder[21,22] three assumptions are made:

1. Thermodynamic properties are all single-valued functions of $P$, $T$, and $z$.

2. During an irreversible process associated with changing $z$ there is an increase in entropy $S$ given by

$$TdS_{irr} = dQ + Adz$$

where $A$ is a thermodynamic state function called the affinity.
3. For all realizable changes $Adz > 0$.

Irreversibility is allowed for by the introduction of the term $Adz$ and assumption (3) means that, in equilibrium, $A = 0$. It should be clear from the preceding discussions that from the kinetic viewpoint the glass transformation is not a unique phenomenon. It is only one type of relaxation process. Others are to be found in viscoelastic behavior (often closely related to the glass transformation), electric polarization (which may or may not be related to glass transformation), magnetic polarization, chemical reactions, and so on.[23] The approach via irreversible thermodynamics is equally applicable to all.

By applying the above definitions and assumptions to the relationship $V = V(P, T, z)$ we obtain

$$\frac{dT_g}{dP} = \frac{\Delta\kappa}{\Delta\alpha} \tag{12.4.1}$$

where $\Delta\kappa = \kappa_{liq} - \kappa_g$ and $\Delta\alpha = \alpha_{liq} - \alpha_g$.

Similarly, from the relationship $H = H(P, T, z)$ or $S = S(P, T, z)$,

$$\frac{dT_g}{dP} = \frac{TV\Delta\alpha}{\Delta C_p} \tag{12.4.2}$$

where $\Delta C_p = C_{p\,liq} - C_{pg}$. We note straight away that both equations predict that $T_g$ increases with pressure, as is observed experimentally (see Section 12.1). Equations (12.4.1) and (12.4.2) yield

$$\frac{\Delta C_p \Delta\kappa}{TV\Delta\alpha^2} = 1 \tag{12.4.3}$$

Equations analogous to (12.4.1), (12.4.2), and (12.4.3) can also be derived from the usual treatment of an equilibrium second-order thermodynamic transition; in that context they are sometimes called the Ehrenfest equations.[24] Their appearance here should not be construed as meaning that the glass transformation is a true second-order thermodynamic transition — the conditions for that conclusion are much more restrictive. Before considering equations (12.4.1)–(12.4.3) in more detail it is worth mentioning some other results of the thermodynamics.

Instead of using the undefined parameter $z$, variables more suited to practical problems could be chosen and incorporated in the formalism. In calorimetric experiments on glass the most convenient parameter is the fictive temperature $T_F$, which in the present terms would be defined by

$$A(P, T_F, z) = 0 \quad \text{with } T_F = T_F(P, z) \quad (12.4.4)$$

If we introduce a linear approximation and use the above definition, we derive

$$A(P, T, z) = A(P, T_F, z) + (T - T_F)\left(\frac{\delta A}{\delta T}\right)_{P, z} + \cdots$$

$$= (T - T_F)\left(\frac{\delta A}{\delta T}\right)_{P, z} + \cdots \quad (12.4.5)$$

With appropriate thermodynamic relationships, such as

$$dQ = dE + P dV = T dS_{irr} - A dz$$

a variety of results can be derived. For instance,

$$T dS_{irr} = A dz = - V \Delta\alpha(T - T_F)dP + \left(\frac{\Delta C_p}{T_F}\right)(T - T_F)dT_F \quad (12.4.6)$$

At constant pressure this gives

$$dS_{irr} = \Delta C_p \left(\frac{1}{T_F} - \frac{1}{T}\right) dT_F \quad (12.4.7)$$

From the second law it follows that

$$\frac{dT_F}{dt} \gtreqless 0 \quad \text{according as } T - T_F \gtreqless 0 \quad (12.4.8)$$

This means that $dS_{irr}/dT \to 0$ as $T_F \to T$, *in whatever direction*. In other words, the approach of a glass to equilibrium is a consequence of the entropy principle just as in any other system, but equilibrium can be approached from temperatures either above or below $T_F$. This is, in effect, a description of the stabilization phenomena described in Figure 12.10. A fictive pressure $P_F$ can also be defined by

$$A(P_F, T, z) = 0 \quad \text{with } P_F = P_F(T, z)$$

Using the same linear approximation, it can be shown that

$$\frac{T - T_F}{P - P_F} = -\frac{VT\Delta\alpha}{\Delta C_p} = \frac{\Delta\kappa}{\Delta\alpha} \qquad (12.4.9)$$

Thus a sudden isobaric change in temperature $\delta T$, which leaves $T_F$ unchanged, is thermodynamically equivalent to a pressure increment of

$$\delta P = -\frac{\Delta C_p}{TV\Delta\alpha}\,\delta T \qquad (12.4.10)$$

This can become quite large (thousands of atmospheres) and is effectively the driving force tending to stabilize the glass.

Most of the discussion of the thermodynamics of glass has revolved around the validity or otherwise of equations (12.4.1)–(12.4.3). The lack of reliable data makes it difficult to establish whether or not equation (12.4.3) is valid. Generally, however, the experimental evidence[21] is that

$$\frac{\Delta C_p \Delta\kappa}{TV\Delta\alpha^2} > 1$$

This quantity attains a value in the region of 1.2–2.4 in Se, for instance, and approximately 1.6 in a borosilicate glass. Equations (12.4.1)–(12.4.3) are a direct consequence of the assumption that a single ordering parameter can describe the thermodynamic state of a glass (together with $P$ and $T$, of course). The fact that equation (12.4.3) is not obeyed means that a single parameter is inadequate, i.e., that two glasses brought to the same $V$ and $T$ by different routes are *not* necessarily in the same thermodynamic state. Either of equations (12.4.1) or (12.4.2) could still be independently valid, however, and a closer consideration of their implications can give some insight into the factor(s) determining the glass transition.[25] They are, as emphasized before, both the result of a one-parameter description using an ordering parameter.

An alternative description of glass formation is to introduce an "excess function," which is a function of temperature and pressure, and to suppose that the liquid ↔ glass transformation occurs when the excess function goes to zero, or becomes constant. For instance, the notion of unoccupied, or "free," or excess volume $V_e$ has featured prominently in many discussions of the glass transformation, especially in polymers. The glass transition is then equated with the condition $V_e \to 0$ or constant. Thus if we write[25]

$$dV_{liq} = dV_g + dV_e$$
$$= V\alpha_g dT - V\kappa_g dP + V\Delta\alpha dT - V\Delta\kappa dP \qquad (12.4.11)$$

in a glass (i.e., a "fast" experiment; $DN \gg 1$) only the first two terms contribute to $dV_e$. One could equally well postulate, however, that at the glass transformation an excess entropy $S_e \to 0$ (or constant). The question of the adequacy of a single ordering parameter then becomes equivalent to asking if the excess thermodynamic variables are functions of a single parameter only, i.e.,

$$V_e = V_e(z), \quad S_e = S_e(z), \quad H_e = H_e(z); \quad z = z(P, T)$$

Put in this way $z$ is a redundant variable since constant $V_e$, $S_e$, and $H_e$ implies constant $z$, and vice versa. Thus $V_e$, say, could be adopted as an independent variable to characterize the glass and the sufficiency of a single ordering parameter is then equivalent to the conditions

$$H_e = H_e(V_e), \quad S_e = S_e(V_e), \quad V_e = V_e(P, T)$$

If $P$ and $T$ are changed so as to keep $V_e$ constant, then

$$dV_e = \left(\frac{\partial V_e}{\delta T}\right) dT + \left(\frac{\delta V_e}{\delta P}\right) dP$$

$$= V\Delta\alpha dT - V\Delta\kappa dP$$

$$= 0 \qquad \text{at constant } V_e$$

Therefore $dP = (\Delta\alpha/\Delta\kappa)dT$.

When applied to the glass transition this is identical with equation (12.4.1).[25] These arguments therefore imply that if $V_e$ determines $T_g$, then

$$\frac{dT_g}{dP} = \frac{\Delta\kappa}{\Delta\alpha}$$

i.e., equation (12.4.1). Similar reasoning shows that,[25] if $S_e$ or $H_e$ determine $T_g$, then

$$\frac{dT_g}{dP} = \frac{TV\Delta\alpha}{\Delta C_p}$$

i.e., equation (12.4.2). Reliable experimental data are again scarce but results pertaining to a variety of glasses, including $S_e$, boric oxide, and several vitreous organic polymers, show that the latter equality holds, but not the former.[25,26]

One must conclude, therefore, that a one-parameter description is inadequate but that the glass transformation occurs when $S_e$ (or $H_e$)

reaches a critical value, i.e., when excess entropy or enthalpy associated with molecular motion in the liquid becomes "frozen-in." It is not possible by these arguments to distinguish between the two variables.

It is convenient to conclude this section with brief comments on the thermodynamic status of the glass transition. If either of equations (12.4.1) or (12.4.2) is valid (but not both), this implies that on the free energy $(G)-P-T$ surface of a liquid there is an infinite set of points, each of which defines a *distinct* glass.[25] If both equations (12.4.1) and (12.4.2) hold, i.e., equation (12.4.3) is valid, there are lines of constant $V_e$, $S_e$, $H_e$, and $z$ along the $G-P-T$ surface of the liquid, and glasses derived from any point along these lines would be identical. Goldstein[25] points out that only if the lines of constant $V_e$, etc., coincided with lines of constant relaxation time could equations (12.4.1)–(12.4.3), as applied to glass, be considered as equivalent to the Ehrenfest equations for a true second-order thermodynamic transition. If the time scale of the experiment is altered, however, the location of the transition on the $G-P-T$ surface is changed. One must still conclude therefore that, as normally observed, the glass transformation is a relaxation process.

### 12.4.3. A Finite Limit to the Glass Transition Temperature

The question of whether a thermodynamically equilibrium transition would occur if the cooling rate were made slower and slower is equivalent to asking whether there is a finite limit to $T_g$, i.e., is there a $T_{g\ \text{limiting}} > 0$ K?

It is first relevant to observe the trend of configurational (or excess) entropy $S_{\text{config}}$ for a supercooled liquid and glass. Many years ago Simon calculated $S_{\text{config}}$ from calorimetric data (latent and specific heats) for some low-temperature organic glasses and typical data for glycerol are shown in Figure 12.13.[27] Note first that in the glassy state $(T \le T_g)$, $S_{\text{config}}$ has a constant value of about 5 eu and the curve apparently extrapolates down to 0 K with little change in $S_{\text{config}}$. This is not a violation of the third law, however, as the glass is not in thermal equilibrium. The constant $S_{\text{config}}$ of about 5 eu is the entropy frozen-in to the glass, because the relaxation times for configurational rearrangement exceed experimental times. Thus much of the configurational disorder (entropy) of the liquid relative to the crystalline solid remains trapped in the glass. This observation implies that there is not an "ideal" glassy state with a *unique* structure; the structure is whatever the liquid structure happens to be when the glass solidifies at $T_g$. Anderson[28] and Goldstein[29] have interpreted this in terms of a potential energy $V(x)$ vs. configuration space diagram with a large number of more or less equivalent energy valleys, separated by "mountain passes" whose height is variable within a certain range, but almost all of which become impassable much below $T_g$; see Figure 12.14.

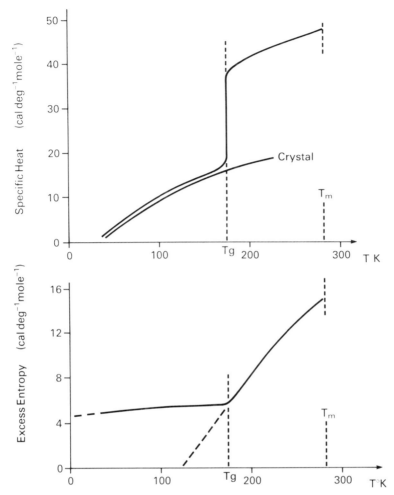

Figure 12.13. The upper diagram is a plot of experimental values of the specific heat of vitreous glycerol ($T < T_g$) and supercooled glycerol ($T > T_g$), with data for the crystal shown for comparison. The lower graph is the entropy–temperature diagram for the formation of vitreous glycerol calculated from the specific-heat data (after Wilks[27]).

Of particular significance in the present context is the observation that $S_{config}$ for the supercooled liquid (operationally, the excess over the crystalline entropy) seems to extrapolate to zero at a finite temperature $T_0$ at a relatively small temperature difference below the normally attainable $T_g$ (this is simply another way of stating the problem discussed in Section 12.3.3 and Figure 12.12). For many molecular and polymeric glasses $T_g - T_0$ is only about 40–60 K.

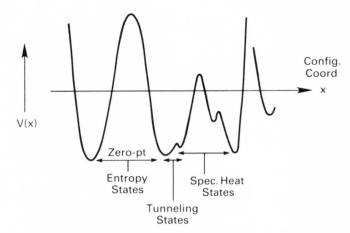

Figure 12.14. A conceptual model of the glassy state in terms of a potential energy vs. configuration space diagram. "Valleys," which are more or less equivalent in energy, are separated by "mountain passes" of variable height (after Anderson[28]).

It will be recalled from clause (3) at the end of Section 12.1.2 that over a wide range of temperature the viscosity of many glass-forming liquids obeys the empirical equation

$$\eta = C \exp[E_\eta / k (T - T_0)] \qquad (12.4.12)$$

In fact, most transport properties of glass-forming liquids, such as fluidity (inverse viscosity), diffusion, conductance, and dielectric relaxation, seem to follow an equation of identical form, so that the characteristic relaxation time $\tau$ associated with a particular property is given by

$$\tau = \tau_0 \exp[A / (T - T_0)] \qquad (12.4.13)$$

Equation (12.4.13) seems to be uniquely and universally applicable to glass-forming liquids, and its particular feature is that as $T \to T_0$ the slope becomes almost infinitely steep so that all physical processes are essentially frozen-out before $T_0$ is reached. Thus $T_0$ is an experimentally inaccessible singularity but it can be estimated by extrapolation of data from higher temperatures. The striking fact is that *where calorimetric data* (i.e., as in Figure 12.13) *and relaxational data are both available, the two values of $T_0$ seem to coincide.*[28,29] The temperature $T_0$ has therefore been interpreted as a limiting value of $T_g$ as cooling rates are made slower and slower. It has also been suggested that $T_0$ is a thermodynamic parameter at which an equilibrium glass transition would presumably occur if it were experimentally attainable.[30] These are very speculative suggestions, however, and to date no formalized theories have been developed, although the correlation between the calorimetric and relaxational values of $T_0$ appears to be established.[28-30]

## 12.5. Models of the Glass Transition

### 12.5.1. Free Volume

Various models of the glass transition have been proposed, but those based on the idea of "free volume" have featured most prominently. The first free-volume theories were proposed by Turnbull and Cohen,[31-33] and they have been developed subsequently by others.[34] A particularly simple approach given by Gee[26] is outlined here.

The total volume of a liquid is divided into:

1. A part "occupied" by the molecules (or atoms),
2. A part in which the molecules are free to move.

The contribution (2) is termed "free volume" and is assumed to be shared communally by a process in which "holes" of varying size and location are continually moving and being redistributed. As the liquid is cooled, both occupied and free volumes contract. A glass is then distinguished from a liquid in two ways:

1. In the glassy state the free volume remains constant, independent of temperature.
2. The redistribution of free volume no longer occurs; the "holes" remain fixed in the positions they occupied when the glass was formed.

The glass–liquid relationship can then be expressed as follows. Let $\alpha_{liq}$ be the expansion coefficient of the liquid (i.e., of occupied + free volume) and $\alpha_g$ the expansion coefficient of the glass (i.e., of occupied volume only), where $\alpha_{liq} - \alpha_g = \Delta\alpha$ is the expansion coefficient of free volume.

The volume–temperature relationships are illustrated schematically in Figure 12.15, and with the *total* volume $V_g$ at $T_g$ as a reference point the free volume $V_f$ at $T > T_g$ is given by

$$V_f = V_{fg} + V_g \Delta\alpha (T - T_g) \tag{12.5.1}$$

where $V_{fg}$ is the free volume of the glass.

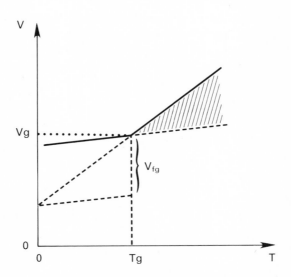

Figure 12.15. A schematic volume–temperature diagram for a glass, illustrating the idea of free volume. In this particular picture the free volume in the glassy state, $V_{\mathrm{fg}}$, is fixed and corresponds to the actual volume of the glass minus the volume of the liquid extrapolated to absolute zero but corrected for the temperature dependence of the volume (i.e., expansion) of the *glass* (after Gee[26]).

The same arguments can be applied to the effects of pressure, and with

$$\Delta\kappa = \kappa_{\mathrm{liq}} - \kappa_{\mathrm{g}}$$

where $\kappa_{\mathrm{g}}$ is the compressibility of occupied volume, the combined effects of $T$ and $P$ on the *free volume of the liquid* can be expressed in the form

$$V_{\mathrm{f}} = V_{\mathrm{fg}} + V_{\mathrm{g}}[\Delta\alpha\,(T - T_{\mathrm{g}}^0) - P\Delta\kappa] \tag{12.5.2}$$

where $T_{\mathrm{g}}^0$ is the glass transition temperature at $P = 0$.

If $T_{\mathrm{g}}$ is associated with the temperature at which the free volume of the liquid tends to $V_{\mathrm{fg}}$, then it follows that

$$\frac{T_{\mathrm{g}} - T_{\mathrm{g}}^0}{P} = \frac{\Delta\kappa}{\Delta\alpha} \tag{12.5.3}$$

It is noteworthy that this is essentially the same result as equation (12.4.1), namely

$$\frac{dT_{\mathrm{g}}}{dP} = \frac{\Delta\kappa}{\Delta\alpha}$$

Several arguments have been used to estimate $V_{fg}$. For example, if the volume obtained by extrapolating the liquid volume to $T = 0$ K is equated to the occupied volume at $T = 0$ K then, approximately,

$$V_{occ}(T = 0) = V_g(1 - \alpha_{liq} T_g) \tag{12.5.4}$$

The total volume of the glass at $T = 0$ K is $V_g(1 - \alpha_g T_g)$, therefore

$$V_{fg} = V_g(1 - \alpha_g T_g) - V_g(1 - \alpha_{liq} T_g)$$

$$= V_g \Delta \alpha T_g$$

i.e.,

$$\frac{V_{fg}}{V_g} = T_g \Delta \alpha \tag{12.5.5}$$

This has been measured for a number of glasses, particularly polymeric glasses, and experimentally $T_g \Delta \alpha$ varies from about 0.08 to 0.13, i.e., *at $T_g$, about 11% of the total volume is free volume.*

Note that free-volume theories are basically thermodynamic models. The liquid → glass transition occurs when an excess function of the liquid — free volume in this case — falls below some critical value and becomes constant (or zero) (see Section 4.4.2).

### 12.5.2. Vibrational Models of the Glass Transition

Glass formation is traditionally discussed by relating glass to its parent liquid phase and this is obviously a natural approach. There have also been attempts to relate the glass transition to vibrational characteristics of the solid. This is a "solid-state" approach to the problem and it predicts a close correlation between the glass transformation temperature $T_g$ and the Debye temperature *of the amorphous solid.*[35] It is rationalized on the basis that, near $\theta_D$, phonon–phonon interactions become dominant and there is a high probability of cooperative phonon interactions generating sufficient local energy to allow "molecules" to cross the energy barrier preventing their translational motions. The principal experimental evidence is that, for a wide variety of glasses (such as Se, $As_2S_3$, $As_2Se_3$, Au–Ge–Si alloys, various silicates, and $ZnCl_2$), the *molar* heat capacity approaches the limiting Dulong–Petit value $3nR$ at $T_g$.[35,36] Boric oxide ($B_2O_3$) is an exception; $C_p$ is only approximately $2nR$ at $T_g$, indicating $\theta_D > T_g$.

A simpler rationalization of the $T_g \sim \theta_D$ rule for network glasses is to consider the glass transition as a bond-breaking process.[37] For translational motion of nearest neighbors some critical energy $E$ must be supplied to nearest-neighbor bonds. The probability of attaining $E$ is small until $T$ reaches values where phonons of sufficient energy become available to cause the population of the higher vibrational levels of nearest-neighbor bonds. But this is the temperature $\theta_D$ at which $\nu_D = kT/h$, where $\nu_D$ is presumably to be identified with the nearest-neighbor bond frequency.[37] The failure of boric oxide to follow the $\theta_D \sim T_g$ rule is explained [36] on the assumption that *intramolecular* bonds between the two-dimensional sheets of $BO_3$ triangles are weak compared with the B–O bonds. In other words, a glass may have several characteristic temperatures associated with various types of bonds and $T_g$ may correspond to the lowest, or one of the lower of these.

The basic premises for this approach are not well established, however, and Schnaus *et al.*[37] point to a notable inconsistency. The two compounds $As_2S_3$ and $As_2Se_3$ have the same structure and their values of $T_g$ are nearly the same (445–462 K for $As_2S_3$ and 436–450 K for $As_2Se_3$). In both cases $C_p \to 3nR$ as $T \to T_g$, but from $C_p$–$T$ curves Schnaus *et al.* obtain

$$\theta_D = 450 \pm 50 \text{ K} \qquad \text{for } As_2S_3$$

and

$$\theta_D = 350 \pm 50 \text{ K} \qquad \text{for } As_2Se_3$$

Thus for $As_2Se_3$, $\theta_D$ is appreciably *less* than $T_g$. Schnaus *et al.* also find that the ratio $\theta_D(As_2S_3)/\theta_D(As_2Se_3)$ is roughly equal to the ratio of the onset frequencies of As–S and As–Se vibrational bands in the infrared, and also to the square root of the ratio of reduced masses for the two types of bond. The bond force constants are therefore not very different. Schnaus *et al.* point out that it is probably not surprising that there is no correlation betweeen characteristic vibrational temperatures and the glass transition.[37,38] In this solid-state transition model the glass transition depends primarily on the *depth* of a potential well $E$, which is related to the transition frequency by

$$\nu = \nu_0 \exp(-E/kT)$$

where $\nu$ is a bond vibrational frequency and will usually be $10^{12}$–$10^{13}$ s$^{-1}$. At the glass transition $\nu$ is normally approximately $10^{-2}$ s$^{-1}$ and is little affected by $\nu_0$. On the other hand, characteristic vibrational temperatures are related to $\nu_0$ and this depends strongly on the *shape* of the potential-energy curve near its minimum. The shape and depth of the potential-energy well are not necessarily related and hence there is no reason to expect a strong correlation between $\theta_D$ and $T_g$.

### 12.5.3. Structural Models

A variety of structural models have been suggested, particularly making use of computer calculations of molecular dynamics.[39] Although not strictly a model of the glass transition, an interesting calculation of the configurational entropy of fused silica has been proposed by Bell and Dean.[40] They constructed a random network of $SiO_2$ by considering the number of ways in which an $SiO_4$ tetrahedron could be added to a growing surface. At typical surface sites the number of possibilities tends to be limited for various reasons:

1. The Si–O–Si angles must be kept in a particular range.
2. One must ensure that further tetrahedra can be added at a later stage.
3. In particular, the necessity of avoiding configurations that represent overlapping molecules severely restricts the number of possibilities.

On adding a single tetrahedron, the following situations must be taken into account:

1. If it bonds *triply* to the existing structure, there is one attitude.
2. If it bonds *doubly* to the existing structure, one or two attitudes are possible.
3. If it bonds singly to the existing structure, it would be unusual for there to be more than four possible attitudes.

At the most, therefore, the corresponding entropies are $k \ln 1$, $k \ln 2$, and $k \ln 4$. The maximum entropies occur if it is assumed that all possible attitudes are equally likely. To achieve an isotropic network it is necessary to ensure that the number of free bonds per unit area of surface remains roughly constant and it is also assumed therefore that triple, double, and single bonding situations occur with relative probabilities $(1 - \varepsilon)$, $(1 - \varepsilon)$, and $(\varepsilon/2)$, where $\varepsilon$ is a constant such that $0 < \varepsilon < 1$. Hence

$$S_{\text{config}} = R \left[ \frac{\varepsilon}{2} \ln 1 + (1 - \varepsilon) \ln 2 + \frac{\varepsilon}{2} \ln 4 \right]$$

$$= 1.38 \text{ cal deg}^{-1}\text{mol}^{-1} \tag{12.5.6}$$

This result was obtained assuming 1, 2, and 4 possible attitudes for triple, double, and single bonding situations. These are upper limits and experience suggests that $1$, $\frac{3}{2}$ and $\frac{5}{2}$ would be more realistic estimates. Consequently

$$S_{\text{config}} = R \left[ \frac{\varepsilon}{2} \ln 1 + (1 + \varepsilon) \ln \frac{3}{2} + \frac{\varepsilon}{2} \ln \frac{5}{2} \right]$$

$$= 0.8\text{–}0.9 \text{ cal deg}^{-1}\text{mol}^{-1} \tag{12.5.7}$$

depending on the value of $\varepsilon$. Experimental values of the configurational entropy of used silica are in the range 0.67–1.08 cal deg$^{-1}$mol$^{-1}$.

### 12.5.4. Statistical Thermodynamics

Statistical thermodynamic treatments have been developed for *polymeric* glasses and the first such model was proposed by Gibbs and Di Marzio[41] based on the definition of an equilibrium transition temperature $T_0$ at which the configurational entropy is zero. This temperature can be regarded as the limiting value of $T_g$ as the cooling rate is made slower and slower (see Section 12.4.3 and Figure 12.13). The configurational entropy $S_{config}$ is calculated by counting the number of ways in which $n_x$ linear molecules, each $x$ segments long, can be placed on a diamond lattice of coordination number $z = 4$, together with $n_0$ holes. The restrictions imposed on the placing of a molecule on the lattice are incorporated in the hindered rotation, which is expressed as the "flex energy" $\Delta\varepsilon$, and $\varepsilon_h$, which is the energy of formation of a hole. The flex energy is the energy difference between the potential-energy minimum of the located bond and the potential minima of the remaining $(z - 2)$ possible orientations available on the lattice. In the case of polyethylene, for example, the *trans* position is considered the most stable and the *cis* positions are the flexed orientations with $\Delta\varepsilon$ the energy between the ground (stable) and flexed states. The details vary with the nature of the polymer molecule. The quantity $\varepsilon_h$ is a measure of the cohesive energy. The configurational entropy is derived from the partition function describing the location of polymer molecules and holes.

As the temperature drops toward $T_0$ the number of available configurational states in the system decreases, until at $T_0$ the system possesses only one degree of freedom and this condition leads to[41]

$$\frac{S_{config}(T_0)}{n_x k T_0} = 0 = \phi\left(\frac{\varepsilon_h}{kT_0}\right) + \lambda\left(\frac{\Delta\varepsilon}{kT_0}\right) + \frac{1}{x}\ln\left\{[(z-2)x+2]\frac{(z-1)}{2}\right\} \quad (12.5.8)$$

where

$$\phi\left(\frac{\varepsilon_h}{kT_0}\right) = \ln\left(\frac{\varepsilon_h}{S_0}\right)^{[(z/2)-1]} + \frac{f_0}{f_x}\ln\left(\frac{f_0}{S_0}\right)$$

and

$$\lambda\left(\frac{\Delta\varepsilon}{kT_0}\right) = \frac{x-3}{x}\ln\left[1 + (z-2)\exp(-\Delta\varepsilon/kT)\right.$$
$$\left. + (\varepsilon/kT)\frac{(z-2)\exp(-\Delta\varepsilon/kT)}{1+(z-2)\exp(-\Delta\varepsilon/kT)}\right]$$

The fractions of occupied and unoccupied sites are $f_x$ and $f_0$, respectively, and $S_0$ is a function of $f_x$, $f_0$, and $z$.

The weaknesses of the Gibbs and Di Marzio theory are:

1. A polymeric molecule of zero stiffness would have a $T_g$ of 0 K.
2. $T_g$ is essentially independent of molecular interactions.

The temperature $T_0$ is calculated to be about 50 K below normally attainable values of $T_g$ and it is not, of course, an experimentally measurable quantity. In this theory $T_0$ is regarded as a true thermodynamic (equilibrium) *second-order* transition temperature and it is essentially the same $T_0$ as discussed in Section 4.4.3.

# References

1. G. W. Morey, *The Properties of Glass*, 2nd edn., Reinhold, New York (1954).
2. G. O. Jones, *Glass*, 2nd edn. (revised by S. Parke), Chapman and Hall, Science Paperbacks, London (1971).
3. Cited by J. Wong and C. A. Angell, in: *Glass Structure by Spectroscopy*, Chap. 1, p. 36, Marcel Dekker, Inc., New York (1976).
4. S. U. Dzhalilov and K. I. Rzaev, *Phys. Status Solidi* **20**, 161 (1967).
5. A. E. Owen, in: *Electronic and Structural Properties of Amorphous Semiconductors* (P. G. LeComber and J. Mort, eds.), Chap. 4, pp. 161–190, Academic Press, New York (1973).
6. J. M. Stevels, in: *Handbuch der Physik* (S. Flugge, ed.), Vol. 13, p. 510, Springer-Verlag, Berlin (1962).
7. A. Eisenberg, *J. Phys. Chem.* **67**, 1333 (1963).
8. D. Turnbull and M. H. Cohen, in: *Modern Aspects of the Vitreous State* (J. D. Mackenzie, ed.), Chap. 3, pp. 38–62, Butterworths, London (1960).
9. H. Rawson, *Inorganic Glass Forming Systems*, Chap. 3, pp. 31–45, Academic Press, New York (1967).
10. C. N. R. Rao and K. J. Rao, *Phase Transitions in Solids*, Chap. 4, pp. 82 ff, McGraw-Hill Book Company, New York (1978).
11. S. Sakka and J. D. Mackenzie, *J. Non-Cryst. Solids* **6**, 145 (1971).
12. P. T. Sarjeant and R. Roy, *Mater. Res. Bull.* **3**, 265 (1968).
13. H. A. Davies, *Phys. Chem. Glasses* **17**, 159 (1976).
14. R. O. Davies and G. O. Jones, *Adv. Phys.* **2**, 370 (1953).
15. R. W. Douglas, in: *Proc. Int. Conf. on Physics of Non-Crystalline Solids* (*Delft, 1964*) (J. A. Prins, ed.), p. 397, North-Holland Publishing Company, Amsterdam (1965).
16. D. Turnbull and M. H. Cohen, in: *Modern Aspects of the Vitreous State* (J. D. Mackenzie, ed.), Vol. 1, Chap. 3, Butterworths, London (1960).
17. W. Kauzmann, *Chem. Rev.* **43**, 219 (1948).
18. J. M. Stevels, *J. Non-Cryst. Solids* **6**, 307 (1971).
19. T. A. Litovitz, in: *Non-Crystalline Solids* (V. D. Frechette, ed.), p. 252, Wiley and Sons, New York (1960).
20. I. L. Hopkins and C. R. Kurkjian, in: *Physical Acoustics* (Warren P. Mason, ed.), Vol. II, Part B, p. 91, Academic Press, New York (1965).
21. a. R. O. Davies and G. O. Jones, *Proc. R. Soc. London, Ser. A.* **217**, 26 (1953). b. R. O. Davies and G. O. Jones, *Adv. Phys.* **2**, 370 (1953).

22. I. Prigogine and R. Defay, *Chemical Thermodynamics*, Longmans, London (1954).

23. R. O. Davies, in: *Non-Crystalline Solids* (V. D. Frechette, ed.), p. 232, Wiley (1960).

24. A. B. Pippard, *The Elements of Classical Thermodynamics*, Cambridge C.U.P. (1957).

25. M. Goldstein, *J. Chem. Phys.* **39**, 3369 (1963).

26. G. Gee, *Contemp. Phys.* **11**, 313 (1970).

27. See, for example, J. Wilks, *The Third Law of Thermodynamics*, Chap. V, University Press, Oxford (1961).

28. P. W. Anderson, in: *Ill-Condensed Matter* (R. Balain, R. Maynard, and G. Toulouse, eds.), Lectures on Amorphous Systems, Course 3, pp. 161–261, North-Holland Publishing Company, Amsterdam (1979).

29. M. Goldstein, in: *Faraday Symposia No. 6, Molecular Motion in Amorphous Solids and Liquids* (F. C. Tompkins, ed.), p. 7, The Chemical Society, London (1972).

30. C. N. R. Rao and K. J. Rao, *Phase Transitions in Solids*, pp. 144–147, McGraw-Hill Book Company, New York (1978).

31. D. Turnbull and M. H. Cohen, *J. Chem. Phys.* **29**, 1049 (1958).

32. M. H. Cohen and D. Turnbull, *J. Chem. Phys.* **31**, 1164 (1959).

33. D. Turnbull and M. H. Cohen, *J. Chem. Phys.* **34**, 120 (1961).

34. G. S. Grest and M. H. Cohen, *Adv. Chem. Phys.* **48**, 454 (1981).

35. C. A. Angell, *J. Am. Ceram. Soc.* **51**, 117, 125 (1968).

36. J. S. Haggerty, A. R. Cooper, and J. H. Heasley, *Phys. Chem. Glasses* **9**, 47 (1968).

37. U. E. Schnaus, C. T. Moynihan, R. W. Gammon, and P. D. Macedo, *Phys. Chem. Glasses* **11**, 213 (1970).

38. C. T. Moynihan, P. B. Macedo, I. D. Aggarwal, and U. E. Schnaus, *J. Non-Cryst. Solids* **6**, 322 (1971).

39. C. A. Angell, J. H. R. Clarke, and L. V. Woodcock, *Adv. Chem. Phys.* **48**, 397 (1981).

40. R. J. Bell and P. Dean, *Phys. Chem. Glasses* **9**, 125 (1968).

41. J. H. Gibbs and E. A. Di Marzio, *J. Chem. Phys.* **28**, 373 (1958).

# Magnetism in Amorphous Solids

## J. M. D. Coey

This chapter aims at providing an overview of collective magnetism in noncrystalline solids. It is descriptive in outlook, and sets out to relate some of the key ideas of amorphous magnetism to well-established concepts of magnetism in crystalline solids, stressing the points where something new emerges because of the lack of a crystal lattice. For example, there are many amorphous ferromagnets yet no amorphous antiferromagnets are known to exist. Why not? In the area of amorphous and disordered solids, magnetism continues to play its traditional role of introducing concepts and posing problems that can be formalized in terms of simplified yet not completely unrealistic models, whose usefulness extends far beyond explaining the magnetic properties of solids. Past examples of this sort have been in the areas of phase transitions, elementary excitations, two-level systems, and physics in one or two dimensions. Another important facet is the practical potential of amorphous magnets that provides a ready justification for working in the field. The scope for applications of these materials will be discussed briefly. However, the bulk of the chapter is devoted to a discussion of magnetic order in amorphous solids, and the influence of a nonperiodic lattice on its ingredients. Beside the magnetic ground states themselves, topics of interest are defects, excitations, and the manner in which the magnetically ordered state breaks up with increasing temperature. For a more detailed discussion and bibliography on the subjects raised here, the reader is referred to recent monographs and review articles.[1-6]

**J. M. D. Coey** · Department of Pure and Applied Physics, Trinity College, Dublin 2, Ireland.

## 13.1. Types of Disorder

Conceptually, it is possible to distinguish three types of disorder in a solid. These distinctions are illustrated in Figure 13.1 for two-dimensional networks. By distorting a perfect crystal (Figure 13.1a) in such a fashion as to introduce bonds of different lengths making different angles with each other, it is possible to destroy the periodic structure (Figure 13.1b). This *bond disorder* is perhaps the simplest variety, as the bond-disordered network remains topologically equivalent to the crystal. We note that some bond disorder is present even in a real crystal at finite temperatures due to thermal displacements of the atoms from their equilibrium lattice sites. These thermal displacements do not destroy the underlying periodicity because there remains in the hot crystal a large probability of finding the atoms close to their lattice sites.

A much stronger type of disorder is the *topological disorder* shown in Figure 13.1c. The network there includes four-, five-, seven-, and eight-membered rings distributed at random among the six-membered ones.

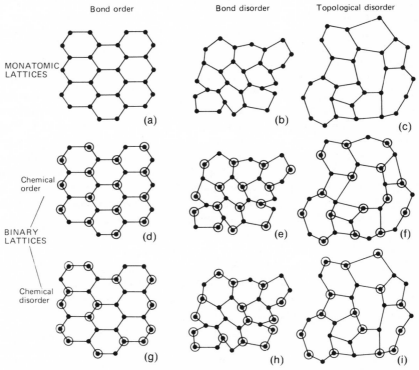

Figure 13.1. Types of disorder on two-dimensional monatomic and binary lattices.

Topological disorder also results when the number of bonds at each atom deviates from three. It necessarily includes bond disorder, but a topologically disordered network cannot be distorted back into a crystal. Some degree of topological disorder may be essential for forming any amorphous solid.

Figure 13.1d shows the structure of an ordered binary AB alloy. It may be disordered while retaining its crystallinity (Figure 13.1g), and this *chemical disorder* is of fundamental importance in metallurgy. In a perfectly random solid solution $A_x B_{1-x}$, the probability of an A atom being surrounded by $Z$ B atoms on the $N$ nearest-neighbor sites is

$$P_N(Z) = \frac{N!}{Z!(N-Z)!} (1-x)^{N-Z} x^Z \tag{13.1.1}$$

If the disorder is incomplete, the average number of B neighbors will be different from $(1-x)N$, and a short-range order parameter can be defined. A special case, when $x \to 0$, is the dilute crystalline alloy. The positions of A atoms there are like those of atoms in a gas and, in a sense, they are more disordered than either Figure 13.1b or c, where each atom in the solid (or liquid) has a certain number of nearest neighbors within a narrow range of interatomic distances. Bond disorder may be imposed on chemically ordered or disordered lattices (Figure 13.1e or h). In Figure 13.1e each A atom is surrounded by three B nearest neighbors, but the chemical order is absent in Figure 13.1g, h, and i. Finally, when topological disorder is imposed on a binary network, chemical order may be retained (Figure 13.1f) or destroyed (Figure 13.1i). Metallic magnetic glasses are almost invariably multicomponent alloys, of type (h) or (i) in Figure 13.1. They involve bond disorder, some chemical disorder, and probably topological disorder as well. Amorphous transition metals are of the type shown in Figure 13.1c. In amorphous binary compounds, the number of bonds formed by each component may be constant, yet topological disorder can result from a distribution of even-membered rings (Figure 13.1f). The magnetic lattice, however, is of type (c) in Figure 13.1 when only one component carries a moment.

Bond and chemical disorder in a magnetic material introduce a distribution in the magnetic moments and exchange coupling between interacting pairs of atoms. In the case when the B atom in an $A_x B_{1-x}$ alloy is nonmagnetic, chemical disorder leads us to the idea of *percolation*. At a certain critical concentration $x_P$, there appear infinite continuous interaction paths joining atoms, and many of the A atoms belong to the bulk cluster. The rest belong to small, isolated clusters. Any sort of magnetic long-range order is only possible for $x > x_P$, because all the atoms belong to isolated clusters when $x < x_P$. The value of $x_P$ depends critically on the

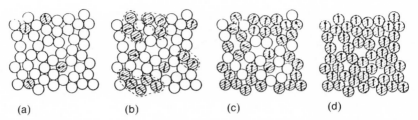

(a)              (b)              (c)              (d)

Figure 13.2. Illustration of the magnetic order as a function of concentration in an amorphous magnet with ferromagnetic nearest-neighbor exchange. (a) the dilute limit; (b) $x < x_p$; (c) $x > x_p$; (d) the concentrated limit.

range of the interaction, but for nearest-neighbor coupling it is of order $2/N$. Figure 2 illustrates the idea. Percolation has been extensively studied on crystalline lattices of type (g) in Figure 13.1,[7] and the fraction of A atoms in the bulk cluster falls to zero at $x_P$ in a manner perfectly analogous to the order parameter at $T_c$ for second-order phase transition $P_\infty \sim (x - x_P)^\beta$. Provided the bond interaction in a magnetic model is always ferromagnetic, regardless of bond length, bond disorder will not influence the percolation concentration. The value of $x_P$ depends on the degree of chemical and topological disorder. The disordered A–B alloy with magnetic A atoms is an example of the *site* percolation problem. Alternatively, one might imagine breaking *bonds* at random, and the percolation threshold will occur for a fraction $y_P$ of broken bonds. Quite generally, $y_P > x_P$. In magnetic systems it may be useful to think of bond percolation when the interaction strength depends critically on bond length.

## 13.2. Effects of Disorder

The basic requirements for magnetic order in a solid are (i) the existence of magnetic moments associated with unpaired electrons on the atoms, and (ii) an interaction to couple them together. The electrostatic fields acting at the atomic sites also have a profound effect, influencing the orientation of the atomic moments through spin–orbit coupling, thereby creating magnetic anisotropy. The various types of disorder modify each of these three key factors, which will now be considered in turn.

### 13.2.1. Magnetic Moments

A magnetic moment exists on a free atom whenever the atom has unpaired electrons. Any atom with an odd number of electrons must carry one. However, these unpaired outer electrons in solids are usually employed in the formation of covalent bonds or enter broad bands where their strong paramagnetism is destroyed. Exceptions are the atoms of the transition series where the unpaired electrons reside in an inner shell, and therefore cannot participate fully in the bonding.

A satisfactory theory exists for the magnetic moments of transition ions (including the rare earths) when an integral number of electrons are localized on the ion core.[8] This theory is applicable to many insulating and semiconducting compounds of the $3d$-series and practically all $4f$-materials except a few alloys involving elements such as Ce, which can have two different charge states in their compounds.

The theory of localized magnetism of noninteracting transition ions treats the Hamiltonian

$$\mathcal{H} = \mathcal{H}^c + \mathcal{H}^{so} + \mathcal{H}^{ef} \tag{13.2.1}$$

where $\mathcal{H}^c$ represents the Coulomb interactions among $d$- or $f$-electrons, which are responsible for strong electron–electron correlation resulting in coupling of spin and orbital angular momenta of the individual electrons according to Hund's rules to give resultants $S$ and $L$; $\mathcal{H}^{so}$ and $\mathcal{H}^{ef}$ represent the much weaker spin–orbit and electrostatic field interactions, whose relative magnitudes are inverted for the $3d$- and $4f$-series. The electrostatic field is generally referred to as the "crystal field," but this terminology is evidently inappropriate for noncrystalline solids. For the rare earths, the spin–orbit interaction

$$\mathcal{H}^{so} = -\lambda L \cdot S \tag{13.2.2}$$

is of order $10^4$ K, and $L$ and $S$ couple according to Hund's rules to give a resultant $J$, which is a good quantum number for the $4f$-series. A perturbation of order 100 K is caused by the interaction of the electrostatic field due to the atom's environment with the asymmetric charge distribution in the unfilled $f$-shell. Its main effect is to introduce local magnetic anisotropy but the atomic magnetic moment at low temperatures may also be reduced from its free-ion value. The electrostatic field interaction is of order $10^4$ K in the $3d$ transition series because, unlike the $4f$-shell, the $3d$-shell is not an inner shell, well shielded by outer valence electrons from the electrostatic field created by neighboring ions. Its effect is to quench partially, or completely, the orbital angular momentum. Spin–orbit coupling, of order $10^2$–$10^3$ K, then serves to mix the electrostatic energy levels slightly and give a Lande $g$ factor a little different from the spin-only value of 2. It also brings about local magnetic anisotropy. In crystals, where the local anisotropy is the same for each atom, the contributions add to give magnetocrystalline anisotropy, but in an amorphous solid they must add in a random way.

In reality, the wave functions of transition metal orbitals containing the unpaired electrons in compounds will be partially mixed with those of the ligands to form molecular orbitals. This covalency reduces the unpaired $3d$- or $4f$-occupancy below the purely ionic value, to an extent which depends on the metal–ligand overlap integrals and the energies of the atomic orbitals. The effect of noncrystallinity in an ionic solid will be to replace the fixed crystal field and overlap integrals of the crystal with a distribution of electrostatic field, which will be of such low symmetry as to remove all orbital degeneracy, except Kramers degeneracy, and a distribution of overlap integrals that results in slightly different population of the magnetic orbitals from one site to the next. These effects modify the orbital and spin moments of the ion, respectively. The latter effect may be examined in $S$-state ions, such as $Mn^{2+}$, $Fe^{3+}$, $Eu^{2+}$, or $Gd^{3+}$. It will be unimportant for rare-earth ions because the $4f$-shell is so small and well shielded by outer electrons that the overlap integrals are negligible.

Figure 13.3 shows the distributions of magnetic hyperfine fields acting at the nuclei for different types of magnetic glasses. The hyperfine field is roughly proportional to the magnetic moment, so the curves reflect the

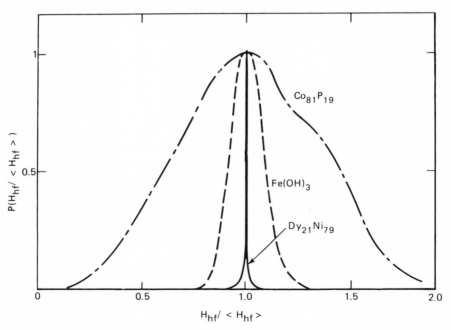

Figure 13.3. Hyperfine field distributions measured by Mössbauer spectroscopy or NMR at $^{161}$Dy in an amorphous rare-earth metallic alloy $Dy_{21}Ni_{79}$, at $^{59}$Co in a transition-metal metallic alloy $Co_{81}P_{19}$ and at $^{57}$Fe in a transition-metal insulating compound $Fe(OH)_3 \cdot 0.9\ H_2O$.[9]

moment distributions. Data were taken at $T$ equal to about zero, so thermal effects are absent. The hyperfine field distribution is extremely narrow for the rare earth Dy, with a relative width of only 1%. It is moderately broad, about 10%, for the insulating ferric compound due to variations in the Fe–O bond length.

Inevitably, direct overlap of $3d$- or $4f$-wave functions will lead to formation of narrow bands in any solid. In a one-electron picture the bands will conduct electricity if partially filled, but in fact electron–electron interactions may be so strong as to open up a correlation gap in the effective density of states.[10] Many transition-metal compounds, which should be narrow-band metals according to classical band theory, are really Mott insulators because of correlation. As the overlap increases, the "Mott–Hubbard" subbands broaden and the moment is reduced. Ultimately the bands cross at the metal–insulator transition where the greatly reduced localized moment becomes itinerant. As the bands broaden further, the moments are entirely destroyed and the metal ultimately becomes a Pauli paramagnet when they are broad enough for correlations to be neglected entirely. The simplest model Hamiltonian for a metal exhibiting magnetic effects is that of Hubbard,

$$\mathcal{H} = \mathcal{H}^{\mathrm{T}} + \mathcal{H}^{c} \tag{13.2.3}$$

where $\mathcal{H}^{\mathrm{T}} = \Sigma_{ij\sigma} t_{ij} C^{+}_{i\sigma} C_{j\sigma}$ is the term allowing electron transfer from one site to the next which gives the one-electron band structure. Here $t_{ij}$ is related to $\Delta$, the one-electron bandwidth in the tight-binding approximation, by $\Delta = Z t_{ij}$ ; $\mathcal{H}^{c}$ is the Coulomb correlation interaction, written most simply for a single, nondegenerate half-filled $s$-band as $\Sigma_{i} U n_{i\uparrow} n_{i\downarrow}$ where $U = \langle e^{2}/r_{12} \rangle$ represents the average intra-atomic Coulomb interaction of two electrons on the same site. The band structure and magnetic moment at $T = 0$ for the Hubbard model as a function of $\Delta/U$ are illustrated schematically in Figure 13.4. The ground state for the half-filled $s$-band is antiferromagnetic.

Magnetism in real $3d$-metals is greatly complicated by the degeneracy of the $d$-bands and their overlap with the $4s$-band. In the solid, $d$-orbitals of different symmetry will overlap to varying extents, giving bands of different widths. The magnetism of iron, for example, has a partly localized and partly itinerant character, which has so far defied accurate calculation.

Early attempts to explain the nonintegral magnetic moments in transition metals were based on the concept of a spin-polarized $3d$-band, overlapping with the $4s$-band, as shown in Figure 13.5. The occupancy of the $\uparrow$ and $\downarrow$ $d$-bands determines the moment, and this will depend on the total number of $3d/4s$ electrons, which changes by one on passing from one $3d$-element to the next in the periodic table, or on alloying different

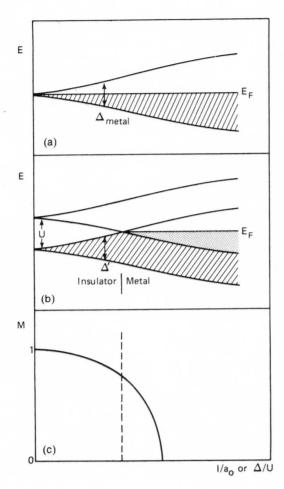

Figure 13.4. Schematic band structure of an array of $1s^1$ ions as a function of the inverse of their separation, calculated (a) without and (b) with intra-atomic correlations; (c) variation of the magnetic moment per atom calculated for the same array of ions at $T = 0$ (after Hubbard[11] and Cyrot[12]).

elements. The moments of alloys predicted by filling the rigid bands of this model are often in fairly good agreement with observation. Moments of alloys with nonmagnetic metals can also be explained within the rigid-band model by invoking charge transfer. The model has also been recast by Malozemoff *et al.* to account for the systematic variation of the atomic magnetic moments in a series of amorphous ferromagnetic alloys and leads[13] to the following equation:

$$M = 2(N_d^\uparrow + N_{sp}^\uparrow) - Z \qquad (13.2.4)$$

where $N_d^\uparrow$ is taken to be 5 for iron and elements to its right which are

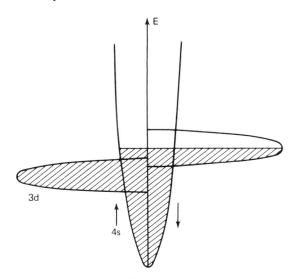

Figure 13.5. Spin-polarized band scheme for explaining the magnetic moments of $3d$-metals and alloys.

assumed to have full spin up $d$-bands, and 0 for elements to its left. $N_{sp}^{\uparrow}$ is supposed to be 0.3, and $Z$ is the normal chemical valence (e.g. 3 for Y, 4 for Si, 8 for Fe ...). The quantities are averaged over the alloy composition. Equation (13.2.4) provides a rough account of the concentration dependence of the magnetic moment in a number of binary amorphous alloys, quite an achievement for a model that depends only on counting the number of electrons, and takes no account of atomic structure or nearest-neighbor environments. Although there is little direct experimental evidence in favor of a large splitting of $\uparrow$ and $\downarrow$ bands and photoemission experiments on alloys do not usually support a rigid band picture, an equation like (13.2.4) can be justified if the Fermi level is pinned at a minimum in the density of states of $\uparrow$ or $\downarrow$ $d$ bands.

In any case, the $3d$-metals are near the limit for the appearance of magnetism, and their moments are all quite sensitive to changes in the overlap integrals, as shown in Figure 13.4c. Even iron, which has the best-developed moment of them all, can be rendered nonmagnetic by alloying with elements which broaden the $3d$-density of states sufficiently. There will generally be a much greater spread of magnetic moments in amorphous metals due to the variability of nearest-neighbor distances than is found in amorphous insulators. This fact was illustrated for amorphous $Co_4P$ in Figure 13.3. In some cases the spread can be large enough to permit coexistence of magnetic and nonmagnetic atoms of the same element on different sites.

It has long been known that magnetic and nonmagnetic atoms of the same element can coexist on different sites of certain crystalline solid solutions. A model was first developed for iron impurities in $Nb_{1-x}Mo_x$ by Jaccarino and Walker,[14] who postulated that iron in these alloys was nonmagnetic when surrounded by less than seven molybdenum nearest neighbors. Mixing of the iron $d$-electrons with the conduction electrons of niobium neighbors broadens the local density of states, and if there are enough of them the broadening is such that the transition element can no longer sustain a moment.

One of the attractive features of amorphous metals is that binary alloys $A_x B_{1-x}$ can often be prepared over a wide concentration range where no crystalline solid solutions exist. Particularly interesting from our present viewpoint are alloys where pure A is magnetic but B is not. The appearance of magnetism around a critical concentration $x_c$ has been examined in a number of binary systems involving a $3d$-element such as Fe, Co, or Ni together with an early transition metal from group IIIB or IVB or a metalloid from group IIIA or IVA. As an illustration of the evolution of electronic structure with $x$, Figure 13.6 shows the magnetic, nonmagnetic, and superconducting zones in the a-$M_x Zr_{1-x}$ systems. The disappearance of magnetism occurs at roughly similar concentrations for a variety of other alloys of Fe, Co, or Ni, although not all of them become superconducting at low $3d$-transition metal concentrations. A list of critical concentrations for binary alloys is given in Table 13.1[49] together with the prediction of equation (13.2.4).

A theoretical model of some relevance in this context is that of Anderson for a single impurity atom in a broad-band metal.[15] A simple treatment of the Anderson model has been given by Mott.[10] The result of mixing the impurity wave functions with the conduction band is to produce a local density of states $N(E)$ for each of the two spin states, separated by the energy $U$ when there is no overlap. This is shown in Figure 13.7. The

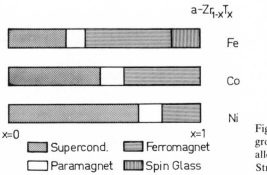

Figure 13.6. Electronic and magnetic ground states in amorphous $Zr_{1-x}M_x$ alloys (after Z. Altounian and J. Ström-Olsen).

**Table 13.1. Critical Concentrations for Binary Fe, Co, and Ni Amorphous Alloys (%)**

| III | | | | IV | | | | V | | | |
|---|---|---|---|---|---|---|---|---|---|---|---|
| | Fe | Co | Ni | | Fe | Co | Ni | | Fe | Co | Ni |
| B | 38 | — | — | Si | 39 | — | — | P | 45 | 68 | 75 |
| Y | 36 | 45 | 80 | Ge | 40 | — | — | Sb | 43 | — | — |
| La | — | 42 | — | Sn | 35 | 51 | — | Nb | 56 | 59 | — |
| | | | | Ti | 43 | 69 | — | Ta | 71 | — | — |
| | | | | Zr | 38 | 64 | 85 | | | | |
| | | | | Hf | 45 | 37 | — | | | | |
| | | | | Th | 43 | — | — | | | | |
| Eq. (13.2.4) | 48 | 60 | 80 | | 57 | 68 | 85 | | 63 | 73 | 88 |

number of unpaired impurity electrons $n = n_\uparrow - n_\downarrow$ is given by

$$n = \nu(E_F + \tfrac{1}{2}nU) - \nu(E_F - \tfrac{1}{2}nU) \qquad (13.2.5)$$

where $\nu(E) = \int_0^E N(E)dE$. By expanding equation (13.2.4) as a power series for small $n$, we find that

$$n = nUN(E_F) + \tfrac{1}{24}(nU)^3 N''(E_F) + \cdots \qquad (13.2.6)$$

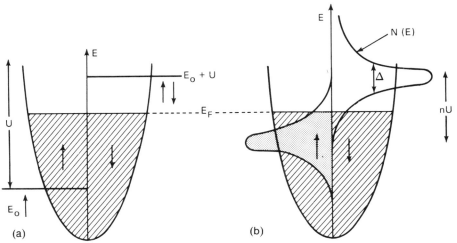

Figure 13.7. Local density of states for a $3d$-impurity in an $s$-band metal. (a) is the case where there is no mixing of the wave functions of the impurity level at $E_0$ with the conducting band, while (b) shows the result of such mixing.

so that $n = \{24[UN(E_F) - 1]/ - N''(E_F)U^3\}^{1/2}$. Since $N''(E_F)$ is negative, a moment will form provided

$$U > 1/N(E_F) \tag{13.2.7}$$

$N(E_F)$ is proportional to $1/\Delta$, the inverse width of the local density of states, and this depends in turn on the overlap energy integral between the impurity electrons and the conduction band. If $N(E_F)$ were to vary smoothly with some parameter, such as concentration, then the magnetic moment on the impurity, just below the critical concentration $x_c$ where the moment disappears, would vary as $(x - x_c)^{1/2}$.

### 13.2.2. Exchange Interactions

The appearance of a magnetically ordered structure, such as ferromagnetism, requires an interaction to couple the magnetic moments together. The problem of the exchange interaction can be separated from that of the existence of the moments themselves provided they are essentially localized in character. This is certainly true of insulators, and is a fair approximation in many metals, although a fraction of the $3d$-moment can often be attributed to interatomic exchange. It has been found that exchange interactions between two spins usually tend to align them parallel or antiparallel. This tendency is represented phenomenologically by the isotropic Heisenberg interaction for localized moments $g\mu_B S$,

$$\mathcal{H}_{ij} = -\mathcal{I}_{ij}\mathbf{S}_i \cdot \mathbf{S}_j \tag{13.2.8}$$

where $\mathcal{I}_{ij}$ is the exchange constant between spins at site $i$ and site $j$. It is positive for ferromagnetic coupling and negative for antiferromagnetic coupling. Typical values for the interaction between two spins lie in the range 1–100 K. Other interactions exist that favor a perpendicular configuration for the pair. These may be represented by the Dzyloshinskii–Moriya (DM) term $\mathcal{D}_{ij}\mathbf{S}_i \times \mathbf{S}_j$, but they are weak by comparison with the Heisenberg interaction. The DM interaction is zero by symmetry for certain spin configurations in crystals, but no such restraints apply in amorphous or disordered solids.[16]

The origin of exchange coupling lies in the electrostatic interaction between electrons of either spin on different sites. Various exchange mechanisms exist — direct exchange, superexchange via ligands, indirect exchange via conduction electrons [Ruderman–Kittel–Kasuya–Yoshida (RKKY) interaction] — but all depend sensitively on the distance between the interacting electrons. Attempts to calculate the exchange constant between localized spins in equation (13.2.8) from first principles have led to much too small values, but nonetheless $\mathscr{I}_{ij}$ can be accepted as a phenomenological parameter depending on $r_{ij}$, and in the case of superexchange on the bond angle. Though $\mathscr{I}_{ij}$ or $\mathscr{D}_{ij}$ is usually assumed to be isotropic, this is a gross simplification. For the $4f$-series, $S$ in (13.2.8) is replaced by $J$, while an effective value is obtained for $3d$-metals by equating $\mu_i$, the atomic moment, to $g\mu_B S$.

The direct exchange mechanism is most important in $3d$-transition metals. The Slater–Néel diagram in Figure 13.8a indicates roughly how the exchange varies with distance between magnetic shells in the $3d$-metals.

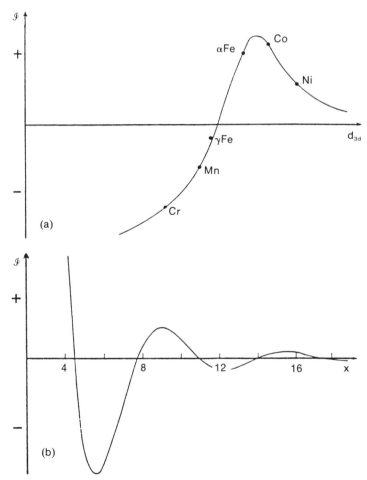

Figure 13.8. Distance-dependence of exchange interactions. (a) schematic variation of direct exchange constant as a function of distance between $3d$-shells in the $3d$-transition metals; (b) the RKKY interaction, $\mathscr{I}(r)$.

The value of $\mathscr{I}$ changes sign for $\gamma$Fe at 2.55 Å.[17] It is clear that the distribution of interatomic separations in a noncrystalline solid can lead to a distribution of exchange interactions, which may sometimes include interactions of both signs. The same is true of the other exchange mechanisms as well. Superexchange is negative for a 180° metal–ligand–metal bond, but a weaker positive interaction may occur for bond angles near 90°.[18] The RKKY interaction oscillates in sign as a function of $r$, as shown in Figure 13.8b. It arises from polarization of the conduction band by localized spins and it is the most important mechanism for $4f$-metals. Assuming a free-electron model for the conduction electrons, calculation gives[19]

$$\mathscr{I}(r) \propto \frac{-18\pi Q^2 \mathscr{I}_{sf}^2}{E_F} \left( \frac{x \cos x - \sin x}{x^4} \right) \tag{13.2.9}$$

where $x = 2k_F r$, $\mathscr{I}_{sf}$ is the interaction between the localized spins and the conduction electrons, $E_F$ and $k_F$ are the Fermi energy and wave vector, respectively, and $Q$ is the charge of the ion core.

The probability of finding an exchange interaction of a given magnitude and sign may be represented on a $P(\mathscr{I})$ diagram. For a crystal the diagram consists of one or more delta functions, but the disorder in an amorphous solid will broaden the peaks, even to the extent that interactions of both signs may be included in the distribution, as indicated in Figure 13.9. The distribution of exchange in an amorphous solid can be modified experimentally by applying pressure to decrease the interatomic distances, or by hydrogenation, which increases them but also modifies the conduction electron concentration and density of states.[20]

In a solid, it is useful to replace the sum of equation (13.2.8) over all pairs of sites $i$, $j$ by a sum over sites $i$, assuming that the interaction of any atom with all its neighbors may be replaced by an effective field — the molecular field approximation. The moment $\boldsymbol{\mu}_i$ associated with the spin $S_i$ is $g\mu_B S_i$, and its interaction with a magnetic field $\mathbf{H}_{eff}$ is given by

$$\mathscr{H}_i = -g\mu_B \mathbf{S}_i \cdot \mathbf{H}_{eff} \tag{13.2.10}$$

where $\mathbf{H}_{eff}$ is defined by the average,

$$\mathbf{H}_{eff} = \frac{1}{g\mu_B} \sum_j \mathscr{I}_{ij} \langle \mathbf{S}_j \rangle \tag{13.2.11}$$

The molecular field approximation therefore consists of replacing $\sum_{ij} \mathscr{I}_{ij} \mathbf{S}_i \cdot \mathbf{S}_j$ by $\sum_i (\sum_j \mathscr{I}_{ij} \langle \mathbf{S}_j \rangle) \cdot \mathbf{S}_i$. It is an extremely useful approximation and forms the basis of much of the theory of collective magnetism.[8] In

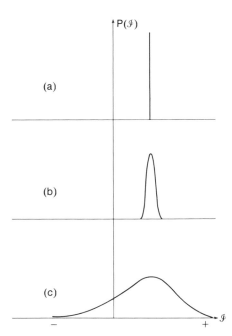

Figure 13.9. Probability of finding a given value of exchange (a) in a crystal, and (b) and (c) in amorphous solids with differing amounts of disorder.

amorphous magnets, the effective field will certainly have a probability distribution of magnitude, and in all cases except ferromagnets where $P(\mathscr{I})$ is predominantly or exclusively positive it will have a probability distribution in direction as well.

Exchange interactions are not the only ones which couple magnetic moments. The classical *dipolar interaction*

$$\mathscr{H}_{ij}^{\text{dip}} = \frac{\boldsymbol{\mu}_i \cdot \boldsymbol{\mu}_j}{r_{ij}^3} - \frac{3(\boldsymbol{\mu}_i \cdot \mathbf{r}_{ij})(\boldsymbol{\mu}_j \cdot \mathbf{r}_{ij})}{r_{ij}^5} \qquad (13.2.12)$$

tends to align the two moments $\boldsymbol{\mu}_i$ and $\boldsymbol{\mu}_j$ parallel, along the line joining their sites, $\mathbf{r}_{ij}$. Like the RKKY interaction, the dipolar interaction varies as $1/r^3$, but it is much weaker than exchange in most materials, except for some insulating compounds of the rare earths. Dipolar interactions between two moments are of order 1 K and are quite anisotropic since they depend on the orientation of the moments relative to $\mathbf{r}_{ij}$. The net dipolar field is identically zero at sites in a lattice having cubic symmetry. In noncrystalline solids it will be distributed both in magnitude and direction. Despite their weakness, dipolar interactions can sometimes influence significantly the ordered magnetic structures in noncrystalline solids.[21]

### 13.2.3. Anisotropy

The electrostatic field created by the surroundings at an atomic site in a crystal has the point symmetry of the site. Its interaction with $3d$-electrons is stronger than either spin–orbit coupling or exchange, and it acts to remove some or all of the orbital degeneracy of the free ion. Certain preferred directions are then imposed on the ionic moment in the ground state by spin–orbit coupling, and this is the microscopic origin of magneto-crystalline anisotropy. For the $4f$-electrons, spin–orbit coupling is much greater than the electrostatic field interaction, so magnetocrystalline anisotropy is created by the latter directly. The interaction can be expressed by the single-ion Hamiltonian[22]

$$\mathcal{H}^{\text{ef}} = \sum_{n=0}^{n'} \sum_{m=-n}^{n} \langle J \| \theta_n \| J \rangle \langle r^n \rangle (1 - \sigma_n) K_n^m A_n^m \hat{O}_n^m \qquad (13.2.13)$$

where $n$ is even and $n \geq |m|$. Terms up to $n = 4$ are required for $d$-ions and terms up to $n = 6$ for $f$-ions. The number of terms in equation (13.2.13) is greatly reduced by symmetry in crystals, with an appropriate choice of axes. For instance, there are no second-order ($n = 2$) terms on sites of cubic symmetry. The values of $A_n^m$ are known as crystal-field parameters, and involve sums over the surrounding ionic charge distribution and a conduction electron contribution when necessary. The quantities $\hat{O}_n^m$ are Stevens' operator equivalents, which are combinations of the angular-momentum operators, while $\langle J \| \theta_n \| J \rangle$ are reduced matrix elements, $K_n^m$ are constants tabulated by Hutchings, $(1 - \sigma_n)$ is a shielding factor, and $\langle r^n \rangle$ is the average of the electron position over the $3d$- or $4f$-wave functions.

No simplification of equation (13.2.13) by symmetry will occur in a noncrystalline solid. The electrostatic interaction may be imagined to give a complicated three-dimensional energy contour, according to the direction of the orbital moment, with no particular symmetry except an inversion center. Reversing the direction of the orbital motion does not change the energy. One axis however must be lower in energy than all the rest, and this is the justification for the model of Harris, Plishke, and Zuckermann (HPZ),[23] who represent the electrostatic field interaction in noncrystalline solids by the single-ion Hamiltonian

$$\mathcal{H}_i^{\text{ef}} = - D_i S_{z_i}^2 \qquad (13.2.14)$$

where $D_i$ has a distribution in magnitude and defines easy axes $z_i$, which differ in direction at each site. The approximation is illustrated by Figure 13.10. Equation (13.2.14) is equivalent to the leading $\hat{O}_2^0$ term in equation (13.2.13). Attempts have been made to improve the approximation by including the other second-order term $O_2^2$, or by retaining a group of terms which represent a quasicrystalline environment in the amorphous solid, which will be differently oriented at each site.[2]

Figure 13.11 provides a summary of the effects of disorder, contrasting the ingredients of magnetism in a perfect crystal and in a noncrystalline solid.

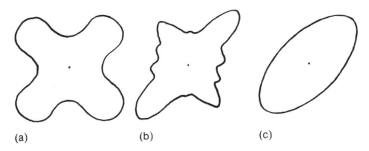

(a)                (b)                (c)

Figure 13.10. Schematic single-ion anisotropy energy surfaces: (a) a crystal, (b) a noncrystalline solid, and (c) the surface assumed in the HPZ model, equation (13.2.13).

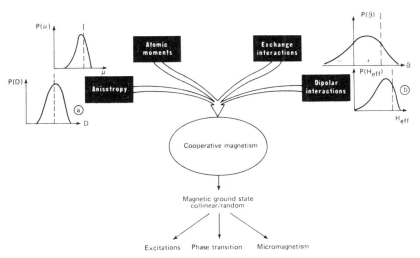

Figure 13.11. Summary of the ingredients of cooperative magnetism, showing how unique values of $\mu$, $D$, and $H_{\text{eff}}$ in the crystal (dashed lines) are replaced by probability distributions for their magnitudes in amorphous solids. Not included on the diagram are (a) random isotropic distribution of the directions of the local anisotropy axes, and (b) possible random distribution of the directions of $H_{\text{eff}}$. Some of the resulting subjects of interest in amorphous magnetism are also shown.

## 13.3. Collective Magnetic Order

### 13.3.1. Ferromagnetic and Antiferromagnetic Interactions

Ferromagnetic order requires the presence of magnetic moments, and a positive exchange interaction to couple them together. It is not incompatible with a noncrystalline structure provided the concentration of magnetic atoms exceeds the percolation threshold, the distribution of exchange is essentially positive, and random anisotropy is negligible compared with exchange coupling. Ferromagnetism has been well known in chemically disordered crystals for a long time, but it was not until 1960 that Gubanov pointed explicitly to the possibility of amorphous ferromagnetism, and extensive experimental study of amorphous ferromagnets has been going on since about 1970. Chemical and bond disorder may modify the magnitude of the atomic moments and bond disorder will bring about a distribution of exchange interactions, but neither effect will destroy the ferromagnetic ground state provided $\mathscr{I}$ remains positive and $x > x_P$. There are no fundamental differences in principle between crystalline and amorphous ferromagnets. In certain circumstances the phase transition at the Curie point may be smeared out by disorder, and magnetic and nonmagnetic atoms can sometimes coexist in the ferromagnetically ordered phase. These effects, like many others in amorphous magnetism, are also found in chemically disordered crystals.

Nevertheless, there are often large differences in the values of the magnetic parameters compared with crystals of the same composition. For example, crystalline $YCo_3$ has its Curie point at 300 K while the amorphous form has a larger magnetic moment and would become paramagnetic only above 600 K.[24] The Curie point of the amorphous phase cannot actually be attained because it crystallizes first. In other cases, such as $Fe_2B$, the amorphous form has the lower ordering temperature.[25] Other features of amorphous ferromagnets are that the $M(T)$ curves tend to be flatter than for crystals, and the Bloch $T^{3/2}$ law for the low-temperature variation of the magnetization due to spin waves applies over a wider range of reduced temperature. Bulk anisotropy may be much less than in corresponding crystalline phases.

In contrast to ferromagnetism, where disorder does not necessarily induce much change in the phenomenon itself, antiferromagnetic interactions on a disordered lattice can produce results that are quite different from those in crystals. The concept of *frustration*[25] is not entirely unfamiliar in crystals. It is present to a limited degree on the fcc lattice, and on the two-dimensional triangular lattice. Frustration arises with antiferromagnetic interactions whenever the geometry of the lattice is such that the neighbors of a given atom are themselves neighbors of each other. It is therefore impossible to find a configuration for the spins where all the interactions are simultaneously satisfied. The idea is most simply presented by considering the three-, four-, and five-membered rings of Figure 13.12. With Ising spins, $S_z = \pm 1$, the lowest-energy configurations have energy $-0.33\mathscr{J}$, $-\mathscr{J}$, and $-0.6\mathscr{J}$ per bond for antiferromagnetic pair interactions, but all have the lowest possible energy, $-\mathscr{J}$ per bond, with ferromagnetic coupling. Frustration occurs with antiferromagnetic interactions on odd-membered rings. A related feature is the degeneracy of the lowest-energy state, 6, 2, and 10 for the three rings with negative $\mathscr{J}$, but 2 for all three when $\mathscr{J}$ is positive. Increased degeneracy (or near-degeneracy when the bond interactions are unequal) accompanies frustration. The classical limit is also quite instructive. Bond energies for classical spins are higher than for Ising spins, $-0.5\mathscr{J}$, $-\mathscr{J}$, and $0.81\mathscr{J}$ for the three rings respectively, but frustration is still present in the odd-membered ones, and it leads to noncollinear lowest-energy states as shown in Figure 13.12.

Frustration also arises on even lattices, such as the square lattice, when

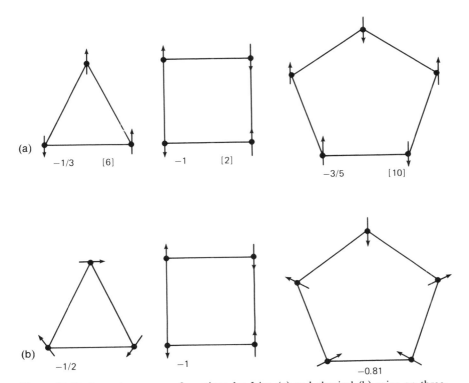

Figure 13.12. Lowest-energy configurations for Ising (a) and classical (b) spins on three-, four-, and five- membered rings. Energy per bond (in units of $\mathscr{J}$) and the degree of degeneracy for Ising spins is shown under each ring.

positive and negative exchange bonds are distributed at random.[26] A square or "plaquette" is frustrated if there is an odd number of antiferromagnetic interactions (1 or 3). The degree of frustration depends on the relative proportion of the two signs of interaction.

No frustration is introduced by chemical disorder alone in an unfrustrated crystal with antiferromagnetic nearest-neighbor exchange interactions, but if the interaction is longer range and at least partly antiferromagnetic (such as RKKY), the lattice will be frustrated. Bond disorder of the unfrustrated crystalline lattice will introduce frustration if the change in length of some of the bonds is sufficient to change the sign of $\mathscr{I}$ from negative to positive. Topological disorder is always prone to frustration when it involves odd-membered rings. In any case, whenever frustration is present it will be accompanied by degeneracy or near-degeneracy of the ground state. Many different configurations of the spin system are close in energy, although to pass from one of these states to another involves going through intermediate higher-energy states, so they are separated by energy barriers. The free energy surface in spin configuration space is corrugated and pitted with many local minima.

### 13.3.2. Ideal Magnetic Structures

From the considerations in the preceding paragraph, it is plausible that frustration in a random lattice will result in random noncollinear magnetic structures. Frustration introduced by competing exchange interactions in crystals sometimes results in helimagnetic order,[27] but long-range helimagnetism is unlikely in a noncrystalline solid. One approach would be to consider the random noncollinear structures in amorphous solids as superpositions of spirals of variable phase and propagation direction. Frustration arises in noncrystalline solids from antiferromagnetic interactions on lattices with particular topological disorder (odd-membered rings), and also when competing exchange interactions of both signs are present on any disordered lattice. In a broader sense, we can also consider amorphous solids with significant single-ion anisotropy as being frustrated. It is impossible to fully satisfy the exchange and single-ion interactions at each site because they favor different orientations of the spins. The energy of the compromise ground state will be higher than that given by the sum of the interactions, and a number of alternative configurations will exist that have almost the same energy, but are separated by energy barriers of different heights. Competition between single-ion anisotropy and exchange in a noncrystalline solid will lead to some sort of a random noncollinear structure even if $\mathscr{I}$ is everywhere positive. The same competition in some crystals, including certain rare-earth metals, also produces spiral or periodically modulated spin arrangements.

Turning, for a moment, from general concepts to concrete experimental data, some typical initial magnetization curves for a selection of amorphous magnets are shown in Figure 13.13. These curves are all taken at a temperature of 4.2 or 1.6 K, close to $T = 0$, and they show the reduced magnetization $M/M_0$, where $M_0$ is the value corresponding to parallel alignment of all atomic moments. The maximum field, 150 kOe, is large by laboratory standards but much less than the dominant interactions (exchange for $YCo_3$, $YFe_3$, and $FeF_3$, single-ion anisotropy for $DyNi_3$). These materials each contain only one magnetic species and so there is a single magnetic subnetwork. (Nickel in $DyNi_3$ does not bear a moment.) The applied field is corrected for demagnetizing effects. $YCo_3$ saturates very easily, like a typical ferromagnet. It obviously has a simple collinear

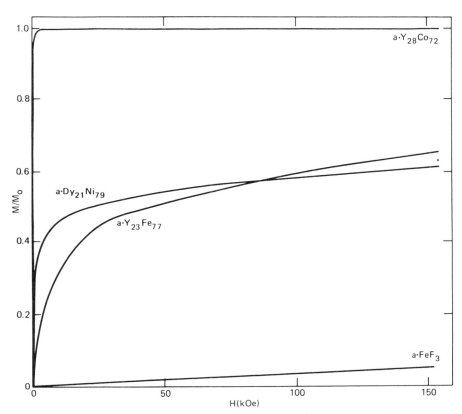

Figure 13.13. Typical low-temperature magnetization curves for an amorphous ferromagnet, a-$Y_{28}Co_{72}$; two asperomagnets, one exchange dominated, a-$Y_{23}Fe_{77}$, and the other anisotropy dominated, a-$Dy_{21}Ni_{79}$, and a speromagnet a-$FeF_3$ (after Coey[9]).

structure and any antiferromagnetic interactions or random anisotropy introduce negligible perturbations of the ferromagnetic order. The field induces only very weak magnetization in $FeF_3$, the opposite extreme where the susceptibility is essentially independent of the field. Apart from a tiny remanence $M_r$ ($M_r/M_0$ is approximately $10^{-3}$), the magnetization curve of amorphous $FeF_3$ resembles that of a crystalline antiferromagnet. By contrast, for $DyNi_3$ and $YFe_3$ the field quite easily induces a large moment of approximately one-half the saturation magnetization, but the approach to saturation in higher fields is slow, and it would be necessary to apply fields of order 1000 kOe to align the spins completely. Both materials show remanence at low temperatures such that $M_r/M_0 > 10^{-1}$. Clearly, in the magnetized state at least, they have magnetic structures with a large net magnetization that nevertheless fall far short of the ferromagnetic saturation value.

As an aid to discussing magnetism in noncrystalline solids it is useful now to define some broad categories of magnetic order encountered in amorphous materials. They fall into two groups: collinear and random. The essential feature of magnetic order in every case is understood to be that a component of the magnetic moment of each atom is constant in time, i.e., $\bar{S}_i \neq 0$, where $\bar{S}_i$ denotes the time average of the moment, which may have spin and orbital contributions, along $z_i$, the local quantization axis. Pictorial definitions at $T = 0$ for materials with one magnetic *subnetwork* are illustrated in Figure 13.14, a subnetwork being a chemical "sublattice," defined as the ensemble of atoms which carry a magnetic moment and have similar magnetic interactions. For the examples in Figure 13.13, the subnetworks are composed of the cobalt, iron, or dysprosium atoms in $Y_{28}Co_{72}$, $FeF_3$, or $Y_{23}Fe_{77}$, and $Dy_{21}Ni_{79}$. Figures 13.14a and b represent collinear structures having, respectively, the maximum and zero net moment, corresponding to *ferromagnetic* and *antiferromagnetic* order. Figures 13.14c and d represent random noncollinear structures, one with a substantial net magnetization, the other with none. We will refer to them as *asperomagnetic* and *speromagnetic*, respectively. In the speromagnetic structure,[28] the moments are distributed at random in directions with no preferred orientation. The formal definition is that $-1 \leqslant \bar{S}_i \cdot \bar{S}_j / |\bar{S}_i| |\bar{S}_j| \leqslant 1$, but $\langle \bar{S}_i \cdot \bar{S}_j \rangle = 0$ for $r_{ij} \geqslant a$, where $\langle \ \rangle$ denotes the average over all pairs with separation equal to $r_{ij}$ and $a$ is the average nearest-neighbor distance. The definition admits the possibility of very short-range averaged correlation at the level of the first- or second-neighbor shells, such as may exist when the nearest-neighbor interactions are antiferromagnetic, but there are no correlations on average at longer distances. By contrast, there is some long-range preferred orientation of the moments in an asperomagnet (Figure 13.14d) such that $0 < \langle \bar{S}_i \cdot \bar{S}_j \rangle < \langle |\bar{S}_i|^2 \rangle$.

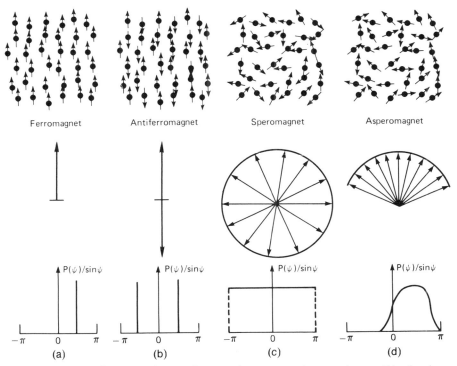

Figure 13.14. Possible one-subnetwork magnetic structures in amorphous solids showing schematically the spatial distribution of moment directions and the corresponding normalized angular probability distributions $P(\psi)/\sin \psi$.

Any definition of a magnetically ordered structure is only valid within a *domain*, so some consideration must be given to the length scale over which the definition may be applied. An iron bar usually has no net magnetization if no field is applied, so the averaged spin correlation $\langle \bar{\mathbf{S}}_i \cdot \bar{\mathbf{S}}_j \rangle$ is zero for $r_{ij}$ much greater than the ferromagnetic domain size while it is equal to $\langle |\bar{\mathbf{S}}_i|^2 \rangle$ within the domain. Hence the definition of ferromagnetic order applies only within the domain, whose size is usually greater than one micron. The concept of domain size has no obvious meaning for a speromagnet, but for an asperomagnet it can be defined as the distance beyond which the averaged spin correlation $\langle \bar{\mathbf{S}}_i \cdot \bar{\mathbf{S}}_j \rangle$ tends to zero. The definition is useful provided this distance is much greater than the interatomic spacing, although the physical explanation of domain formation may not necessarily be the same as in a ferromagnet, where domains are the result of minimizing the free energy of the sample, including the interaction with its own dipolar field.

Of the four structures in Figure 13.14 there is direct experimental evidence of all except the amorphous antiferromagnet, formally defined by the relations

$$\bar{\mathbf{S}}_i \cdot \bar{\mathbf{S}}_j = \pm |\bar{\mathbf{S}}_i| \, |\bar{\mathbf{S}}_j|; \quad \langle \bar{\mathbf{S}}_i \cdot \bar{\mathbf{S}}_j \rangle = 0 \qquad \text{for } r_{ij} \gg a$$

These structures are for materials with one magnetic subnetwork, but amorphous materials also exist with two magnetic subnetworks. All known examples are alloys of rare-earth and $3d$-transition metals, where each chemical subnetwork is composed of atoms of one sort or the other. The dominant interaction is ferromagnetic within the $3d$-subnetwork, but the $3d$–$4f$-coupling depends on the rare-earth element. It is doubtful whether one can ever distinguish two magnetic subnetworks on a geometric rather than a chemical basis in noncrystalline solids. If this were feasible, then one could have an amorphous collinear antiferromagnet. In crystals, of course, it is possible. The spinel structure, for instance, has two sublattices formed by cation sites with tetrahedral (A) and octahedral (B) oxygen coordination, and the principal magnetic interaction is A–B coupling because the neigboring cations of an atom on one sublattice all belong to the other. While it is surely possible to conceive of a noncrystalline structure with this property (a spinel with bond disorder would do), it is improbable that examples really exist.

Several classes of two-subnetwork structures may be distinguished. Possible collinear structures are *ferromagnetic* or *ferrimagnetic*, according to whether the subnetworks are coupled parallel or antiparallel. If one or both subnetworks possess a random, noncollinear structure but nevertheless have a net magnetization, the structure will be termed *sperimagnetic*.[29] Figure 13.15 illustrates the definitions. If both possess random noncollinear structures with no net moment, the structure is once again *speromagnetic*.

Compared with regular crystals, the novel feature of magnetism in amorphous solids is the possibility of finding "ordered" magnetic structures, where the atomic moments are fixed in directions that are essentially random. Order in this sense means that there are no significant fluctuations in the average moment directions on an experimental times scale. Temporal but not spatial fluctuations are suppressed as the spins "freeze" on cooling the sample sufficiently, but there is little evidence that this occurs at a well-defined conventional phase transition. It seems to be a continuous process, like the glass transition in a quenched liquid. By contrast, a phase transition quite analogous to that in crystals is found in many amorphous ferromagnets and ferrimagnets. Furthermore, the random, noncollinear structures (speromagnetic, asperomagnetic, and sperimagnetic) have magnetic excitation spectra, which may be quite different compared with those of crystalline or amorphous collinear magnets.[30]

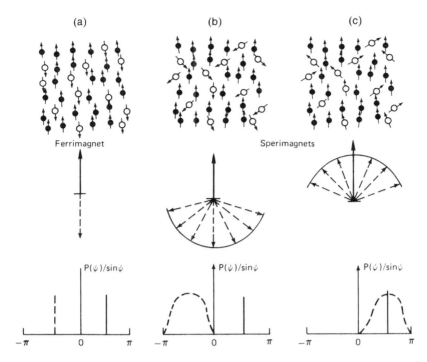

Figure 13.15. Possible two-subnetwork magnetic structures in amorphous solids showing schematically the spatial distribution of moment directions and the corresponding normalized angular probability distributions $P(\psi)/\sin\psi$. Solid and dashed lines represent moments for two chemical species.

### 13.3.3. Spin Glasses

In the broadest sense of the term, all the random noncollinear amorphous magnets can be described as *spin glasses*. They do not exhibit the usual magnetic phase transition, the magnetic order is random to some degree, and there are indications that the ground state is not unique. Originally applied by B. R. Coles to dilute crystalline alloys such as $Au$Fe or $Cu$Mn containing about 1% of magnetic impurities, the term "spin glass" no longer has any precise, generally-accepted significance. It has been applied by various authors to materials, which are magnetically dilute or concentrated, metallic or insulating, crystalline or amorphous, when they possess some magnetic characteristics in common with the canonical examples. A common feature of all these systems appears to be a degree of frustration arising from competing interactions in the presence of a disordered structure. It also denotes a class of theoretical models, which have some relevance in explaining these characteristics. Sometimes "spin glass" has been used in the sense of an ideal magnetic structure, similar to speromagnetism, except that not even short-range correlations are allowed; $\bar{\mathbf{S}}_i \neq 0$; $\langle \bar{\mathbf{S}}_i \cdot \bar{\mathbf{S}}_j \rangle = 0$ for *all* $r_{ij}$.

The canonical spin glasses, dilute alloys of transition metals in noble metal hosts, whose disorder has more in common with a gas than a liquid, show the following characteristic magnetic properties.[31]

1. A sharp peak appears in the low field (about 1 Oe) ac susceptibility at a temperature $T_f$, which usually increases with increasing measuring frequency.
2. These features become rounded in higher ac or dc fields.
3. Near $T_f$, irreversible behavior including remanence and coercivity appear.
4. The temperature derivative of the magnetization measured in small fields with decreasing temperature shows a discontinuity at $T_f$.
5. The remanence, of order a few percent of the collinear saturation magnetization at $T = 0$, decays with time, varying approximately linearly as a function of ln $t$, or as $t^{-1/n}$ where $n$ is approximately 10.
6. Remanence measured for the zero-field cooled state (isothermal remanent magnetization) is generally less than that obtained by cooling through $T_f$ with the field already applied (thermoremanent magnetization). The two become equal when the applied field is sufficiently large.
7. The magnetization of the zero-field cooled sample depends both on time and the thermal history of the sample.
8. The magnetic entropy at $T_f$ is only a fraction of the total magnetic entropy of disorder.
9. The paramagnetic Curie temperature $\theta_p$, extrapolated from susceptibility data taken well above $T_f$, is close to zero.
10. Various magnetic properties, including spin freezing temperature and magnitude of the remanence, scale with the concentration of magnetic impurities.

Together, this behavior constitutes the spin-glass syndrome. Some of the properties, like the ln $t$ decay of the remanence, $\theta_p \simeq 0$, or field dependence of the susceptibility peak, occur also for other sorts of magnetism, so in practice it is necessary to observe several of these characteristics before classifying any material as a spin glass. No consensus has yet evolved on the relative significance of the above factors, but perhaps the most telling features are those related to the irreversible susceptibility. The magnetization measured in the field cooled and zero-field cooled states of a typical ferromagnet, a typical antiferromagnet, and a canonical spin glass are compared in Figure 13.16. The behavior of

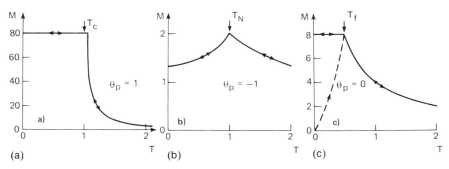

Figure 13.16. Comparison of the magnetization measured in a small applied field for (a) a ferromagnet, (b) an antiferromagnet, and (c) a spin glass. The dashed line indicates the zero-field cooled behavior of the spin glass, which evolves with time. The curves are schematic and the units arbitrary, but the same average magnitude of the exchange interaction is chosen for all three. Values of the paramagnetic Curie temperature in $\chi = C/(T - \theta_p)$ are indicated.

random, noncollinear amorphous magnets is closest to that of the canonical spin glass. The first theoretical models of spin glasses were based on the idea that the magnetic interaction in chemically disordered dilute alloys must be of the oscillatory RKKY type, so that the $P(\mathscr{I})$ distribution was taken as a gaussian, centered at $\mathscr{I} = 0$.

The existence was postulated of a ground state, where each spin "freezes" into a direction that is randomly oriented, isotropic, and different for every spin. Taking $q = \lim_{t \to \infty} \langle \mathbf{S}_i(0) \cdot \mathbf{S}_i(t) \rangle$ as the order parameter, it was shown that the system settles into the postulated ground state below $T_f$. Within a novel form of mean-field theory, a phase transition with a kink in the susceptibility was found at $T_f$.[32] Later studies have uncovered the enormous difficulties inherent in the theoretical investigation of the spin-glass problem. Some cherished concepts, such as linear-response theory, have had to be abandoned, and it seems that nonergodic behavior prevails at low temperature. Spin-glass theory is a prolific and controversial subject.[33]

### 13.3.4. Summary

A summary of the types of collective magnetic order that may be found in crystalline solids and in magnetic glasses is given in Figure 13.17. The most important determining factors are shown around the border of each part of the figure, and they are linked to the various magnetic structures by solid or dashed lines according to whether they are essential ingredients or simply possible contributing factors. In each part of the diagram, collinear structures are shown in the upper half of the figure, and noncollinear ones in the lower half.

a) Crystalline solids

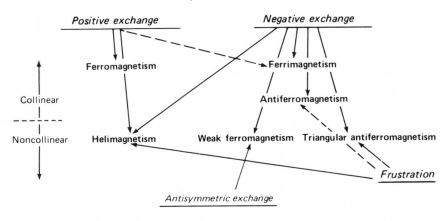

b) Noncrystalline solids

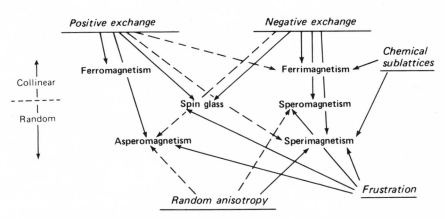

Figure 13.17. Relation between the various forms of collective magnetic order and the important contributing factors in (a) crystalline and (b) noncrystalline solids.

## 13.4. Applications of Amorphous Magnets[2]

The uses to which crystalline magnetic materials are put may be roughly divided into three broad areas,[34,35] indicated in Figure 13.18. Iron-based alloys with soft magnetic properties for electromagnetic power applications (motors, generators, transformers, inductors) represent by far the greatest volume of production. There is a comparable market for high-value materials employed in digital and analog data storage and recording, and for a variety of other electronic and magnetomechanical applications. The third sector, hard magnetic materials for permanent magnets, represents 10–15% of the total market. We note that roughly half the materials used are metallic alloys and half are insulators. Ferrites are mainly used in data storage and recording, but also feature in high-frequency electronic and power components, and as ceramic permanent magnets.

Practically all useful magnetic materials possess a spontaneous magnetic moment at ambient temperatures. In metals this arises directly from ferromagnetic coupling of the atomic moments, while in insulators the dominant coupling is generally antiferromagnetic superexchange and a net ferrimagnetic moment only arises if the crystal structure incorporates two unequal cation sublattices. From the discussion of Section 13.3 it appears that topological disorder in amorphous compounds makes it impossible to define sublattices, and antiferromagnetic interactions are satisfied as far as possible by a random, noncollinear structure, which includes marked antiparallel correlations of the nearest-neighbor spins. Frustration depresses the magnetic ordering temperature far below that expected from the exchange strength of the individual bond interactions. An amorphous insulating compound, which is magnetically ordered at room temperature, has yet to be discovered: their spin freezing transitions are usually below 100 K. For these reasons there is no prospect of replacing ferrites for magnetic applications by their noncrystalline counterparts.

The requirement of a spontaneous net moment restricts us to amorphous materials with predominantly ferromagnetic interactions. Although

Figure 13.18. Schematic division of the world market for magnetic materials. The pie represents roughly $5 \times 10^9$ per annum. The relative importance of metals (alloys) and insulators (ferrites) are indicated in each sector.

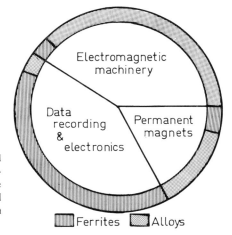

there are a few insulating and semiconducting materials in this category,[36] the need for a high Curie temperature limits the selection to amorphous metallic alloys containing a high proportion of Fe, Co, or Ni, or some combination of these three. Many such ferromagnetic alloys can be prepared as glasses by melt-spinning. Ferrimagnetic and sperimagnetic amorphous alloys of the rare earths with Fe, Co, or Ni also have potential use in devices. The dominant coupling is ferromagnetic, but the rare-earth transition–metal coupling is antiferromagnetic for the second half of the rare-earth series.

At a qualitative level there are really no novel bulk magnetic properties of ferromagnetic amorphous metals that do not exist in crystalline metals. Even the lack of magnetocrystalline anisotropy, which one might have expected from the random addition of the single-ion contributions, does not preclude hysteresis. Very soft ferromagnetic metallic glasses are available, but their properties tend ultimately to be limited by strains, which create bulk anisotropy via magnetostriction. Nonmagnetostrictive glasses with excellent permeability and minimal hysteresis superior to the best supermalloy do exist, but they are rich in cobalt whose cost restricts their application to electronic components and specialized devices.

The advantages conferred by the glassy state are essentially quantitative. A great degree of chemical substitution is possible, far beyond that tolerated in crystalline solid solutions. This allows properties to be tailored over a wider range. But the decisive advantage of metallic glasses for a specific application usually lies in the combination of good magnetic properties with another, nonmagnetic factor such as hardness, high resistivity, corrosion resistance, or ease of fabrication. At present, magnetic applications are thought to offer the best economic prospects for metallic glasses.[37] A brief summary of the main uses proposed for amorphous magnets follows.

### 13.4.1. Electromagnetic Power Apparatus

Iron-based glasses with low hysteresis losses are being evaluated for widespread use in transformers, motors, and inductors. Prototype distribution transformers with glassy metal cores (e.g., $Fe_{81}B_{13.5}Si_{3.5}C_2$) are now in service in the U.S. and Japan.[37,38] The main advantages of metallic glasses are their high resistivity and ease of fabrication in thin strips, which minimize eddy-current losses. A drawback is their lower saturation induction (16 kG), compared with silicon steel (12 kG), which means that larger transformers are needed. Nevertheless, there are genuine prospects for worthwhile savings on the $2 \times 10^9$ lost annually in the world's distribution transformers by changing to metallic glass in the cores. Almost as much is lost every year in electric motors, and there too prototype motors with metallic glass stators are being developed.[37,38] They are more difficult to fabricate from glassy ribbons than toroidal cores, and adoption of efficient motors by individual consumers would require a better appreciation of the total cost of ownership. Another promising area is in high-energy pulsed power applications where large inductors are employed in magnetic switches, and in pulse compressors used for firing high-power lasers and accelerators.[37]

### 13.4.2. Electronic and Other Devices

Many of the electronics applications exploit the good power losses and high permeability of certain magnetic glasses in the audiofrequency ranges, and up to frequencies of order $10^5$ Hz. Inductive heads for audio and video recording made from cobalt-based glasses possess good magnetic properties and superior wear resistance.[39] They are marketed in Japan. Other uses are in switched-mode power supplies, magnetic amplifiers, and a variety of other devices.[40]

The original commercial application of soft magnetic glasses was the simplest, magnetic shielding.[37] New, very-high-permeability shields manufactured from nonmagnetostrictive Co glasses [such as $Co_{66}Fe_4(MoSiB)_{30}$] are capable of efficiently shielding extremely weak fields and are insensitive to stress.[40]

Another group of applications exploits the magnetoelastic properties of compositions with substantial magnetostriction. Saturation linear magnetostriction of about 30 ppm is available in iron-based glasses, and values an order of magnitude greater are found in some amorphous rare-earth iron alloys. These materials may be used in transducers. Other metallic glasses show invar or elinvar characteristics[41] (i.e., volume or elastic moduli are independent of temperature over a range of temperature where the alloy is ferromagnetically ordered). The most striking manifestation of bulk anisotropy in the magnetic properties of amorphous alloys is the $\Delta E$ effect.[42] Young's modulus is hypersensitive to a magnetic field applied perpendicular to the initial domain orientation direction in a glassy soft ferromagnetic ribbon. The value of $E$ may change by as much as 30% in a field of less than 10 Oe. The effect on the velocity of sound has been exploited in magnetically tunable surface acoustic wave delay lines.[43] Products range from a computerized go game to directable antenna arrays.

### 13.4.3. Perpendicular Recording Media

Thin films of amorphous ferrimagnetic or sperimagnetic alloys with bulk anisotropy perpendicular to their planes have been used for permanent or erasable data storage. Bit packing density is limited by demagnetizing effects when the magnetization lies in the film plane so, for high-density recording, media with their magnetization perpendicular to the plane must be used. Such anisotropy may be induced during film preparation by controlling parameters, such as substrate bias, for sputtered films. However, its microscopic orgin is not completely clear, but possible explanations[44] include preferred orientation of atom pairs relative to the film plane, aligned columnar voids or inhomogeneities, or a built-in macroscopic strain arising from the unidirectional accretion of an evaporated or sputtered thin film.

The most intensively developed and sophisticated system of perpendicular magnetic recording is magnetic bubble recording,[45] where mobile reversed cylindrical domains represent the bits. Prototype memory devices have been built from amorphous ferrimagnetic GdCoMo alloys, but these alloys are best suited for bubbles in the 0.25–0.6 μm range. The present generation of commercial bubble memories use larger bubbles in epitaxially grown crystalline garnet films.

A more recent development, where amorphous alloys are better placed to become the preferred materials, is thermomagnetic recording of beam-addressable memory. Bits are written by locally heating the film with an electron or laser beam, provoking a change in magnetization of the spot as the temperature rises either to the compensation point or to the Curie point of the alloy. Thermomagnetic contact printing works the same way. Information can then be read out using the magneto-optic Faraday or Kerr effect. A good signal-to-noise ratio is obtained because of a reasonably large polar Kerr rotation angle, and the absence of grain boundaries in amorphous films. An erasable system of real-time video recording on a GdCo disk using compensation point laser writing has been described lately.[46]

### 13.4.4. Permanent Magnets

Another recent development has been the discovery that partially amorphous RFe alloys (R = Pr, Nd, Sm) prepared by slow melt spinning can show useful hard magnetic properties,[47] rivaling in some respects those of crystalline $SmCo_5$. These results have stimulated interest in the magnetic consequences of metastable microstructures formed at intermediate quench rates, and led to the discovery of the new $Nd_2Fe_{14}B$ phase[48] which seems set to revolutionize permanent magnet applications — probably the first time that work on amorphous materials has led to a technological breakthrough involving their crystalline counterparts.

ACKNOWLEDGMENT. It is a pleasure to thank Prof. Kishin Moorjani for his collaboration on many aspects of the work on which this text is based.

# References

1. K. Handrich and S. Kobe, *Amorphe Ferro- und Ferrimagnetika*, Physik-Verlag, Weinheim (1980).
2. K. Moorjani and J. M. D. Coey, *Magnetic Glasses*, Elsevier, Amsterdam (1984).
3. Chapters by E. M. Gyorgy and R. Alben, J. I. Budnick and G. S. Cargill, in: *Metallic Glasses*, American Society of Metals, Metals Park, Ohio (1978).
4. Chapters by J. Durand in: *The Magnetic, Chemical and Structural Properties of Glassy Metallic Alloys* (R. Hasegawa, ed.), CRC Press, Boca Raton (1982) and *Glassy Metals II* (H. J. Guntherodt and H. Beck, eds.), Springer-Verlag, Berlin (1983).
5. R. W. Cochrane, R. Harris, and M. J. Zuckermann, *Phys. Rep.* **48**, 1 (1978); J. J. Rhyne, in: *Handbook on the Physics and Chemistry of Rare Earths* (K. A. Gschneider and L. Eyring, eds.), Ch. 16, North-Holland, Amsterdam (1978).
6. F. E. Luborsky, in: *Ferromagnetic Materials* (E. P. Wolfarth, ed.), Vol. 1, Ch. 2, North-Holland, Amsterdam (1980) and chapters in *Amorphous Metallic Alloys* (F. E. Luborsky, ed.), Butterworth, London (1982).
7. D. Stauffer, *Phys. Kep.* **54**, 1 (1979).
8. D. H. Martin, *Magnetism in Solids*, Iliffe, London (1967).
9. J. M. D. Coey, *J. Appl. Phys.* **49**, 1646 (1978).
10. N. F. Mott, *Metal–Insulator Transitions*, Taylor and Francis, London (1974).
11. J. Hubbard, *Proc. R. Soc. London, Ser. A* **273**, 238 (1963); **277**, 237 (1964); **280**, 401 (1964).
12. M. Cyrot, *J. Phys. (Paris)* **33**, 125 (1972); *Philos. Mag.* **25**, 1301 (1972).
13. T. Mizoguchi, AIP Conf. Proc. **34**, 286 (1976); A. P. Malozemoff, A. R. Williams and V. L. Moruzzi, *IEEE Trans Magn* **MAG-19** 1983 (1983); *Phys. Rev* **B29**, 1620 (1984); ibid. **B30**, 6565 (1984).
14. V. Jaccarino and L. R. Walker, *Phys. Rev. Lett.* **15**, 258 (1965).
15. P. W. Anderson, *Phys. Rev.* **124**, 41 (1961).
16. P. M. Levy, C. Morgan-Pond, and A. Fert, *J. Appl. Phys* **53**, 2168 (1982).
17. W. Kummerle and U. Gradmann, *Solid State Commun.* **24**, 33 (1977).
18. J. B. Goodenough, *Magnetism and the Chemical Bond*, Wiley–Interscience, New York (1963).
19. D. C. Mattis, *The Theory of Magnetism I*, Springer-Verlag, Berlin (1981).
20. J. M. D. Coey, D. Ryan, D. Gignoux, A. Liénard, and J. P. Rebouillat, *J. Appl. Phys.* **53**, 7804 (1982); H. Fujimori, K. Nakanishi, H. Hiroyoshi, and N. S. Kazama, *J. Appl. Phys.* **53**, 7792 (1982).
21. W. Nägele, K. Knorr, W. Prandl, P. Convert, and J. L. Bueroz, *J. Phys C* **11**, 3295 (1978).
22. M. F. Hutchings, *Solid State Phys.* **16**, 227 (1966).
23. R. Harris, M. Plishke, and M. J. Zuckerman, *Phys. Rev. Lett.* **31**, 160 (1973).
24. K. H. J. Buschow, M. Brouha, J. W. M. Biesterbos, and A. G. Dirks, *Physica B* **91**, 261 (1977); K. H. J. Buschow, *J. Appl. Phys.* **53**, 7713 (1982).
25. N. A. Blum, K. Moorjani, T. O. Poehler, and F. G. Satkiewicz, *J. Appl. Phys.* **53**, 2074 (1982); C. L. Chien and K. M. Unruh, *Phys. Rev. B* **24**, 1556 (1981).
26. G. Toulouse, *Commun. Phys.* **2**, 115 (1977).
27. A. Herpin, *Théorie du Magnétisme*, Presses Universitaires de France, Paris (1968).
28. J. M. D. Coey and P. W. Readman, *Nature* **246**, 445, 476 (1973).
29. J. M. D. Coey, J. Chappert, J. P. Rebouillat, and T. S. Wang, *Phys. Rev. Lett.* **36**, 1061 (1976).
30. J. M. D. Coey and S. von Molnar, *J. Phys. (Paris)* **39**, L327 (1978).
31. Recent reviews include P. J. Ford, *Contemp. Phys.* **23**, 141 (1982); J. A. Mydosh and G. J. Nieuwenhuys, in: *Ferromagnetic Materials* (E. P. Wolfarth, ed.), Vol. 1, p. 71, North-Holland, Amsterdam (1980); R. Rammal and J. Souletie, *Proceedings of 1981 Les Houches*

*Summer School*; J. Joffrin, in: *Ill-Condensed Matter* (R. Balian, R. Maynard, and G. Toulouse eds.), p. 63, North-Holland, Amsterdam/World Scientific, Singapore (1979).

32. S. F. Edwards and P. W. Anderson, *J. Phys. F* **5**, 965 (1975).

33. P. W. Anderson, in: *Ill-Condensed Matter* (R. Balian, R. Maynard, and G. Toulouse, eds.), p. 159, North-Holland, Amsterdam/World Scientific, Singapore (1979); G. Parisi, in: *Disordered Systems and Localization* (C. Castellani, C. di Castro, and L. Peliti, eds.), p. 107, Springer-Verlag, Berlin (1981); G. Toulouse, in: *Congrès de la Société Française de Physique*, p. 3, Editions de Physique, Paris (1982).

34. I. S. Jacobs, *J. Appl. Phys.* **50**, 7294 (1979).

35. U. Enz, in: *Ferromagnetic Materials* (E. P. Wolfarth, ed.), Vol. 3, p. 1, North-Holland, Amsterdam (1982).

36. J. M. D. Coey, E. Devlin, and R. J. Gambino, *J. Appl. Phys.* **53**, 7810 (1982); F. J. Litterst, *J. Phys. (Paris)* **36**, L197 (1975).

37. C. H. Smith, *IEEE Trans. Magn.* **MAG-18**, 1376 (1982).

38. L. A. Johnson, E. P. Connell, D. J. Bailey, and S. M. Hegyi, in: *IEEE PES Meeting*, Paper 81TD 641–0, Minneapolis (1981).

39. K. Shiiki, S. Otomoto, and M. Kudo, *J. Appl. Phys.* **52**, 2483 (1981).

40. H. Warlimont and R. Boll, *J. Magn. & Magn. Mater.* **26**, 97 (1982).

41. K. Fukamichi, T. Masumoto, and M. Kikuchi, *IEEE Trans. Magn.* **MAG-15**, 1404 (1979).

42. B. S. Berry, in: *Metallic Glasses*, Ch. 7, American Society of Metals, Metals Park, Ohio (1978).

43. D. C. Webb, D. W. Forester, A. K. Ganguly, and C. Vittoria, *IEEE Trans. Magn.* **MAG-15**, 1410 (1979).

44. G. S. Cargill and T. Mizoguchi, *J. Appl. Phys.* **50**, 3570 (1979).

45. A. H. Eschenfelder, *Magnetic Bubble Technology*, Springer-Verlag, Berlin (1980).

46. Y. Togami, *IEEE Trans. Magn.* **MAG-18**, 1233 (1982).

47. J. J. Croat, *IEEE Trans. Magn.* **MAG-18**, 1442 (1982).

48. Proceedings of the Workshop on Nd–Fe–B Permanent Magnets, Brussels, 25 October 1984 (I. V. Mitchell, ed.), Brussels (1984).

49. J. M. D. Coey, *IEEE Trans. Magn.* **MAG-20**, 1278 (1984).

# Properties of Two-Level Systems

## W. A. Phillips

### 14.1. General Introduction

#### 14.1.1. Heat Capacity and Thermal Conductivity

The easiest way to introduce this subject is by showing some early results on the thermal properties of vitreous silica and other amorphous solids. The heat capacity[1] $C$ and thermal conductivity[1] $\kappa$ are shown as functions of the absolute temperature $T$ in Figures 14.1 and 14.2. Also shown in these figures are $C$ and $\kappa$ for a similar crystalline material, $\alpha$-quartz, which can be used to review the behavior expected on the basis of simple theories.

The heat capacity of $\alpha$-quartz varies as $T^3$ below 10 K, as expected in line with Debye theory, which predicts that in the long-wavelength limit the density of phonon states $g(\omega)$ varies quadratically with the phonon frequency $\omega$; $g(\omega) = A\omega^2$ if the velocity of sound $v_s$ is a constant, where $\omega/q = v_s$, with $q$ the phonon wave vector. At higher temperatures, $C$ increases more rapidly than $T^3$ as a result of the phonon dispersion, which increases $g(\omega)$.

The temperature variation of the thermal conductivity, $\kappa \propto T^3$, can be most easily explained by means of the simple kinetic equation

$$\kappa = \tfrac{1}{3}Cv_s\lambda \tag{14.1.1}$$

**W. A. Phillips** · Cavendish Laboratory, Department of Physics, University of Cambridge, Madingley Road, Cambridge CB3 0HE, England.

where $\lambda$ is the phonon mean free path. At low temperatures phonons are scattered by defects in the crystal, or by the surfaces of the sample, so that $\lambda$ is independent of $T$. The thermal conductivity $\kappa$ is therefore proportional to $C$ and hence to $T^3$. Above 10 K the reduction in $\lambda$ by phonon–phonon scattering leads to the fall in $\kappa$.

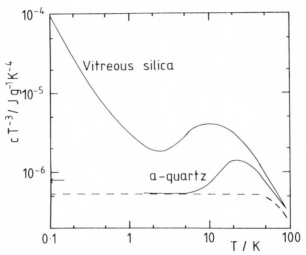

Figure 14.1. Heat capacity of vitreous silica and crystalline quartz as a function of temperature $T$ (after Jones[1] and Zeller and Pohl[1]).

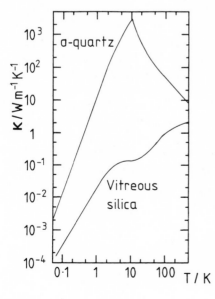

Figure 14.2. Thermal conductivity of vitreous silica and crystalline quartz (after Jones[1] and Zeller and Pohl[1]).

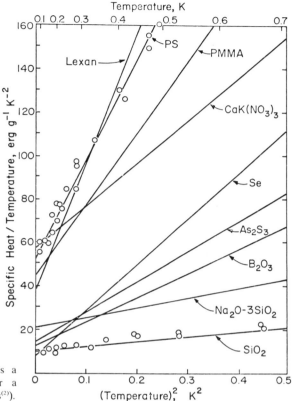

Figure 14.3. Heat capacity as a function of temperature for a range of glasses (after Stephens[2]).

These ideas are well known, and serve to emphasize the peculiarity of the results in the glass. The magnitude of $C$ varies roughly as $T$ below 1 K, and at 0.1 K it is about two orders of magnitude greater in the glass than in the crystal. Below 1 K $\kappa$ varies as $T^2$, in the range 4 to 30 K it increases only slightly with $T$, and then increases at higher temperatures toward a value similar to that of quartz.

These results were soon supplemented by measurements on a wide range of other amorphous solids, such as oxide glasses, chalcogenide glasses, polymers, and amorphous metals. In all these materials similar behavior was observed, as can be seen for a representative sample[2] in Figures 14.3 and 14.4.

### 14.1.2. Early Theories

The universality of the phenomena and the simplicity of the (idealized) temperature dependences, $C$ proportional to $T$ and $\kappa$ proportional to $T^2$,

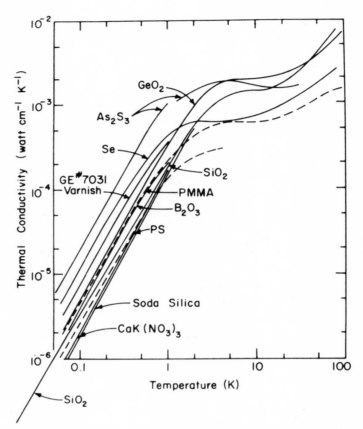

Figure 14.4. Thermal conductivity as a function of temperature for a range of glasses (after Stephens[2]).

proved to be very attractive for theorists. A large number of different theories were proposed in a short space of time. Some of these were based on realistic descriptions of the amorphous state, and are worth describing briefly in order to show how ideas from other branches of the study of amorphous solids can be applied to this field.

The first and perhaps most obvious explanation was in terms of electron states.[3] In the amorphous state the sharp distinction between energy bands and energy gaps is blurred, and it was suggested that the specific heat was a result of the almost constant density of localized electron states at the Fermi level, giving a linear temperature-dependent heat capacity just as electrons do in a metal. However, it turns out that the density of these localized states is much too small to explain the heat capacity. In silica, for example, the density of states deduced from the heat capacity is approximately $10^{46}$ $J^{-1}m^{-3}$ ($10^{21}$ $eV^{-1}cm^{-3}$) while optical measurements imply a density of states less than $10^{43}$ $J^{-1}m^{-3}$. After all, vitreous silica is almost perfectly transparent as can be shown by its use in optical fibers.

A second explanation involved the damping or scattering of phonons in a glass. This is based on the well-established idea that the lack of translational symmetry prevents the use of a wave vector $q$ to describe the vibrational modes except at low frequencies, where a continuum description should hold. A mode of well-defined $q$ will therefore decay as it propagates in the glass. Fulde and Wagner[4] suggested a specific mechanism for this decay, based on structural relaxation, and suggested that this could explain not only the thermal conductivity but also (through the consequent broadening of the spectral response function) the heat capacity. However, subsequent experiments, which will be described later, did not provide evidence for their mechanism, although the explanation of the thermal conductivity resembles current ideas.

The third and final example is a model for the thermal conductivity that is based on the scattering of sound waves by inhomogeneities in the glass structure. This is Rayleigh scattering by the local variations in the velocity of sound. However, two problems arise with this explanation.[5] The first concerns the frequency or $q$ dependence of the effect, as $\omega^4$, $\omega^2$, and $\omega$ dependences of the scattering rate have been suggested, and the second the magnitude of the fluctuations. Density fluctuations estimated thermodynamically or from light-scattering experiments are too small, but there is no way of estimating the local variations of sound velocity. Such scattering is undoubtedly important above 1 K, but probably not at lower temperatures.

Many other theories have been suggested, but in most cases these theories are inconsistent with what is known of the amorphous state and with the more precise experiments which will now be described.

### 14.1.3. Acoustic Experiments

At 1 K the frequency of a thermal phonon, given by $\hbar\omega \sim kT$, is about 30 GHz. The properties of such phonons can be investigated directly by means of thermal Brillouin scattering.[6] An incident photon excites or absorbs a phonon: measurements of the frequency shift at a particular scattering angle give the frequency and wave vector of the phonon, and the width of the Brillouin line gives the inverse phonon lifetime. Measurements show that both transverse and longitudinal acoustic phonons exist in a glass, and that there is no dispersion at these frequencies (i.e., the sound velocity is identical in $SiO_2$ with that measured at lower frequencies). The

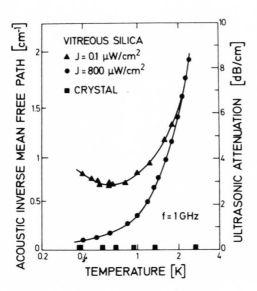

Figure 14.5. Ultrasonic attenuation as a function of temperature at high and low powers (after Hunklinger and Schickfus in Phillips[7]).

mean free path of these acoustic phonons is at least ten wavelengths at room temperature, and becomes larger as the temperature is reduced. Taken together with the thermal measurements these results show that *additional* excitations are present at low energies; they contribute directly to the heat capacity and scatter the existing phonons.

Acoustic measurements at lower frequencies, 100 MHz to 1 GHz, have provided a large amount of detailed information on the nature of these excitations. Many of these will be described later, but the results[7] shown in Figure 14.5 are particularly important. Below 1 K the attenuation increases with decreasing temperature if the acoustic intensity is low, just as expected on the basis of the thermal conductivity. The values of the phonon free path deduced from the acoustic measurements are consistent with those calculated from the thermal conductivity. However, the most valuable aspect of these results from the standpoint of investigating the low-frequency excitations is the observed saturation of the attenuation at high acoustic intensities. The implication of these experiments is that the excitations must be represented by two-level systems and not, for example, by harmonic oscillators. This point will be considered later in more detail.

### 14.1.4. Tunneling States

A specific form of the two-level system is provided by a tunneling state. Indeed, this model was suggested before the discovery of saturation and predicted the effect. The basic idea is that, in contrast to a crystalline solid where the position of each atom is determined by symmetry, the amorphous solid contains atoms or groups of atoms which are equally happy to sit in either of two local potential minima. It has been claimed that this is an essential property of the glass, representing the additional entropy frozen in at the glass transition $T_g$. If the barrier is not too large the atom can tunnel from one minimum to the other (just as in the ammonia molecule).

The state can be represented as in Figure 14.6, and the quantum-mechanical treatment of this double-well potential will now be considered in some detail.

The calculation of the energy levels of a particle in a double-well potential $V$ of the form shown in Figure 14.6a usually starts with the solution of the single-well problem shown in Figure 14.6b. The choice of these two basis states is known as the *well, nondiagonal*, or *localized* representation. Each state is the ground state of the appropriate harmonic potential $V_1$ or $V_2$, both of which are shown continued as dotted lines in Figure 14.6a. The Hamiltonian can be written as

$$H = H_1 + (V - V_1) = H_2 + (V - V_2) \qquad (14.1.2)$$

where $H_1$ and $H_2$ are the individual Hamiltonian operators. In this

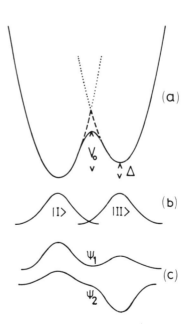

Figure 14.6. The double-well potential, together with the wave functions in the localized and diagonal representations.

representation the Hamiltonian matrix becomes

$$\begin{pmatrix} E_1 + \langle 1 | V - V_1 | 1 \rangle & \langle 1 | H | 2 \rangle \\ \langle 2 | H | 1 \rangle & E_2 + \langle 2 | V - V_2 | 2 \rangle \end{pmatrix} \qquad (14.1.3)$$

To a good approximation each term $\langle i | V - V_i | i \rangle$ can be neglected in comparison with $E_i$. If the zero of energy is chosen as the mean of the two ground-state energies $E_1$ and $E_2$, expression (14.1.3) can be written in the form

$$\frac{1}{2} \begin{pmatrix} -\Delta & -\Delta_0 \\ -\Delta_0 & \Delta \end{pmatrix} \qquad (14.1.4)$$

where $\Delta_0$ is defined as

$$\Delta_0 = -2\langle 1 | H | 2 \rangle \qquad (14.1.5)$$

the negative sign being introduced because the matrix element $\langle 1 | H | 2 \rangle$ is negative. We note that only if the wells are identical, apart from a relative displacement in energy, is the $\Delta$ of expression (14.1.4) identical to that of Figure 14.6a.

The quantity $\Delta_0$ can be evaluated for specific potentials. For two identical three-dimensional harmonic oscillators with $\Delta = 0$ and with an overall potential shown by the dashed continuation in Figure 14.6a,[8]

$$\Delta_0 = \hbar\omega_0 \left[ 3 - \left( \frac{8V_0}{\pi\hbar\omega_0} \right)^{1/2} \right] e^{-2V_0/\hbar\omega_0} \qquad (14.1.6)$$

where $V_0$ is the minimum energy barrier between the two wells, and $\hbar\omega_0$ is equal to $2E_1$ or $2E_2$. Since for our purposes (and indeed for the localized representation to be useful) $V_0 \gg \hbar\omega_0$, equation (14.1.6) becomes

$$\Delta_0 = -2\hbar\omega_0 \left( \frac{2V_0}{\pi\hbar\omega_0} \right)^{1/2} e^{-2V_0/\hbar\omega_0}$$

$$= -4 \left( \frac{2V_0^3\hbar^2}{md^2} \right)^{1/4} e^{-(2mV_0/\hbar^2)^{1/2}d/2} \qquad (14.1.7)$$

where $m$ is the mass of the particle and $d$ the separation of the two minima. This value of $\Delta_0$ is just twice that found for the equivalent problem in one dimension.

As an alternative example, the solution of Mathieu's equation for a rigid rotator in a twofold symmetric potential gives,[9] in the same limit $V_0 \gg \hbar\omega_0$, the approximate result

$$\Delta_0 \simeq -8\hbar\omega_0 \left(\frac{2V_0}{\pi\hbar\omega_0}\right)^{1/2} e^{-4V_0/\hbar\omega_0} \tag{14.1.8}$$

where again $V_0$ is the height of the barrier separating the two wells, and $\omega_0$ is the angular frequency of small oscillations within a single minimum.

Although both results imply a similar exponential dependence of $\Delta_0$ on $V_0$ the numerical relationships are different, and the use of a particular form of equation for $\Delta_0$ cannot be justified unless, as is often the case in crystals, the microscopic symmetry of the potential is known. For this reason an expression of the form

$$\Delta_0 = -\hbar\Omega e^{-d(2mV_0/\hbar^2)^{1/2}} \tag{14.1.9}$$

where $\hbar\Omega$ is an energy roughly equal to $\hbar\omega_0$, is usually adequate in the case of amorphous solids. It is noteworthy that $\langle 1 | H | 2 \rangle$ is negative in these examples because the negative contribution from $V - V_2$ (for example) in equation (14.1.2) overwhelms the positive contribution from $H_2$.

The matrix (14.1.4) can be diagonalized to obtain the eigenstates, the *true*, *diagonal*, or *energy* representation. The eigenfunctions, shown in Figure 14.6c, have energies $\pm E/2$, where

$$E^2 = (\Delta^2 + \Delta_0^2) \tag{14.1.10}$$

and are usually written in the analytic forms

$$\psi_1 = |1\rangle\cos\theta + |2\rangle\sin\theta \tag{14.1.11}$$

$$\psi_2 = |1\rangle\sin\theta - |2\rangle\cos\theta \tag{14.1.12}$$

where $\tan 2\theta = \Delta_0/\Delta$; $\psi_2$ is the lower energy state because $\Delta_0$ (as defined here) is positive.

The functions $\psi_1$ and $\psi_2$ defined by equations (14.1.11) and (14.1.12) are orthonormal only to the extent that the overlap term $\langle 1 | 2 \rangle$ can be put equal to zero. As far as normalization is concerned this presents no difficulty, as an additional multiplicative factor can be readily calculated, but may lead to more important problems in the calculation of matrix elements. Any matrix element involving the overlap of $\langle 1 |$ and $| 2 \rangle$ must be evaluated carefully through a correct choice of orthonormal states.

It is worth pointing out the existence of a wide range of notation for equation (14.1.10). Not only have several symbols been used for the various energies in this equation but the same symbols have been used for half the energies. Occasional discrepancies of powers of two in equations cited in the literature can often be traced to this difference in definition.

For this, as for any problem involving two energy levels, there is a formal analogy with the problem of a spin $\frac{1}{2}$ particle in a magnetic field. The Hamiltonian matrix (14.1.4) can be rewritten in terms of spin operators, although here too there is a choice of notation between the Pauli spin matrices

$$\sigma_x = \begin{pmatrix} 0 & 1 \\ 1 & 0 \end{pmatrix}, \quad \sigma_y = \begin{pmatrix} 0 & -i \\ i & 0 \end{pmatrix}, \quad \sigma_z = \begin{pmatrix} 1 & 0 \\ 0 & -1 \end{pmatrix} \quad (14.1.13)$$

and the spin $\frac{1}{2}$ operators defined by $S_i = \frac{1}{2}\sigma_i$. After diagonalization the Hamiltonian can be written in the obvious form

$$H = \tfrac{1}{2}E\sigma_z \qquad (14.1.14)$$

The advantages of this analogy will be seen later when it is used to interpret nonlinear and coherent effects in the interaction of tunneling states with acoustic and electric fields.

In many cases the results of experiments can be interpreted in terms of a simpler model, which ignores the detailed origin of the two levels and uses the energy $E$ as the only parameter, in contrast to the tunneling model, which uses both $\Delta$ and $\Delta_0$. This model is often known as the *two-level system* model, in contrast to the *tunneling* model described earlier.

The second step in the discussion of tunneling states in amorphous solids involves an evaluation of the distribution functions for the parameters $\Delta_0$ and $\Delta$, the obvious choice of independent variables. Both parameters are expected to take a wide range of values in the amorphous solid. In the case of the asymmetry $\Delta$ it can be argued that the distribution function must be symmetric as positive and negative values of $\Delta$ are equally likely, and further that the scale of energy variations will be determined by the thermal energy available at the effective temperature characterizing the "frozen liquid" state. Since $T_g$ is between 300 and 1000 K for most glasses, this will be of the order of 0.05 eV. Below 1 K the thermal energy is $10^{-4}$ eV or less, so that the low-temperature properties are sensitive to the center of a broad symmetric distribution. The distribution function $f(\Delta)$ can therefore be taken to be a constant.

A less firmly based argument can be used to derive the form of $g(\Delta_0)$, where $\Delta_0$ varies exponentially with the barrier height and separation. If the exponent is assumed to vary smoothly on the scale of $kT_g$, the form of $g(\Delta_0)$ is determined by the exponential dependence of $\Delta_0$ on the barrier parameters:

$$g(\Delta_0) \propto \frac{1}{\Delta_0} \qquad (14.1.15)$$

Some care should be taken not to assume that this form of $g(\Delta_0)$ is always (or even sometimes) valid. There is little direct experimental evidence in support of this precise form and indeed other slightly modified forms have been suggested. However, the use of relationship (14.1.15) allows a preliminary comparison to be made between experiment and theory.

The density of states $n(E)$ can be calculated from $f(\Delta)$ and $g(\Delta_0)$. The details of this calculation involve a treatment of the singularities introduced by integrating expression (14.1.15) and will be discussed later in connection with time-dependent heat capacities, but as might be expected the result is a slowly varying logarithmic function of energy. The origin of the "linear" temperature-dependent term can be seen if this slight energy dependence is neglected, so that $n(E) = n_0$, a constant. The heat capacity is

$$C(T) = n_0 \int_0^\infty \left(\frac{E}{2kT}\right)^2 \operatorname{sech}^2 \frac{E}{2kT} \, dE \qquad (14.1.16)$$

where the integrand represents the response of a single two-level system. Integration gives

$$C(T) = \frac{\pi^2}{6} n_0 k^2 T \qquad (14.1.17)$$

a result that can be used to estimate $n_0$.

## 14.2. Phenomenological Theory

### 14.2.1. Transition Probabilities and Relaxation Times

Transitions between the states $\psi_1$ and $\psi_2$ occur through the perturbation of the potential well of Figure 14.6a by a photon or phonon with energy $\hbar\omega = E$. This perturbation can change $\Delta$ or $\Delta_0$ (or both) but, of the two, changes in $\Delta$ are much more important.

The reason for this is twofold. The first is that the wavelength of the perturbing electric or strain field is much greater than the separation of the wells. As in the electric-dipole approximation in semiclassical radiation theory this leads to a perturbing potential which is essentially antisymmetric, equivalent to a change in $\Delta$ and not in $\Delta_0$. Second, the matrix elements, calculated in the localized basis $|1\rangle$ and $|2\rangle$, are relatively much smaller for a symmetric perturbation. (The use of the words symmetric and antisymmetric is of course not exactly correct, because the potential well of Figure 14.6a is not symmetric. It is, however, a useful approximation, identifying perturbations that tend to change $\Delta$ or $\Delta_0$ separately.)

The second effect can be illustrated by comparing the matrix elements for two perturbing potentials $Ax$ and $Bx^2$ for the one-dimensional double harmonic oscillator of Figure 14.6a with $\Delta = 0$. The appropriate quantity for comparison is the ratio of the matrix element to the change in energy of each well ($Ad/2$ and $Bd^2/4$ in the two cases, where $d$ is the separation of the minima). For the antisymmetric perturbation this ratio is $\pm 1$ for the diagonal matrix elements and zero for the off-diagonal. In the case of the symmetric perturbation $Bx^2$ the diagonal elements are equal in both magnitude and sign and so give simply a shift in the zero of energy, while the ratio of the off-diagonal matrix elements to $Bd^2/4$ is $(\hbar\omega_0/4V)\exp(-2V_0/\hbar\omega_0)$. The value of $V_0/\hbar\omega_0$ is considerably greater than unity for the low-temperature application of this model, and so the off-diagonal terms in the $|1\rangle$, $|2\rangle$ basis can be neglected. [It is noteworthy that for $\Delta = 0$ the functions $\psi_1$ and $\psi_2$ defined by equations (14.1.11) and (14.1.12) are orthogonal, and so no particular precautions need be taken in the calculation.]

The result has general validity. If the potential wells are not equivalent, or if $\Delta$ is not equal to zero, the antisymmetric perturbation will give off-diagonal terms and the symmetric perturbation will give, in addition to the off-diagonal terms, unequal diagonal terms. However, all these matrix elements are proportional to a factor of the general form $\exp(-2V_0/\hbar\omega_0)$ and so will be relatively unimportant. The perturbation to be included in the Hamiltonian (14.1.4) is therefore diagonal in the basis $|1\rangle$, $|2\rangle$. Using the transformation defined by equations (14.1.11) and (14.1.12), the perturbation in the $\psi_1$, $\psi_2$ basis has the form

$$\begin{pmatrix} \cos 2\theta & \sin 2\theta \\ \sin 2\theta & -\cos 2\theta \end{pmatrix} \quad \text{or} \quad \begin{pmatrix} \Delta/E & \Delta_0/E \\ \Delta_0/E & -\Delta/E \end{pmatrix}$$

The interaction Hamiltonian can therefore be written in terms of the Pauli operators as

$$H_{\text{int}} = \left(\frac{\Delta}{E}\sigma_z + \frac{\Delta_0}{E}\sigma_x\right) p_0 \cdot F + \left(\frac{\Delta}{E}\sigma_z + \frac{\Delta_0}{E}\sigma_x\right)\gamma e \qquad (14.2.1)$$

in the presence of an electric field $F$ and a strain field $e$; $p_0$ and $\gamma$ are defined as $\frac{1}{2}\partial\Delta/\partial F$ and $\frac{1}{2}\partial\Delta/\partial e$, respectively, and are therefore also equal to the electric and elastic dipole moments of the equivalent classical potential. The vector character of $F$ is here preserved, but the quantity $\gamma e$ has been written as an average over orientations (it can also be averaged over polarizations, although transverse and longitudinal models are usually considered separately). The off-diagonal term $\sigma_x$ produces transitions between $\psi_1$ and $\psi_2$ while the diagonal term $\sigma_z$ changes their relative energies.

In the simpler two-level system model the relationship between the diagonal and off-diagonal terms is ignored. The interaction Hamiltonian becomes

$$H_{int} = (\tfrac{1}{2}D\sigma_z + M\sigma_z)e + (\tfrac{1}{2}\mu\sigma_z + \mu'\sigma_x)F \qquad (14.2.2)$$

where the diagonal and off-diagonal terms are specified independently.

The interaction between tunneling states of two-level systems and photons can conveniently be described by the use of the Einstein coefficients. Consider first the interaction of two-level systems with thermal phonons. Each two-level system is continually absorbing and emitting thermal phonons. The rate equation for the probability $p_1$ of finding the system in the ground state $\psi_1$ can be written

$$\frac{dp_1}{dt} = -p_1 B\rho(E) + p_2[A + B\rho(E)] \qquad (14.2.3)$$

where $A$ and $B$ are phonon Einstein coefficients and $\rho(E)$ is the phonon energy density (per unit volume) evaluated at an energy $E$ equal to the two-level system energy; $\rho(E)$ is given in terms of the density of states $g(E)$ by

$$\rho(E) = \frac{Eg(E)}{e^{E/kT} - 1} \qquad (14.2.4)$$

In thermal equilibrium $dp_1/dt = 0$, and since $p_1 + p_2 = 1$

$$A/B = Eg(E) \qquad (14.2.5)$$

For small departures from equilibrium equation (14.2.3) defines a relaxation time $\tau$, where

$$\tau^{-1} = [A + 2B\rho(E)] \qquad (14.2.6)$$

which can be rewritten, using equation (14.2.4), as

$$\tau^{-1} = A \coth(E/2kT) \qquad (14.2.7)$$

$1/A$ is the natural lifetime of the system at absolute zero, and at any temperature $\hbar\tau^{-1}$ is the uncertainty in the energy $E$.

The analysis can be continued to calculate the rate at which phonons are scattered by the two-level systems. If the density of states of the two-level systems is $n(E)$ per unit volume, the change in the phonon energy density is given by

$$\frac{\partial\rho(E)}{\partial t} = En(E)\,\frac{\partial p_1}{\partial t} \tag{14.2.8}$$

so that using equation (14.2.3)

$$\frac{d\rho(E)}{dt} + n(E)BE(p_1 - p_2)\rho(E) = n(E)EAp_2$$

The phonon lifetime is given by

$$\tau_{ph}^{-1} = n(E)BE(p_1 - p_2) = \frac{n(E)A}{g(E)}\tanh\left(\frac{E}{2kT}\right) \tag{14.2.9}$$

The coefficient $B$ can be calculated starting from equation (14.2.11) using a derivation equivalent to that of the corresponding optical problem. However, unlike the optical case, the form chosen for the density of states $g(E)$ must be specified for phonons, as must the polarization. The Debye approximation can be used at temperatures of 1 K and below, so that the phonon density of states has a quadratic dependence on energy. Further, the phonon polarization can be classified as either longitudinal $l$ or transverse $t$. For a single polarization $\alpha$, $B$ is given by

$$B = \frac{\pi M_\alpha^2}{\hbar\rho_0 v_\alpha^2} \tag{14.2.10}$$

where $\rho_0$ is the density of the solid and $v_\alpha$ is the velocity of sound for polarization $\alpha$. The value of $M$, as discussed in connection with equation (14.2.1), is an average over orientations. The relaxation times can now be written for the *two-level system* as

$$\tau^{-1} = \sum_\alpha \frac{M_\alpha^2}{v_\alpha^5}\frac{E^3}{2\pi\rho_0\hbar^4}\coth\left(\frac{E}{2kT}\right) \tag{14.2.11}$$

and

$$\tau_{ph}^{-1} = \frac{\pi M_\alpha^2}{\hbar\rho_0 v_\alpha^2}\,n(E)E\tanh\left(\frac{E}{2kT}\right) \tag{14.2.12}$$

where both longitudinal and transverse contributions have been included in the expression for $\tau^{-1}$, but $\tau_{ph}^{-1}$ is written for a single polarization $\alpha$.

For the *tunneling states* these equations must be modified to take account of the explicit form of $M_\alpha$. The tunneling state relaxation time is

$$\tau^{-1} = \sum \frac{\gamma_\alpha^2}{v_\alpha^5} \frac{\Delta_0^2 E}{2\pi\rho_0 \hbar^4} \coth\left(\frac{E}{2kT}\right) \tag{14.2.13}$$

but the expression for the phonon scattering time is more complicated because, for a given phonon energy $E$, each tunneling state scatters phonons at a rate determined by $M_\alpha^2$, proportional to $\Delta_0^2$. In an amorphous solid, as opposed to a crystal, there will in general be a wide range of local environments and hence a range of values of $\Delta_0$. The exact expression for $\tau_{ph}^{-1}$ involves the distribution function for $\Delta_0$ but for the moment it is sufficient to write

$$\tau_{ph}^{-1} = \frac{\pi\gamma_\alpha^2}{\hbar\rho_0 v_\alpha^2} \bar{n}(E)E \tanh(E/k2T) \tag{14.2.14}$$

where $\bar{n}(E)$ is an effective density of states. In principle, the coupling parameter $\gamma_\alpha$ also varies from tunneling state to tunneling state, and it too should be represented by an average, although this can be incorporated into $\bar{n}(E)$.

One important difference between the two-level system model and the tunneling-state model should be noted. As described in Section 14.1.4, the specific heat can be calculated from $n(E)$. In the two-level system model the same parameter enters directly into the expression for $\tau_{ph}^{-1}$, but in the tunneling model this is not so, and the relationship between the heat capacity and $\tau_{ph}^{-1}$ depends on an unknown distribution function.

A description of absorption and emission in terms of the Einstein coefficients and semiclassical radiation theory is obviously oversimplified, and it is important to consider the extent to which it is valid. The semiclassical approach can be replaced by a quantized field calculation without changing the results. More important is the neglect in the Einstein treatment of coherence between the wave functions of the two energy levels. The two-level system is here characterized by two parameters, the occupation probabilities, instead of the three that a full quantum treatment requires. This limitation means that the Einstein approach cannot provide a detailed explanation of nonlinear and coherent effects (including higher-order transitions involved in Raman scattering). It does, however, give an accurate description of one-phonon or photon emission and absorption, although even simple scattering needs to be treated more carefully.

## 14.2.2. Heat Capacity

The results developed in Section 14.2.1 can be used to discuss the dependence of the heat capacity on the time scale $t_0$ of the measurement. It was mentioned in connection with the derivation of $n(E)$ that the integration of $g(\Delta_0)$ needs care. The upper limit to $\Delta_0$ is obviously given by $E$, but in principle the lower limit is not well defined but depends on the way in which $n(E)$ is measured. The time scale $t_0$ of the experiment determines the minimum value of $\Delta_0$ through equation (14.2.13) for the relaxation time. Defining a distribution function $g(E, \tau)dEd\tau$, and writing it in terms of $f(\Delta)$ and $g(\Delta)$ gives [using equation (14.2.13)]

$$g(E, \tau)dEd\tau = f(\Delta)g(\Delta_0)\frac{\Delta_0 E}{2\Delta\tau}\,dEd\tau \qquad (14.2.15)$$

A minimum relaxation time $\tau_{min}$ is defined by putting $\Delta_0 = E$ in equation (14.2.13). If $f(\Delta)$ is constant, $g(\Delta_0) \propto 1/\Delta_0$ gives

$$g(E, \tau) = \frac{A}{\tau(1 - \tau_{min}/\tau)^{1/2}} \qquad (14.2.16)$$

Equation (14.2.16) can in turn be used to give the density of states

$$n(E) = A \int_{\tau_{min}}^{t_0} \frac{d\tau}{\tau(1 - \tau_{min}/\tau)^{1/2}} = A \ln(4t_0/\tau_{min}) \qquad (14.2.17)$$

The tunneling model therefore predicts not only that there is a wide distribution of relaxation times but also that the heat capacity depends on the time scale $t_0$ on which it is measured. Of course, the precise forms of equations (14.2.16) and (14.2.17) depend on the form chosen for $g(\Delta_0)$.

Over the last few years considerable effort has been devoted to experimental verifications of these predictions. The existence of a wide range of relaxation times has been demonstrated conclusively, but until recently the measurements of the heat capacity on short time scales have been unreliable and contradictory. The two most recent experiments indicate the kind of information that can now be obtained.

If a sample of glass at a temperature $T_1$ is suddenly connected to a thermal reservoir at temperature $T_0$ $(T_0 > T_1)$ by a thermal link of conductance $\alpha$, in the absence of a distribution of relaxation times the temperature of the sample will decrease exponentially with time. As the temperature of the sample approaches $T_0$, the temperature–time curves can be superposed as shown in Figure 14.7. The observed behavior is quite different: as shown[10] in Figure 14.8 the temperature varies much more slowly with time than in Figure 14.7.

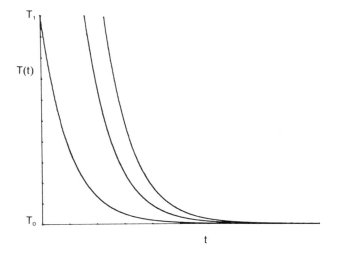

Figure 14.7. Exponential decays expected on cooling from different initial temperatures to $T_0$ if no relaxation effects are present.

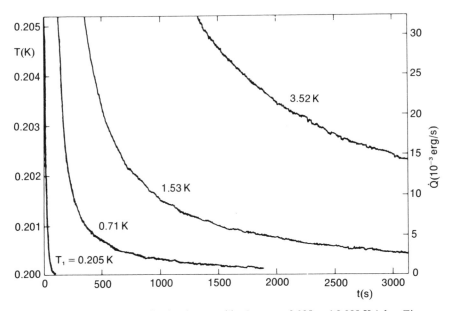

Figure 14.8. Thermal relaxation in vitreous silica between 0.205 and 0.200 K (after Zimmermann and Weber[10]).

These results can be understood on the basis of an idealized model of the experiment. The rate at which heat is evolved from the sample is given by $\alpha \dot{T}$ and at long times this will be the result of the slow evolution of energy from tunneling states with long relaxation times. Ordinary phonons relax rapidly and can be neglected except insofar as they define the temperature. Using the distribution function defined in equation (14.2.16) the instantaneous rate of energy loss is

$$\dot{Q} = \int_{kT_0}^{kT_1} \int_{\tau_{min}}^{\infty} \frac{E}{\tau} g(E, \tau) e^{-t/\tau} dE d\tau \qquad (14.2.18)$$

This can be integrated if $t \gg \tau_{min}$ (which is certainly true) to give

$$\dot{Q} = \frac{Ak^2}{2t} (T_1^2 - T_0^2) \qquad (14.2.19)$$

Equation (14.2.19) agrees well with the results and illustrates an important aspect of the amorphous state. Amorphous solids contain a large range of relaxation times and under very general assumptions [such as those outlined in connection with equation (14.2.15)] this leads to nonexponential relaxation. In particular, integration of functions similar to that used in equation (14.2.13) leads to relaxation as $1/t$.

The main conclusion of the experiments shown in Figure 14.8 is that there is a broad distribution of long $(50 \text{ s} < \tau < 5000 \text{ s})$ relaxation times, in agreement with low-frequency acoustic measurements. A probe of much shorter relaxation times is provided by heat-pulse propagation. A pulse of heat is applied to one side of a thin glass slide, and the temperature is monitored as a function of time at the other. By comparing the temperature–time response with solutions of the diffusion equation, a value for the heat capacity on a time scale corresponding to the thermal-diffusion time can be measured. Values of this characteristic time varying from 1 μs to a few seconds have been obtained by using slides of different thicknesses. The results of two of these studies[11] are shown in Figure 14.9, where the decrease in heat capacity is clear. However, this figure demonstrates equally well the uncertainty of the results. At least five sets of experiments have been performed, with inconsistent results, although all show the existence of relaxation time effects.

### 14.2.3. Phonon Scattering

The phonon scattering time defined in Equation (14.2.14) can be applied both to thermal conductivity and to acoustic experiments. In general the acoustic experiments are used to measure the parameter $\bar{n}\gamma^2$

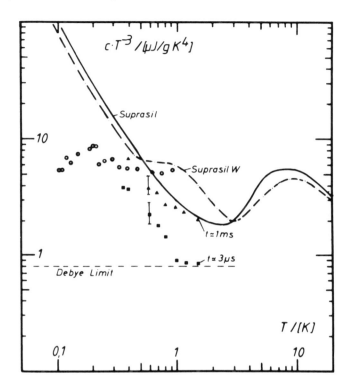

Figure 14.9. Heat capacity of vitreous silica measured on different time scales (after Meissner and Spitzmann[11]).

(and $v_s$), which can then be used to calculate the thermal conductivity below 1 K.

The temperature dependence of $\kappa$ follows from application of the dominant phonon approximation to equations (14.1.1) and (14.2.14). This approximation states that at a temperature $T$ the relevant phonons are those for which $\hbar\omega \sim kT$. The temperature dependence of $\tau_{ph}$ is obtained by putting $E = \hbar\omega \sim kT$, so $\tau_{ph} \propto 1/T$. Together with the $T^3$ variation of $C$ this gives the $T^2$ dependence of $\kappa$. A more careful treatment involves not only the use of the real energy-dependent density of states $\bar{n}$ in equation (14.2.14) but also a generalization of equation (14.1.1) to include an integration over all phonon frequencies. Such a calculation reproduces the experimental results more exactly, and shows why the observed temperature dependence of $\kappa$ is $T^{1.9}$ and not $T^2$.

Equation (14.2.14) has been extensively tested by means of acoustic experiments at low powers. At acoustic frequencies the condition $\hbar\omega \ll kT$

is satisfied at all but the lowest temperatures so that $l_{ph}^{-1} \propto \omega^2/T$, the attenuation increasing at lower temperatures as shown in Figure 14.5. In fact the complete expression (14.2.14) has been shown to give a good description of the low power attenuation, as shown[12] in Figure 14.10.

Associated with this attenuation is a temperature-dependent acoustic velocity. Applying the Kramers–Kronig equations to the acoustic case gives a value for this contribution $\Delta v$ to the velocity given by

$$\Delta v = \frac{P}{\pi} \int_0^\infty \frac{v \tau_{ph}^{-1} d\omega'}{\omega^2 - \omega'^2} \qquad (14.2.20)$$

Relative to a reference temperature $T_0$ the change in velocity is given in the limit $\hbar\omega \ll kT$, using equation (14.2.14), by

$$v(T) - v(T_0) = \frac{\bar{n}\gamma^2}{\rho v} \ln(T/T_0) \qquad (14.2.21)$$

Experimental results,[13] as illustrated in Figure 14.11, confirm this behavior and give values for the product $\bar{n}\gamma^2$. The logarithmic dependence of the velocity is a characteristic feature of tunneling states and has been used to show the existence of such states in materials, such as amorphous metals, where the contribution of tunneling states to the thermal properties is difficult to identify.

Perhaps the most important feature of the acoustic experiments is the phenomenon of saturation. This can be readily understood on the basis of the Einstein treatment. Implicit in equation (14.2.4) is the fact that transitions are induced by phonons from a thermal distribution. At larger acoustic powers this will no longer be the case, and $\rho(E)$ in equation (14.2.3) will be determined by the acoustic phonon flux. As this increases spontaneous transitions become relatively unimportant, and the populations of the two levels will both tend to one-half. Stimulated emission processes therefore compensate for the absorption, and there is very little

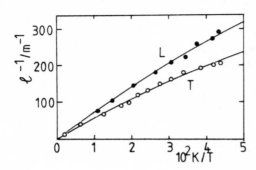

Figure 14.10. Temperature dependence of the measured phonon free path. The solid curves are $l^{-1} = l_0^{-1} \tanh(\hbar\omega/2kT)$ (after Golding *et al.*[12]).

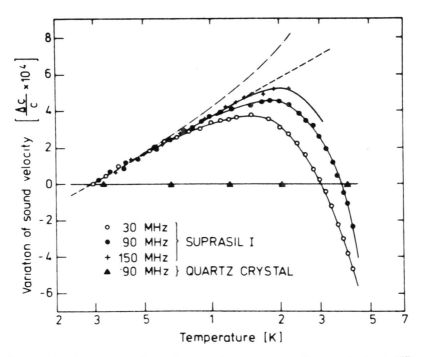

Figure 14.11. Acoustic velocity as a function of temperature in silica (after Piché et al.[13]).

net attenuation. An estimate of the acoustic power at which this occurs can be obtained by equating the thermal-phonon and acoustic-phonon concentrations.

Although this appears straightforward one point requires clarification. The thermal-phonon concentration can be expressed per unit frequency (or energy) range, but a calculation of the corresponding acoustic-phonon concentration requires a knowledge of the effective bandwidth. This can be determined either by the length of the acoustic pulse (i.e., an experimental broadening) or by the intrinsic lifetime of the tunneling state. This has been mentioned previously in connection with equation (14.2.7), and will be discussed in much more detail later. At this point it is sufficient to say that most information is obtainable in the continuous-wave limit, where the effective bandwidth is intrinsic. In this case, if the number of acoustic phonons per unit frequency range is equated to the number of thermal phonons, the critical energy density $E_c$ is given by

$$\frac{E_c}{\hbar\omega T_2^{-1}} = \frac{\omega^2}{\pi v^3}\frac{kT}{\hbar\omega}$$

where $T_2^{-1}$ is the line width of the tunneling state and the final factor comes from the low-frequency expansion of the Bose factor. This gives

$$E_c = \frac{\omega^2 kT}{\pi v^3 T_2}$$

or equivalently for the power density $J_c$ (W m$^{-2}$),

$$J_c = \omega^2 kT / \pi v^2 T_2 \qquad (14.2.22)$$

The results shown[7] in Figure 14.5 imply a value of roughly 10 ns for $T_2$, but this is not the value predicted by equation (14.2.13) in general. Interaction between tunneling states gives a line width $T_2^{-1}$ greater than that predicted from the relaxation time $T_1$.

### 14.2.4. Dielectric Absorption

There is an exact parallel between the absorption of electromagnetic and acoustic waves. The dielectric absorption varies in the same way as the acoustic attenuation, and the velocity of light in the glass (or the dielectric constant) varies logarithmically with temperature. Results in vitreous silica[14] are shown in Figure 14.12.

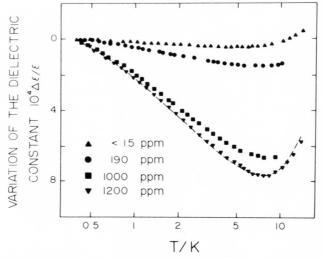

Figure 14.12. Temperature dependence of the dielectric constant for vitreous silica containing different concentrations of OH (after Hunklinger and Schickfus[14]).

The main difference between the electromagnetic and acoustic results lies in the sample dependence of the results. In the acoustic case all samples of $SiO_2$, for example, give the same value for $\bar{n}\gamma^2$, but the measurements of dielectric constant represented in Figure 14.12 show an effect proportional to the concentration of water (or hydroxyl ions) in the silica. This is not unexpected in view of the large dipole moment of hydroxyl, but is clear evidence of the role of impurities in some of these phenomena.

### 14.2.5. Thermal Expansion

Although the range of experimental data is limited, the thermal expansion $\beta$ of glasses at low temperatures reflects the other unusual thermal properties. Results in vitreous silica[15] are shown in Figures 14.13 and 14.14. The Grüneisen parameter used to describe the expansion is defined from the experimental results as $\beta V/\chi_T C_V$ where $\chi_T$ is the isothermal compressibility. This parameter balances the strong temperature dependencies of $\beta$ and $C_V$ to give a quantity which varies slowly with temperature and which is of order $+1$ in ordinary crystalline solids. Microscopically, it is a measure of the variation of the frequencies of the excitations with volume, $-\partial \ln \omega/\partial \ln T$, as can be seen from a quasi-harmonic treatment. The vibrational frequencies are assumed to vary with

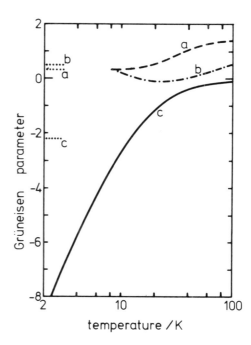

Figure 14.13. Grüneisen parameter as a function of temperature for (a) KCl, (b) crystalline germanium, and (c) vitreous silica. The dotted lines give the limiting value calculated from the pressure derivatives of the sound velocities (after Wright[15]).

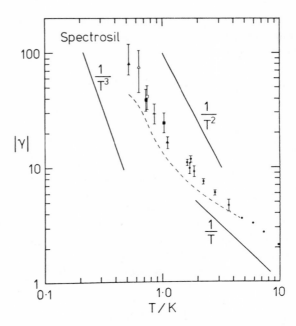

Figure 14.14. Grüneisen parameter as a function of temperature below 10 K (after Wright[15]).

volume but not explicitly with temperature, so that the entropy $S$ can be written as $S[\omega(V)/T]$. Then

$$\left(\frac{\partial S}{\partial V}\right)_T = \frac{1}{T}\, S'\left(\frac{\partial \omega}{\partial V}\right)_T$$

and

$$\left(\frac{\partial S}{\partial T}\right)_V = -\frac{\omega S'}{T^2}$$

Using $C_V = T(\partial S/\partial T)_V$ and rearranging gives

$$\left(\frac{\partial S}{\partial V}\right)_T = -\frac{C_V}{\omega}\left(\frac{\partial \omega}{\partial V}\right)_T$$

Now

$$(\partial S/\partial V)_T = -(\partial P/\partial V)_T(\partial T/\partial V)_P$$

so

$$\beta V/\chi_T C_V = -\partial \ln \omega/\partial \ln V$$

This means that an explanation of the large negative Grüneisen parameter at 1 K or below requires a demonstration that the total energy of the tunneling states *decreases* dramatically as the volume of the solid is decreased.

Although no detailed explanation has been attempted it is possible to see two ways in which this might occur. First, the tunnel splitting $\Delta_0$ is exponentially dependent on the barrier height and the separation of the two potential minima. If these parameters are sensitive to volume changes (as they are in the case of tunneling defects in crystals) a large change in $\Delta_0$, and hence in the energy of the tunneling state, will result. The main problem with this explanation is that for most states of a given energy $E$, $\Delta_0 \ll E$.

The second possible explanation involves the large coupling of these states to strain fields. A typical coupling constant of 1 eV gives a value of $10^4$ for $\partial \ln \omega / \partial \ln V$. Although this value for a single state is enormous, states are at first sight equally likely to increase in energy as to decrease, and it is difficult to see how a large resultant value can be obtained.

## 14.3. Dynamics of Two-Level Systems

### 14.3.1. Quantum Treatment

The Einstein analysis presented in Section 14.2.1 neglects an important aspect of the quantum-mechanical theory of two-level systems, the coherence between the wave functions describing the two states. A simple analysis will be used in this section to show that additional detailed information can be obtained from echo experiments, which explore this coherence.

If an alternating electric field is applied to a two-level system the Hamiltonian is of the form

$$H = H_0 + H_1(t)$$

where $H_0$ has eigenfunctions $\psi_1$ and $\psi_2$ satisfying

$$H_0\psi_1 = -\frac{E}{2}\psi_1 \quad \text{and} \quad H_0\psi_2 = +\frac{E}{2}\psi_2$$

and $H_1(t) = -2e\varepsilon_0 \cos \omega t$. The general solution of

$$H\Psi = i\hbar \frac{\partial \Psi}{\partial t}$$

can be written in the form

$$\Psi = c_1(t)\psi_1 e^{i\omega_0 t/2} + c_2(t)\psi_2 e^{-i\omega_0 t/2} \tag{14.3.1}$$

where $\hbar\omega_0 = E$. Substituting this solution in Schrodinger's equation gives

$$\left. \begin{array}{c} H_{11}c_1(t) + H_{12}e^{-i\omega_0 t}c_2(t) = i\hbar\dot{c}_1(t) \\[2mm] \text{and} \qquad\qquad\qquad\qquad\qquad\qquad\qquad\qquad \\[2mm] H_{21}e^{i\omega_0 t}c_1(t) + H_{22}c_2(t) = i\hbar\dot{c}_2(t) \end{array} \right\} \tag{14.3.2}$$

where

$$H_{ij} = \langle \psi_i \,|\, H_1(t) \,|\, \psi_j \rangle$$

In the amorphous solid, containing a broad distribution of frequencies $\omega_0$, some states will always be in resonance for any driving frequency $\omega$: the frequency of the electric field selects a small number of two-level systems from the distribution. The most important terms on the left of equations (14.3.2) are therefore those for which $\omega = \omega_0$ and which give a time-independent contribution to $\dot{c}_1$ and $\dot{c}_2$. The rapidly varying terms can be neglected just as they are in the analogous discussion of magnetic resonance. In addition, since from equation (14.2.13) the strongest coupling between tunneling states of energy $E$ and photons (or phonons) occurs in the symmetric case when $\Delta_0 = E$, the discussion can be limited to these almost symmetric states for which $H_{11}$ and $H_{22}$ are very small. With these assumptions equations (14.3.2) become

$$H_1^0 c_2 = i\hbar\dot{c}_1 \quad \text{and} \quad H_1^0 c_1 = i\hbar\dot{c}_2 \tag{14.3.3}$$

where $H_1^0 = -\langle \psi_1 \,|\, x\mathscr{E}_0 \,|\, \psi_2 \rangle = -\mathscr{E}_0 p'$, defining the induced dipole moment $p'$. Combining the two equations (14.3.3) gives equations for $c_1$ and $c_2$ of the form

$$\ddot{c}_1 = -\frac{\mathscr{E}_0^2 p'^2}{\hbar^2}\, c_1 = -\omega_1^2 c_1 \tag{14.3.4}$$

with solutions $e^{\pm i\omega_1 t}$.

In all experiments involving coherence the sample must initially be in a well-defined state, and in amorphous solids this is achieved by keeping the sample as cold as possible (10 mK). A basic experiment consists of applying two pulses of rf electric field, the first of length $\tau$ and the second, a time $t_0$ later, of length $2\tau$. If the two-level system is initially in the state $\psi_1$ the appropriate solution during the first pulse is

$$\Psi(t) = \psi_1 \cos \omega_1 t + \psi_2 \sin \omega_1 t \tag{14.3.5}$$

The pulse length (or the magnitude of $\varepsilon_0$) is chosen so that $\omega_1\tau = \pi/4$, or

$$\frac{2\mathscr{E}_0 p'\tau}{h} = \frac{\pi}{2} \tag{14.3.6}$$

(a $\pi/2$-pulse) so that at the end of the pulse the state of the system is described by

$$\psi(\tau) = \frac{1}{\sqrt{2}}(\psi_1 + \psi_2) \tag{14.3.7}$$

For the symmetric case, the combination (14.3.7) corresponds to a state that instantaneously has a large dipole moment, in contrast to the states $\psi_1$ or $\psi_2$, which do not. This large dipole moment can be detected experimentally.

During the time interval between the pulses the evolution of the state function can be followed in the usual way for time-independent problems, so that at time $t_0$ after the first pulse ($t_0 \gg \tau$)

$$\psi(t_0) = \frac{1}{\sqrt{2}}(\psi_1 e^{+i\omega_0 t_0/2} + \psi_2 e^{-i\omega_0 t_0/2}) \tag{14.3.8}$$

The second pulse acts exactly as the first, except that the phase change $\omega_1 t$ during the pulse is $\pi/2$ instead of $\pi/4$. Choosing solutions of equation (14.3.4) to match equation (14.3.8) and redefining the origin of $t$ gives

$$\Psi(t) = \frac{1}{\sqrt{2}}[(\psi_1 \cos\omega_1 t + \psi_2 \sin\omega_1 t)e^{i\omega_0 t_0/2} + (\psi_1 \sin\omega_1 t + \psi_2 \cos\omega_1 t)e^{-i\omega_0 t_0/2}]$$

$$0 < t < 2\tau \tag{14.3.9}$$

At the end of the pulse

$$\Psi(t_0 + 2\tau) = \frac{1}{\sqrt{2}}(\psi_2 e^{i\omega_0 t_0/2} + \psi_1 e^{-i\omega_0 t_0/2}) \tag{14.3.10}$$

and at a later time $t$, again taking $t \gg \tau$,

$$\Psi(t + t_0) = \frac{1}{\sqrt{2}}(\psi_2 e^{i\omega_0(t_0-t)/2} + \psi_1 e^{-i\omega_0(t_0-t)/2}) \tag{14.3.11}$$

Equation (14.3.11) shows that at a time $t_0$ after the second pulse the system is again in the state

$$\Psi = \frac{1}{\sqrt{2}} (\psi_1 + \psi_2)$$

with a large dipole moment. More importantly this is true for all values of $\omega_0$. In an amorphous solid, as a result of the finite lifetime of the states and the finite length of the exciting pulse, the electric field will inevitably excite states with a range of energies or frequencies $\omega_0$. The effect of this is to give a very sharp "echo" $t$ when equation (14.3.11) is averaged over a range of $\omega_0$. It is interesting that the broad distribution of energies is responsible for this sharpness of the echo.

### 14.3.2. Echo Experiments

This analysis assumes that the phase coherence of the state function is maintained for the complete time $2t_0$. Any disruption of this phase coherence will prevent a tunneling state from contributing to the final echo, and therefore the experiment provides a way of measuring the dephasing time, known as $T_2$. The echo amplitude in this simple picture is expected to decrease as $\exp(-2t_0/T_2)$, and $T_2$ can be determined by measuring the echo amplitude as a function of $t_0$.

These experiments, in which two pulses of acoustic or electric fields are used to measure $T_2$, are known as spontaneous echo experiments. In the ideal photon case two electric-field pulses of length $\tau$ and $2\tau$ (typically 1 μs) at a frequency $\omega/2\pi$ of about 1 GHz are applied to a sample held at a temperature $T$ of a few mK, where $\hbar\omega > kT$. This low temperature is needed in order to prepare the system in the ground state $\psi_1$. The amplitude of the echo is measured as a function of the electric-field amplitude $\mathscr{E}_0$ at fixed pulse separation $t_0$, and also as a function of $t_0$ at fixed $\mathscr{E}_0$. A typical echo pattern is shown[16] in Figure 14.15.

Results typical of electric echo experiments are shown[17] in Figure 14.16. The maximum in the echo amplitude as a function of $\mathscr{E}_0$ corresponds from equation (14.3.6) to $\pi/2$- and $\pi$-pulses or $\mathscr{E}_0\tau p'/\hbar = \pi/4$, and can be used to evaluate $p'$. It is clear from Figure 14.16 that two distinct species of dipole occur in these glasses, and by choosing $\mathscr{E}_0$ appropriately each type can be studied in turn. A number of complications, including the random orientation of the dipoles relative to the field and the problem of determining the local field, make precise values of $p'$ difficult to determine, but values are typically of order $3 \times 10^{-30}$ cm (1 Debye). The corresponding phonon experiment in silica gives values for the coupling constant $\gamma$ of about 1 eV.

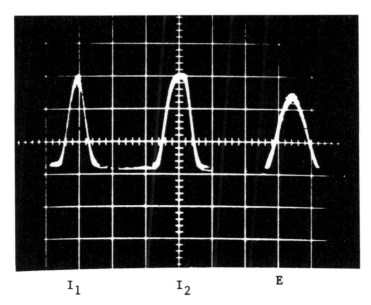

Figure 14.15. Spontaneous electric echo pattern (after Bernard[16]).

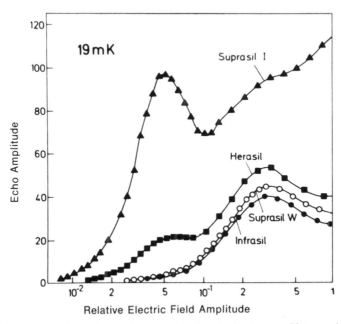

Figure 14.16. Amplitude of the spontaneous electric echo in vitreous-silica samples, plotted as a function of electric field (after Golding *et al.*[17]).

The variation of the echo amplitude with $t_0$ is shown[16] in Figure 14.17 for a pulse area that corresponds to the first peak of Figure 14.16 (which turns out to be the result of OH groups). An exponential variation is found in this case, although in general more complicated patterns of decay are observed. These nonexponential decays may result from a distribution of relaxation times similar to that discussed in Section 14.2.2, but could also be characteristic of the molecular process responsible for $T_2$, to be discussed in Section 14.3.3. The results shown in Figure 14.17 show that $T_2$ is about 70 μs at 10 mK, and varies as $T^{-1}$. Phonon echo experiments show slightly smaller values of $T_2$, 10 μs at 20 mK, with a temperature variation of $T^{-2}$. These phonon experiments are probably observing the second set of states, but the different temperature dependences may well result from different experimental conditions.

More complicated three-pulse sequences of electric and acoustic fields can be used to measure $T_1$ in "stimulated" echo experiments. In one such experiment an initial $\pi$-pluse is followed a time $t_0$ later by $\pi/2$- and $\pi$-pulses, relatively closely spaced. The analysis of this experiment follows that given in Section 14.3.1. The system, initially in the ground state $\psi_1$, is excited coherently to the state $\psi_2$ by the first $\pi$-pulse. During the time $t_0$ a fraction $e^{-t_0/T_1}$ of the systems will relax back to the ground state $\psi_1$. The

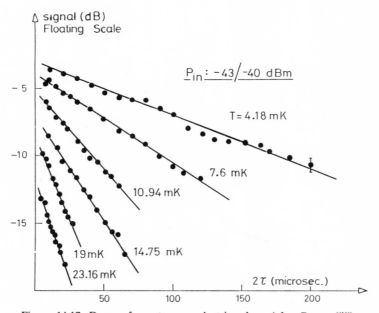

Figure 14.17. Decay of spontaneous electric echoes (after Bernard[16]).

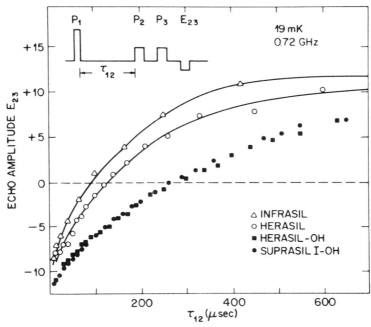

Figure 14.18. Amplitude of stimulated echo as a function of the time between initial and second pulse (after Golding *et al.*[17]).

double-pulse sequence gives a spontaneous echo just as before, but the contribution of those systems which decayed back to $\psi_1$ will be opposite to that of the systems remaining in $\psi_2$. The amplitude of the echo will therefore be proportional to $1 - e^{-t_0/T_1}$ and measurements as a function of $t_0$ can be used to measure $T_1$. Both photon and phonon measurements in vitreous silica give results in agreement with equation (14.2.13), although this experiment is also affected by the wide distribution of tunneling-state relaxation times. Results of stimulated photon experiments are shown[17] in Figure 14.18. Measured values of $T_1$ are up to ten times larger than those of $T_2$.

### 14.3.3. Spectral Diffusion

A dephasing time $T_2$ much shorter than an energy relaxation time $T_1$ implies, by analogy with magnetic resonance, interaction between the two-level systems. This interaction could take two forms. The term proportional to $\sigma_x$ in the Hamiltonian gives rise to a process in which two systems exchange energy by mutual excitation and de-excitation with exchange of a resonant phonon. In a glass, however, the density of resonant two-level systems lying within the bandwidth of a pulse is too small for this to be an effective dephasing process. Much more effective is the nonresonant process involving $\sigma_z$ in which a transition of one system, equivalent to a reorientation of an elastic dipole, gives rise to a strain field, which affects the energy of a neighbor. Since all systems with energy less than about $2k_B T$ undergo such transitions the number of effective interactions is much greater.

A semiquantitative explanation of the effect can be given by means of a simple physical picture. Each system experiences a fluctuating local strain field, which in equilibrium gives rise to a fluctuating energy $\Delta E_0$, as shown in the uper half of Figure 14.19. The initial field pulse, of duration less than the time scale of the fluctuations, selects from the distribution a subset, which instantaneously has energy $E_0 = \hbar\omega_0$. Because the mean energies of systems within this subset differ, the energies gradually spread out over a range determined by the fluctuating fields. This is illustrated in the lower half of Figure 14.19b. At times long compared to the average relaxation time the width of the energy distribution of the subset reaches $\Delta E_0$, and is then independent of time.

A value for $\Delta E_0$ can be evaluated from the strength of the coupling between the two-level systems and the phonons. The elastic dipole moment

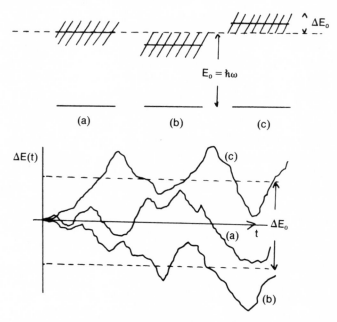

Figure 14.19. Representation of broadening arising from spectral diffusion.

is $\gamma$ so that the interaction energy is

$$\delta E \sim \frac{\gamma^2}{\rho v^2 r^3} \tag{14.3.12}$$

Replacing $1/r^3$ by the concentration of thermally excitable two-level systems gives

$$\Delta E_0 \sim \gamma^2 \bar{n} k_B T / \rho v^2 \tag{14.3.13}$$

In the long-time limit the time interval $t_0$ between the two pulses of a spontaneous echo experiment is much greater than $T_{min}$, defined as the shortest energy relaxation time for two-level systems with energy $E \sim k_B T$. The dephasing time can be derived as the time for which the spread in phase, $\Delta E_0 t / \hbar$, is of order $\pi/2$. This gives

$$T_2^{-1} = \frac{\pi}{2} \frac{\Delta E_0}{\hbar} \simeq \frac{\pi \gamma^2 \bar{n} k_B T}{2 \hbar \rho v^2} \tag{14.3.14}$$

and an exponential decay of the echo amplitude.

In the short-time limit, $t_0 \ll T_{min}$, the width of the energy distribution increases roughly as $\Delta E_0 (1 - e^{-t/T_{min}})$, so

$$\Delta E(t) = \Delta E_0 t / T_{min} \tag{14.3.15}$$

This defines a dephasing time by

$$\frac{1}{\hbar} \int_0^{T_2} \Delta E(t) dt = \frac{\pi}{2}$$

and so

$$T_2^2 = \hbar \pi T_{min} / \Delta E_0 \tag{14.3.16}$$

The decay is nonexponential, varying with $t_0$ as

$$\exp(- \Delta E_0 t_0^2 / \hbar T_{min}) \tag{14.3.17}$$

The dephasing time therefore varies with temperature as $T$ for $t_0 > T_{min}$ and as $T^{-2}$ for $t_0 \ll T_{min}$, using $T_{min} \propto T^{-3}$. Using numerical values for the constants in equation (14.3.14) the value of $T_2$ at 20 mK can be estimated as 20 $\mu$s in vitreous silica if $\bar{n}\gamma^2$ is taken as $10^7$ J m$^{-3}$, a value derived from acoustic velocity measurements of the kind shown in Figure 14.11. This is in reasonable agreement with experiment.

It must be emphasized that many complications have been ignored in this simple argument. In particular the broad distribution of relaxation times $T_1$ complicates the analysis, and a rigorous derivation of equation

(14.3.17) is complicated. However, this analysis does show the underlying physical principles, and illustrates that this spectral diffusion can explain the observed interaction between two-level systems.

### 14.3.4. Saturation and Linewidth Experiments

The phenomenon of spectral diffusion can be probed by direct measurements of the linewidth of two-level systems, and indeed such experiments provided the first evidence that interaction between the systems is important. The ideal experiment uses a large-intensity initial pulse to saturate a subset of two-level systems, followed by a second weaker pulse, which monitors the recovery to thermal equilibrium. This second pulse may have the same frequency $\omega_0$ as the first, in which case the time delay between the two is varied, or can be used to examine the frequency dependence. Interpretation of the various experiments depends critically on the relationship between the length $\tau$ of the saturating pulse, and the relaxation times $T_1$ and $T_2$.

Figure 14.20 shows[18] saturation recovery experiments in vitreous silica, where $\tau_{12}$ is the time delay between the pulses, both at frequency $\omega_0$. The temperature is sufficiently low for $T_2$ to be comparable to $\tau$ (0.06 μs), and both are much less than $T_1$. The initial pulse saturates a subset of

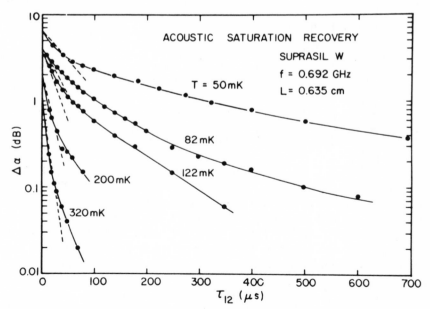

Figure 14.20. Attenuation of a weak probing pulse applied a time $\tau_{12}$ after a saturating pulse (after Golding and Graebner[18]).

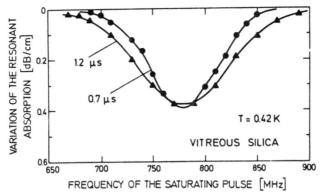

Figure 14.21. Attenuation of weak probing pulse as a function of saturation-pulse frequency for two values of saturating-pulse length (after Arnold *et al.*[19]).

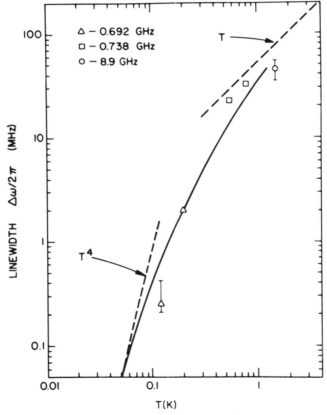

Figure 14.22. Linewidth obtained from saturation experiments plotted as a function of temperature (after Golding and Graebner[18]).

two-level systems in a frequency band of width $\hbar/\tau$ at $\omega_0$, comparable to the linewidth $\Delta E_0$ (Figure 14.19). Spectral diffusion does not therefore have a large effect, and the second pulse monitors the recovery of equilibrium through energy relaxation, and hence measures $T_1$. The dashed lines in the figure correspond to a one-phonon process with $\gamma = 1.4$ eV. At longer times the recovery shows a much slower time variation, reminiscent of the data shown in Figure 14.8, and may be another illustration of the large range of relaxation times in glasses.

The intrinsic linewidth of the two-level systems can be measured by working in the limit $T_2 < \tau < T_1$, and measuring the saturation produced by an initial pulse of variable frequency $\omega$ at fixed initial frequency $\omega_0$ of the second. (For obvious experimental reasons it is better to keep the frequency of the second detected pulse constant.) Experimental results in vitreous silica are shown[19] in Figure 14.21. The width of the line is much larger than $\hbar/\tau$ or $\hbar/T_1$, giving additional evidence for interaction effects. In addition the dependence of linewidth on pulse length is direct evidence for the time evolution of the linewidth implied by spectral diffusion: the longer the pulse the broader the line (Figure 14.19). A detailed comparison of linewidth results[18] with the predictions of spectral-diffusion theory is shown in Figure 14.22.

## 14.4. Microscopic Interpretations

### 14.4.1. Impurity Effects

It is clear that a wide range of thermal, acoustic, and electrical measurements can be interpreted in terms of two-level systems, or, more specifically, in terms of a tunneling model. The vast majority of the nonthermal experiments have been performed on vitreous silica, and have established many important parameters: the density of states, the dipole moment, and the phonon coupling constant are all known. Much less is known about the properties of these states in other amorphous systems such as organic polymers, chalcogenide glasses, or even other oxide glasses. Glassy metals have, however, been extensively studied and will be discussed in Section 14.4.2.

Although experiments can be interpreted in terms of a general tunneling model, such an approach leaves unanswered two important questions: what is the microscopic nature of the states, and why do (almost?) all amorphous solids contain them? In particular, one of the more important questions concerning the low-temperature properties of glasses is whether the effects are intrinsic, in the sense that they would occur in a fully coordinated random-network model of a glass, or whether defects, in the form of incomplete coordination, or water, must be present. To answer this it is helpful to look at the results[20] shown in Figure 14.23. Samples of $As_2S_3$ were prepared with increasing care and decreasing impurity concentrations. The concentrations of impurity in the two purest samples, determined by mass spectroscopy, were about $10^{24}$ m$^{-3}$ of Ge and less than $10^{23}$ m$^{-3}$, respectively. Even so, both these samples showed an excess heat capacity, with the order-of-magnitude reduction in impurity concentration halving the excess heat capacity. The corresponding additional entropies can be evaluated and, on the assumption that two-level systems are involved, lead to *minimum* numbers of contributing states which are between $10^{23}$ m$^{-3}$ and $10^{24}$ m$^{-3}$ in both cases. It was concluded from the lack of proportionality between impurity level and heat capacity that an "intrinsic" term would be left after removal of all impurities, although this is a slightly dangerous conclusion which depends crucially on the reliability of the chemical analysis. Measurements of $\kappa$ in the two samples point to an intrinsic effect, since the values are the same within experimental error.

Other evidence is ambiguous. Measurements in the more highly coordinated amorphous solids As and Ge show a much smaller excess heat

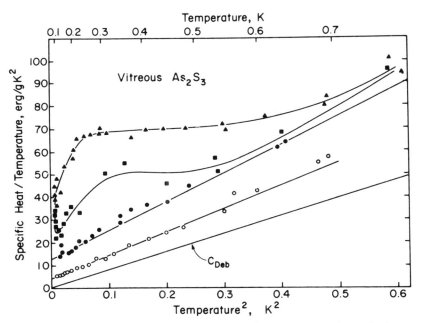

Figure 14.23. Heat capacity of $As_2S_3$. The heat capacity decreases as the purity increases (after Stephens[20]).

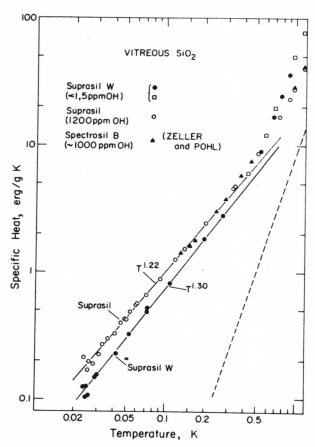

Figure 14.24. Heat capacity of vitreous silica containing 1 and 1200 ppm OH (after Lasjaunias *et al.*[(22)]).

capacity, although the interpretation of these results is complicated by sample inhomogeneities. It appears, however, that the density of tunneling states is less in these more rigid networks than in the glasses containing twofold coordinated atoms or groups of atoms.

The effect of one particular impurity, the hydroxyl group OH, has been studied in some detail. These studies[(21)] not only identify one particular impurity which can change the low-temperature properties, but also help to build up a picture of the microscopic structure of one possible tunneling state as discussed in Section 14.4.3. Figure 14.24 shows the effect of OH on the heat capacity[(22)] of $SiO_2$: a sample containing 1200 ppm of OH has a heat capacity larger by about 30% than a sample containing less than 1.5 ppm.

### 14.4.2. Metallic Glasses

The existence of free electrons in amorphous metals makes it difficult to deduce the presence of tunneling states from the thermal properties. By estimating the electronic contribution to the thermal conductivity, using the Weideman–Franz ratio and the measured electrical conductivity, the phonon contribution to $\kappa$ can be shown to resemble that of amorphous insulators. However, the most direct thermal evidence comes from measurements on amorphous superconductors, as shown[23] in Figure 14.25 for zirconium-palladium. Below the transition temperature $T_c$ of 2.5 K the electronic contribution to the heat capacity vanishes exponentially, and the resulting linear term at very low temperatures can be identified with tunneling states. Below $T_c$ heat is carried only by phonons and, as shown[23]

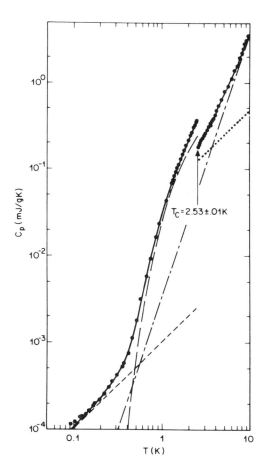

Figure 14.25. Heat capacity of superconducting amorphous $Zr_{70}Pd_{30}$ (after Graebner et al.[23]).

in Figure 14.26, the temperature dependence of $\kappa$ is very similar to that found in vitreous silica. The magnitude of $\kappa$ in $Zr_{70}Pd_{30}$ is smaller, but if the results are used to calculate $\bar{n}\gamma^2$ using equations (14.1.1) and (14.2.14) a value of approximately $5 \times 10^{46}$ J m$^{-3}$ is found, half that of vitreous silica or NiP but typical of other amorphous metals.

As in insulating glasses, acoustic measurements provide the most direct evidence for two-level systems in metallic glasses. The most straightforward experiment, in which effects of saturation can be ignored, is the measurement of the temperature variation of the acoustic velocity. Results in NiP are shown[24] in Figure 14.27. A logarithmic variation of velocity with temperature is observed, as in silica, although the slope is smaller by an order of magnitude and is significantly different for longitudi-

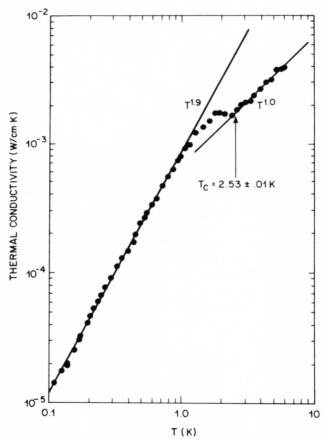

Figure 14.26. Thermal conductivity of superconducting amorphous $Zr_{70}Pd_{30}$ (after Graebner et al.[23]).

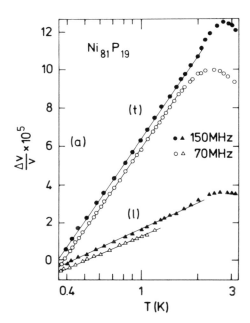

Figure 14.27. Variation of acoustic velocity in amorphous $Ni_{81}P_{19}$ (after Lohneysen[24]).

nal and transverse waves. A value of $\bar{n}\gamma^2$ equal to $3 \times 10^{46}$ J m$^{-3}$ can be calculated from either slope (unlike silica, where $\bar{n}\gamma^2$ is larger by a factor of two for longitudinal waves). This should be compared with the value of about $15 \times 10^{46}$ J m$^{-3}$ estimated from the phonon contribution to $\kappa$. It is clear that unlike the insulating glasses, where in general the values of $n\gamma^2$ derived from thermal and acoustic measurements are in good agreement, there is a major discrepancy in amorphous metals.

Further experimental evidence emphasizes that the tunneling-state theory cannot be applied directly to metals. Two-pulse saturation experiments, as described in Section 14.3, have been completely unsuccessful even at temperatures of 10 mK. At this temperature $T_1$ in silica is 200 μs (at a frequency of 1 GHz) but the failure of attempts to measure $T_1$ in PdSiCu imply that it is less than 25 ns under these conditions. If $T_1$ were governed by one-phonon processes in this glassy metal, the value of the coupling constant $\gamma$ would need to be at least an order of magnitude larger than in silica. This is clearly inconsistent with the thermal results.

This rapid relaxation is a result of direct interaction of the tunneling states with the electrons. Electrons are inelastically scattered from tunneling systems in a process formally equivalent to nuclear spin relaxation in metals. The magnitude of $T_1$ can be derived from the Fermi expression for transition probabilities, so that the probability per unit time of an electron exciting a tunneling state is

$$\omega_{ij} = \frac{2\pi}{\hbar} K^2 g(\varepsilon_F)$$

where $K$ is the electron-tunneling state matrix element and $g(\varepsilon_F)$ is the density of electronic states evaluated at the Fermi level. If the energy splitting $E$ of the tunneling state is less than $k_B T$, then the number of electrons capable of inducing the transition is $g(\varepsilon_F)k_B T$, and the time between transitions $1/T_1$ is given by

$$\frac{1}{T_1} \approx \frac{2\pi}{\hbar} [Kg(\varepsilon_F)]^2 k_B T \tag{14.4.1}$$

For $E > k_B T$, the $k_B T$ factor is replaced by one obtained from integrating over the Fermi factors to give

$$\frac{1}{T_1} = \frac{2\pi}{\hbar} [Kg(\varepsilon_F)]^2 E \coth \frac{E}{2k_B T} \tag{14.4.2}$$

Estimating $K$, the difference in potential seen by the electrons in the two configurations of the tunneling state, as 1 eV, and taking $g(\varepsilon_F)$ as 0.1 $(\text{eV})^{-1}$ per atom gives $T_1$ about 10 ns at 10 mK, consistent with experiment.

This short relaxation time not only explains the absence of phonon echoes but also implies that the acoustic power needed to saturate the tunneling states will be much higher than in vitreous silica. The interaction with electrons provides another channel through which equilibrium of the tunneling states can be maintained, and the saturating acoustic power should be larger by the ratio of electron and phonon scattering rates. Experimental results, verifying this prediction, are shown[25] in Figure 14.28.

The magnitude of $T_1$ has an indirect effect on $\kappa$. In addition to the resonance scattering of phonons by tunneling states, it is also possible to identify a relaxation contribution. The strain field of a nonresonant phonon of frequency $\omega$ perturbs the equilibrium of a tunneling state through the coupling $\gamma$. The tunneling state returns to equilibrium with a relaxation time $T_1$ so that at low frequencies $\omega T_1 \ll 1$ equilibrium is always maintained. At high frequencies $\omega T_1 \gg 1$ the tunneling state does not have time to respond to the strain field, but for $\omega T_1$ near unity the response will lag behind the strain. This phase difference gives rise to a relaxation loss, which can be formally described by means of the Debye equations, written

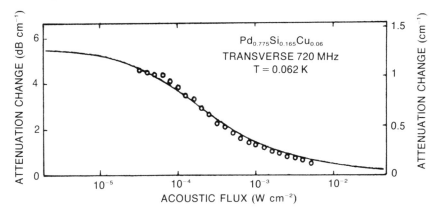

Figure 14.28. Attenuation as a function of intensity in the amorphous alloy Pd–Si–Cu (after Doussineau et al.[25]).

in slightly simplified form and using $\tau$ in place of $T_1$:

$$l_{\text{rel}}^{-1} = \frac{\bar{n}\gamma^2}{\rho v^3 k_B T} \int_0^\infty \text{sech}^2 \frac{E}{2k_B T} \int_{\tau_{\min}}^\infty \frac{\omega^2 \tau}{1 + \omega^2 \tau^2} g(\tau) d\tau \qquad (14.4.3)$$

$$\frac{\Delta v}{v} = \frac{\bar{n}\gamma^2}{2\rho v^2 k_B T} \int_0^\infty \text{sech}^2 \frac{E}{2k_B T} dE \int_{\tau_{\min}}^\infty \frac{g(\tau)}{1 + \omega^2 \tau^2} d\tau \qquad (14.4.4)$$

Physically, the first factor in equation (14.4.3) is the contribution from a simple elastic dipole, the first integral excludes all states that have $E > k_B T$ and so remain in thermal equilibrium, while the second integral describes the relaxation. In general, the first integral gives a factor $k_B T$, and the dominant factor in $g(\tau)$ is $1/\tau$, following the argument leading to equation (14.1.15). Equations (14.4.3) and (14.4.4) become, in this simplified form,

$$l_{\text{rel}}^{-1} = A \int_{\tau_{\min}}^\infty \frac{\omega^2}{1 + \omega^2 \tau^2} d\tau = A\omega \left( \frac{\pi}{2} - \tan^{-1} \omega\tau_{\min} \right) \qquad (14.4.5)$$

$$\frac{\Delta v}{v} = B \int_{\tau_{\min}}^\infty \frac{1}{(1 + \omega^2 \tau^2)\tau} d\tau = B \ln \left( \frac{\omega^2 \tau_{\min}^2}{1 + \omega^2 \tau_{\min}^2} \right) \qquad (14.4.6)$$

If $\omega\tau_{\min} > 1$,

$$l^{-1} = A/\tau_{\min}, \quad \Delta v/v = -B/\omega^2 \tau_{\min}^2 \qquad (14.4.7)$$

and if $\omega\tau_{\min} < 1$,

$$l^{-1} = A\omega, \quad \Delta v/v = B \ln(\omega^2 \tau_{\min}^2) \qquad (14.4.8)$$

In insulating glasses $\tau_{min} \simeq aT^{-3}$, where $a$ is $10^{-8}$ sK$^3$ so that for typical acoustic frequencies $\omega\tau_{min} = 1$ at about 2 K. In this temperature region the rapid variation $\Delta v/v = -BT^6/a^2\omega^2$ gives rise to the "turnover" of the velocity shown in Figure 14.11, and $l^{-1} = AT^3/a$ to the upturn in attenuation shown in Figure 14.5 at higher temperatures.

In metallic glasses the relaxation time $\tau$ (or $T_1$) is much shorter, with $\tau_{min} \simeq bT^{-1}$ where $b \simeq 10^{-10}$ sK in the low-temperature regime where electronic processes dominate. Phonon scattering becomes important at a temperature $T_0$ given by $bT_0^{-1} = aT_0^{-3}$, or about 10 K. At an acoustic frequency of 1 GHz, $\omega\tau_{min}$ equals 1 at approximately 100 mK. This means that the logarithmic slope measured in an acoustic experiment contains contributions from both resonance and relaxation terms, as does the measured attenuation. Since the two contributions are of different sign in the velocity, but add in the attenuation, the values of coupling constant deduced from the velocity variation are significantly smaller than those deduced from the acoustic attenuation or thermal conductivity. Detailed calculations of the relaxation contribution have been made in PdSiCu, and agree well with experiment.

The effects of electrons in determining $T_1$ are dramatically confirmed by acoustic measurements in amorphous superconductors. Attenuation as a function of temperature is shown[26] in Figure 14.29 for Pd$_{30}$Zr$_{70}$. These

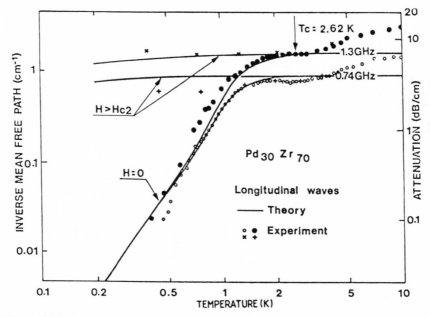

Figure 14.29. Attenuation of sound in superconducting amorphous Pd$_{30}$Zr$_{70}$ (after Arnold *et al.*[26]).

results can be explained by noting that from equation (14.4.8) the attenuation $l^{-1}$ is not sensitive to the details of the relaxation process if $\omega\tau \ll 1$, but depends only on frequency. In this limit, satisfied above $T_c$ in $Pb_{30}Zr_{70}$, the attenuation is constant even though the phonon and electron contributions to the tunneling-state lifetimes are varying as $T^{-3}$ and $T^{-1}$ respectively. Below $T_c$ the number of effective electrons drops as a result of the superconducting energy gap, but in the amorphous superconductor, where the direct electron–phonon interaction does not give significant attenuation, this is not immediately effective. Only when $T_1$ has increased to such an extent that $\omega T_1 > 1$ does the attenuation drop following equation (14.4.7). This will occur at approximately $T_c/2$, where the electron contribution to $T_1$ becomes negligible in comparison to the phonon. Below $T_c/2$ the attenuation will decrease as $T^3$, characteristic of phonon processes.

The most complete microscopic description of a possible intrinsic tunneling state comes from results on Nb/Zr alloys. Although crystalline, this material is disordered in the sense that it contains small amounts of the $\omega$-phase in a matrix of the $\beta$-phase. In a 20% Nb sample isolated regions of $\omega$-phase, about 5 Å in diameter, are formed at a concentration of about $10^{25}$ m$^{-3}$. Results for the heat capacity and $\kappa$[27] show that glassy behavior can be obtained in this disordered crystal and, furthermore, that this behavior can occur where the disorder is localized to very small regions separated by about 100 Å.

The microscopic picture of the states responsible for the additional heat capacity and the $T^2$ variation of $\kappa$ is reasonably well established in the Nb/Zr alloy. The $\beta$–$\omega$ transformation occurs by moving two atoms simultaneously through a distance of about 0.5 Å as shown[27] in Figure 14.30. The potential for each atom can be described by a double potential well, and it appears on the basis of electron microscopy, inelastic neutron, and Mössbauer measurements that a large range of potential barriers $V$ must exist in the solid. Such a microscopic state could easily exist in the glassy metals and is exactly of the form suggested by the tunneling model.

### 14.4.3. OH in Vitreous Silica

The effect of hydroxyl, OH, on the low-temperature properties of vitreous silica has been investigated in a wide range of experiments comparing commercial "water-free" and "wet" samples. An example is given in Figure 14.24. The various experiments can be related and linked to form a consistent picture of one particular impurity tunneling state.

The chemistry of OH in silica has been extensively studied in connection with the growth of "wet" oxide films on silicon and with the optical absorption arising from OH stretching vibrations, of particular

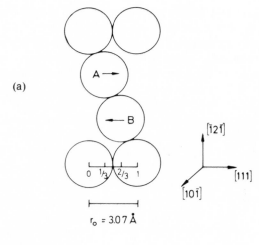

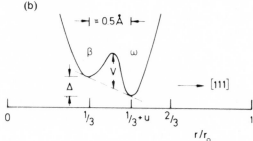

Figure 14.30. Schematic representation of the $\beta-\omega$ phase transition in Nb/Zr (after Lou[27]).

importance in applications of optic fibers. The general conclusion of these studies is that the chemical reaction $Si-O-Si + H_2O \rightarrow 2SiOH$ gives rise to OH groups chemically bonded to the silica network. The intensity of the fundamental OH stretching vibration is proportional to OH concentration, but the width of the absorption line is constant indicating that, in general, interactions between OH groups can be neglected. However, the absorption line is broad ($\Delta\nu/\nu$ about 5%) indicating that interactions between OH and the silica host *are* important. Each OH group can therefore be treated independently of the others, moving in a potential that varies considerably from site to site.

In organic compounds an OH group bonded to a carbon atom can rotate in a twofold (phenols) or threefold (tertiary alcohols) symmetric potential, although the ideal symmetry is of course modified in the solid state by intermolecular interactions. A similar picture can be used for OH in silica with the choice of "symmetry" left to experiment: because the interactions between OH and its surroundings are large, there is no *a priori* reason for preferring the threefold potential given by the tetrahedral coordination of the Si atom.

Figure 14.31 shows[21] that the effect of 1200 ppm OH on the dielectric loss of vitreous silica is to produce a broad low-temperature relaxation peak. The physical interpretation of relaxation losses is as described in the last section, but in this case the information is sufficiently complete to allow the distribution function $g(\tau)$ to be derived. At high temperatures a classical theory can be used to deduce the distribution of energy barriers from $g(\tau)$, and this is turn can be used to derive $g(\Delta_0)$. It turns out that the real $g(\Delta_0)$ differs from equation (14.1.15) by having fewer small values of $\Delta_0$, and corresponds to the proton "rotating" in an approximately twofold symmetric potential.

Knowledge of $g(\Delta_0)$ together with the dipole moment $p_0$ of equation (14.2.1) allows the temperature variation of the velocity of light, shown in Figure 14.12, to be calculated. The departure of $g(\Delta_0)$ from $1/\Delta_0$ means that the logarithmic slope, given by equation (14.2.21), depends on measuring frequency: the available data can be well-fitted by the calculation. In addition the density of states can be calculated, following equations (14.2.15)–(14.2.17), and this leads to a calculated value of the heat capacity that agrees with the difference between the two sets of data shown in Figure 14.24. This is shown[21] in Figure 14.32. Finally, the idea of OH as a discrete tunneling state is consistent with photon echo experiments, although precise numerical calculations are difficult.

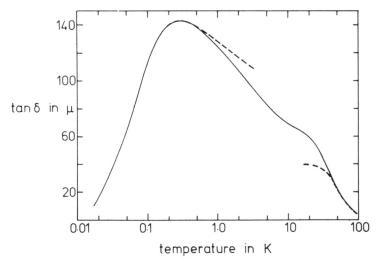

Figure 14.31. The contribution of 1200 ppm OH to the dielectric loss of vitreous silica (after Phillips[21]).

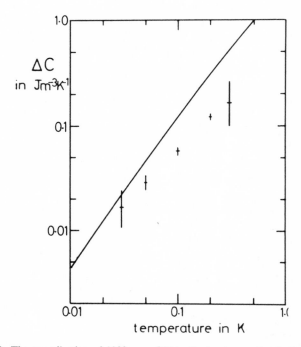

Figure 14.32. The contribution of 1200 ppm OH to the heat capacity of vitreous silica. The line refers to calculated values (after Phillips[21]).

This detailed analysis of one kind of impurity is useful in that it shows how impurities can contribute to the low-temperature properties but it is equally clear, from Figures 14.24 and 14.16, that other, possibly intrinsic, two-level systems are present. The nature of these intrinsic states is unknown, although a number of possibilities have been proposed and will be discussed in the next section.

### 14.4.4. The Structural Origin of Tunneling States

Sections 14.1, 14.2, and 14.3 have described the theory of the tunneling model and how it accounts for a variety of experiments on amorphous solids at low temperatures. This section attempts to identify the possible atomic scale origins of the tunneling states in simple covalently bonded inorganic glasses such as vitreous silica, elemental amorphous semiconductors Se, As, P, Ge, and Si, and chalcogenide glasses such as $Ge_xSe_{1-x}$, $As_2S_3$, and $As_2Se_3$. The structure of all these materials seems to be best described by the continuous random network (CRN) model, and models for the origin of double-well potentials usually involve atomic motion without bond-breaking. The double well is often associated with a twofold coordinated atom, e.g., oxygen in silica, or with similar displacements to those observed in transitions between two crystalline forms. There is a strong case for the existence of double-well potentials from the ultrasonic attenuation peak at around 50 K and from the onset of large atomic movement near the glass transition $T_g$.

In a defect-free CRN, each atom forms the same number of covalent bonds with its neighbors (often 8-$N$ where $N$ is the group in the periodic table). If more than one type of atom is present, chemical ordering is possible. X-ray diffraction and other structural studies demonstrate that long-range order is absent, but short-range order is often similar to that of the corresponding crystal. The radial distribution function (rdf) gives the coordination number and demonstrates that the bond length is fixed, variations arising chiefly from thermal vibrations. On the other hand, the bond angle generally has a spread of 10–20° about a mean value. The rdf does show broad peaks at larger distances, but often features present in the crystalline form are completely absent. In the CRN model, this is achieved by allowing the dihedral angle (defining rotation about a bond) to take a continuous spectrum of values.

The tunneling model attributes the low-temperature behavior of glasses to the existence of double-well potentials. From their energy possible combinations of the parameters $V$, $d$, and $m$ (barrier height, width, and mass of tunneling species) can be deduced. For example, an oxygen atom tunneling 0.1 nm through a potential barrier of height $8 \times 10^{-22}$ J ($5 \times 10^{-3}$ eV or 60 K) gives a state with energy equivalent to 1 K, assuming that the vibration frequency $\Omega$ is given by the Debye frequency ($\theta_D = 495$ K in vitreous silica). A state of energy 0.05 K requires a barrier of twice the height.

Some of the proposed origins of the double-well potentials were originally advanced in connection with the ultrasonic attenuation peak at 50 K in vitreous silica. Anderson and Bömmel[28] proposed that the relaxing unit was an oxygen atom "flipping" between two equilibrium positions transverse to the Si–O–Si bond. This change does not preserve the O–Si–O bond angles at the two silicon atoms involved, and so the true equilibrium configuration must involve a local rearrangement of the network to minimize the strain energy of the bonds. This model is attractive for the majority of amorphous solids, which contain twofold coordinated atoms. However, the threefold coordinated materials a-As and a-P both show a similar peak in the ultrasonic attenuation[29] and a small excess low-temperature heat capacity although the structure of these two materials is much less perfect than that of most bulk glasses. Low-temperature heat-capacity experiments give no clear evidence for a linear term in amorphous Ge[30] and surface-wave attenuation measurements are consistent with the absence of double-well potentials:[31] in a-Si the logarithmic sound-velocity variation indicates that $\bar{n}\gamma^2$ is at least a factor of 12 lower than in an $SiO_2$ film.

Estimates of the total number of double-well potentials from the 50 K ultrasonic attenuation peak lead to typical values of 1 per 100 atoms in $SiO_2$, a-As, a-P, and many other materials.[29] The required potential barrier in vitreous silica is obtained from an initial Si–O–Si angle of 174°. The experimental rdf shows around 1% of the Si–O–Si bond angles in this range, in good agreement. In threefold coordinated materials, if the model applies, the pyramidal unit must be rather flatter than usual, with the bond angles larger than their usual values, probably close to 120°.

Vukcevich[32] developed a model for vitreous silica that also invokes a potential function, which possesses two minima separated by a barrier. Whole $SiO_4$ tetrahedra can assume one of two equilibrium positions, the fraction in each varying with temperature. The model seems difficult to reconcile with the random-network model, as the four Si atoms at the centers of the neighboring tetrahedra need to be situated so that a rotation of the central tetrahedron changes all four Si–O–Si bond angles in the same sense, but is based on the structural changes seen in the $\alpha-\beta$ displacive transition in quartz at 545 °C.

Mon and Ashcroft[33] gave essentially the same argument for the origin of double-well potentials. They also connected the occurrence of the amorphous state and crystal polymorphism with the existence of low-temperature thermal anomalies.

A useful step would be to search CRN models by computer for double-well potentials. Smith[34] used a fourfold coordinated model representing a-Ge and a-Si. One atom at a time was moved and the energy calculated, but the surrounding network was not allowed to relax. Many large barriers and large asymmetries were found (4 per atom) but the numbers at low $\Delta_0$ and $\Delta$ were of the right order of magnitude to give a typical excess specific heat. Similar, but more realistic calculations would give a very useful guide to the microscopic nature of the tunneling systems.

A variety of electric and optic measurements on amorphous semiconductors suggests a broad division into two types of material. Both possess mobility edges separating extended from localized states in the valence and conduction bands, and the Fermi level is pinned at around midgap, i.e., it does not shift with temperature or on the addition of impurities. This suggests a density of localized defect states in the band gap. The first group of materials includes a-Ge and a-Si prepared in the absence of hydrogen and oxygen and at low temperatures shows an esr signal in the dark and dc conductivity proportional to $\exp(-T/T_0)^{1/4}$, where $T_0$ is a constant. This law can be derived for variable-range hopping of electrons between localized states at $E_F$. Both observations suggest that defect states at the Fermi level are singly occupied at absolute zero.

The second group encompasses the chalcogenide glasses; most work has been performed on $As_2S_3$, $As_2Se_3$, and Se. Variable-range hopping and a dark esr signal are not observed. There are, therefore, no single electrons at $E_F$ to give a linear heat capacity in chalcogenide (and oxide) glasses. However, an esr signal (and optical absorption at midgap) can be induced by irradiation with slightly sub band-gap light, and removed by annealing above typically 100 K or by irradiation with infrared radiation at around half the band gap. These materials also show photoluminescence at about half the band gap when irradiated with band-gap light.

These and other experimental results were brought together and explained using a simple model by Street and Mott[35] based on a spin-pairing idea of Anderson.[36] The model assumes the presence of defects in amorphous materials above their $T_g$, in much the same way as a thermal-equilibrium concentration of vacancies and other defects exists in crystals. These defects are "frozen in" during quenching or deposition. The simplest defect imaginable is a "dangling bond" where the coordination is not satisfied at one atom in an otherwise perfect random network. Such a structure can be constructed with a variety of local environments and the dangling-bond orbital can be unoccupied ($D^+$), singly occupied ($D^0$), or doubly occupied ($D^-$). This notation of Street and Mott is used to specify the charge state of the defect. In a-Si and a-Ge, the centers in the gap are singly occupied ($D^0$) giving rise to the esr signal. In chalcogenides, however, Street and Mott proposed that equal numbers of $D^+$ and $D^-$ states are energetically favored over $D^0$, accounting for the absence of unpaired electrons. The correlation energy, usually positive due to electron–electron repulsion favoring singly occupied levels, is effectively negative due to lattice relaxation around the $D^+$ defect. The stability of the charged-defect states is confirmed by theoretical studies and chemical-bond arguments, which can also predict the coordination and position in the band gap. Other support for the existence of undercoordinated and overcoordinated atoms has come from the observation of small sharp features in the infrared and Raman spectra of several amorphous solids, notably silica and a-As.

One modification to this model is to suppose that oppositely charged defects do not occur at random but form preferentially in close proximity, gaining electrostatic energy and producing a neutral center (the "intimate valence alternation pair" or IVAP of Kastner[37]). In a model for the ac conductivity of chalcogenide glasses based on thermally activated hopping of electron pairs from $D^-$ to $D^+$, Elliott[38] found that pairs of defects only 0.5 nm apart dominate, in contrast to the 10 nm mean separation. As discussed below, it is likely that if electron tunneling between charged defects is responsible for the formation of the two-level systems, the defects need to be closely spaced.

At least one experiment on $As_2S_3$ reveals interesting correlations between charged defects and two-level systems.[39] This used electric echoes to probe tunneling states in $As_2S_3$. After irradiation with band-gap light (2.41 eV) the echo signal is reduced but can be progressively restored by annealing at increasingly higher temperatures or by irradiation with midgap (1–2 eV) radiation. The magnitude of the echo signal therefore corresponds to the population of the $D^+$ (or $D^-$) center; excitation to $D^0$ reduces the echo signal.

However, the link between charged defects and two-level systems is uncertain. Single-electron hopping, either thermally activated or by phonon-assisted tunneling, is excluded at low temperature owing to the large energy required (about half the band gap) to transfer the electron to another site. Hopping of two electrons together is therefore the only possibility, but the rate for phonon-assisted tunneling is vanishingly small unless the separation between sites is of order 0.1 nm, corresponding to the IVAP. Structural relaxation around the initial and final positions of the electron pair therefore involves essentially the same atoms, and the transition can be equally well described in terms of a potential barrier to atomic motion. The transition may involve interchange of $D^+$ and $D^-$ centers as in the Elliott model, or the other bond rearrangements. Alternatively, the coordination of each atom may remain fixed and the charged defect (presumably an undercoordinated atom in this case) merely makes the lattice locally less rigid allowing a double potential well to form.

While the electric-echo experiments[39] in $As_2S_3$ clearly demonstrate a link between the charged defect centers and two-level systems, a structural model based on atomic tunneling still appears preferable. The existence of potential barriers of the required height and width seems very plausible based on current views of the glass transition: tunneling states are a necessary consequence of their existence. Two-level systems also occur in polymers and amorphous metals: the only common feature is structural disorder. It is attractive to believe that this results generally in double potential wells.

# References

1. D. P. Jones, Thesis (1982) after R. C. Zeller and R. O. Pohl, *Phys. Rev. B* **4**, 2029 (1971).
2. R. B. Stephens, *Phys. Rev. B* **8**, 2896 (1973).
3. D. Redfield, *Phys. Rev. Lett.* **27**, 730 (1971).
4. P. Fulde and H. Wagner, *Phys. Rev. Lett.* **27**, 1280 (1971).
5. D. P. Jones, N. Thomas, and W. A. Phillips, *Philos. Mag. B* **38**, 271 (1978).
6. W. F. Love, *Phys. Rev. Lett.* **31**, 822 (1973).

7. W. A. Phillips, ed., *Amorphous Solids: Low Temperature Properties*, Springer-Verlag, Berlin (1981).
8. E. Merzbacher, *Quantum Mechanics*, 2nd edn., John Wiley and Sons, New York (1970).
9. M. Abramowitz and I. A. Stegun, *Handbook of Mathematical Functions*, Dover, New York (1970).
10. J. Zimmermann and G. Weber, *Phys. Rev. Lett.* **46**, 661 (1981).
11. M. Meissner and K. Spitzmann, *Phys. Rev. Lett.* **46**, 265 (1981).
12. B. Golding, J. E. Graebner, and R. J. Shutz, *Phys. Rev. B* **14**, 1660 (1976).
13. L. Piché, R. Maynard, S. Hunklinger, and J. Jäckle, *Phys. Rev. Lett.* **32**, 1426 (1974).
14. S. Hunklinger and M. v Schickfus, *J. Phys. C* **9**, L439 (1976).
15. O. B. Wright, Thesis, University of Cambridge (1982).
16. L. Bernard, Thèse, Grenoble (1979).
17. B. Golding, M. v Schickfus, S. Hunklinger, and K. Dransfield, *Phys. Rev. Lett.* **43**, 1817 (1979).
18. B. Golding and J. E. Graebner, in: *Amorphous Solids: Low Temperature Properties* (W. A. Phillips, ed.), Springer-Verlag, Berlin (1981).
19. W. Arnold, C. Martinon, and S. Hunklinger, *J. Phys. C* **6**, 961 (1978).
20. R. B. Stephens, *Phys. Rev. B* **13**, 852 (1976).
21. W. A. Phillips, *Philos. Mag. B* **43**, 747 (1981).
22. J. C. Lasjaunias, A. Ravex, M. Vandorpe, and S. Hunklinger, *Solid State Commun.* **17**, 1045 (1975).
23. J. E. Graebner, B. Golding, R. J. Schutz, F. S. L. Hsu, and H. S. Chen, *Phys. Rev. Lett.* **39**, 1480 (1977).
24. H. v Lohneysen, *Phys. Rep.* **79**, 161 (1981).
25. P. Doussineau, P. Legros, A. Levelut, and A. Robin, *J. Phys. (Paris)* **39L**, 265 (1978).
26. W. Arnold, P. Doussineau, C. Frénois, and A. Levelut, *J. Phys. (Paris) Lett.* **42**, L289 (1981).
27. L. F. Lou, *Solid State Commun.* **19**, 335 (1976).
28. O. L. Anderson and H. E. Bommel, *J. Am. Ceram. Soc.* **38**, 125 (1955).
29. K. S. Gilroy and W. A. Phillips, *J. Non-Cryst. Solids* **35–36**, 1135 (1980).
30. H. v Lohneysen and H. J. Schink, *Phys. Rev. Lett.* **48**, 1121 (1982).
31. M. v Haumeder, U. Strom, and S. Hunklinger, *Phys. Rev. Lett.* **44**, 84 (1980).
32. M. R. Vukcevich, *J. Non-Cryst. Solids* **11**, 25 (1972).
33. K. K. Mon and N. W. Ashcroft, *Solid State Commun* **27**, 609 (1978).
34. D. A. Smith, *Phys. Rev. Lett.* **42**, 729 (1979).
35. R. A. Street and N. F. Mott, *Phys. Rev. Lett.* **35**, 1293 (1975).
36. P. W. Anderson, *Phys. Rev. Lett.* **34**, 953 (1975).
37. M. Kastner, *J. Non-Cryst. Solids* **31**, 223 (1978).
38. S. R. Elliott, *Philos. Mag.* **37**, 553 (1978).
39. D. L. Fox, B. Golding, and W. H. Haemmerle, *Phys. Rev. Lett.* **49**, 1356 (1982).

# Defects in Amorphous Semiconductors

## E. A. Davis

### 15.1. Introduction

The concept of point defects in noncrystalline solids is not an easy one to grasp for anyone unfamiliar with the subject. One's first impression might be that the structure of a glass or an amorphous thin film is totally defective!

The clue to an appreciation of the nature of the defects in these materials lies in the realization that, at least for covalently bonded solids, there is an amorphous analog to the atomic structure of the perfect single crystal, namely the continuous random network (CRN). In a CRN every atom is bonded to the required number of nearest neighbors to satisfy its valency, the bond lengths and bond angles being very similar to those in the crystal. Disorder arises simply by removing any restrictions, other than that required to achieve complete connectivity, on the relative orientation of second-neighbor bonds. Short-range order is thereby preserved while periodicity in the structure is completely lost.

The simplest defect that can then be envisaged in a CRN is that of an atom whose valency is *not* satisfied, i.e., one which is undercoordinated or overcoordinated. Such an atom will then carry either a charge or a spin, both of which will be sufficiently localized in space to justify description of the site as a point defect.

E. A. Davis · Department of Physics, University of Leicester, Leicester LE1 7RH, England.

A CRN model may contain *odd*-membered rings of atoms, not normally present in the crystalline phase. If the material contains two kinds of atom, such rings will necessarily contain bonds between like atoms, a feature again not normally present in the crystal. These "wrong" bonds can also be considered as defects but are only usefully classified as such if their density is considerably less than that of the bonds between unlike atoms.

Several other types of defects can occur in amorphous solids. For example, Figure 15.1 illustrates several possible arrangements for a tetrahedrally coordinated structure in which undercoordinated atoms (dangling bonds) lie in proximity to each other. A vacancy in a crystal composed of atoms of valency $N$ may be regarded as $N$ undercoordinated atoms in the same vicinity. In a CRN there would seem no reason to favor such high-order symmetric defects and it appears more likely that defects involving less than $N$ undercoordinated atoms will occur. Indeed the evidence is that, at least for a-Si and a-Ge, the single isolated dangling bond is the dominant defect.

Some mention should be made of hydrogenated a-Si. It is now well established that incorporation of H neutralizes most of the dangling bonds, reducing the density of levels in the gap and permitting doping. However, normally, much more hydrogen than is required for this purpose is incorporated and other defect centers (such as Si–H–Si) could then become important.

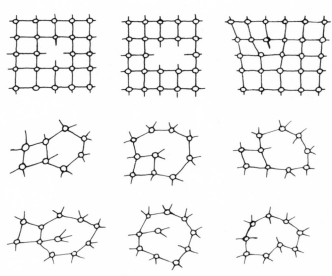

Figure 15.1. Defects in crystals and CRN models of amorphous materials corresponding to dangling bonds in various configurations. The structure on the top right corresponds to a dislocation, i.e., a row of dangling bonds perpendicular to the plane shown.

In chalcogenides the concept of the single dangling bond is also important. However, it will be seen in Section 15.3 that pairing of electrons at such sites is favored. This leads to the defects having strikingly different behavior as far as many properties are concerned.

## 15.2. Positive- and Negative-U Defects

The levels in the energy gap associated with dangling bonds in tetrahedrally bonded amorphous semiconductors are believed to be as illustrated schematically in Figure 15.2a. The lower of the two bands ($E$) corresponds to single occupancy by an electron, in which configuration the defect is neutral and carries a spin. The upper band of levels ($E + U$) corresponds to double occupancy and lies at a "correlation energy" $U$ higher. When the defect contains two electrons it has no net spin and is negatively charged.

The variation of the Fermi energy $E_F$ with electron occupancy $n$ is illustrated below the density-of-states diagram. Such a variation can be obtained in practice by, for example, doping of hydrogenated a-Si. For the situation shown the occupancy is less than one per dangling bond and $E_F$ lies in the lower band. The density of levels at $E + U$ available for double

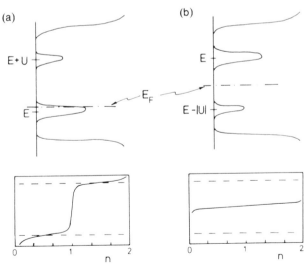

Figure 15.2. Density of states and variation of Fermi level with electron occupancy $n$ for a semiconductor with defects having (a) a positive correlation energy and (b) a negative correlation energy.

occupancy is drawn equal to the density of single occupied states. The density of singly occupied levels can be determined by measuring the strength of the ESR signal associated with the free spins. For a-Si:H, the behavior of this with doping confirms the general picture presented. For chalcogenides, no free spins are detected. This leads to the conclusion that electrons are paired at any defect sites present. The Coulomb repulsion associated with having two electrons at the same site must be outweighed by a negative term in the total energy due to electron–phonon interaction, which leads to configuration changes in the local atomic structure. The centers are therefore associated with an effective correlation energy $U$, which is negative.

The energy-level diagram for negative-$U$ centers is illustrated in Figure 15.2b. The levels corresponding to double occupancy lie below those corresponding to single occupancy. As the electron occupancy $n$ is increased from zero, the electrons enter the defect sites in pairs. The density of lower levels is equal to the number of electrons introduced and this band is therefore always full. Just as for an intrinsic semiconductor, $E_F$ lies midway between the two levels; at a finite temperature, some occupancy of the upper level occurs at the expense of the lower level. This "pinning" of $E_F$ between the two levels is maintained at all values of $n$ between 0 and 2 except near the two extremes and is the reason why chalcogenide semiconductors cannot be made strongly n- or p-type by doping.

## 15.3. Defects in Chalcogenides

A specific model for negative-$U$ defects in chalcogenides is illustrated in Figure 15.3a for the case of a-Se. Consider two dangling bonds at the ends of the selenium chain. When they contain a single electron, the defects are neutral and will be designated $D^0$. Transfer of an electron from one chain end to the other will lead to the creation of two charged defects $D^+$ and $D^-$. It is proposed that the reaction

$$2D^0 \rightarrow D^+ + D^-$$

is exothermic, the necessary lowering in energy being provided by local structural rearrangements. On a configurational-coordinate diagram (Figure 15.3b) the positive correlation energy $U$ associated with two electrons at $D^-$ in the absence of configurational charges, becomes negative ($U_{eff}$) after lattice relaxations. The "chemical" reason for the exothermic nature of the reaction is that, at $D^-$, an extra bond with a neighboring chain can be formed by utilizing the normally nonbonding lone-pair electrons associated

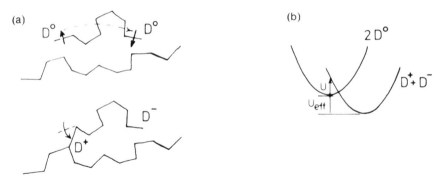

Figure 15.3. (a) Illustration of the transfer of an electron from one chain end to another creating two charged defects $D^+$ and $D^-$. The reaction is assumed to be exothermic, the $D^+$ defect forming a threefold coordinated atom. (b) The same reaction on a configurational-coordinate diagram. The positive correlation energy $U$ associated with two electrons at one site is turned into an effective negative correlation energy $U_{eff}$ because of the configurational changes.

with each Se atom. The coordination of Se atoms at $D^+$ is therefore three, in contrast to that at $D^-$ where it is one and to that at a normally bonded Se atom where it is two. It is proposed that the lattice distortion at $D^-$ is negligible, at $D^+$ it is considerable, and at $D^0$ it is intermediate. Accurate calculations of the energy levels associated with the three charge states of the defect are difficult and the principal sources of evidence that the above model is appropriate for chalcogenides are experimental. Luminescence in particular is a powerful technique for studying the levels.

## 15.4. Luminescence from Chalcogenides

As a general rule, luminescence from chalcogenides occurs in a band peaking at an energy close to that of half the optical band gap, the integrated intensity being greatest when excitation is by photons of energy corresponding to the tail of the optical absorption edge (see Figure 15.4). Since there is no corresponding absorption at the same energy, the implication is that the luminescence is Stokes shifted.

An explanation for these results is illustrated in Figure 15.5a. The defect involved has an energy level just above the top of the valence band (it is in fact associated with the $D^-$ state referred to in Section 15.3). Its excitation requires photons of energy comparable to the band-gap energy as illustrated by (i) process A — creation of an electron–hole pair with subsequent rapid trapping of the hole, or by (ii) process A' — direct excitation of an electron in the defect level. Following excitation by either of these processes, the lattice surrounding the defect distorts leading to a new energy level for the defect that lies near midgap (process B). This level corresponds to the $D^0$ state of the defect. Subsequent recombination of the electron with the hole at the defect then leads to the radiative transition (process C) and the $D^0$ center is reconverted to $D^-$. The difference between the absorption and luminescence energies is thereby to be regarded as a Stokes shift and can be represented on a configurational-coordinate diagram as shown in Figure 15.5b.

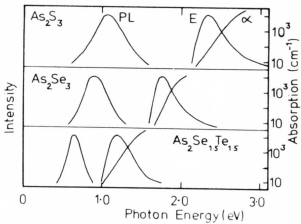

Figure 15.4. Low-temperature photoluminescence spectra (PL), excitation spectra (E), and optical absorption edges ($\alpha$) for amorphous (glass) chalcogenides. The luminescence band occurs at roughly half the energy gap (which can be taken approximately as the energy corresponding to $\alpha = 10$ cm$^{-1}$).

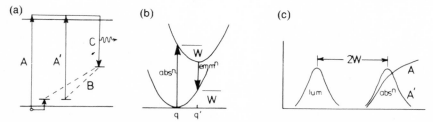

Figure 15.5. (a) Photogeneration and luminescence processes. (A) Photoexcitation across the gap followed by rapid trapping of the hole at the defect center. (A') Direct excitation of the defect center. (B) Relaxation of the center. (b) Configurational-coordinate diagram illustrating absorption and emission processes. The distortion energy is $W$ and the Stokes shift is $2W$. (c) Lineshapes of luminescence and absorption in the case of direct (A') and continuum (A) excitation.

An important experimental observation is that the luminescence efficiency is greatest at low temperatures. The temperature dependence is believed to arise from a competing nonradiative mechanism, which occurs if the initially excited electron escapes from the vicinity of the center. Should this occur the centers are then left in a metastable state and, as long as they remain in this condition, are unavailable for the primary excitation processes A or A'. Such a situation would then be expected to give rise to "luminescence fatigue," i.e., a decrease in the luminescence intensity during prolonged photoexcitation. One also expects "photoinduced absorption," with a threshold energy corresponding to excitation of an electron into or out of the metastable center, and "photoinduced ESR" since the $D^0$ center carries a spin. All of those effects have been observed. A block diagram of the various steps is illustrated in Figure 15.6.

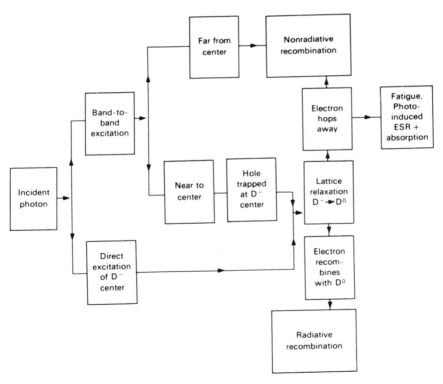

Figure 15.6. Block diagram of the various steps in photogeneration and recombination mechanisms for amorphous chalcogenides. In principle $D^-$ could be exchanged for $D^+$ and the roles of electrons and holes reversed.

## 15.5. Luminescence from a-Si:H

In contrast to the luminescence observed in chalcogenides which involves defect states, the principal luminescence band in a-Si:H arises from radiative transitions between conduction- and valence-band tail states. The role of defects on this band is to provide a nonradiative path for recombination and the band-edge luminescence (at about 1.35 eV) is strongly quenched when the defect density exceeds about $10^{17}$ cm$^{-3}$. However, in samples with a defect density equal to approximately $10^{18}$ cm$^{-3}$, a weaker luminescence band centered at about 0.9 eV is observed. It is believed that the transition involved is that of an electron, which has been trapped at a dangling bond (the upper level of Figure 15.2a) to a valence-band tail state. This can occur following across-the-gap excitation (which produces band-edge luminescence also) or by direct absorption into the dangling-bond state using subbandgap radiation. From the width of the defect-related luminescence band, a weak electron–phonon coupling involving a distortion energy $\sim 0.1$ eV has been inferred.

## 15.6. Group-V Materials

The normal atomic coordination in amorphous group-V elements is three and as such leads to a structure that is more flexible than Ge or Si, but less flexible than Se. In addition the electronic structure differs from either class, the uppermost valence band being composed primarily of $p$-bonding electrons with lone-pair $s$-electrons lying lower in energy. The question then arises as to whether the defect states in those materials exhibit positive- or negative-$U$ characteristics.

Photoluminescence in a-As occurs with emission in a band centered close to midgap energy (Figure 15.7) and there is a concomitant photoinduced ESR signal and optical absorption similar to those found for chalcogenides. However, other properties suggest Ge(Si)-like features. The material exhibits an ESR signal in the dark that is weakly temperature-dependent. Another feature that distinguishes a-As from chalcogenides is the existence of a doubly peaked photoexcitation spectrum.

The above properties have been incorporated into a model for defect states in a-As, which has some features of both Ge-like and chalcogenide-like materials. In essence the model is based on the concept of dangling bonds, many of which react to form negative-$U$ D$^+$ and D$^-$ defect states but some, because of topographical constraints, do not react in this manner but instead remain as unpaired dangling bonds. It is the latter which give rise to the ESR signal. The Fermi energy is however pinned by the D$^+$ and D$^-$ defects. Experiment and theory have been combined to give a fairly

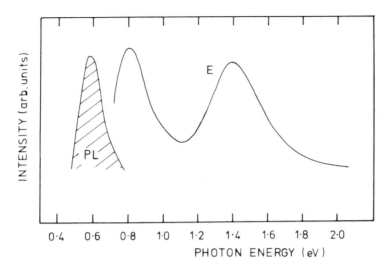

Figure 15.7. Low-temperature photoluminescence (PL) and excitation spectra (E) for amorphous As. The optical absorption edge (not shown) cuts down through the high energy (1.4 eV) peak in the excitation spectrum.

detailed picture of the energy levels in the gap arising from the defects. Relative to chalcogenides, the levels lie at higher energies in the gap, a feature that explains the photoexcitation spectrum shown in Figure 15.7.

Amorphous Sb, As, and P form an interesting sequence in that there is a gradual change from amorphous Ge-like to amorphous chalcogenide-like properties. This is a consequence of the average bond angles, which range from 96° to 102° and make the formation of negative-$U$ defects more likely in a-P and less likely in a-Sb.

## 15.7. Conclusions

Defects in the major classes of amorphous materials are seen to be associated in all cases with dangling bonds, but the way they manifest themselves in experimental studies depends very much on whether electron pairing at such sites is favored. If it is, then negative-$U$ states form in the gap with striking consequences for electrical and optical properties.

In this review only the simplest early experiments on luminescence and ESR have been presented. It should however be appreciated that, in recent years, more sophisticated techniques, such as light-induced ESR, time-resolved photoluminescence (TRL), optically detected magnetic resonance (ODMR), and spin-dependent photoconductivity, have been brought to bear on the problems. The bibliography contains references to some of these studies. As might be anticipated, some of the earlier suggestions have been verified while others have been brought into question. In particular, the question of whether $D^+$ and $D^-$ centers combine to form "intimate valence alternation pairs" in certain materials is not resolved.

It is certain that studies of defects in amorphous materials, which have such relevance to device applications, will continue to occupy the scene for many years.

# Bibliography

## General

N. F. Mott and E. A. Davis, *Electronic Processes in Non-Crystalline Materials*, 2nd edn., Oxford University Press (1979).

E. A. Davis, in: *Topics in Applied Physics*, Volume 36: *Amorphous Semiconductors* (M. H. Brodsky, ed.), Springer-Verlag, Berlin (1979), p. 41.

D. Adler, in: Proceedings of the 9th International Conference on Amorphous and Liquid Semiconductors, *J. Phys.* (*Paris*) **42**, C4, 3 (1981).

J. Robertson, *Phys. Chem. Glasses* **23**, 1 (1982).

R. A. Street, *Adv. Phys.* **25**, 397 (1976).

## Amorphous Si:H

W. E. Spear, *Adv. Phys.* **26**, 811 (1977).

I. G. Austin, T. S. Nashashibi, P. G. LeComber, and W. E. Spear, *J. Non-Cryst. Solids* **32**, 373 (1979).

T. M. Searle, T. S. Nashashibi, I. G. Austin, R. Devonshire, and G. Lockwood, *Philos. Mag. B* **39**, 389 (1979).

R. A. Street and D. K. Biegelsen, *J. Non-Cryst. Solids* **35** and **36**, 651 (1980).

R. A. Street, D. K. Biegelsen, and J. C. Knights, *Phys. Rev.* **24**, 969 (1981).

R. A. Street, *Adv. Phys.* **30**, 593 (1981).

H. Dersch, J. Stuke, and J. Beichler, *Phys. Status Solidi B* **105**, 265 (1981).

H. Dersch, J. Stuke, and J. Beichler, *App. Phys. Lett.* **38**, 456 (1981).

R. A. Street, *Phys. Rev. Lett.* **49**, 1187 (1982).

R. A. Street, D. K. Biegelsen, and J. Zesch, *Phys. Rev. B* **25**, 4334 (1982).

R. A. Street, *Phys. Rev.* **26**, 3588 (1982).

R. A. Street, *Philos. Mag. B* **46**, 263 (1982).

S. P. Depinna, B. C. Cavenett, I. G. Austin, T. M. Searle, M. J. Thompson, J. Allison, and P. G. LeComber, *Philos. Mag. B* **46**, 473 (1982).

S. P. Depinna, B. C. Cavenett, T. M. Searle, and I. G. Austin, *Philos. Mag. B* **46**, 501 (1982).

W. M. Pontuschka, W. W. Carlos, and P. C. Taylor, *Phys. Rev. B* **25**, 4362 (1982).

D. V. Lang, J. D. Cohen, and J. P. Harbison, *Phys. Rev. B* **25**, 5285 (1982).

D. C. Allan and J. D. Joannopoulos, *Phys. Rev. Lett.* **44**, 43 (1982).

W. E. Spear and P. G. LeComber, in: *The Physics of Amorphous Silicon and its Applications* (J. Joannopoulos and G. Lucovsky, eds.), Springer-Verlag, Berlin (1983).

## Group-V Materials

G. N. Greaves, S. R. Elliott, and E. A. Davis, *Adv. Phys.* **28**, 49 (1979).

S. R. Elliott and E. A. Davis, *J. Phys. C* **12**, 2577 (1979).

W. B. Pollard and J. D. Joannopoulos, *Phys. Rev. B* **19**, 4217 (1979).

E. A. Davis, *J. Phys. (Paris) C* **4**, 855 (1981).

S. Depinna and B. C. Cavenett, *Solid State Comm.* **40**, 813 (1981); *Solid State. Comm.* **43**, 25 (1982).

# Index